Electron Microscopy of Materials

MATERIALS RESEARCH SOCIETY SYMPOSIA PROCEEDINGS   VOLUME 31
ISSN 0272 - 9172

Volume 1—Laser and Electron-Beam Solid Interactions and Materials Processing, J.F. Gibbons, L.D. Hess, T.W. Sigmon, 1981

Volume 2—Defects in Semiconductors, J. Narayan, T.Y. Tan, 1981

Volume 3—Nuclear and Electron Resonance Spectroscopies Applied to Materials Science, E.N. Kaufmann, G.K. Shenoy, 1981

Volume 4—Laser and Electron-Beam Interactions with Solids, B.R. Appleton, G.K. Celler, 1982

Volume 5—Grain Boundaries in Semiconductors, H.J. Leamy, G.E. Pike, C.H. Seager, 1982

Volume 6—Scientific Basis for Nuclear Waste Management, S.V. Topp, 1982

Volume 7—Metastable Materials Formation by Ion Implantation, S.T. Picraux, W.J. Choyke, 1982

Volume 8—Rapidly Solidified Amorphous and Crystalline Alloys, B.H. Kear, B.C. Giessen, M. Cohen, 1982

Volume 9—Materials Processing in the Reduced Gravity Environment of Space, G.E. Rindone, 1982

Volume 10—Thin Films and Interfaces, P.S. Ho, K.N. Tu, 1982

Volume 11—Scientific Basis for Nuclear Waste Management V, W. Lutze, 1982

Volume 12—In Situ Composites IV, F.D. Lemkey, H.E. Cline, M. McLean, 1982

Volume 13—Laser-Solid Interactions and Transient Thermal Processing of Materials, J. Narayan, W.L. Brown, R.A. Lemons, 1983

Volume 14—Defects in Semiconductors II, S. Mahajan, J.W. Corbett, 1983

Volume 15—Scientific Basis for Nuclear Waste Management VI, D.G. Brookins, 1983

Volume 16—Nuclear Radiation Detector Materials, E.E. Haller, H.W. Kraner, W.A. Higinbotham, 1983

Volume 17—Laser Diagnostics and Photochemical Processing for Semiconductor Devices, R.M. Osgood, S.R.J. Brueck, H.R. Schlossberg, 1983

Volume 18—Interfaces and Contacts, R. Ludeke, K. Rose, 1983

Volume 19—Alloy Phase Diagrams, L.H. Bennett, T.B. Massalski, B.C. Giessen, 1983

Volume 20—Intercalated Graphite, M.S. Dresselhaus, G. Dresselhaus, J.E. Fischer, M.J. Moran, 1983

Volume 21—Phase Transformations in Solids, T. Tsakalakos, 1984

Volume 22—High Pressure in Science and Technology, C. Homan, R.K. MacCrone, E. Whalley, 1984

Volume 23—Energy Beam-Solid Interactions and Transient Thermal Processing, J.C.C. Fan, N.M. Johnson, 1984

Volume 24—Defect Properties and Processing of High-Technology Nonmetallic Materials, J.H. Crawford, Jr., Y. Chen, W.A. Sibley, 1984

MATERIALS RESEARCH SOCIETY SYMPOSIA PROCEEDINGS   VOLUME 31

Volume 25—Thin Films and Interfaces II, J.E.E. Baglin, D.R. Campbell, W.K. Chu, 1984

Volume 26—Scientific Basis for Nuclear Waste Management VII, G.L. McVay, 1984

Volume 27—Ion Implantation and Ion Beam Processing of Materials, G.K. Hubler, O.W. Holland, C.R. Clayton, C.W. White, 1984

Volume 28—Rapidly Solidified Metastable Materials, B.H. Kear, B.C. Giessen, 1984

Volume 29—Laser-Controlled Chemical Processing of Surfaces, A.W. Johnson, D.J. Ehrlich, H.R. Schlossberg, 1984

Volume 30—Plasma Processing and Synthesis of Materials, J. Szekely, D. Apelian, 1984

Volume 31—Electron Microscopy of Materials, W. Krakow, D. Smith, L.W. Hobbs, 1984

Volume 32—Better Ceramics Through Chemistry, C.J. Brinker, D.R. Ulrich, D.E. Clark, 1984

Volume 33—Comparison of Thin Film Transistor and SOI Technologies, H.W. Lam, M.J. Thompson, 1984

Volume 34—Physical Metallurgy of Cast Iron, H. Fredriksson, 1985

MATERIALS RESEARCH SOCIETY SYMPOSIA PROCEEDINGS   VOLUME 31

Electron Microscopy of Materials

Symposium held November 1983 in Boston, Massachusetts, U.S.A.

EDITORS:

William Krakow
David A. Smith
IBM, Thomas J. Watson Research Center, Yorktown Heights,
New York, U.S.A.

Linn W. Hobbs
Department of Materials Science and Engineering, Massachusetts
Institute of Technology, Cambridge, Massachusetts, U.S.A.

NORTH-HOLLAND
NEW YORK • AMSTERDAM • OXFORD

©1984 by Elsevier Science Publishing Co., Inc.
All rights reserved.

This book has been registered with the Copyright Clearance Center, Inc. For further information, please contact the Copyright Clearance Center, Salem, Massachusetts.

Published by:

Elsevier Science Publishing Company, Inc.
52 Vanderbilt Avenue, New York, New York 10017

Sole distributors outside the United States and Canada:

Elsevier Science Publishers B.V.
P.O. Box 211, 1000 AE Amsterdam, The Netherlands

Library of Congress Cataloging in Publication Data

Main entry under title:

Electron microscopy of materials.
    (Materials Research Society symposia proceedings, ISSN 0272-9172; v.31)
    Sponsored by the Materials Research Society.

    Includes indexes.
    1. Materials—Microscopy—Congresses. 2. Electron microscopy—Congresses. I. Krakow, William. II. Smith, David A. III. Hobbs, Linn W. Materials Research Society. V. Series.
TA417.23.E43 1984   620.1'1'0287   84-10245
ISBN 0-444-00897-7

Manufactured in the United States of America

Contents

Preface xi

SECTION I - CURRENT TRENDS IN ELECTRON MICROSCOPE CHARACTERIZATION TECHNIQUES

Electron Microscopy at Atomic Resolution... 1
  R. Gronsky

An Overview of Analytical Electron Microscopy 11
  D.B. Williams

Computer Interfaced Electron Microscope 23
  Y. Kokubo, S. Moriguchi, J. Hosoi, E. Watanabe and J. Nash

Microanalysis with High Spatial Resolution 33
  P.E. Batson, C.R.M. Grovenor, D.A. Smith and C. Wong

Image Processing for Electron Microscope Investigations
of Materials.. 39
  William Krakow

Combination of CBD and High Resolution with 400 kV 57
  S. Suzuki, T. Honda, Y. Kokubo and Y. Harada

Recent Development in Hitachi Transmission
Electron Microscope.. 63
  Shigeto Isakozawa, Isao Matsui, Shoji Kamimura and Akira Tonomura

Electron Microscopy and Analysis at Higher Voltages:
The Philips EM430 300kV Microscope 71
  Peter N. Hagemann and J.S. Fahy

Determination of the Specific Site Occupation of Rare Earth in $Y_{1.7}Sm_{0.6}Lu_{0.7}Fe_5O_{12}$ Thin Films by the Orientation Dependence of Characteristic X-Ray Emissions 79
   Kannan M. Krishnan, Peter Rez, Raja Mishra and Gareth Thomas

SECTION II - SEMICONDUCTING MATERIALS

TEM/EBIC Investigations of Structural Defects in Polycrystalline Solar Cells..85
   D.G. Ast, B. Cunningham and R. Gleichmann

Dopant Site Location by Electron Channeling in Ion Implanted Silicon..97
   S.J. Pennycook, J. Narayan and O.W. Holland

High Resolution TEM Studies of Defects Near Si-SiO$_2$ Interface..105
   J.H. Mazur and J. Washburn

Effects of Microstructure and Chemical Composition on EBIC Contrast in HEM Solar Cell Silicon 111
   T.D. Sullivan

Electron Microscopy of Semiconductors Reconstructed Surfaces...........117
   Pierre M. Petroff

Intrinsic Point Defects and Diffusion Processes in Silicon.....................127
   T.Y. Tan

Structural Analyses of Metal/GaAs Contacts and Ge/GaAs and AlAs/GaAs Heterojunctions...143
   T.S. Kuan

Structure of Thermally-Induced Microdefects in Czocharlski Silicon 153
   F.A. Ponce and S. Hahn

Crystallinity, Morphology, and Conductivity of Boron-Doped Microcrystalline Silicon 159
   G. Rajeswaran, J. Tafto, R.L. Sabatini and P.E. Vanier

Interfacial Reactions of Nickel Films on GaAs 165
   L.J. Chen, and Y.F. Hsieh

TEM Observation of Defects in InGaAsP and InGaP Crystals
on GaAs Substrates Grown by Liquid Phase Epitaxy.........................171
   O. Ueda, S. Isozumi, S. Komiya, T. Kusunoki, and I. Umebu

SECTION III - SURFACES AND INTERFACES

Reflection Electron Microscopy and Diffraction from Crystal Surfaces 177
   J.M. Cowley

Interpretation of the Atomic Surface Structure of Ag_2O on
(111) Au Thin Films 189
   William Krakow

Formation and Transformation of Amorphous Silicide Alloys...............201
   K.N. Tu, T. Tien and S.R. Herd

The Characterisation of Interfacial Dislocation Structures....................211
   W.A.T. Clark

Dislocations, Steps and Interphase Boundary Migration.......................223
   D.A. Smith

Microdiffraction Studies of Small Gold Metallic Particles....................233
   Miguel Jose-Yacaman, Alfredo Gomez, and Krystyna Truszkowska

High Resolution Study of the Relationship Between Misfit
Accommodation and Growth of $Cu_{2-x}S$ in CdS 241
   T. Sands, J. Washburn and R. Gronsky

Epitaxy of Bcc and Hexagonal Metals on Fcc(001) Substrates.............247
   C.R.M. Grovenor and D.A. Smith

STEM Microanalysis of Hazed Polycrystalline Silicon Layers...............255
   G.J.C. Carpenter and D.C. Houghton

Misfit Dislocation in the Interface Between the Metallic
and Insultating Phases in Cr Doped V_2O_3 261
   Nobuo Otsuka and Hiroshi Sato

SECTION IV - CERAMIC MATERIALS

The Structure of Grain Boundaries and Phase Boundaries in Ceramics 267
   C.B. Carter

Applications of High Resolution Electron Microscopy in Ceramics Research 279
   G. Van Tendeloo

The Defect Structures of Fe_9S_{10} 291
   Thao Nguyen and Linn W. Hobbs

TEM Dislocations in Sapphire (α-Al_2O_3) 303
   K.P.D. Lagerlöff, T.E. Mitchell, and A.H. Heuer

TEM Observations of Grain Boundaries in Ceramics 317
   M. Rühle

Fault Structures in CVD Silicon Nitride 325
   T.M. Shaw, J.W. Steeds and D.R. Clarke

High Resolution TEM Studies of β-Alumina Type Structures 331
   K.J. Morrissey, Z. Elgat, Y. Kouh and C.B. Carter

The $\gamma \longrightarrow \alpha$ Phase Transformation in Al_2O_3 337
   D.S. Tucker and J.J. Hren

Calcium Site Occupancy in $BaTiO_3$ 345
   H.M. Chan, M.P. Harmer, M. Lal, and D.M. Smyth

Characterisation of Portland Cement Hydration by Electron Optical Techniques 351
   Karen L. Scrivener and P.L. Pratt

Ordered Defect Flouride Compounds in ZrO_2 Alloys 357
   S. Farmer, V. Lanteri, J. Hangas, T.E. Mitchell and A. Heuer

Author Index 369

Subject Index 371

Preface

This proceedings contain the papers which were presented at the symposium on the "Electron Microscopy of Materials" held in Boston, Massachusetts, November 14-17, 1983, and sponsored by the Materials Research Society. The symposium provided an interdisciplinary forum for scientists engaged in the use of electron microscope characterization techniques to study materials. This year four specific topics or subdivisions were chosen which include: Current Trends in Electron Microscopy Characterization Techniques, Semiconducting Materials, Surfaces and Interfaces and Ceramic Materials. The proceedings are therefore divided into these four parts. One of the distinguishing features of this symposium was the emphasis on new advances in both electron microscopes and problem solving which here-to-fore was unreachable. Substantial progress has been made in a variety of materials areas as evidenced in many of the papers of this proceedings. Also, many innovations were presented in the area of electron microscope equipment.

In order to speed publications of the proceedings, manuscripts were submitted by symposium participants in camera-ready form. The manuscripts were reviewed for format style and modified accordingly.

A large part of the success of the symposium was due to financial contributions from both electron microscope manufacturers and corporations. We would like to specifically acknowledge International Business Machines Corporation, JEOL U.S.A., Inc., Hitachi Scientific Instruments and Philips Electronic Instruments for their generous assistance. Finally, we would like to extend our sincere thanks to Mrs. Lorraine Miro for her secretarial services before and after the Conference.

                                                          W. Krakow
                                                          D.A. Smith
                                                          L.W. Hobbs

SECTION I

Current Trends in Electron Microscope Characterization Techniques

ELECTRON MICROSCOPY AT ATOMIC RESOLUTION

R. Gronsky
National Center for Electron Microscopy
Materials and Molecular Research Division
Lawrence Berkeley Laboratory
Berkeley, California   94720

ABSTRACT

The direct imaging of atomic structure in solids has become increasingly easier to accomplish with modern transmission electron microscopes, many of which have an information retrieval limit near 0.2nm point resolution. Achieving better resolution, particularly with any useful range of specimen tilting, requires a major design effort. This presentation describes the new Atomic Resolution Microscope (ARM), recently put into operation at the Lawrence Berkeley Laboratory. Capable of 0.18nm or better "interpretable" resolution over a voltage range of 400 kV to 1000kV with $\pm$ 40° biaxial specimen tilting, the ARM features a number of new electron-optical and microprocessor-control designs. These will be highlighted, and its atomic resolution performance demonstrated for a selection of inorganic crystals.

INTRODUCTION

In its most common mode of operation, the transmission electron microscope (TEM) produces images of thin specimens by an amplitude-contrast mechanism. Such images are formed by the utilization of a small objective aperture to admit only one scattered "beam" from the diffraction spectrum of the object through the microscope optics (Fig. 1(a) and (b)). Contrast under these conditions results from the spatial variation of the intensity distribution contained within the chosen beam, and image resolution is determined by the extent to which the sampled scattering event is localized within the specimen.

Alternatively, thin specimens can be imaged in the TEM by a phase-contrast mechanism. This requires admitting more than one beam from the diffraction spectrum of the object through the microscope optics (Fig. 1(c)), and with proper setting of the objective lens current, the phase variations between the chosen beams are made to produce image contrast. Furthermore since these phase shifts are localized at the individual scattering species, image resolution is determined by the extent to which the complete diffraction spectrum is included in the imaging aperture, if all beams are included, atoms are resolved.

Although there have been no limitations on resolution imposed by instrumentation for quite some time in amplitude-contrast imaging, only the most modern TEM's have been successfully used for phase-contrast imaging of fine scale structure. Fortunately the sources of experimental difficulties which continue to prevent atomic resolution by this technique are also quite well known.

One set of problems has to do with beam-specimen interactions. In the phase-contrast imaging method, the specimen is frequently considered an integral part of the imaging optics since its only effect on the incident beam should be an alteration in phase [1]. This mandates that

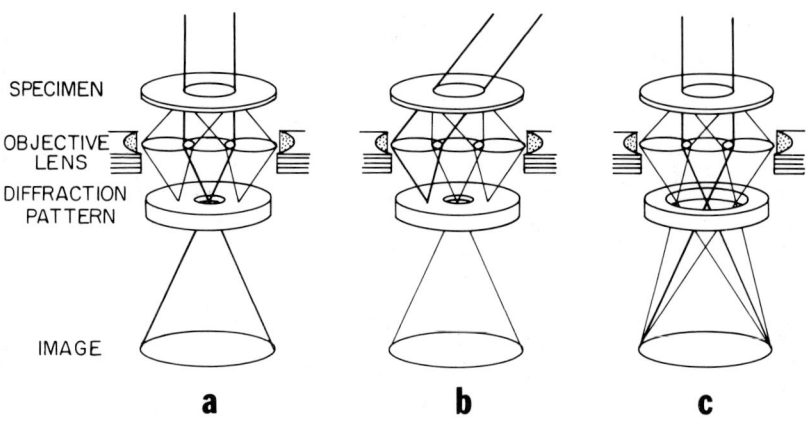

FIG. 1. Ray diagram showing conditions for (a) bright field amplitude-contrast, (b) dark-field amplitude and (c) phase-contrast imaging.

the object be very "thin", with the most severe constraints placed on preparing specimens of high atomic number. Other limitations include specimen orientation, irradiation damage, both ionization and knock-on, and contamination under the illuminating beam from both the microscope vacuum environment and specimen-borne adsorbates.

Another set of problems stems from electron optics. Much like the specimen, the objective lens in the TEM affects the electron beam through an overall phase distortion. This phase change (X) in the diffraction spectrum is a function of scattering angle or position ($\underset{\sim}{u}$) in reciprocal space and can be written [2]

$$X(\underset{\sim}{u}) = \frac{2\pi}{\lambda} [C_s \frac{\lambda^4 u^4}{4} + \Delta z \frac{\lambda^2 u^2}{2}]$$

where λ is the electron wavelength, C_s the spherical aberration coefficient of the objective lens and Δz the defocus of the objective lens. Improving the contrast transfer characteristics of the objective lens therefore basically amounts to reducing λ and C_s while controlling Δz to match phases for the largest number of "beams" in the scattering distribution [3]. Attention must also be given to source coherence [4] and chromatic aberration [5] since these, too, serve to truncate the number of beams which can be effectively used during phase-contrast imaging.

ACHIEVING ATOMIC RESOLUTION

These combined effects impose limitations on the interpretation of phase-contrast images which are presently more severe than those imposed on their formation. Specifically, the point-to-point resolution of a microscope for directly interpretable images is established by the Scherzer defocus condition [6] whereby the phases of the electron waves scattered by the specimen are uniformly controlled over the entire range of scattering vectors from $|\underset{\sim}{u}|=0$ to

$$|\underset{\sim}{u}| = [0.66 \, C_s^{1/4} \, \lambda^{3/4}]^{-1}.$$

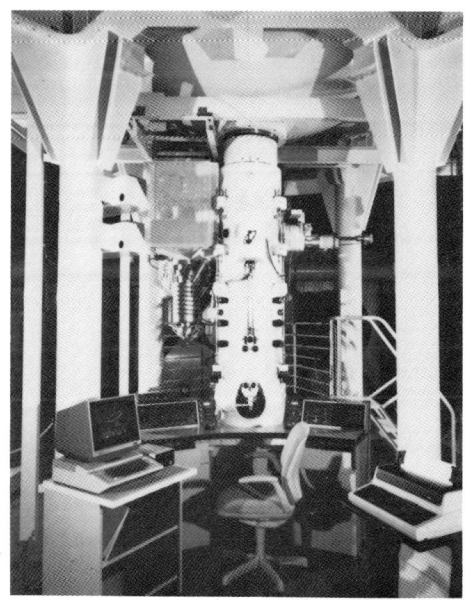

FIG. 2. Console level view of the ARM. Large ion pump mount is visible on the upper left of the column; diagonal tube attached to left of specimen chamber is mass spectrometer head. Specimen exchange rod is attached horizontally to the right, and the manual controls for the goniometer descend the front of the column. The terminal at the lower right addresses all memory locations, digital console display and text field on micrograph negatives. Console controls are efficiently interfaced to the CPU, which reduces their total number.

FIG. 3. Side view of one half of the ARM vibration isolation system. The ten pneumatic isolators are located just below the console floor level, each carrying an average load of 27,130 lbs.

FIG. 4. Microcomputer and monitor used for specimen manipulation by keystroke control. Specimen position is indicated by digital readout and graphics display. Lower lines on monitor represent lens pole piece as height (z) reference.

Another resolution limit is set however by the ultimate cut-off of spectra resulting from instrumental instabilities. This "information retrieval limit" [7] is characterized by fine detail in the image beyond the Scherzer limit where the phases of the electron waves scattered by the specimen are not uniformly controlled. For this reason the image is no longer "directly" interpretable; it must be viewed with accurate knowledge of the complete contrast transfer characteristics of the objective lens.

An approach to achieving atomic resolution in electron microscopy is therefore one of recognizing the present limitations of phase-contrast imaging and carefully interpreting or enhancing high resolution images in conjunction with computer simulations that account for instrumental shortcomings [8]. Alternatively, another approach is one of improving existing instrumentation. This was the course adopted by the recently completed Atomic Resolution Microscope (ARM) project [9,10] which is highlighted in the remainder of this review.

INSTRUMENTATION

The ARM (Fig. 2) is a high voltage electron microscope which has been designed for a point-to-point resolution consistently at or better than 0.18 nm over its entire 400 kV to 1000 kV accelerating potential range. Consequently the microscope can be tuned to a voltage which is below the threshold for knock-on damage in any specimen of interest and used to directly image its contiguous atom structure. Voltage is generated by a Cockroft-Walton system using pressurized Freon insulation, with AC and DC columns concentrically disposed about the evacuated accelerating tube, all housed within a single tank. Both ripple and stability are maintained within 1×10^{-6} per minute using highly sensitive feedback compensation circuity. The instrument also has a unique vacuum system to control contamination, and employs two sputter ion pumps of capacity 1000 ℓ/sec each, backed by titanium sublimation pumps and turbomolecular pumps to maintain a pressure in the low 10^{-8} torr range throughout the electron-optical column and the accelerator tube.

To remove mechanical instabilities the microscope has been attached to a 100 ton inertial block and mounted on ten pneumatic isolators (Fig. 3) capable of dynamic leveling to within ± 0.001 inch from a preset value. The natural vibrational frequency of this system is 0.64 Hz in the horizontal plane and 1.15 Hz in the vertical.

Special attention has also been given to beam brightness in the ARM. At 1000 kV and 25 μA current, the LaB$_6$ cathodes in this microscope yield 10^8 amps/cm^2-str., enabling 2 sec. photographic exposure times at ∿300,000 x magnification. This level of brightness is furthermore preserved through the viewing glass (made under special subcontract by Nikon Optics) due to its tailored transmission coefficient which is matched to the wavelength of the photons emitted from the screen phosphor.

The key to the variable-voltage, atomic resolution performance of the ARM is its top entry objective stage, which, in addition to ± 40° biaxial tilting, incorporates a height (z) control to alter specimen position along the optic axis over a 2 mm range within the objective lens. Using the z - control to focus the specimen, the microscope can be operated at the appropriate objective lens current which maintains a constant $C_s \lambda$ product for any accelerating voltage. This principle is in fact put into operation automatically on the ARM; the instantaneous orientation of the specimen is furthermore displayed via a graphics software package on the monitor screen (Fig. 4).

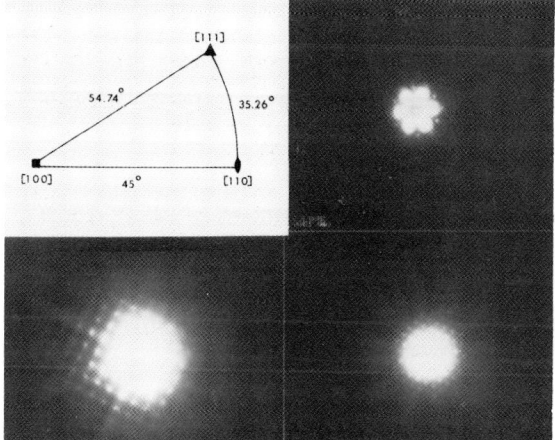
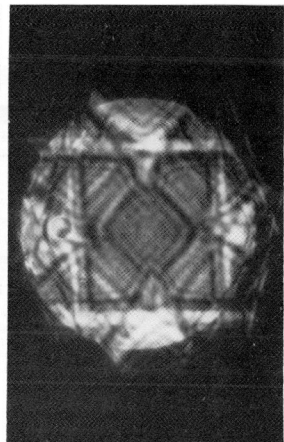

FIG. 5. Results of ARM tilting experiment showing access to full crystallographic unit triangle of a silicon specimen.

FIG. 6. Central disc of CBED pattern from [011] silicon specimen.

PERFORMANCE

The versatility of the high resolution, high angle goniometer of the ARM is demonstrated in Fig. 5. Moreover the z-control stage enables the specimen to be lowered within the lens pole piece to such an extent that the pre-field of the objective can be used as a probe-former. In this way, convergent beam electron diffraction (CBED) patterns can also be recorded from the same specimen areas imaged at high resolution (Fig. 6), providing greater precision in diffraction studies through the analysis of higher order Laue zone (HOLZ) lines [12].

Another effect of the height adjustable stage is its influence upon the optical properties of the objective lens. In general, the lens spherical aberration coefficient decreases with increasing lens excitation, although the exact functional variation is not a simple one [13]. At higher excitation, the lens focal length shortens, making it necessary to drop the specimen deeper into the focusing field. The most serious complication of this process is that the lens is physically constricted nearer the pole piece gap, and this in turn reduces the maximum tilting range of the specimen goniometer. In the ARM, the specimen in its lowest z position is "restricted" to an otherwise generous ± 25° tilting range.

However the primary purpose of the height control is to optimize lens performance. An example of its proper use is shown in Fig. 7, the computed phase contrast transfer function for the ARM at 1000 kV operation with the specimen now located at an optimum 1.9 mm above its deepest immersion position. This calculation reveals that with proper care the ARM is theoretically capable of 0.13 nm point-to-point resolution.

Experimentally, the contrast transfer function of the ARM has been measured by the optical diffractogram method [4]. The 1 MeV results are shown in Fig. 8, where the point-to-point resulution is demonstrated to be 0.16 nm.

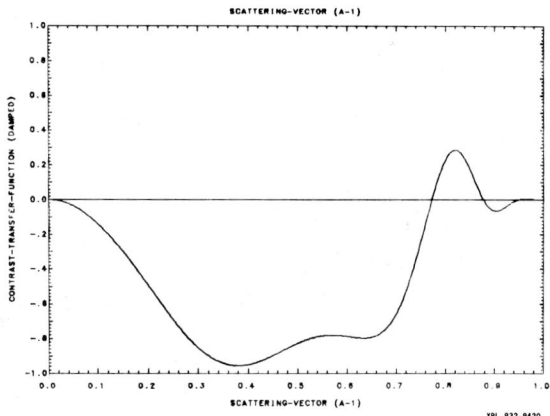

FIG. 7. Computed phase contrast transfer function for the ARM, assuming 1000 kV accelerating voltage, C_s = 2.3 mm, C_c = 3.4 mm, defocus = -52 nm (Scherzer condition), and a disc-shaped source with a divergence half angle of 0.5 mrad and 2 eV energy spread. At this imaging condition, the phases for all scattered waves are uniformly controlled over the widest range of scattering angles. The theoretical resolution limit is determined by the first crossover of the zero axis, here shown to be 0.13 nm.

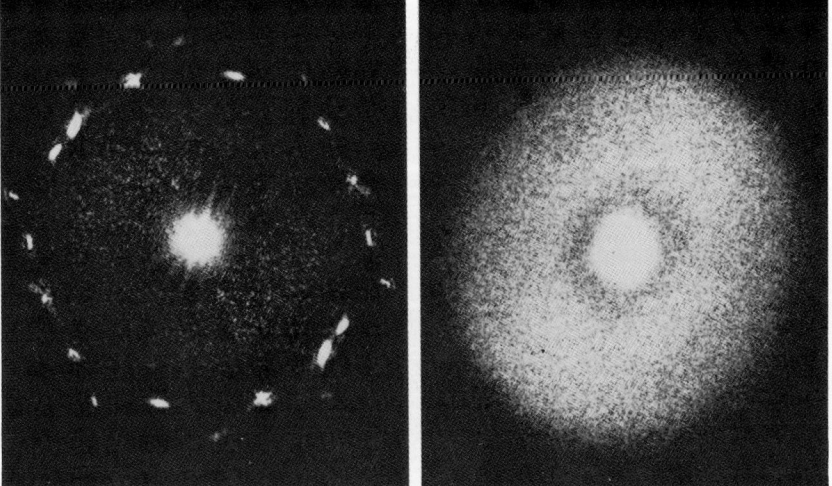

FIG. 8. Optical diffractograms from high resolution images of (a) gold particles showing discrete 0.23, 0.20 and 0.14 nm reflections (left) and (b) amorphous Si (right) at Scherzer defocus, 1000 kV. The diffractogram halo in (b) is continuous out to a resolution limit of 0.16 nm.

APPLICATIONS

The ARM accepts common 3 mm diameter specimens which have been thinned to electron transparency by standard techniques and screened for suitability in another TEM. Three cartridges can be simultaneously loaded into the specimen airlock and individually withdrawn by a turret mechanism for rapid insertion into the goniometer stage. Once in position, focusing and astigmatism correction are carried out while viewing the image directly on the screen phosphor. Through-focus series are facilitated by infinite-turn, continuously variable potentiometer control of the objective lens current, with a minimum focus increment of 0.3 nm. In addition, the stigmator coils are fed by six independent CPU channels, permitting different reference settings to be stored in memory for rapid recall. Operation is further simplified by digital readout of all relevant lens and deflector coil excitation values at the console. Specific data concerning film number, operator code, magnification or camera length, and a 14 character text assigned by a user terminal are also recorded in the margin of each negative, of which there are 50 per camera load.

Two typical applications, one to an aluminum alloy, one to an ionic crystal, are presented in the last two figures.

SUMMARY

Current developments in atomic resolution microscopy are progressing rapidly on two fronts. One is the production of high precision microscopes in the medium ranges of accelerating potential (350 kV to 400 kV) which have the benefit of improved resolution and the advantage of a more compact size, increasing accessibility. The other is the

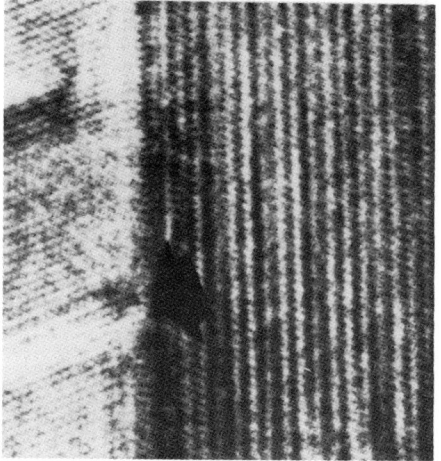

FIG. 9. (a) (top) Atomic resolution image of a four atom layer γ' platelet in an Al-Ag alloy. The long particle dimension is 15 nm.
(b) (bottom) Atomic resolution image of an impingement event between γ' platelet and large ordered γ plate in an Al-Ag alloy. The impingement region again shows four atom layer stacking (arrow), demonstrating the role of Shockley partial dislocations in the growth process. Here the ledge mechanism is observed to be operable at the single atomic plane level.

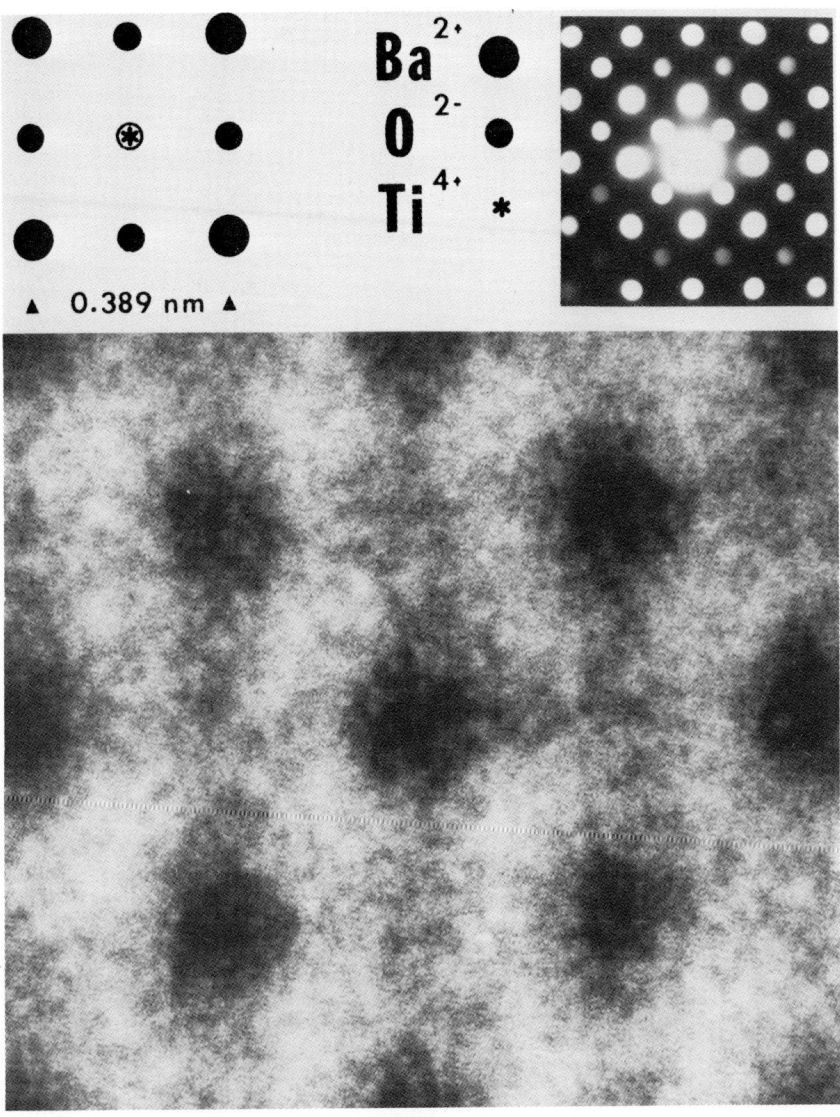

FIG. 10. Ionic positions, electron diffraction pattern and atomic resolution image of a very thin section of barium titanate in an [001] zone axis orientation. The specimen was prepared by crushing in an agate mortar and suspending the powdered crystallites on a holey carbon support film. It was necessary to use the goniometer to tilt one of the crystallites which extended beyond the carbon film edge for this image, recorded at 1000 kV and -60 nm defocus.

extended use of image enhancement through real time video acquisition, digitization and computer processing for better signal-to-noise statistics. The latter approach obviously has higher potential for universal acceptance since it is not microscope-limited.

It should also be noted that there is essentially parallel activity in the further development of the scanning transmission electron microscope (STEM), which was intentionally omitted from this brief review. Although scanning beam systems continue to lag fixed beam systems in imaging quality, they offer outstanding advantages for complementary analyses, most notably in spatially-resolved spectroscopy.

Finally, it is the intention of the ARM project to continue to urge the field of electron microscopy toward higher resolution while providing the field of materials science with a superior characterization tool [15]. Electron microscopy at atomic resolution is now both possible and extremely promising for materials research.

ACKNOWLEDGEMENTS

The author wishes to thank Dr. W. Krakow and the Program Committee for their kind invitation to present this review.

The Atomic Resolution Microscope is a User Facility supported by the Director, Office of Energy Research, Office of Basic Energy Sciences, Materials Science Division of the U.S. Department of Energy under Contract No. DE-AC03-76SF00098.

REFERENCES

1. S. Horiuchi, Ultramicroscopy 10, 229 (1982).
2. J.W. Goodman, Introduction to Fourier Optics, McGraw-Hill, New York (1968).
3. J.M. Cowley and A.F. Moodie, Proc. Phys. Soc. 76, 378 (1960).
4. J. Frank, Optik. 38, 519 (1973).
5. P.L. Fejes, Acta Cryst A33, 109 (1977).
6. O. Scherzer, J. Appl. Phys. 20, 20 (1949).
7. J.C.H. Spence, Experimental High Resolution Electron Microscopy, Clarendon Press, Oxford (1981).
8. W.O. Saxton, Computer Techniques and Image Processing in Electron Microscopy, Academic Press, New York (1978).
9. R. Gronsky in 38th. Annual Proc. Electron Microscopy Soc. Amer., San Francisco, G.W. Bailey (ed.), p. 2 (1980).
10. R. Gronsky and G. Thomas in 41st Annual Proc. Electron Microscopy Soc. Amer., Phoenix, G.W. Bailey (ed.), p. 310 (1983).
11. H. Watanabe, T. Honda, K. Tsuno, H. Kitajima, S. Katoh, Y. Baba, H. Kobayashi, N. Yoshimura, T. Itoh, Y. Harada, S. Sakurai, Y. Noguchi, and T. Etoh in Proc. Seventh Int. Conf. on High Voltage Electron Microscopy, Berkeley, R.M. Fisher, R. Gronsky and K.H. Westmacott (eds.), p. 5 (1983).
12. J.W. Steeds in Introduction to Analytical Electron Microscopy, J.J. Hren, J.I. Goldstein and D.C. Joy (eds.), Plenum, New York, p. 387 (1979).
13. K. Tsuno and T. Honda, Optik 64, 367 (1983).
14. F. Thon, Z. Naturforsch 20a, 154 (1965).
15. A User's Guide to the National Center for Electron Microscopy is available upon request from M. Moore, NCEM, Lawrence Berkeley Laboratory, Berkeley, California 94720.

AN OVERVIEW OF ANALYTICAL ELECTRON MICROSCOPY

D. B. WILLIAMS
Department of Metallurgy & Materials Engineering, Whitaker Lab #5
Lehigh University, Bethlehem, PA 18015

ABSTRACT

Analytical electron microscopy techniques comprising imaging, chemical analysis and microdiffraction are described together with details of the instrumentation required. Using the analytical electron microscope (AEM), the materials scientist can gain combined chemical and crystallographic information with a spatial resolution and sensitivity not available in other imaging instruments. Examples of the application of the AEM to determine solute distribution and crystal structure data are given.

INTRODUCTION

The AEM permits characterization of the chemistry and structure of an electron-transparent specimen using spectroscopic and microdiffraction techniques. Specifically, the AEM gives improved spatial resolution and sensitivity of chemical analysis when compared with more conventional electron-microscope based techniques, e.g. electron probe microanalysis. Similarly, microdiffraction in the AEM offers higher spatial resolution and increased accuracy of crystal structure determination compared with conventional selected area diffraction in the TEM and other common forms of diffraction analysis. When combined with the high resolution imaging and defect analysis capability of a modern TEM, the AEM is the most powerful and indispensable instrument available for the comprehensive characterization of materials.

Given the range of information obtainable in the AEM, an overview article must, of necessity, paint the broad picture while relying on specific references to give more detailed aspects, and two texts are currently available [1,2]. Although AEM is equally relevant to the biological sciences, the examples and techniques discussed here will emphasize the study of crystalline materials, and current limitations of the technique will be noted.

INFORMATION AVAILABLE IN THE AEM

Typical AEMs operate at ~ 100 kV although higher voltage instruments up to ~ 400 kV are now available. When such a high energy beam of electrons strikes a thin-foil specimen, many signals are generated, as shown in Fig. 1a. In theory, all these signals are accessible, but in practice only certain ones are routinely studied. They are (see Fig. 1b): a) transmitted or scattered electrons for imaging, b) x-rays and energy loss electrons for microanalysis and c) the scattered electron distribution (diffraction pattern) for crystal structure determination. Before we study each of these three groups of signals, a brief word on instrumentation is necessary.

AEM INSTRUMENTATION

In order to realize the spatial resolution capabilities of AEM the electron beam has to be confined to a fine probe as in a conventional SEM while retaining the high kV of a TEM beam. However, without the capability of

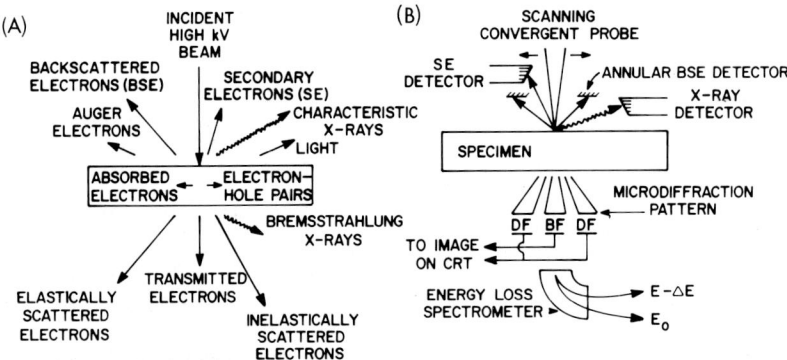

FIG. 1-(a) Signals generated when a high kV electron beam is incident on a thin foil specimen; (b) Typical selection of signals detected in a TEM/STEM.

scanning a fine probe, it is impossible to form images of the region of interest. Hence AEMs are usually based on a scanning transmission (STEM) instrument, although recent improvements in condenser lens optics mean that modern TEMs are capable of generating the necessary fine probes (≤ 10 nm) for AEM while also permitting conventional TEM imaging with beams > ~ 50 μm diameter, at the turn of a control. Under such circumstances a scanning system is unnecessary except in the relatively rare cases when scanning images offer an advantage over static TEM images. (See the section on imaging for a brief discussion of this.)

Since the AEM is primarily a signal-generating instrument the brightness (β) of the electron source is of paramount importance. Because of this the first STEMs to be developed [3] were ultra-high vacuum (UHV) instruments equipped with a field emission gun (FEG) offering values of β of ~ $10^9 A\ cm^{-2} sr^{-1}$. Such instruments are incapable of operating in a conventional TEM mode and can only image using a scanning beam. They are therefore termed 'dedicated STEMs.' While offering high resolution imaging, equivalent to high resolution TEMs, and excellent microanalytical capabilities, dedicated STEMs have, until recently, lacked the electron-optical flexibility of TEM-based instruments. Without any post-specimen lenses (just as a SEM lacks imaging lenses) the microdiffraction capability of a dedicated STEM is poor, compared with the alternative TEM/STEMs. For a detailed comparison of the instruments see reference [4]. In a TEM/STEM it is customary to use a LaB_6 source since this usually gives sufficient brightness (β ~ $10^6 A\ cm^{-2} sr^{-1}$) for all the signals of interest to be detected, and quantified where necessary, without the UHV requirement of a FEG. TEM/STEMs are far more common than dedicated STEMs, and Fig. 1b shows schematically how such an instrument can be used to form images and diffraction patterns, and how x-ray and electron energy loss data can be simultaneously acquired.

IMAGING IN THE AEM

Imaging in the AEM is versatile insofar as any of the signals detectable (see Fig. 1a) can be used (in theory) to modulate a CRT, giving an appropriate scanning image. The most obvious signals are the transmitted and scattered electrons which can be used to form bright field (BF) and dark field (DF) images respectively. Similarly SE and BSE signals are accessible and therefore the AEM can act as a high resolution, high kV SEM, although this capability is rarely used. This is probably because of the restrictions in specimen size in a TEM/STEM compared with a conventional SEM.

For the materials scientist using a STEM there is nothing to be gained from using scanning BF and DF images of crystalline specimens. This is because, in order to recreate in STEM the dynamical contrast information in a BF or DF TEM image, the theorem of reciprocity [5] has to be satisfied. In practice this means that a very small STEM detector collection angle ($< 10^{-4}$ rads) has to be employed [6]. Such images are very noisy (see Fig. 2). In addition the quality of a scanning image, even on a high resolution CRT, does not approach the quality of a conventional image on a photographic plate [7]. Therefore, in the study of crystalline materials the STEM image from an AEM serves mainly as an aid in positioning the probe on a feature that is to be analyzed by spectroscopy or through microdiffraction (see subsequent sections).

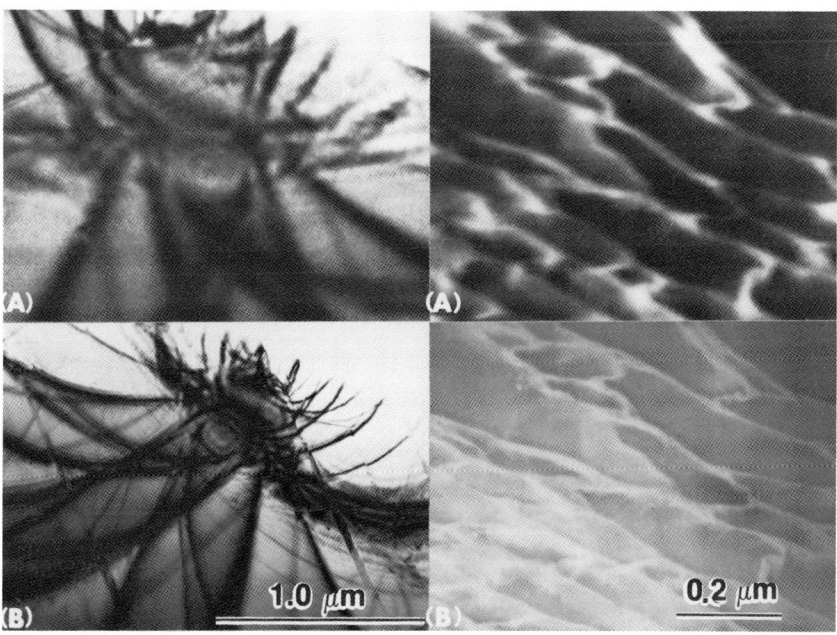

FIG. 2-(a) STEM BF image showing dynamical contrast from a bend center in stainless steel. Note poor contrast and high noise level when compared with a TEM BF image of the same region (b).

FIG. 3-(a) STEM annular DF image of butadiene cell structure in a rubber-modified polymer matrix. Note the increased contrast when compared with a TEM DF image of the same region (b).

The only case where a STEM image offers an improvement over TEM is when considering DF imaging of amorphous specimens, which exhibit mass-thickness (amplitude) contrast. The increased collection efficiency using an annular DF detector compared with an objective aperture in TEM, means an improvement in signal to noise and a consequent improvement in the quality of information in the image (Fig. 3). In addition, because the signal is digitized, conventional SEM signal-processing operations can be performed. This results, in effect in increased contrast/resolution over a TEM at a given thickness, or increased penetration compared with TEM for a given contrast level.

Therefore imaging of unstained polymeric or biological specimens becomes a possibility and improvements of ~ 5X over TEM penetration have been reported [8].

CHEMICAL ANALYSIS IN THE AEM

X-ray Spectrometry

X-ray spectrometry is invariably carried out using an energy dispersive spectrometer (EDS) in the stage of the AEM, usually positioned to look down on the specimen with a take-off angle (angle from plane of specimen to the detector axis) of $> \sim 20°$. This position minimizes corrections of the spectral data for absorption [9], and fluorescence [10], maximizes the characteristic peak to bremsstrahlung background ratio [11] and minimizes the detection of artifacts due to stray radiation interacting with the specimen [2] and the microscope stage [12].

An EDS is used because it is more efficient than a crystal spectrometer, easily fits in the confines of the AEM stage and collects the whole spectrum simultaneously, usually permitting quantitative analysis after acquiring a spectrum for ~ 60-100s. The drawback to EDS is that the Be window, which protects the liquid-N_2 cooled Si crystal from contamination, absorbs low energy x-rays thus restricting those elements that can be detected to $Z \geq 11$ (Na). Advances in detector technology, both in terms of using ultra-thin windows to allow the low energy x-rays to penetrate, and creating low noise-level electronic amplification now permits detection of C K_α x-rays [13] and possibly boron. Such instrumentation is not easily interfaced to a STEM but it is considered that the engineering difficulties will soon be overcome, leaving only Li and Be as elements of interest to materials scientists that cannot be detected using EDS.

A typical EDS output from an AEM is shown in Fig. 4. This was obtained from a thin foil of stainless steel in ~ 60s. A major advantage of EDS in the AEM is that a semi-quantitative analysis accurate to $\leq \pm 10\%$ is often possible, simply by measuring the principal peak heights (above the approximately linear background) with a ruler. (Except at very low energies (≤ 1.5 keV) the background can be approximated to a straight line over small (1-2 keV) energy ranges; background subtraction routines are, therefore, trivial.) Given that the specimen is thin ($< \sim 200$ nm) the only times this procedure will give higher errors than $\sim \pm 10\%$ is if there is significant absorption of a certain x-ray (which only occurs in a few systems of interest [9]) or if the x-ray peaks are of very low energy (≤ 1.5 keV) when the EDS detector itself influences the peak height [14]. Therefore from Fig. 4, it can be reasonably concluded that the specimen is Fe ~ 26% Ni 13% Cr. A full quantitative analysis gives Fe 25.4% Ni 13.6% Cr [2].

To achieve full quantification it is necessary to apply the so-called 'Cliff-Lorimer ratio equation' [15]:

$$\frac{C_A}{C_B} = k_{AB} \frac{I_A}{I_B} \qquad (1)$$

which for a thin foil binary system (A,B) relates the compositions in wt% (C) to the characteristic x-ray intensities (I) through a proportionality factor k_{AB}, which (as can be deduced from the preceding argument) is usually close to unity. Measurement of k_{AB} can be carried out using appropriate standards [16], or by calculation from first principles [17]. The former approach is tedious but more accurate ($\leq \pm 3\%$) while the latter is more error prone ($\sim \pm 15\%$ [18] but can be applied to any elements A,B. Usually a fixed element B is considered (either Si [15] or Fe [19] have been used) and lists of k_{AB} factors are available in the literature [1,2,15,16,19].

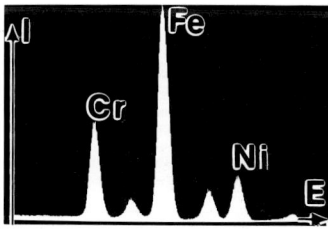

FIG. 4-Typical EDS spectrum from a thin specimen of stainless steel in the AEM. The spectrum is a plot of x-ray intensity I vs energy E. Characteristic (K_α) peaks from Cr, Fe and Ni are indicated on a slowly decreasing background (bremsstrahlung) intensity. The unidentified peaks are K_β peaks from the same elements.

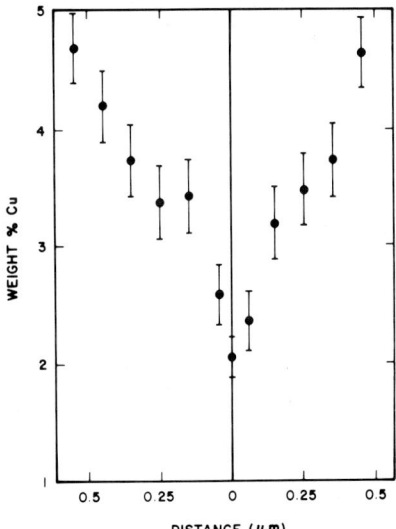

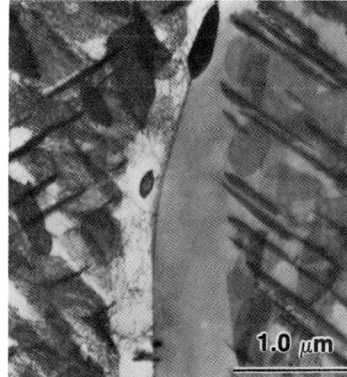

FIG. 5-(a) Cu solute profile across a grain boundary precipitate-free zone (b) in Al-4% Cu aged 0.5 hrs. at 250°C to produce θ on the grain boundary and θ' in the matrix (courtesy J. R. Michael).

Equation (1) only applies if absorption or fluorescence of x-rays is negligible within the specimen. In the few cases where this assumption is invalid, suitable corrections for both absorption [9] and fluorescence [10] are available. These corrections modify equation (1) and are usually incorporated in the software program for thin-foil analysis in the computer attached to the EDS system.

Apart from quantification of a single spectrum, as in Fig. 4, it is obviously possible to generate composition profiles, around a grain boundary for example. Fig. 5 shows a profile of the copper concentration around a grain boundary exhibiting θ precipitation in Al-4% Cu. In order to determine such profiles it is necessary to know a) the minimum mass fraction or detectability limit (C_{DL}) that can be detected and b) the spatial resolution of microanalysis. The former can be determined for element B in A using a simple formula, assuming (a) Gaussian statistics in the characteristic peak, and (b) that the specimen composition is known [20]. The equation is:

$$C_{DL(B)} = \frac{3(2\ I_B^b)^{\frac{1}{2}}}{I_A - I_A^b}\ C_A \qquad (2)$$

where I_A is the intensity of the characteristic peak from element A and I^b is the background intensity under the characteristic peak from element B

(I_B^b) or $A(I_A^b)$. Experiments on homogeneous Cu-Mn solid solution of known composition indicate that C_{DL} for Mn in Cu is ~ 0.08 wt% (800 ppm)[21]. While this may not seem very sensitive, because the analyzed volume is very small the actual mass detected is ~ 10^{-18} gms and theory [22] indicates that ~ 10^{-20} gms should be detectable under ideal conditions. An example of the sensitivity of EDS in the AEM is given in Fig. 6 which shows the detection of less than a monolayer of Bi segregated to a grain boundary in Cu.

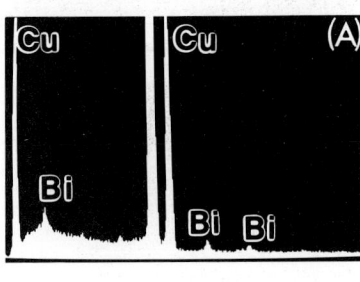 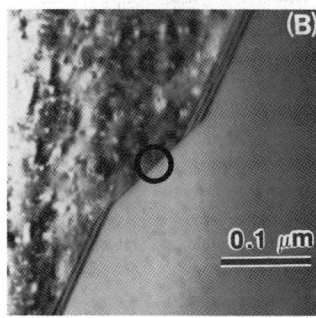

FIG. 6-(a) EDS spectrum showing the presence of Bi (M and L characteristic peaks) in a Cu-0.009% Bi alloy. The Bi is segregated to the grain boundary in the region circled in (b).

The spatial resolution of the analysis depends on both the specimen and the AEM itself. The latter determines the electron beam size and current necessary to generate sufficient x-rays for analysis, while the specimen characteristics determine how much this electron probe will spread, primarily through elastic interactions. Various theoretical models have been used, as reviewed by Newbury [23], and many experimental studies performed to compare the theories. In practice, the simplest model [17] assuming a single electron scattering event gives a reasonable prediction of the beam spreading b through a thin foil of thickness t(cm):

$$b = 625 \frac{Z}{E_0} \left(\frac{\rho}{A}\right)^{\frac{1}{2}} t^{3/2} \text{ cm} \qquad (3)$$

where Z is the atomic number, ρ the density (gm cm^{-3}), A the atomic weight, and E_0 the accelerating voltage (kV). While more sophisticated models may be better in certain circumstances [23], this equation usually suffices to give a very good first approximation to the spatial resolution of x-ray microanalysis.

Electron Energy Loss Spectrometry (EELS)

EELS analyzes the energy distribution of electrons transmitted through a thin-foil specimen. The energy loss spectrum can be divided conveniently into three parts: those electrons that have lost no energy (zero loss), those that have lost < 50 eV (low loss) and those that have lost > 50 eV (high loss). The low-loss electron signal is primarily electrons that have generated plasmon oscillations in conduction band electrons. In Al and Mg base alloys where such losses predominate, some indirect chemical information is available [24]. However EELS is mainly concerned with high-loss electrons, in particular those that have lost more than a certain critical energy E_c required to ionize an inner-shell (K, L or M) electron. In contrast to EDS, EELS is a very efficient technique usually detecting > 75% of the loss electrons [25]. Because of the increase in ionization cross section as Z decreases, EELS is more sensitive to the light elements, such that in an extreme case, EELS of helium is possible [26]. More practically, despite the increased sensitivity to low Z elements, quantification of the data is not routine, for reasons discussed below.

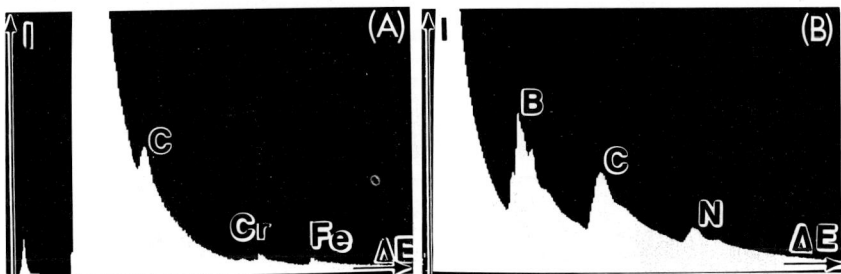

FIG. 7-(a) Typical EELS spectrum from a carbide in stainless steel. The spectrum is a plot of electron intensity I versus energy loss ΔE. The small peak at the left is the zero loss peak and the abrupt intensity increase is a 10,000X gain in the display to show detail in the very low intensity high loss regions. The small peaks superimposed on the rapidly falling background are the ionization loss edges due to C, Cr and Fe. In (b) which is part of a spectrum from BN on a C support film, the increase in background intensity due to having several edges in the spectrum is shown. (Reproduced by permission of Philips Electronic Instruments Inc.)

As shown in Fig. 7a the energy loss spectrum consists of a zero loss peak followed by ionization loss peaks (or edges as they are termed) superimposed on a rapidly falling background. Only ~ 5% of the electrons entering the spectrometer have lost > 50 eV and therefore ionization loss spectrometry for chemical analysis consists of detecting very small edges on a relatively high background. Moreover, the background (due to multiple scattering) increases more rapidly than the edge intensity as specimen thickness increases [25]. Therefore there is an upper limit to specimen thickness for EELS and this may be as small as ~ 50 nm. Intensity drops rapidly with increasing energy loss (ΔE) and therefore losses less than 10^3 eV are usually studied. Very low losses (e.g. LiK at 54 eV, AlL_{23} at 72 eV) have relatively strong edges but are also superimposed on a relatively high background. Furthermore, in multi-element specimens the background intensity for any edge is raised directly depending on how many previous edges exist (see e.g. Fig. 7b). This is because the edge is not a discrete peak as in the EDS spectrum but contains electrons which have lost <u>any</u> amount of energy $\geq E_c$, up to and including E_o, the beam energy. Therefore multi-element spectra have a high background.

Quantification of EELS data is, in theory, straightforward, offering absolute quantification if necessary, or a ratio if preferred. The appropriate equations are [27]

$$N_A = \frac{I_{KA}(\beta,\Delta)}{I_o(\beta,\Delta)} \cdot \frac{1}{\sigma_K(\beta,\Delta)} \quad (4) \qquad \frac{N_A}{N_B} = \frac{I_{KA}(\beta,\Delta)}{I_{KB}(\beta,\Delta)} \frac{\sigma_{KB}(\beta,\Delta)}{\sigma_{KA}(\beta,\Delta)} \quad (5)$$

where N_A is the number of A atoms cm^{-2} contributing to the spectrum, I_{KA} is the intensity in the K edge for element A collected with a spectrometer entrance semiangle (β), and an energy range of integration Δ, I_o is the intensity in the combined zero loss and low loss regions obtained under the same conditions of β and Δ and $\sigma_K(\beta,\Delta)$ is the partial ionization cross section for K shell ionization in element A.

To obtain $I_K(\beta,\Delta)$ the value of β has to be determined and a choice of the integration window Δ made. Egerton [27] has discussed these variables in detail. Secondly the background has to be subtracted. There is no simple way as in EDS, since in EELS the background is always a strongly varying function of the value of the energy loss. In practice analysts have used a curve of the form $A \cdot E^{-r}$ where A and r are constants and E is the energy of the loss electrons [28]. This approach has very specific limitations [29] and in certain circumstances other curve-fitting procedures give better results [30]. To fit the background, the spectrum preceding the edge has to be free of other edges for ~ 50-100 eV. In a multi-component specimen, this is not always possible. Background fitting and subtraction then becomes a matter of estimation.

Nevertheless, given that a background fit can be made and $I_K(\beta,\Delta)$ determined, then quantification requires knowledge of $I_o(\beta,\Delta)$, which can be obtained directly from the spectrum with no background subtraction. The value of $\sigma(\beta,\Delta)$ the partial ionization cross section can be obtained from appropriately modified values of the full cross section [31] or by using a hydrogenic approximation [32]. Errors in either case approach $\pm$ 15% for σ [18] and this value is reflected in a similar error in the value of N or N_A/N_B.

In theory the spatial resolution is governed by the localization of the energy loss interaction and is therefore as small or smaller than for EDS [2]. However there are no experimental studies to support this. Similarly the sensitivity of EELS is considered to be better than EDS but again experimental data are sorely lacking [2].

In summary EELS quantification is not straightforward and is very specimen-dependent. The relative accuracy of quantification is poor when compared with EDS. Furthermore, when ultra-thin window EDS units become available, EELS will perhaps be restricted to microanalysis of those elements with $Z < 5$.

Despite this somewhat gloomy prognosis it should be noted that EELS is not just a microanalysis technique. Within the energy loss spectrum there is information about the specimen thickness [25], the electronic structure of the specimen [33] and the bonding of the atoms in the specimen [34]. Using an EELS it is also possible to energy-filter the electron beam coming through the spectrometer and use the filtered beam to form an image of the specimen. Energy filtering has certain advantages, depending on the specimen, but removing chromatic aberration effects has proved useful in enhancing the contrast in certain specimens [35] as well as in zone axis patterns [36] which are useful displays of the crystallographic symmetry of the specimen. Similarly, if suitable on-line computer analysis is available, chemical images can be obtained [2] in a manner similar to x-ray mapping. It is considered that applications such as these may ultimately be the principal use for EELS, rather than quantitative microanalysis, bearing in mind the current difficulties with spectral manipulation and the requirement for very thin specimens.

MICRODIFFRACTION

In a TEM/STEM, wherever the beam is scanning the specimen, the transmitted and scattered beams are brought back to the same position in the back focal plane of the objective lens. This means a diffraction pattern (see Fig. 1b) is always present here and may be viewed on the TEM screen by appropriate adjustment of the imaging lenses. Because the incident beam is convergent the diffraction pattern consists of discs rather than points and is therefore termed a convergent beam diffraction (CBD) pattern. When the electron probe is positioned on a point of interest in the specimen, in a manner similar to that necessary for x-ray microanalysis, the CBD pattern then comes from a region of the specimen defined by the probe size and the degree of beam spreading, and this may therefore be $\lesssim\sim 50$ nm. This value is an order of magnitude less than the spatial resolution of conventional electron microscope selected area diffraction (SAD) patterns and CBD is therefore a form of microdiffraction. Other microdiffraction techniques such as those due to Riecke [37], Grigson [38] or the Rocking Beam technique [39] can be performed on a TEM/STEM or dedicated STEM. However CBD is the most versatile and widespread in its use, so the discussion will be confined to this method only.

Depending on the thickness of the specimen and the diffraction conditions, the discs in the CBD pattern may be (a) blank, (b) contain parallel (extinction-type) fringes or (c) detailed dynamical scattering information. In the first case (which occurs when the specimen is very thin or away from strong diffracting conditions (Fig. 8a)), no more information is available than in a conventional electron diffraction pattern. Measurement of disc spacings and angles between principal directions permits determination of d-spacings and crystal orientation in the usual manner. The only advantage over SAD in this case is the improvement in spatial resolution that can be achieved. If the specimen is under two-beam conditions, parallel Kossel-Möllenstedt fringes appear in the discs (Fig. 8b), and under certain circumstances these can be used to define the specimen thickness very accurately [40]. If the specimen is on a zone axis the discs contain the extinction fringes which now appear concentric (Fig. 8c). From thick specimens it is also possible (again within strict limitations and experimental conditions as described by Steeds [41]) to observe fine dark lines giving a pattern within the central 000 disc (Fig. 8d). These lines are so-called 'higher order Laue zone' (HOLZ) lines. They are most important because they contain very accurate three-dimensional crystal symmetry information, which permits a full analysis of the point and/or space group of the specimen to be performed [42,43]. They are best observed when the beam is down a zone axis.

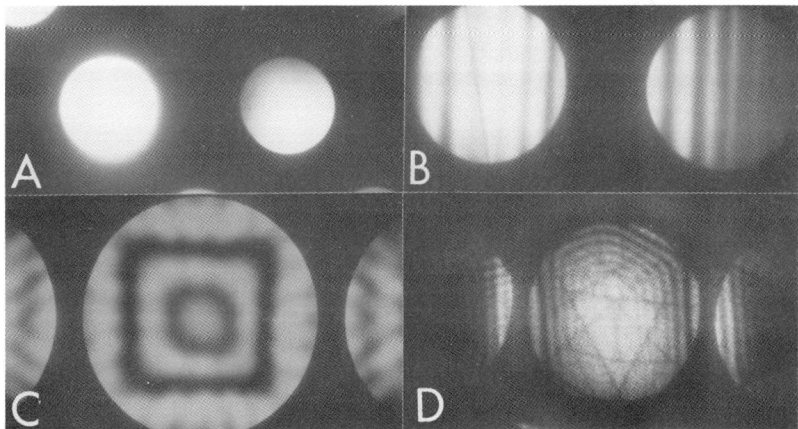

FIG. 8-Aspects of convergent beam microdiffraction patterns. (a) Pattern from very thin sample of stainless steel; (b) Two-beam conditions in a thicker region of the same sample; (c) 000 disc when the specimen in (b) is tilted onto a zone axis; and (d) Defect HOLZ lines in the 000 disc from a very thick region of the specimen (compare the number of extinction fringes in the patterns (c) and (d)).

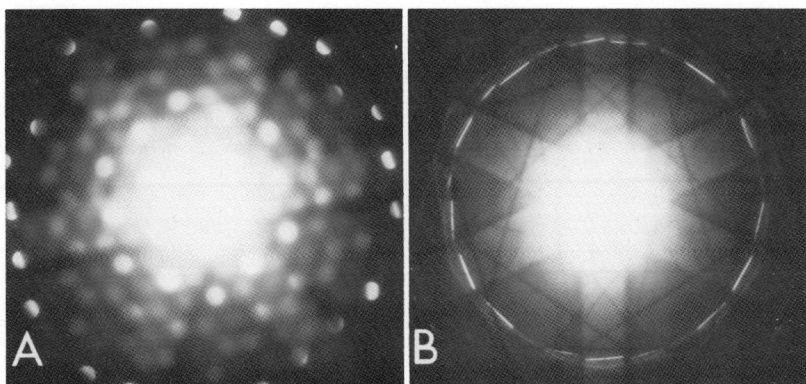

FIG. 9-(a) Diffraction pattern from stainless steel showing diffracted intensity in the first order Laue zone ring of spots. If a thicker sample is used the ring is defined by a series of excess HOLZ lines as shown in (b).

These HOLZ lines are the electron diffraction analog of Kossel lines [42] and are continuous with the Kikuchi lines in the pattern. They arise because, with a convergent beam, it is possible to get reasonable diffracted intensity from planes in the crystal which are not approximately parallel to the electron beam. These planes have reciprocal lattice points in higher order (first, second, etc.) Laue zones than the zero order zone which is all that is seen in conventional SAD patterns. Using modern TEM electron optics it is possible to view the fundamental HOLZ reflections which occur at scattering angles many degrees away from the transmitted beam and a typical example showing a ring of spots corresponding to the first order Laue zone is shown in Fig. 9a. By measuring the radius of this ring, the lattice spacing in the direction parallel to the beam can be determined [42,44] and therefore from a single two-dimensional pattern, a three-dimensional structure can be determined. In thicker specimens the bright lines associated with the HOLZ discs are the excess HOLZ lines (see Fig. 9b), the partners of which account for the fine array of defect lines in the 000 disc, in Fig. 8d.

A combination of all the data in a CBD pattern is extraordinarily detailed information about the crystallography of a minute volume of material. The technique is rapidly becoming the standard method of determining new crystal structures [45] re-defining old (wrong!) structures [46], determining lattice parameter changes with an accuracy of $\sim 2 \times 10^{-4}$ nm [41] and observing symmetry variations on a similar scale [47]. When it is considered that such information is <u>always</u> available on a TEM/STEM, combined with the chemical spectroscopy techniques described above, and that high resolution lattice or structure imaging can be performed on the same instrument, the invaluable role of the AEM in current materials research is obvious.

ACKNOWLEDGEMENT

The author has been funded in his AEM search by grants from the NSF (DMR 81-08308), DOE (DE-ACO2-83ER45016) and NASA (NAG9-45). Their support is gratefully acknowledged.

REFERENCES

1. J. J. Hren, J. I. Goldstein and D. C. Joy, eds., Introduction to Analytical Electron Microscopy (Plenum, New York 1979. Revised edition 1984.)
2. D. B. Williams, Practical Analytical Electron Microscopy in Materials Science (Philips Electronic Instruments, New Jersey 1984).
3. A. V. Crewe, J. Wall and J. Langmore, Science 168, 1338 (1970).
4. I. Y. T. Chan, J. M. Cowley and R. W. Carpenter in: Analytical Electron Microscopy-1981, R. H. Geiss, ed. (San Francisco Press 1981), 107.
5. J. M. Cowley, Appl. Phys. Lett. 15, 58 (1969).
6. D. M. Maher and D. C. Joy, Ultramicroscopy 1, 239 (1976).
7. L. M. Brown, J. Phys. F: Metal Physics 11, 1 (1981).
8. V. E. Cosslett, Phys. Stat. Sol. A55, 545 (1979).
9. D. B. Williams and J. I. Goldstein in: Analytical Electron Microscopy-1981, R. H. Geiss ed. (San Francisco Press 1981), 39.
10. C. Nockolds, M. J. Nasir, G. Cliff and G. W. Lorimer in: Electron Microscopy and Analysis 1979, T. Mulvey, ed. (The Institute of Physics, Bristol 1980), 417.
11. N. J. Zaluzec in Reference [1], 121.
12. D. B. Williams and J. I. Goldstein in: Energy Dispersive Spectrometry (NBS Special Publication No. 604 Washington, D.C., 1981), 341.
13. L. E. Thomas in: Microbeam Analysis-1983, R. Gooley, ed. (San Francisco Press 1983), 70.
14. S. Mehta, J. I. Goldstein, D. B. Williams and A. D. Romig in: Microbeam Analysis-1979, D. E. Newbury, ed. (San Francisco Press 1979), 119.
15. G. Cliff and G. W. Lorimer, J. Microsc. 103, 203 (1975).
16. J. Wood, D. B. Williams and J. I. Goldstein in: Quantitative Microanalysis with High Spatial Resolution, G. W. Lorimer, M. H. Jacobs and P. Doig, eds. (The Metals Society, London 1981), 24.
17. J. I. Goldstein, J. L. Costley, G. W. Lorimer and S. J. B. Reed in: SEM/1977 1, O. Johari, ed. (IITRI, Chicago 1977), 315.
18. D. M. Maher, D. C. Joy, M. B. Ellington, N. J. Zaluzec and P. E. Mochel in: Analytical Electron Microscopy-1981, R. H. Geiss, ed. (San Francisco Press 1981), 33.
19. J. E. Wood, D. B. Williams and J. I. Goldstein, J. Microsc. in press (1983).
20. A. D. Romig and J. I. Goldstein in: Microbeam Analysis-1979, D. E. Newbury, ed. (San Francisco Press 1981), 124.
21. D. B. Williams, J. I. Goldstein and J. R. Michael in: Microbeam Analysis-1982, K. F. J. Heinrich, ed. (San Francisco Press 1982), 21.
22. D. C. Joy and D. M. Maher in: SEM/1977 1, O. Johari, ed. (IITRI Chicago 1977), 325.
23. D. E. Newbury in: Microbeam Analysis-1982, K. F. J. Heinrich, ed. (San Francisco Press 1982), 79.
24. D. B. Williams and J. W. Edington, J. Microsc. 108, 113 (1976).
25. D. C. Joy in Reference [1], 223.
26. R. P. Burgner, O. L. Krivanek and P. R. Swann in: Proc. 40th EMSA Meeting, G. W. Bailey, ed. (Claitors, Baton Rouge 1982), 650.
27. R. F. Egerton, Ultramicroscopy 3, 243 (1978).
28. D. C. Joy, R. F. Egerton and D. M. Maher in: SEM/1979 2, O. Johari, ed. (AMF O'Hare 1979), 817.
29. D. C. Joh and D. M. Maher, J. Microsc. 124, 37 (1981).
30. J. Bentley, G. L. Lehman and P. Sklad in: Analytical Electron Microscopy-1981, R. H. Geiss, ed. (San Francisco Press 1981), 161.
31. M. Isaacson and D. Johnson, Ultramicroscopy 1, 33 (1975).
32. R. F. Egerton, Ultramicroscopy 4, 169 (1979).
33. C. H. Chen, J. Silcox and R. Vincent, Phys. Rev. B12, 64 (1975).
34. B. K. Teo and D. C. Joy, eds., EXAFS Spectroscopy (Plenum, New York 1981).

35. R. F. Egerton, Phil. Mag. 34, 49 (1976).
36. A. Higgs and O. L. Krivanek in: Proc. 39th EMSA Meeting, G. W. Bailey, ed. (Claitors, Baton Rouge 1981), 346.
37. W. D. Riecke, Z. Angew. Physik 27, 155 (1969).
38. C. W. B. Grigson, Rev. Sci. Instrum. 36, 1587 (1965).
39. K. J. van Oostrum, A. Leenhouts and A. Jore, Appl. Phys. Lett. 23, 283 (1972).
40. S. M. Allen, Phil. Mag. A43, 325 (1981).
41. J. W. Steeds in: Quantitative Microanalysis with High Spatial Resolution, G. W. Lorimer, M. H. Jacobs and P. Doig, eds. (The Metals Society London 1981), 210.
42. J. W. Steeds in Reference [1], 387.
43. B. F. Buxton, J. A. Eades, J. W. Steeds and G. M. Rackham, Phil. Trans. Roy. Soc. 281, 181 (1976).
44. M. Raghavan, J. Y. Koo and R. Petkovic-Luton, J. Metals 35, 44 (1983).
45. M. Tanaka, R. Saito and D. Watanabe, Acta Cryst. A36, 350 (1980).
46. K. K. Fung, S. McKernan, J. W. Steeds and J. A. Wilson, J. Phys. C.: Solid State Physics 14, 5417 (1981).
47. M. P. Shaw, R. C. Ecob, A. J. Porter and B. Ralph in: Quantitative Microanalysis with High Spatial Resolution, G. W. Lorimer, M. H. Jacobs and P. Doig, eds. (The Metals Society London 1981), 229.

COMPUTER INTERFACED ELECTRON MICROSCOPE

Y. KOKUBO, S. MORIGUCHI, J. HOSOI, E. WATANABE AND J. NASH*
JEOL Ltd., 1418 Nakagami, Akishima, Tokyo 196, Japan
*JEOL (U.S.A.) Inc., 11 Dearborn Road, Peabody, MA 01960, U.S.A.

ABSTRACT

Some applications of the computer for the electron microscope, in three major areas - 1) control of the microscope, 2) image processing, and 3) structure analysis - are discussed in the present paper.

INTRODUCTION

The papers on application of image processing in electron microscopy can be found as early as mid 1950's. Most of these early applications were made by use of analog circuitry. And it was in the latter half of 1960's that digital processing was practically applied in this field and reported in literature [1]. Since then, digital processing by a computer has been reported very often in a wide range of field [2], including 2D, 3D reconstruction [3] [4] [5] [6] [7]; theoretical image simulation which is restricted to "multislice method" for thin periodic crystals [8] [9]; the same type of image simulation for amorphous specimens [10] [11] [12]; and the interface between amorphous and crystalline materials by wave-optical calculation [13].

INSTRUMENTATION

Fig. 1 shows the block diagram of the JEM-200 TEMSCAN with an interfaced mini-computer. It consists of a 200 kV conventional transmission electron microscope (JEM-200CX), scanning image device (ASID-3D), PDP11/23 mini-computer, 256 kB internal memory and two 10 MB additional disc cartridges with 1 MB floppy disc memory for large mass storage devices. For TEM image acquisition, a high sensitivity silicon intensified visicon is attached to the bottom of the microscope, thus ensuring good image quality on the monitor CRT as low as 1 pA/cm^2 electron density on a TEM screen. Video signal from the silicon visicon are converted into digital signals through a high-speed A/D converter and storaged into frame memory by 512 x 512 pixels with 256 gray level within 1/30 second and also stored into computer memory through DMA. A VTR can also be interfaced to iamge fram memory, to transfer images from frame memory to VTR or from VTR to image frame memory.

In SEM/STEM image acquisition, the computer controls two 12 Bit D/A converters to produce scan signals to drive ASID-3D, which are large as 4,096 x 4,096 pixels elements. A video signal from a photomultiplier or other detectors is converted into a digitized signal through a voltage to frequency (V/F) converter or A/D converter, and thus fed to computer memory. By using an additional several

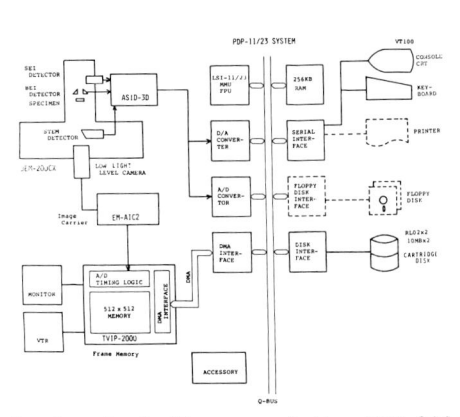

Fig.1. Block diagram of the JEM-200CX TEMSCAN with an interfaced PDP11/23.

Mat. Res. Soc. Symp. Proc. Vol. 31 (1984) © Elsevier Science Publishing Co., Inc.

V/F or A/D converters, it is possible to simultaneously acquire different images or their mixed images of produced by secondary electrons, backscattered electrons, scattered electrons, energy-filtered electrons and transmitted electrons as on-line image processing. Though not mentioned in Fig. 1 a joystick is also interfaced for the purposes below: image mixing ratio control, image brightness and contrast control, gray level control, look up table control, and cursor or marker control for position and hight measurement and other different applications. In our experiment, other interfacings are also carried out, including specimen stage control by a stepping mortor for montage micrographs in the biological field, additional external spectrometer control, and lens control to optimize the electron beam spot size or to optimize the beam convergent angle for X-ray and diffraction analysis.

APPLICATION

A) Electron microscope control

So far, tremendous applications have been carried out as mentioned before. In this section, we shall discuss automatic stigmator control, in ultra high resolution microscopy, hollow cone illumination in TEM dark field microscopy, and line width and height measurement in SEM/STEM respectively.

1] Automated stigmator control

In present electron microscopy, even the direct observation of atomic structure is possible. However, tremendous attention have to be paid to beam axis alignment, optimum objective lens focussing, and stigmator control; it is not too much to say that these jobs require artisan-like techniques. Especially in stigmator control, optical Fourier transformation by a laser beam (optical diffraction method, ODM) is dominant to check the symmetry of the power spectrum of an electron micrograph and repeat this measurement for optimum astigmatism correction. However, this technique requires a lot of time for photographic processing and measurement. A computer replaces this ODM to digital diffraction method (DDM) by using digital Fourier transformation and fix optimum astigmatism correction in a short time. An automatic amorphouse silicon image obtained by 650 k direct magnification with 5,000 Å underfocus is shown in Fig. 2(A), where the image appears to have no astigmatism. Fig. 2(B) shows the power spectrum of the image in Fig. 2(A). From this distorted power spectrum, it is clearly seen that some astigmatism is involved in the image. A binary image of the power spectrum - after space domain filtered - is shown in Fig. 2(C) to measure astigmatic defocus difference. Generally, the astigmatic defocus difference, δ_z is defined by the following formula using the major axis (r_ℓ), the minor axis (r_s) and azimuth α in an ellipse.

$$\delta_z = \frac{n}{\lambda K^2 M^2} \{\frac{1}{(2r_s)^2} - \frac{1}{(2r_\ell)^2}\} + \frac{C_s \lambda^2 K^2 M^2}{2}((2r_\ell)^2 - (2r_s)^2) \quad (1)$$

where K : camera length for Fourier transormation
  C_s: spherical aberration
  λ : wavelength
  M : magnification
  n : order of ring

And correction currents for the X and Y stigmators, I_x, I_y, for astigmatic defocus difference δ_z are given below:

$$\delta I_x = C \cdot \delta_z \cdot \sin (|\theta - \alpha| - \pi/4) \quad (2)$$

$$\delta I_y = C \cdot \delta_z \cdot \cos (|\theta - \alpha + \pi/4| - \pi/4) \quad (3)$$

where C: const (3.62×10^3)
  α: aximuth angle

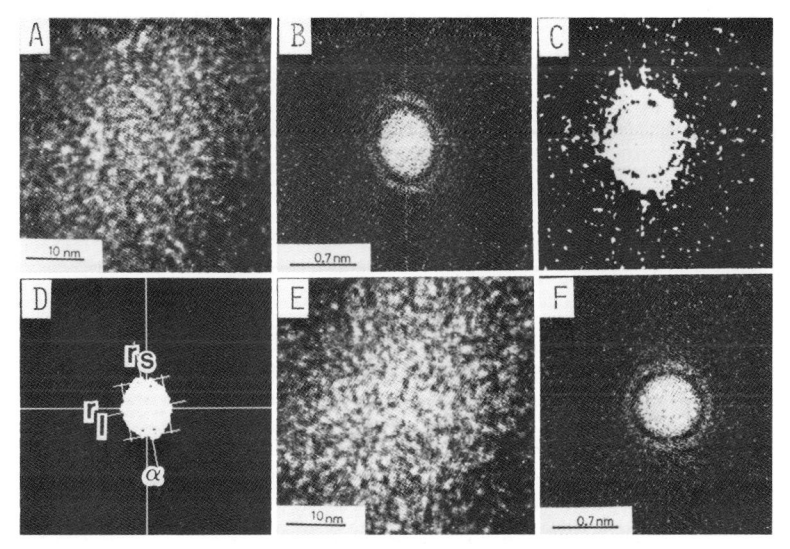

Fig. 2. Automatic stigmator control

Fig. 2(D) also shows the automatic measurement of the major axis (r_ℓ), the minor axis (r_s) and azimth α by a computer for formulas (1), (2) and (3) to set optimum stigmator. The final result of this automatic astigmatism correction for the image is shown in Fig. 2(E), and a power spectrum of the iamge is also shown in Fig. 2(F), ensuring perfect astigmatism correction, unlike Fig. 2(B). As shown here, the digital diffraction method (DDM) using a computer saves tremendous time and labor, and is very effective for automatic stigmator control.

2] Hollow cone inllumination

Hollow cone illumination is one of the well-known techniques of dark field microscopy to get good image contrast. Usually, an annular type condenser aperture is used for electron beam symmetrical tilt illumination (cone illumination), where only electrons scattered within a certain limited angle were used for image forming. However, this mechanical annular aperture does not cover all tilt angles for different types of specimens. Electrically controlled hollow cone illumination [14] was reported in stead of the annular aperture method to expand the application. Fig. 3 shows a block diagram of hollow cone illumination using computer.

Two scan signals as a function of time (t) - one is $R(t)\cos(\theta t)$ and the other is $r(t)\sin(\theta t)$ from computer - are simultaneously fed to the X and Y coils of both the double deflection coil of the microscope and deflection coil of the CRT display. It is clear that $R(t)$ controls tilt angle and $\cos(\theta t)$, $\sin(\theta t)$ control rotation of electron beam. Therefore, by changing the time (t) step by step, the 3D cone illumination of the electron beam is available without changing the beam position on a specimen. Fig. 4 shows a series of typical hollow cone illuminated diffraction

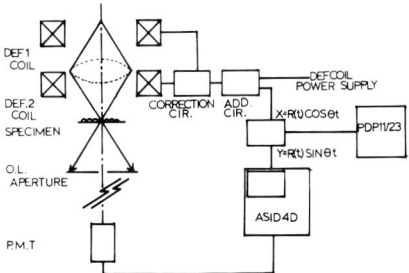

Fig. 3. Block diagram of hollow cone illumination using computer

patterns which correspond to these different tilt angles of a MOS$_2$ specimen on microscope screen. As shown here, it is a little inconvenient to set optimum tilt angle from these picture; therefore, transmitted electrons through the objective aperture are detected by a photomultiplier and fed into a display CRT for brightness modulation as shown in Fig. 3; thus diffraction pattern observation on CRT display is ensured. A series of results of an Au (100) specimen is shown in Fig. 5. At (A) is result when not only sin(θt), cos(θt) but also R(t) scaned full of rang. At (B), (C), (D) and (E) are results when R(t) is adjusted to cover only limits of {200}, {220}, {400} and {420} orientation by using joystick control. In this way, the operator can set the optimum tilt angle which satisfies any bragg's condition while observing a corresponding diffraction pattern on the CRT display using a joystick. Once the optimum tilt angle is fixed, merely switching from the diffraction mode to image mode, makes it possible to find hollow cone dark field image corresponding to that tilt angle on the flourescence screen.

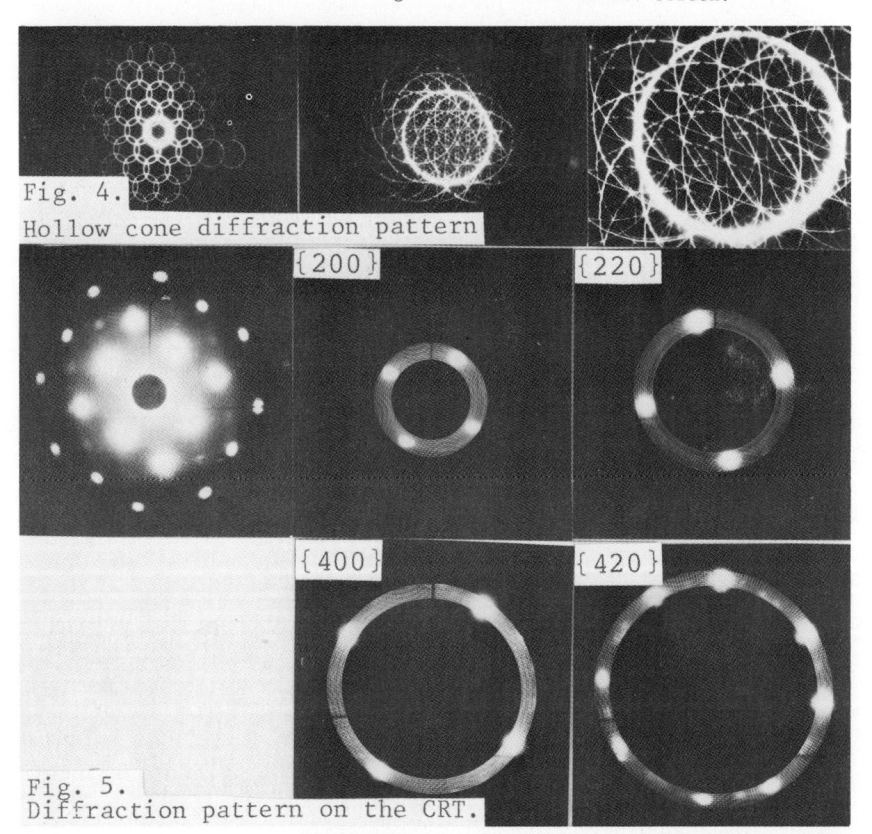

Fig. 4. Hollow cone diffraction pattern

Fig. 5. Diffraction pattern on the CRT.

3] Line width and height measurement

One of big applications of scanning electron microscopy is IC inspection for measurement of line width and height of an IC chip or photoresist. Fig. 6 shows an example of line whidth measurement of an IC specimen. Several cross markers are superimposed on any points of image to measure each position through joystick control and the computer calculates the minimum distance between two of any markers automatically, and prints out the results. The X, Y positions of each cross marker and the distances between the original extreme left marker to each right marker are printed out.

In height measurement in electron microscopy, stereoscopy is well known as a technique using two images obtained with different tilt angles. In this experiment, two images are superimposed by green and red colors on a color coded CRT to view a stereo image through stereo glasses. A relative height meaurement was carried out by combination of relative tilt angle difference, image shift value of two images and image magnification on CRT display. The operator can set a pair of cross markers - one is green and the other is red - on anywhere on the CRT display for point of interest, and control the distance between two cross markers so as to read image shift accurately by joystick control. Fig. 7 shows the result of IC specimen height measurement of four different points A, B, C and D in the image. The X and Y coordinates of each point are listed, and relative height are also printed out. From this result, it is easily understandable that A, B, C and D are located from the surface to above the specimen respectively. In this way, use of a computer or microprocessor allows for simple measurement and calculation.

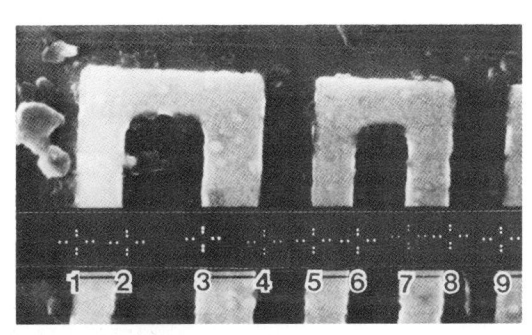

Fig. 6. Line width measurement of an IC specimen.

serial	mag(x 1K)	x1	y1	x2	y2	distance
1 :	1.500	3.059	6.537	4.058	6.563	1.000
2 :	1.500	3.059	6.537	5.674	6.563	2.615
3 :	1.500	3.059	6.537	6.963	6.563	3.904
4 :	1.500	3.059	6.537	7.871	6.563	4.812
5 :	1.500	3.059	6.537	8.781	6.563	5.723
6 :	1.500	3.059	6.537	9.774	6.563	6.715
7 :	1.500	3.059	6.537	10.706	6.563	7.647
8 :	1.500	3.059	6.537	11.749	6.563	8.691

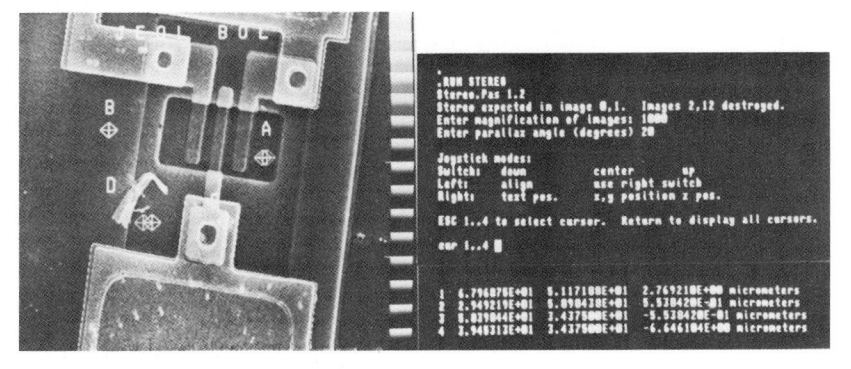

Fig. 7. Height measurement by stereoscopy

B] Image processing
Once SEM/STEM and TEM images are stored into computer memory, many types

of image processings [15], [16] are available, including not only those mentioned before in the section of instrumentation, but also image contrast, and fine structure enhancement. In this section, gray scale expansion, image differentiation, digital filtering and image reconstruction by a combination of Fourier and inverse Fourier transformation are discussed.

1] Gray scale expansion

Intensity hystorams of images are very important to grasp image quality quantitatively. The most basic image contrast enhancement is carried out by setting the optimum lower and upper limits of the intensity hystogram for imaging. Fig. 8(A) shows a TEM image of polyovirus negatively stained with 0.5% uranyl acetate and 3% ammonium molybdate. (B) shows the result of digital diffraction pattern of (A). And (C) shows the result of gray scale expansion; it is seen that the contrast improved and noise also increased. (D) shows a digital diffraction pattern of (C); it is seen that using gray scale expansion high order information was newly included.

Fig. 9 is also another example of a paracrystal tropomyosin specimen negatively stained with 2% uranyl. (A) is an original image, (B) a digital diffraction pattern of (A), (C) a result of gray scale expansion, and (D) a digital diffraction pattern of (C). And it will be seen that high order information is obtained in (D) as shown by arrow, through gray scale expansion. From these two examples, it is rather say that gray scale expansion increases the image contrast but also the noise level. Therefore, this technique is not always effective for improving the image contrast. However, when this technique is combined with another technique which will be discuss latter, good results will be obtained.

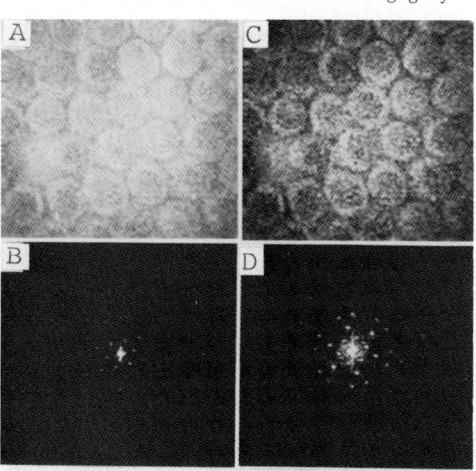

Fig.8. Gray scale expantion for polyovirus specimen.

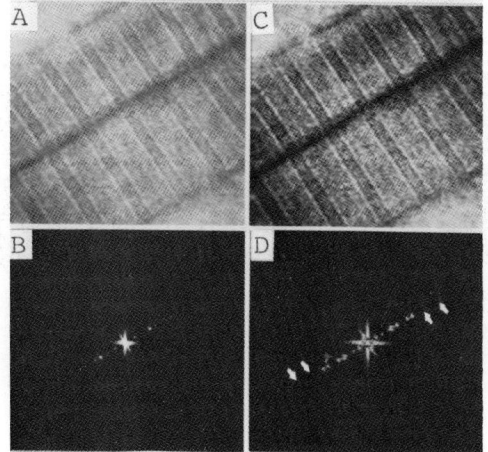

Fig.9. Gray scale expansion for paracrystal tropomyosin.

2] Image differentiation

Image differentiation is one of the well-known techniques for image contrast enhancement and image crisping. Fig. 10(A) shows an original STEM image of a MOS_2 specimen. A result of image differentiation by Sobel kernel is shown in Fig. 10(B) with extremely high contrast. And another result of a combination of an original image and a differentiated image by Sobel is shown in Fig. 10(C), and a result of a combination of an original image and a Laplacian image is shown in Fig. 10(D). Those results indicate that image is improved not only in contrast but also in cripness.

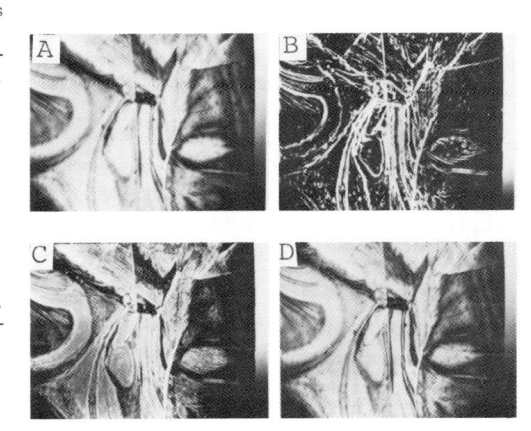

Fig.10. Image differentiation and image mixing.

3] Image filtering

Image filtering is also another technique for image quality improvement by eliminating unnecessary signals such as white noise. A lot of different digital filters are reported, but here, we shall show some results obtained by low pass filter and a median filter. Fig. 11(A) shows the low pass filtered image of Fig. 8(C) - reduced noise component. Because of the low pass filter, the noise component decreased but contrast decreased a little bit. Fig. 11(B) shows the median filtered image of Fig. 8(C). It will be seen that there is some difference between the low pass filter and the median filter, but here without discussing the details we shall proceed to the next subject.

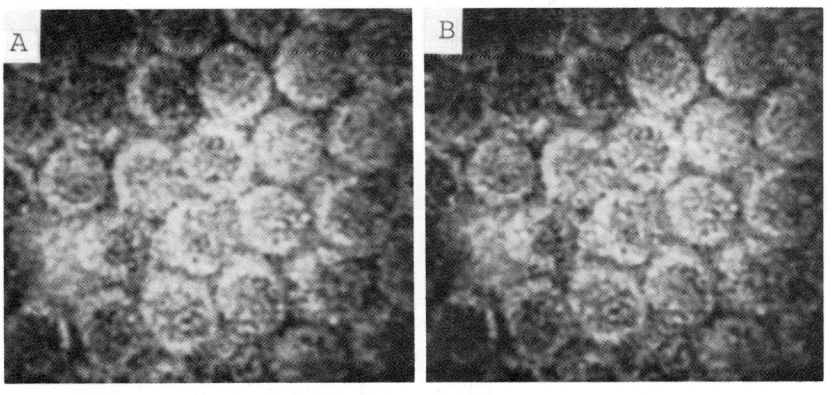

Fig.11. Low pass (A) and median filter image(B)

4] Image reconstruction by Fourier and inverse Fourier transformation

Image reconstruction is possible with the previously discussed combination technique followed by Fourier and inversed Fourier transformation. Image reconstruction is carried out by using only selected spots in a digital diffraction pattern to crisp the image through inversed Fourier transformation. As shown in Fig. 12(A), all spots within a circle are considered for inversed Fourier transformation to the image reconstruction of Fig. 10(A). And the final reconstructed image is shown in Fig. 12(B). Same type of result for paracrystal tropomyosin is shown in Fig. 13(A) and (B); diffraction spots contributed for inversed Fourier transformation are circled in Fig. 13(A), reconstructed image in Fig. 13(B). As seen from both results, the combination of different types of image processing improves image contrast and crispness, discriminates white noise and provides new information.

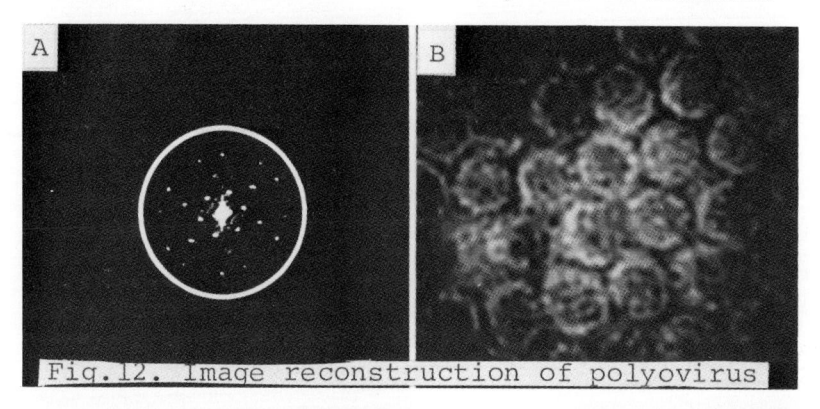

Fig.12. Image reconstruction of polyovirus

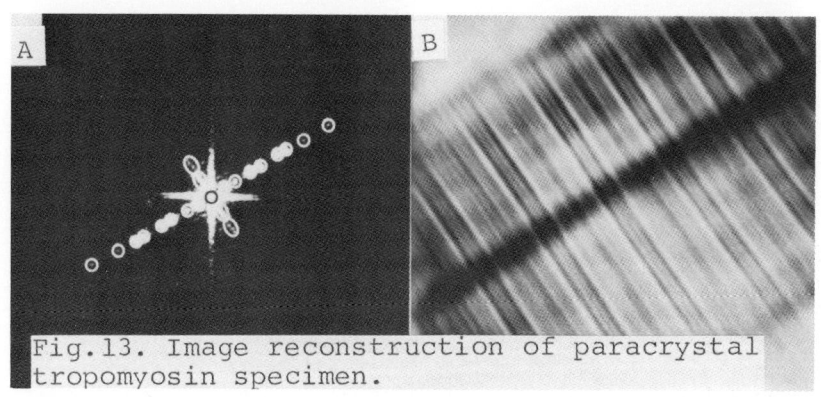

Fig.13. Image reconstruction of paracrystal tropomyosin specimen.

Formerly, these types of image processing were carried out by a combination of a photographic technique and optical digital method using laser beam, however, in this case, there is some possibility of false information being involved in each process, and therefore, real information is not always enhanced. Digital image processing by a computer ensures really acuurate results in an extremely short time by on-line processing.

C] Automated electron diffraction pattern analysis

Electron diffraction pattern analysis, which is an indispensable for structure analysis by electron microscopy, provides information on lattice constant, crystal system, crystal orientation. Formerly, this analysis required a lot of time for photography, measurement of d spacing and indexing process and also demanded crystallographic knowledge of the microscope operator. Combination of an electron microscope with a computer and frame memory, however, now allows this analysis to be carried out with extreme simplicity in a short time by on-line processing.[17] As shown in Fig. 1, a diffraction pattern on a fluorescenct screen is captured by a high-sensitivity silicon visicon and transfered to frame memory, and a computer calculates all different d values, and compare the d value with a crystal file of standard materials stored in memory, and finally prints out the element names, molecular formula, crystal system and lattice orientation of the specimen. A block diagram of software is shown in Fig. 14; it consists of three modules: an acquisition module, analyzing module and index module. An acquisition module has two branches; one is for diffraction pattern acquisition by a high-sensitivity silicon visicon in TEM mode, and the other is for STEM μμ diffraction pattern acquisition by a photomultiplier in STEM. An analysing module has two functions, one is measurement of d spacing from a stored diffraction pattern and the other is to compare these d value with the contents of file, in which the crystal system, molecular formule, d values for typical 650 metallic and alloy elements are included, and identify the element name. Once the specimen is identified, then indexing module carried out for indexing for each diffraction spot or diffraction ring to identify crystal structure. In this system all these processes are carried out fully automatically.

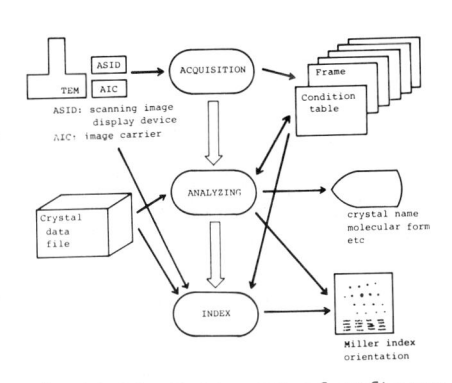

Fig.14 Block diagram of software

Fig. 15 shows results of d value measurement of a MgO specimen; six measured d values are superimposed in the upper left of the diffraction pattern. These d values are compared with the file to retrieve the material name and results are shown in Fig. 16; three different material names with 70 percent existing possibility, which are supposed to correspond to the six d values are listed in the upper left of the pattern. Once material is identified by combination with on X-ray spectrometer or energy analyzer, then indexing is carried out. The final results are shown in Fig. 17; the material name (MgO) and crystal system (FCC) are shown in the upper left and indices which correspond to the individual numbered diffraction spots are also listed on the lower part of the image. Thus, automatic electron diffraction system by computer does not require any crystallographic knowledge, or much operator labor.

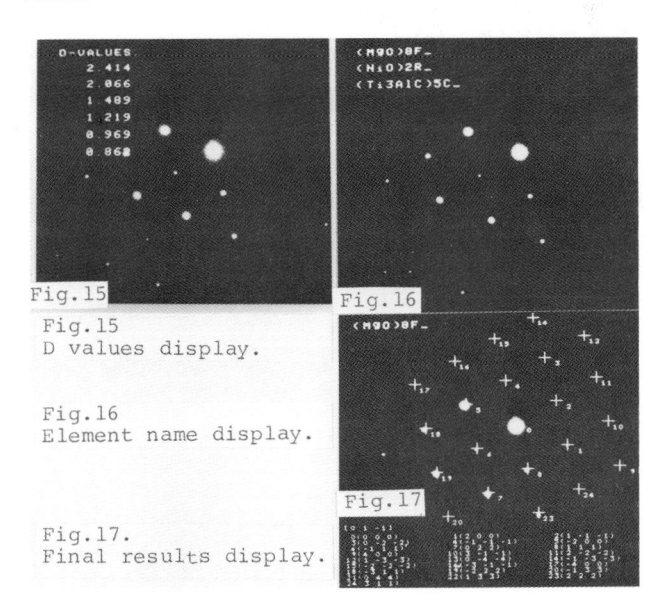

Fig.15
D values display.

Fig.16
Element name display.

Fig.17.
Final results display.

DISCUSSION

The various data shown here are only a handful of our results. It is not too much to say that the application of the computer interfaced electron microscope is really boundless for the major fields of electron microscopy, such as metallurgy, mineralogy, semiconductor research and biology. It will be an important subject for microscope manufactures to make efforts for easy operation and reduction in cost by using a computer. And for manufacturer and researcher, it will also be one of important new subjects to work together to challenge for huge image processing, in order to extract real physical meanings from different types of information and thereby establish standard such as user interface standard, control communication standard between the microscope and computer, high speed bus standard, and software standard.

REFERENCES

1. R. Nathan, Image Processing for Electron Microscopy: I. Enhancement Procedures, in : Advances in Optical and Electron Microscopy, Vol. $\underline{4}$. Eds. R. Barer and V. E. Cosslett (Academic Press New York 1971).
2. P. W. Hawkes, Ed; Computer Processing of Electron Microscope Images (Springer-Verlag New York 1980).
3. L. A. Amos and A. Klug: J. Mol. Biol $\underline{99}$ (1975) 51-73.
4. J. K. Stevens, T. L. Davis, N. Friedman and P. Sterling: Brain Research Review $\underline{2}$ (1980) 265-293.
5. J. Frank, B. Shimkin and H. Dowse: Ultramicroscopy $\underline{6}$ (1981) 210-211.
6. R. Hegerl and A. Altbauer: Ultramicroscopy $\underline{9}$ (1982) 109-116.
7. E. J. Kirkland: Ultramicroscopy $\underline{9}$ (1982) 45.
8. J. M. Cowley and A. F. Moodie: Acta Crystal $\underline{10}$ (1957) 609-619.
9. S. Horiuchi et al: Acta Crystal $\underline{A32}$ (1976) 558-565.
10. W. Krakow: Ultramicroscopy $\underline{4}$ (1979) 55-76.
11. K. Kanaya et al: J. Electron Microscope $\underline{26}$ No. 1 (1977) 1-6.
12. T. Oikawa et al: Ultramicroscopy $\underline{4}$ (1979) 181-191.
13. W. Krakow: Ultramicroscopy $\underline{5}$ (1980) 175-194.
14. W. Krakow and L. A. Howland: Ultramicroscopy $\underline{2}$ (1976) 53.
15. W. O. Saxton: Computer Techniques for Image Processing in Electron Microscopy (Academic Press New York 1978).
16. Y. Kokubo and W. H. Hardy: Ultramicroscopy $\underline{8}$ (1982) 277-286
17. S. Moriguchi et al : 41th Annual Meeting EMSA (1983) 400

MICROANALYSIS WITH HIGH SPATIAL RESOLUTION

P. E. BATSON, C. R. M. GROVENOR, D. A. SMITH, and C. WONG
IBM Thomas J. Watson Research Center, Yorktown Heights, N.Y. 10598

ABSTRACT

Elemental microanalysis, using x-rays and electron energy loss scattering, has been shown to be possible with electron beam probe sizes down to 0.5nm. This paper will discuss some practical problems, such as specimen drift, signal magnitude, and probe-specimen interaction when the probe is made very small. These problems have arisen in two studies: 1) an investigation of As segregation in poly-crystalline Si and 2) imaging of metal spheres with surface and bulk plasmon inelastic scattering.

INTRODUCTION

The Scanning Transmission Electron Microscope equipped with a Field Emission Gun promises to provide practical analysis with a spatial resolution comparable to the 0.5nm resolution of present systems. This is made possible by the large current density (~0.2 nA into a 0.5 nm probe) provided by the field emission source. It must be remembered, however, that this total current is much smaller than the typical 1 μA currents found in SEM and TEM systems. Therefore, the treatment of analytical signals is made more complicated. Secondly, the probe is small and, because the probed system may also be small, the scattering process can be affected by interface structure and probe focussing parameters. It is desirable, therefore, to discuss some of the practical problems which must be addressed in order to realize the potential promised by the techniques.

CHARACTERISTIC X-RAY PRODUCTION

Many straightforward analytical situations arise when multiphase products are formed during annealing of evaporated metal layers. A typical example occurs when Pd is evaporated onto GaAs [1]. Low temperature anneals (~250°C) of Pd films on (110) substrates produce ≈ 50nm crystallites of PdGa embedded in a matrix of a single crystal phase. Diffraction pattern analysis, however, does not unambiguously identify the matrix. Use of the x-ray detector easily establishes a nominal composition of Pd_2GaAs for this matrix. The FEG equipped STEM is therefore important in this case simply because it provides a convenient extension of SEM techniques to systems which are too small for convenient SEM analysis. In addition, since it is a transmission device, it provides a good correlation of structure with the TEM. In the above example, samples were first characterized for structure in the TEM. Interesting areas were then analyzed for composition in the STEM. Since the STEM may be run without regard to detailed setup for these situations where ~3nm spatial resolution is adequate, operator experience and expertise are not critical.

A difficult set of operational problems arise when simple analyses like the above are applied to trace elements; for instance, for dopants in a semiconductor. In this case, we wish to measure concentrations in the 0.1–1.0 atwt.% range with spatial resolutions in the 0.5–20nm range to measure dopant segregation at interfaces. Thus for the example below, at 0.4 atwt.% As in Si implies a needed detection sensitivity of $\sim 8 \times 10^{-21}$g of As (within a 3nm diameter volume through a 50nm foil) -- about 80 atoms. For the 1na beam current, a typical cross section for inelastic scattering of about 10^{-22} cm^2 for the As K_α excitation , we obtain $\sim$ 2c/s at the x-ray detector in the As peak. Clearly then, system background signals must be below 0.1c/s for accurate, reproducible concentration profiles.

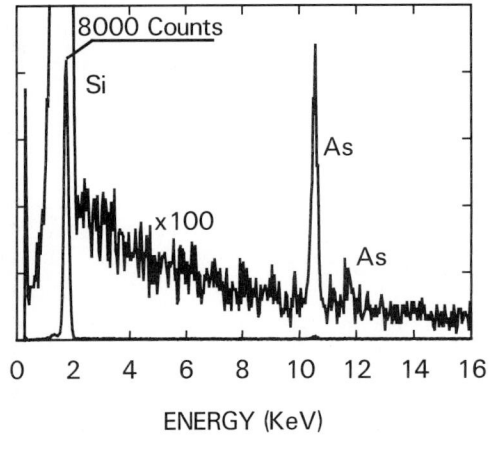

FIG. 1. X-ray spectrum for 0.4 atwt.% As in Si. This was taken with a total count rate of $\sim$ 50c/s and a total counting time of 330 s. Peak to background ratios of 500 are typical in the dedicated STEM.

The example energy dispersive x-ray spectrum shown in Fig. 1 was obtained in the matrix of polycrystalline Si doped to 0.4 atwt.% with As for a study of grain boundary segregation described in detail elsewhere [2]. The spectrum required 330s collection time, eventually accumulating 8000 counts in the Si K_α peak and about 80 counts in the As K_α peak. The bremstrahlung background in the region of the As peak was ~0.02c/s. The important result of the figure is that the peak to background ratio in this Si example was ~500. Typical TEM and SEM systems encounter ratios of ~50. Clearly, trace element analysis would be very difficult in that case. Secondly, notice the absence of spurious characteristic peaks due usually to irradiation of specimen support structure and/or lens surfaces. The extremely clean nature of this spectrum is typical of the dedicated STEM performance due to the particular geometry of the electron beam collimation. In the VG Microscopes Ltd., HB5 STEM, the probe collimation is accomplished prior to the probe forming lenses, at a distance of $\sim$ 25 cm from the specimen and x-ray detector. A second collimation aperture, which does not physically touch the electron beam, is inserted between the first aperture and the specimen. This aperture blocks hard radiation generated at the first collimation aperture. As a result the usual flood of x-ray radiation from the beam aperture is eliminated. Hole count rates due to secondary fluorescence of less than 0.01c/s are therefore typical.

To obtain an x-ray profile across a grain boundary with high spatial resolution, we do the usual thing and introduce an energy window around the As K_α peak. We scan a line segment which is approximately 60nm wide, breaking it into 37 channels 1.8nm wide. Many 0.5s line scans were added together over a long integration time to obtain good statistics. The results for one grain boundary in a sample annealed at 1000°C are shown in Fig. 2. This shows an enhancement of the As signal at a position closely correlated with the image of the grain boundary. The bremstrahlung background intensity determined from Fig. 1 is indicated by the dotted line. The total time to acquire this result was ~1 hour.

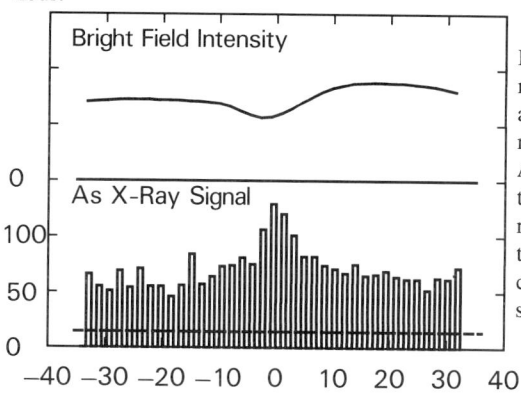

FIG. 2. Bright field and As x-ray signals for a line scan across a single boundary. Typical x-ray count rates were 1c/s in the As K_α peak. The total integration time was ~1 hour. Spatial registry was maintained to better than 0.5nm by periodic cross correlation of the bright field signal with an original.

The largest obstacle to the extension of the simple acquisition of Fig. 1 to the profile in Fig. 2 was the stabilization of the typical ~ 2nm/minute spatial drift rate encountered in the STEM. This results from mechanical and electrical drifts, and possibly from thermal expansion of the specimen under beam heating. To provide information on this drift, the bright field, or in some cases a dark field, signal was used to indicate the position of the boundary structure. This signal was obtained simultaneously with the x-ray signal on a Tracor-Northern TN-4000 multi-channel analyzer. To help alleviate operator fatigue, the structure signal was sampled about 4 times a minute and compared, via a simple cross correlation, with a scan taken at the start of the experiment. This was accomplished in an IBM Series/1 computer interfaced to the TN-4000. Finally an image shift was computed from the correlation peak and applied to the STEM scan amplifiers for a position correction. Registry perpendicular to the grain boundary was maintained to 0.5nm. The system which makes this operation feasible is described elsewhere [3].

It is clear then that systems designed to exploit the high spatial resolution allowed by the FEG for EDX analysis must have well designed probe collimation systems; they must be efficient in their utilization of the limited beam current which is available, and they must include some means to dynamically control the specimen drift. In passing, this particular study revealed ~40% variations in the As grain boundary enhancement for different boundaries in the same sample, indicating the possibility of a segregation dependence on structure. Thus, these studies require several measurements per sample to get a good statistical base. This study investigated ~40 boundaries in 5 samples.

IMAGING WITH INELASTIC SCATTERING

It has been shown recently [4] that imaging in the STEM with inelastically scattered electrons can provide qualitative compositional information on the 1-10nm length scale. We do this by setting energy selecting slits around an inelastic feature of interest and monitor the intensity passed by the slits as the probe is scanned over the sample. Use of the bulk plasmon intensity readily distinguishes regions of uniform dielectric composition with a resolution as good as 1nm. Surface plasmon scattering can emphasize interfaces, revealing information related to the surface quality.

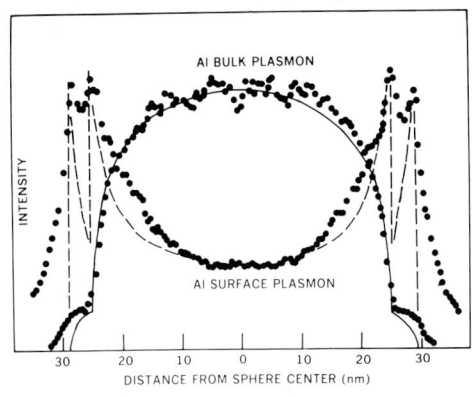

Fig. 3. Inelastic bulk and surface plasmon scattering for scans across a 60nm Al sphere. Comparisons with simple model calculations indicate that multiple scattering to bulk losses must be included even for passage through the 4nm oxide layer.

Figure 3 shows an example of an inelastic scattering line scan through the center of a 60nm diameter Al sphere coated with 4nm of Al_2O_3. Data for both the bulk and surface plasmons are shown. Notice that we may obtain spatial information on a much finer scale than the typical plasmon wavelength of 50-100nm suggests. This results because the inelastic imaging experiment does not sample the plasmon directly, but, rather, displays the change in scattering probability as the probe position is changed. This probability can be quite sensitive to conditions very close to the probe, producing a relatively high resolution technique. For the bulk plasmon scattering in the figure, we have compared the intensity to a simple multiple scattering distribution controlled by Poisson statistics [5].

$$I_p \propto t\sigma_p \, e^{-t\sigma_{TOT}} \qquad (1)$$

where I_p is the observed bulk plasmon intensity, t is the sphere thickness in projection, σ_p is the plasmon single scattering cross section and σ_{TOT} is the scattering cross section for all inelastic processes in the sample. The effect of the surface oxide is included approximately. This result (solid line) is in good agreement with the measured data. If multiple scattering is ignored, the flattened intensity near zero impact parameter cannot be reproduced. Near the metal core edges (at 25nm), this technique identifies the core boundaries with good resolution.

The surface plasmon intensity shows the positions of the major system interfaces. This results mainly because the surface plasmon is most easily excited when the plasmon wave vector lies in a plane parrallel to the incident electron wave vector. Therefore, the scattering intensity follows, roughly, a $1/\cos\theta$ dependence where θ is the angle between the surface normal and the fast electron trajectory. This prediction for the inner and outer surface of the oxide layer is shown as the dashed line in the figure. Of course, this model does not predict scattering of finite impact parameters, so the scattering exterior to the sphere, and the enhanced scattering within the sphere near the exterior interface (near 20 nm), is not predicted [6,7]. The comparable peak heights observed at the metal-oxide and at the oxide-vacuum interfaces are confusing because they suggest that these interfaces support equal amounts of oscillating surface charge. This result is not predicted by more sophisticated theories [8]. A tentative explanation for this behavior may be formed by considering, as above, multiple scattering to bulk losses. When the fast electron follows a trajectory which links the bulk material, it may lose energy to bulk excitations after exciting a surface excitation. The surface loss therefore will be reduced in intensity via multiple scattering in the same way that bulk losses were above. If we realize further that the elastic scattering is reduced by this effect in a similar manner, we may calculate a single scattering probability for surface plasmon excitation:

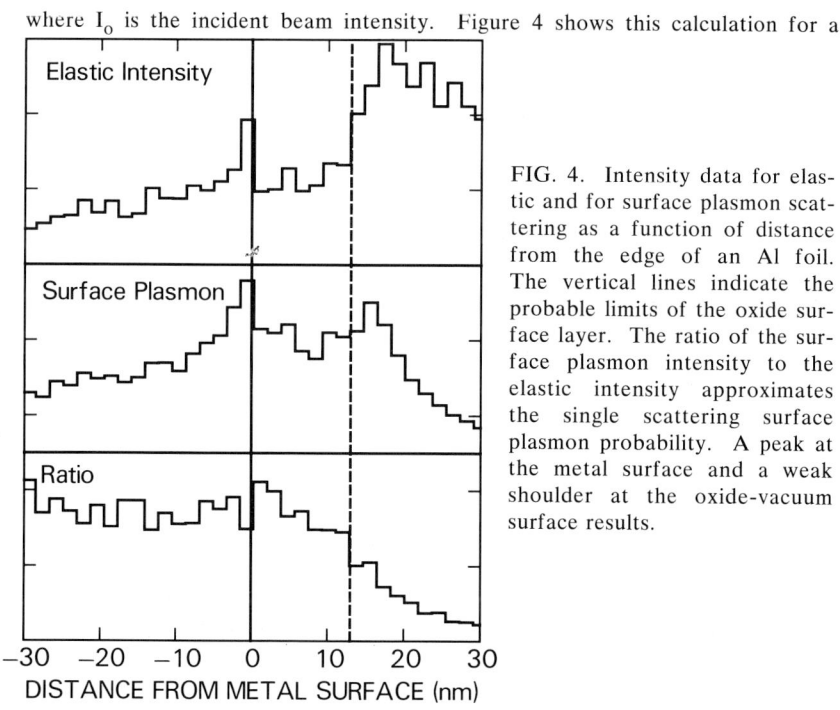

$$P_s = t\sigma_s = \frac{I_s(\text{measured})}{I_{\text{elastic}}} = \frac{I_0 t \sigma_s e^{-t\sigma_{\text{TOT}}}}{I_0 e^{-t\sigma_{\text{TOT}}}} \qquad (2)$$

where I_0 is the incident beam intensity. Figure 4 shows this calculation for a

FIG. 4. Intensity data for elastic and for surface plasmon scattering as a function of distance from the edge of an Al foil. The vertical lines indicate the probable limits of the oxide surface layer. The ratio of the surface plasmon intensity to the elastic intensity approximates the single scattering surface plasmon probability. A peak at the metal surface and a weak shoulder at the oxide-vacuum surface results.

12nm layer of oxide extending beyond the edge of an Al foil. We show in this figure the elastic and inelastic surface plasmon intensities. Notice that the elastic intensity has a peak at the metal surface. Quite possibly in this case, the sharp peak is a focussing (Fresnel) effect. The figure shows also the ratio of the two experimental results following the form of Eq. 2. The main effect of the correction is to increase the apparent scattering in the regions where bulk scattering is dominant. Secondly, the sharp peaks at the boundaries are considerably depressed. The result shows the expected peak at the metal surface with a weak shoulder at the oxide-vacuum surface. This indicates a weak polarization of surface charge on the oxide outer surface, is supported by the detailed theory [8].

CONCLUSIONS

It is clear that the STEM offers unprecedented sensitivity for established techniques and versitility for the creation of new techniques. However, it is also clear that experiments involving trace elements require sophisticated experimental control to deal with the very low signal intensities. In addition, as is true for all new techniques, qualitatively complicated behavior can occur to confuse the interpretation of results, particularly when we approach the spatial resolution limit of the STEM.

REFERENCES

1. P. Oelhafen, J. L. Freeouf, T. S. Kuan, T. N. Jackson, P. E. Batson, J. Vac. Sci. Tech. B1(3), 588 (1983).
2. C.R.M. Grovenor, P. E. Batson, C. Wong and D. A. Smith, in preparation.
3. P. E. Batson and G. Trafas, Ultramicroscopy 8, 293 (1982).
4. P. E. Batson, Ultramicroscopy 9, 277 (1982).
5. D. Misell and A. F. Jones, J. Phys. A2, 540 (1969).
6. M. Schmeitz, J. Phys. C14, 1203 (1981).
7. H. Kohl, Ultramicroscopy 11, 53 (1983).
8. M. Schmeitz, private communication.

IMAGE PROCESSING FOR ELECTRON MICROSCOPE INVESTIGATIONS OF MATERIALS

William Krakow
IBM T.J. Watson Research Center, P.O. Box 218 Yorktown Heights, N.Y. 10598, U.S.A.

ABSTRACT

A time shared television digital image processing system has been developed for on-line electron microscopy and uses a large mainframe computer. The main component of the system is a digital television frame store which has many standard features for digital analysis such as: digitization, zoom and pan, arithmetic and Boolean processors, alphanumeric generators and so on. Images can be acquired at atomic resolution from a TEM, analyzed in real time and hard copy slides made under full computer control. A full range of computer software has been developed or modified from existing software and is generally compatible with IBM Fortran compilers. Some of the areas where extensive menu driven software has been developed are: particle size and feature analysis, algebraic and geometric image manipulations, Fourier analysis, digitization and process control, image contrast correction, text processing, etc. A number of applications areas have been explored which include: the structure of Si/SiO_2 interfaces; nucleation of Au on rocksalt; the formation of hexatic structures from amorphous phases under shear, tension and compression; analysis of atomic surface structure and image motion and the analysis of field ion micrographs of amorphous structures. Several of these areas will be discussed in the context of image processing and materials characterization.

INTRODUCTION

Digital television frame store devices have made it possible to obtain images directly from electron microscopes which are of comparable quality to those obtained on photographic plates. Generally, a limited area is viewed as compared to plates, however, the advantages of real time manipulation of the digitally stored image data often outweigh the restricted field of view. Tasks which often required several hours of photographic preparation (developing and printing) as well as digitization with a slow scan microdensitometer are avoided. There are, of course, a wide range of both commercially available and custom built frame store devices which differ somewhat in hardware capability as well as the type of computer to which they are interfaced. Depending on the type of frame store configuration, it is then possible to develop a wide range of software applications packages which are flexible and can be executed at or near real time modes. Some of these areas which are of importance are: particle size and shape analysis, Fourier analysis, temporal event recognition and automation, image acquisition and manipulation. Of course, the ultimate goal in the case of electron microscopy is to provide full computer control of the important microscope parameters via image information and to present either processed image data or extract useful numerical information from these images. In this paper, I will describe a digital television system and software which has been developed to be run through a large mainframe computer in our laboratory.

In the past few years several groups have interfaced TV rate digital frame stores to high resolution transmission microscopes. Initially their efforts entailed digitizing images from low light level TV pickups with sufficient intensity to store images [1-5]. A considerable amount of variability exists in the TV pickup systems described by these authors; whether fiber optics or lens coupling is employed as well as the type of television pickup tube (isocons, SIT, Newvicon, etc.). Also, the array sizes digitized vary between 128x128x8 bit deep and up to 512x512x8 bits with the latter being compatible with U.S. or European standard line rates. In one case, a direct memory access (DMA) was employed while other systems usually employ one or up to four 8 bit deep frame buffer memories which can acquire data at television rates. Many of the frame store devices have the capability of summing images recursively in real time for image noise reduction using a built in digital arithmetic unit.

Generally, most of the digital frame stores are controlled via a 16 bit minicomputer and hence are somewhat restricted as to maximum memory size available (~128k bytes) for manipulation of images directly in the computer. The earlier work associated with electron microscope was aimed at obtaining the Fourier transform and hence 2-dimensional power spectrum of an image, i.e. the equivalent of a light optical diffraction pattern. These diffractograms can then be used to assess the performance of the microscope in terms of defocus and astigmatism of the objective lens[6]. The bench mark times to obtain the power spectrum described by the various authors range from the slowest of 35 seconds for a 64x64 array to about one second for a 128x128 array using an array processor. In no case does it appear that larger array sizes, of at least 256x256, have been handled in real time, nor have larger 32-bit type computers with ample memory being used for power spectrum analysis. However, some development work at larger array sizes has been stated to be in progress[4]. Also, 256x256 off-line transforms have been reported for use in conjunction with edge smoothing functions of digitized micrographs [7].

Another area where digital frame stores are being used for high resolution microscopy is for instrumental alignment and adjustment purposes. Using covariance between two micrographs, focussing and astigmatism correction have been considered [8,9]. Here essentially one is looking for a null contrast condition (minimum variance) which is characteristic of a condition near exact focus of the microscope ($\Delta f=0$) for phase contrast imaging. Ramping of defocus followed by x and y stigmator ramps are performed using this technique and time estimates for three full iterations was approximately 2 minutes. Preliminary results of variance techniques have been obtained for adjusting beam tilt alignment [10] as well as to correcting for image drift.[11] While it is not clear how viable these techniques will be in practice, they do represent the first attempt to fully automate a high resolution electron microscope. As will soon be shown, another alternative is the possibility of aligning images rapidly with a Boolean function processor attached to a frame store.

Various forms of digital frame stores have been used for on line size and feature analysis; e.g. Bausch and Lomb, Ziess, Lietz, MCI/LINK, etc. Generally, these systems can be coupled to either transmission microscopes at TV data rates and more recently slow scan digital techniques are available for image acquisition from scanning type instruments. The basic functions of these systems are digital data acquisition, windowing, thresholding and shading and finally extracting various forms of primary feature parameters such as areas, cord lengths, axes projections and so on. It is then possible to derive a number of secondary feature parameters describing, in many cases, more exacting measures of various objects geometrical characteristics. These parameters are described extensively in the literature and the reader is referred to one example given as ref. [12] where a wide range of quantitative image analysis procedures are described. It soon becomes apparent upon reviewing the extensive particle analysis literature that very few groups have developed software which can run universally on almost any computer. The software is generally in an assembler language and hence is almost impossible to adapt or modify to new applications. The impetus for streamlining this software was due to limited computer size which is usually some type of 16 bit minicomputer and secondly due to the requirement for very high speed high throughput quality control applications. Many of these problems can be overcome using a more powerful 32 bit computer with an optimizing compiler for Fortran 66 or 77 versions. As will soon be discussed, memory size, processing speed, and adaptability to new feature analysis algorithm are then not severely limited.

TIME SHARED IMAGE PROCESSING ARCHITECTURE AND FACILITIES

A few years ago at the T.J. Watson Research Center it was decided to purchase a digital TV rate frame store device which could display the results of electron microscope image computations or acquire images in monochrome as well as display color coded images of galactic images of plates taken with different spectral wavelengths in real time. It was then decided to interface the frame store to a large IBM mainframe computer, since a reasonably fast interface could be fabricated. This would then alleviate the problem of using a stand alone powerful mini computer and a new operating system, as well as acquiring a large amount of expensive hard disk storage hardware. Also, it was soon determined that little or no applications software was available through frame store vendors in many areas such as particle analysis, Fourier analysis, etc. It was, therefore, decided that frame store functions which could be run via callable Fortran routines were most desirable. Most applications programs could then be easily designed to run from existing software which the present author had already developed and the applications programs supplemented by subroutines to manipulate the frame store. Of course, many new menu driven applications programs were developed which could not be adapted from existing image processing software, and hence the use of a callable frame store subroutine library greatly simplified this task. The advantages of using a mainframe computer, even in a time-shared mode, soon became apparent which include: minimal computer maintenance, a well documented operating system, standard software library packages, large amounts of random access disk storage, high speed numerical calculations on large arrays with 32 bit precision, common use of image processing software from any terminal location, the addition of computer controlled peripherals through the mainframe, multi tasking and so on.

The basic architecture of the time shared image processing system is displayed in Fig. 1. The system is centered about a Grinnell GMR275 full color processor, running at a standard U.S. 525 lines, capable of storing five - 512x512x8 bit deep monochrome images (three used at one time for full color) and four 512x512x 1 bit deep graphic overlays. The frame store has many standard features such as: an 8 bit video digitizer, zoom and pan controls, quad graphic cursors, image analyzer and histogram generator, arithmetic and Boolean processors, alpha numeric character generators, lookup tables, video control routing, internal self diagnostic test proms, and so on. The high speed bidirectional interface to the frame store has both a 16 bit input and output data bus. The communications link with the host computer and the interface is over coaxial cable at rates of 1Mbit/second. This means that the processor system could be located at any site within the research center independent of distance from the host computer. The same coaxial type cabling and controllers are also used for communication with the host via terminals, which can also be at any remote site, to drive the TV processor. A second parallel interface port to the Grinnell system is also available. It can be accessed in a switching mode to allow a local, small computer to control the processor if desired. It is planned that a personal computer fill this function in the immediate future for dedicated laboratory automation of various microscopes requiring precise timing control.

Continuing with Fig. 1, a digitally controlled switcher allows up to 10 video inputs to be individually selected from remote sites, some of which are shown at the bottom of this figure. Also, 10 full color remote RGB (red, green, blue) TV outputs are available and switchable under full computer control. Two of these outputs are always accessible with one being in a central room facility and the other an office, to allow monitoring of current image processing activities. Full color or monochrome (8 bits/color) hard copy can be obtained in a central room facility under computer control via an ASCII line from a precision CRT camera (Matrix 4007) in the form of slides, sheet film, cine' film or polaroid positives and transparencies. Monochrome hard copy can also be obtained off-line on a high resolution font printer capable of generating 140 grey levels with a 512x512 picture being obtained on plain paper in an areas as small as 4 in. x 4 in.. Also hard copy can be obtained on a graphics terminal and hard copy device at 16 grey levels (Tektronix 618 and 4631) when they are attached to a standard IBM computer terminal.

The TV processor is controlled through a virtual machine (VM) computer facility. At this time an IBM 3081 is being employed, with 2M bytes of memory available for routine on line running or up to 4M bytes for special uses. Night time running can have a memory size as large as 16 M bytes. Also available through the computer facility are mass storage media (MSS), which is equivalent to hard disk storage. It is possible to store several hundred byte-packed pictures of 512x512 dimension. Images can be recalled and displayed via the TV frame store using software control even under time sharing in a few seconds. Applications programs for image processing as well as compiled versions are stored on other types of hard disks (minidisks) which are available to a general user in a read only mode. All compiled versions of the interface I/O drivers and subroutines to control the frame store functional logic cards also reside on minidisks.

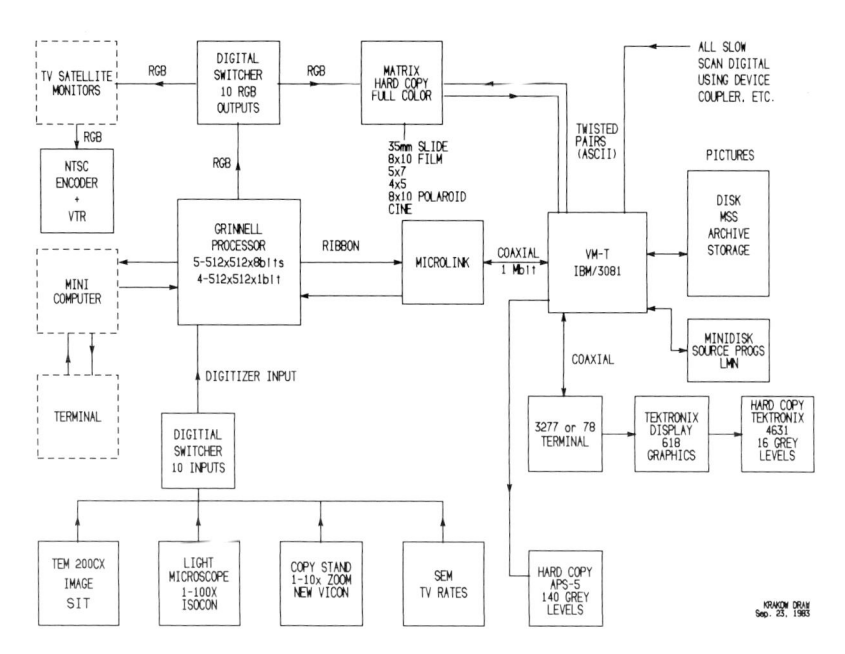

Fig. 1. Block diagram of the time shared image processing system.

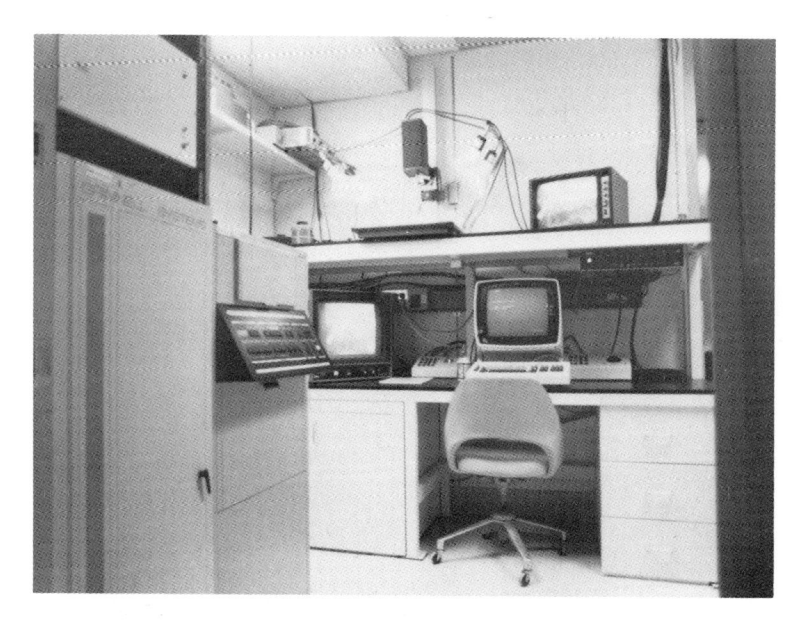

Fig. 2. Central room facility for time shared image processing. Interface and frame store are at the extreme left.

Displayed in Fig. 2. is the central room facility where the frame store (extreme left), the interface (above the frame store) and digital hard copy camera reside (to the right of the frame store). Also contained in this room is an IBM 3277 terminal, RGB monitor and a Newvicon camera on a light table copy stand. Here off-line processing on images which are already stored on the computer system or in the form of photographic plates or prints can be performed independent of the microscopes. This room also contains the digital switching box (to the right of the terminal) for 10 inputs and 10 outputs. A joy stick controller for cursors and track ball unit for panning images are also in this location. One feature of an IBM terminal is the availability of function control keys which can be used instead of joy stick or track ball controls. This feature is particularly desirable for remote site running of the processing system, e.g. from a microscope room.

Finally, as an example of a remote site location for on-line image processing, a JEOL 200CX microscope with a top entry pole piece is displayed in Fig. 3. Here a one stage SIT camera sensitive to 10^{-5} foot-candles of illumination is located below the microscope plate chamber. The camera is coupled to the microscope by a yellow green transmission phosphor screen and optical lens system. A monochrome monitor located above the console allows the direct viewing of the TV pickup an subsequently the video signal is routed to the frame store which is approximately 100 feet away. The processed image is then returned to the microscope room and displayed on a RGB monitor (located to the right of the microscope). Using a pass-through mode, the signals are then sent to a color NTSC video encoder (LENCO CCE-850) and subsequently to a 3/4 inch video tape recorder and NTSC color monitor, which are located to the left of the microscope. Audio dubbing of recorded data is possible during a microscope session. The camera and microscope screens can be protected from outside light sources to avoid the problem of working in total darkness.

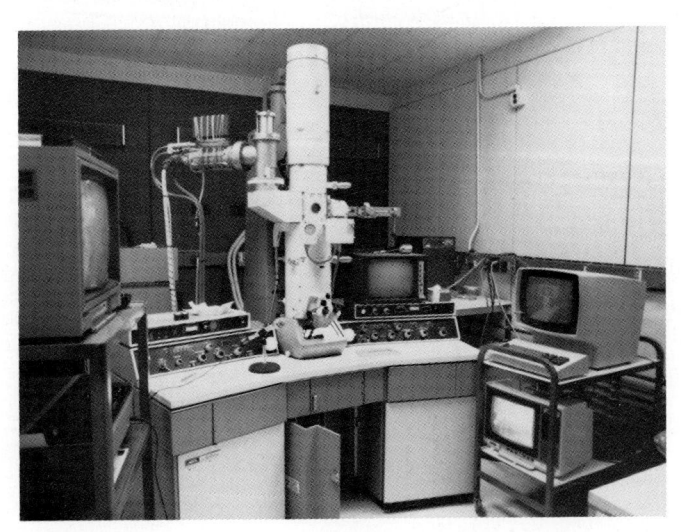

Fig. 3. Microscope and associated image processing facilities.

SOFTWARE APPLICATIONS

Because of the limited space available, it is the intent here to only briefly show several applications areas of image processing where the digital fram store system is employed. In this paper the applications involving full color will not be considered because of the camera-ready style, but will appear in a future publication. Basically, the areas of interest, all of which will now be described, are: devices which reside within the frame store, off-line applications which require only digitization and display, and multifunction menu driven software.

Peripheral Devices Residing in the Frame Store

Perhaps one of the most important areas for frame store use involves the simple task of digitizing imges in real time. Here an 8 bit video digitizer accepts a RS-170 composite video signal and stores the digitized data in one or more display refresh memory banks in real time. Digitized data may directly replace data already stored in memory, or may be recursively added to data already stored in memory. The computer may select to digitize and store a single frame of data, a specified number of consecutive frames, or operate continuously until the computer chooses to stop continuous operation. Two examples of digitization of a high resolution grain boundary image between (110)Si and amorphous Si are given in Fig. 4. Here (4a) is the result of a single frame of video digitization. Fig. (4b) was obtained in the recursive mode with 4 shifted bits, which means it would take 2^4 times as long to replace an image totally as opposed to a single TV frame. It should be noted that (4b) represents a time averaging, hence the noise from insufficient signals is considerably reduced as compared to (4a). Even with a recursive shift factor of three, it has been possible to digitize images at a top microscope magnification of 800,000X of high quality. With this shift, image drift is not significant and the effect of moving the specimen stage does not produce a noticeable time lag of the image which would be evident for higher shift factors (up to 15 bits). It should be noted that image (4b) has been contrast stretched by routing the image, stored in memory, through a video look up table (LUT) which resides in the frame store image function memory. Once the LUT has been loaded an image is then corrected at video rates as digitization proceeds.

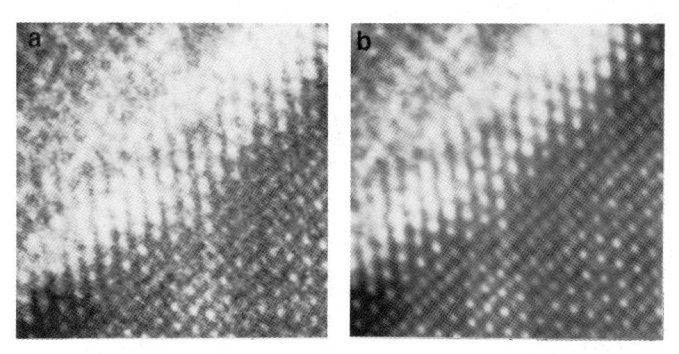

Fig. 4. Digitized images of a (110)Si/amorphous Si interface. a) single frame, (b) time averaged and enhanced.

Another internal function of the digital frame store is the ability to zoom by pixel replication and pan any selected image memory of the frame store where zoom factors of 1,2,4, or 8 are possible. Fig. 5 shows one example of a (110) Au lattice image using an 8x zoom factor. Therefore, an original image on the microscope at ~8000,000X and about the same magnification into the TV pickup, would then effectively be 6.4x10^{6}X. The zoom and pan feature is particularly valuable when aligning different images of a given area in real time to one pixel accuracy either by track ball control or under key board control.

It is often important to determine where the image intensities are most significant for contrast thresholding and stretching. This can be achieved through an image analyzer function which will display the histogram overlay on the output video signal of the full 8 bit dep video picture. It is also possible with this device to read back histogram data to the host computer and even mask limited areas for analysis. Fig. 6 shows one example of a (100)Au lattice image with its overlay histogram displayed which is adjusted to fill the display height presented.

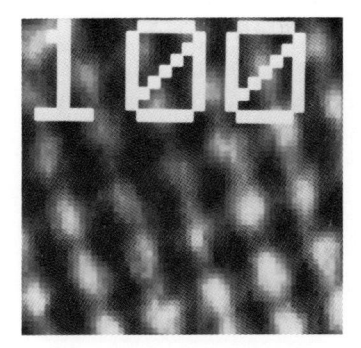

Fig. 5. a) 8x zoom factor applied to a (001) Au lattice image.

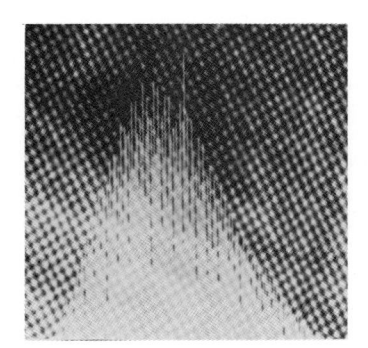

Fig. 6. (001) Au lattice image and its histogram.

Fig. 7. Alphanumeric characters contained in the frame store PROMS.

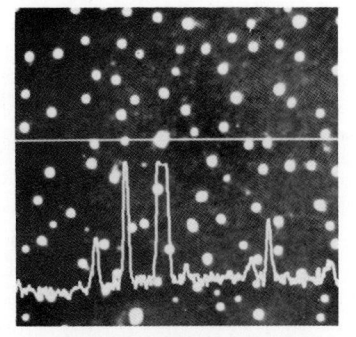

Fig. 8. Densitometer trace of a dark-field image of Au particles.

So far, no information has been given on graphics capability of the digital frame store or alphanumeric capability. Fig. 7 shows the full ASCII character set of 128 characters which are contained on PROMS in the frame store. Many symbols were designed to be of use to electron microscopy applications. Various height and width combinations of letters can be used, as well as various colors to label text material. It should be noted that the frame store is an ASCII device while IBM computers run with EBCDIC format; therefore a software conversion is needed to assign the correct binary code to the frame store from an IBM type terminal. An example of writing vectors is given in Fig. 8, where a line of video data has been read back to the host computer and its intensity plotted as a function of spatial position, i.e. densitometry. Here are 512 interconnected vectors across the screen and their height can range from 0 to 255 units out of the total visible display height of 480 units. Note, word data (16 bits) must be used for display since vectors can range up to a value of 512 which would require two bytes as opposed to pictures requiring one byte/pixel.

Algebraic and Geometric Image Manipulations

The last section considered image manipulations which were at or near television rates directly in the frame store. Here several types of image manipulations will be described where the frame store is used solely for digitization, shipping images to and from the host computer and for display. Most of the image manipulations are performed on the host computer and require the use of floating point arithmetic operations on a 32 bit processor for accuracy. Fig. 9 shows several examples for a 256x256 array of a (110)Si-amorphous Si interface. Here (a) is the original image, (b) is the derivative magnitude (c) is a unidirectional derivative image (y direction), (d) is the second derivative image, (e) is a geometric view of the image if a pin hole camera were looking from bottom to top, (f) is a y modulated image, (g) is a rotated image by 40°, (h) is a compressed image which has been forshortened along the y direction by a factor of three.

The algebraic operations (b) to (d) displayed in Fig. 9 involve simple difference techniques which could be done through an analog hardware device if it were available. Here, however, the computations are accurate and reproducible which may not be the case for an analog method. For the case of geometric manipulations, (e) and (g), interpolation must be employed to accurately preserve image details. While various schemes may exist for performing many of these operations, i.e. analog, hardwired digital, etc., it is important to emphasize that these examples demonstrate image processing done fully under software control on the host computer.

Higher Level Image Processing

It is often advantageous to utilize as many features of the frame store as possible, however, one is often dependent on the host computer and on software subroutines of functions not possible directly with the frame store. In this case one wishes to have menu driven applications programs with a high degree of flexibility. Three such areas will now be discussed which include: particle analysis, Fourier analysis and process control (time lapse control).

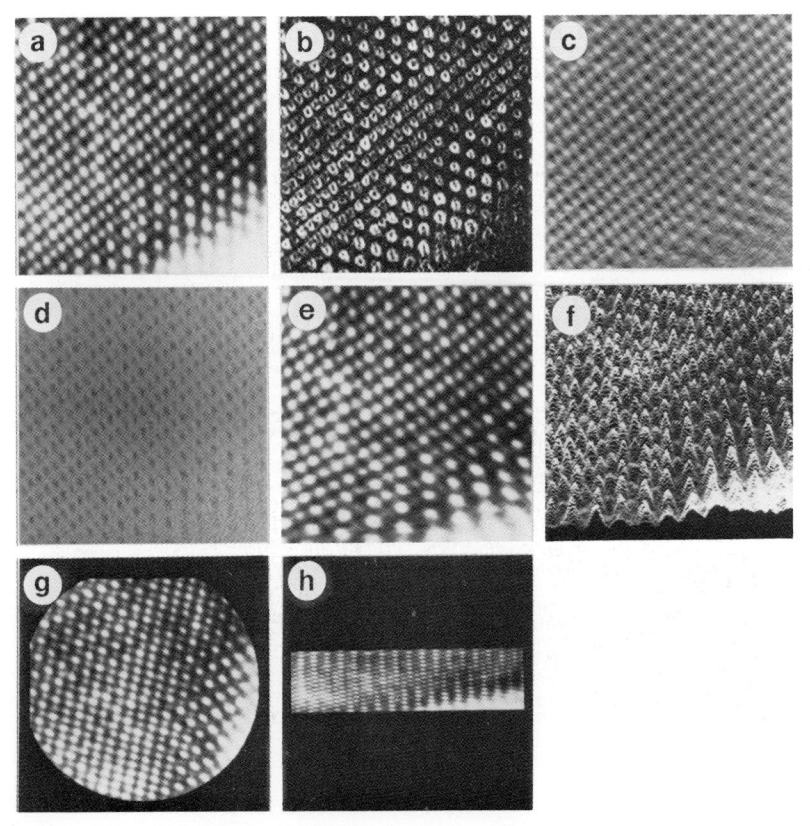

Fig. 9. Algebraic and geometric manipulations of a (110)Si lattice image. a) original, b) first derivative magnitude, c) y direction derivative, d) second derivative, e) perspective view, f) y modulated, g) rotated, 40°, h) vertical compression.

Particle Analysis

Particle analysis is often performed via small minicomputers and often assembler language is used for speed, which does not make it flexible for new applications. It was therefore decided to write the software totally in Fortran and to run off a large mainframe computer, which would provide the ability to analyze very large numbers of particles per field ($>10^3$) with any convoluted shape in reasonable time frames. Basically, the software consists of menu driven programs which allow images to be digitized, thresholded and windowed via frame store functions. Only the portion of the image windowed is then sent back to the host computer for analysis. The analysis is then performed by line segment encoding techniques where the detected objects are put in a vector format reducing the amount of data to be considered by several orders of magnitude. Various types of numerical data are then returned to the computer display screen, such as: number of particles, total area occupied, average number of pixels/particle. Also histogram data is returned to the computer terminal on: particle areas, equivalent spherical diame-

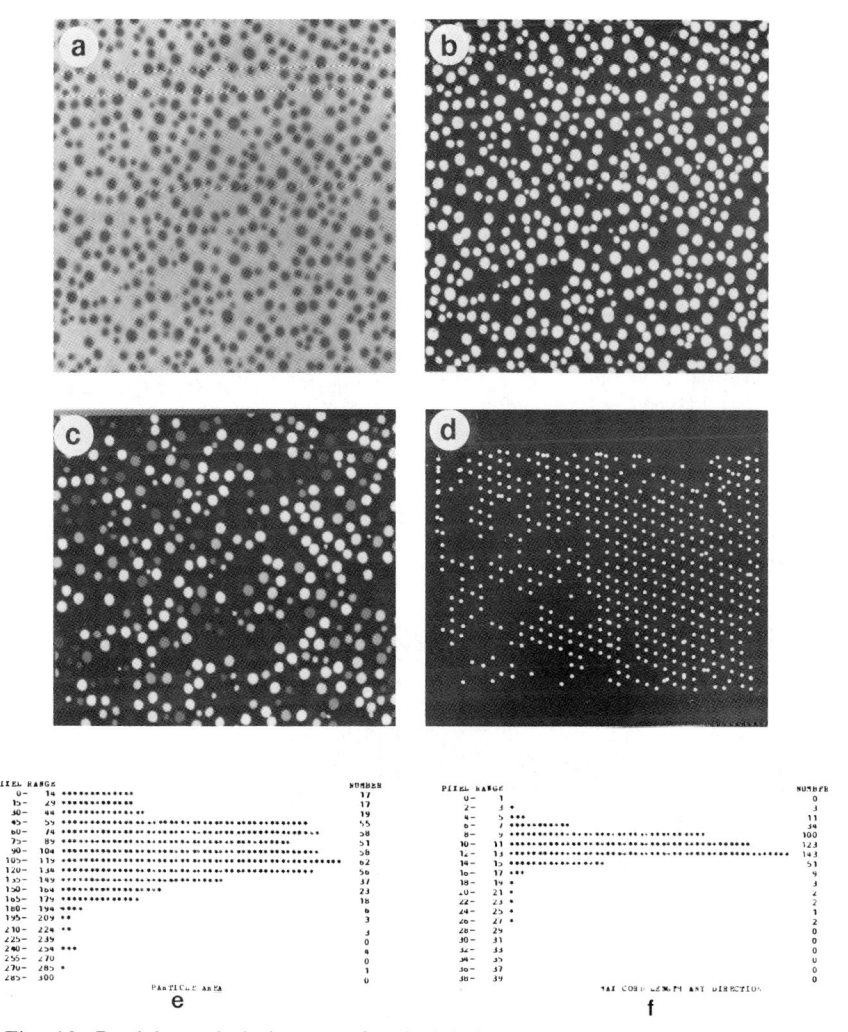

Fig. 10. Particle analysis images, a) original digitized image of Au particles, b) thresholded image of Au, c) bit mapped image of Au, d) centroid image of (110)Si lattice, e) histogram of Au particle areas, f) histogram of Au particles maximum cord lengths.

ters, maximum cord length in any direction, maximum X and Y axis projections, and maximum cord in the X direction. Feature analysis is then performed by using combinations of these primary parameters (e.g. ratio maximum cord length to equivalent spherical diameter). Detected features are displayed on graphic overlay planes and hence do not interfere with the image data. It is also possible to display size distributions by color using the video lookup tables LUT's of the frame store. This topic will be the subject of a future publication dealing with the uses of color in electron microscopy. Also it is possible to perform a perimeter analysis and locate the centroid of objects from the vector encoded data.

Several examples of particle analysis on Au particles nucleated on rocksalt are displayed in Fig. 10. Image (a) shows an original brightfield image of Au particles and (b) is its thresholded result. Image (c) is the result of an analysis of of the Au particle size distributions where different grey levels represent the different size ranges. Image (d) is the result of a centroid analysis on a Si lattice image. Images (e) and (f) are samplings of the area and maximum cord length histogram data from Au particles. All of the steps in performing an analysis through software can be reproduced from the saved state parameters of the program.

Fourier Analysis

In the introduction several different aspects of Fourier analysis were listed where the main intent was to obtain on-line power spectra of images for microscope correction. Generally, the small array sizes employed to reduce computer overhead result in optical transforms which suffered from speckle due to the finite sampling. More efficient and extensive computer methods have been developed in our laboratory for flexible, real time, Fourier analysis via the TV frame store. For these purposes a large Fortran application program contains all the callable subroutines to perform functions from a displayed menu. The main tasks which are performed are (1) digitization, (2) outlining the area to be analyzed with graphic overlays of the frame store. This is referred to as "electronic aperturing" since the computer will use this information when computing the fast Fourier transform (FFT). (3) Reading back image data from the frame store, obtaining the FFT and displaying the logarithmically scaled square modulus of the FFT, i.e. power spectrum, for a 256x256 array. (4) Filtered images can also be obtained by overlaying graphic apertures on the optical diffraction pattern (center or ring). The program contains many additional functions such as: measurement of spacings directly from the diffraction spots of the power spectrum, selection of different images (original, power spectrum or filter), text overlays, hard copy capability, edge grading, aperture size and position selection, scaling of the power spectrum and filtered images.

Results obtained through the digital frame store are given in Fig. 11. Here (a) is an original image of (110) Si with a region outlined by a graphics cursor. Image (b) shows the area sampled and (c) is corresponding optical transform. Note image (b) has edge grading which has eliminated ringing in the transform and permitted more decades of intensity to be displayed. Images (d) and (e) demonstrate the ability to sample very small regions less than a few tens of angstroms in diameter and obtain a useful power spectrum.

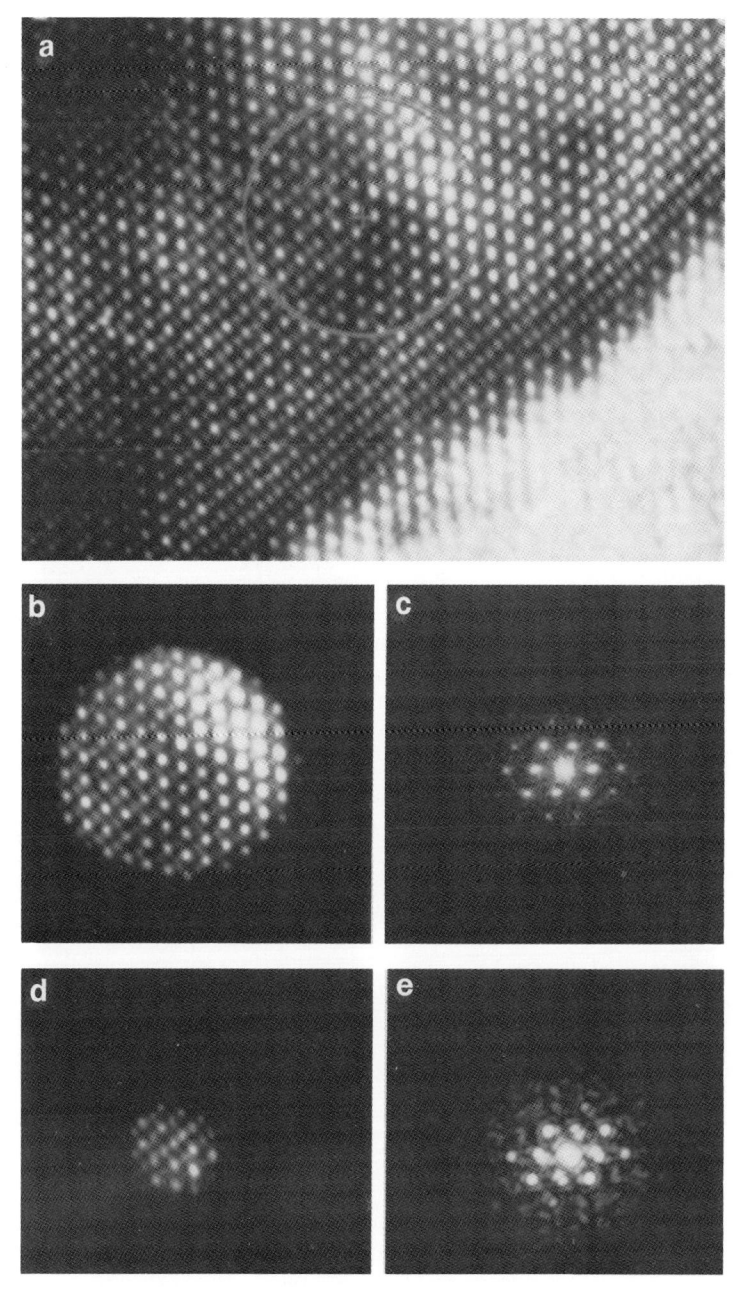

Fig. 11. a) Original field of (110)Si, b) region sampled for Fourier analysis, c) power spectrum of (b), d) smaller sampled region and e) power spectrum of (d).

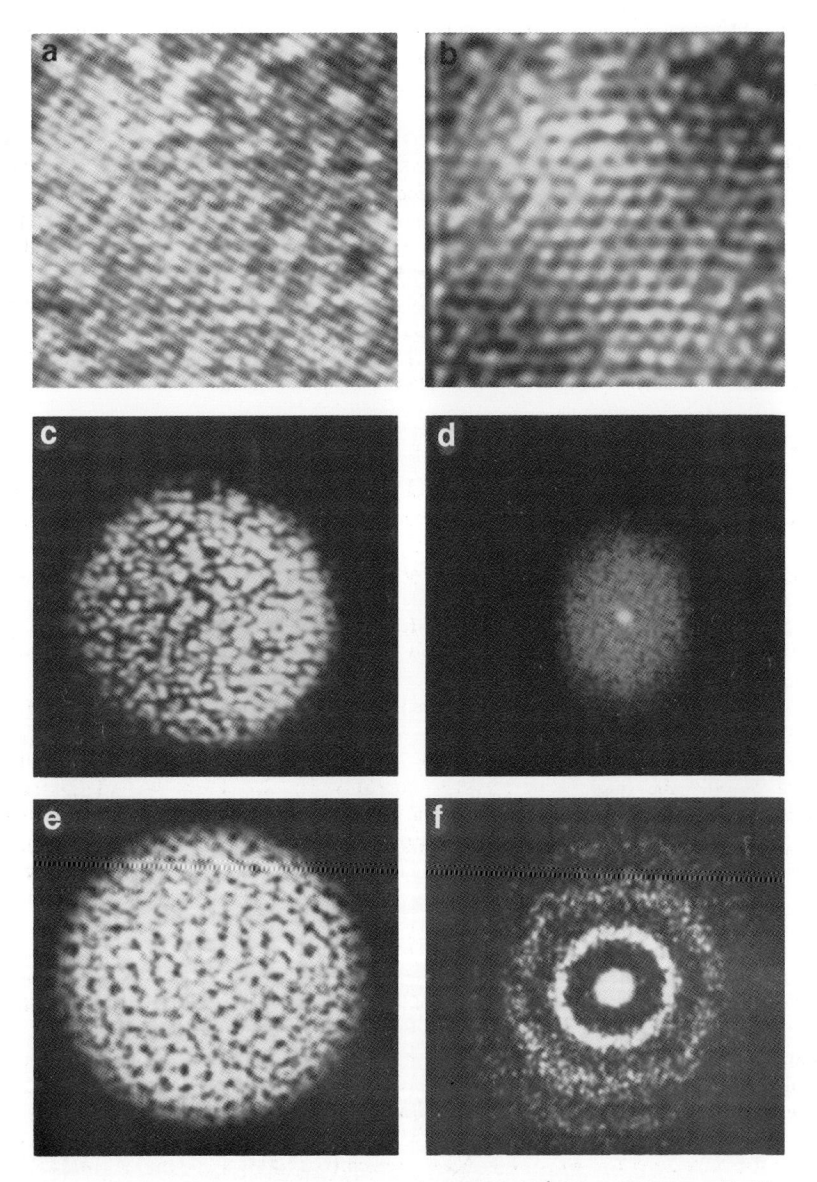

Fig. 12. a)Original image of (111) Au showing 1.43Å lattice fringes, b) filtered image of (a) showing the residual 2.48Å periodicity of the surface atoms. c) Field ion micrograph and d) its power spectrum. e) original model of a relaxed hexagonal close packing of atoms, d) power spectrum of a small region of (f).

Fig. 12 shows examples of image filtering and some new applications areas for Fourier analysis. Image (a) is an original image of a (111) Au film containing predominantly the {220} 1.43Å bulk lattice fringes. If these are removed by filtering in (b) one is then left with the surface lattice periods of 2.48Å which is expected for a monolayer of close packed Au atoms. Figs. 12c and 12d show an analysis of a field ion micro-graph (FIM) of an amorphous material. Here no correlation exists between different image features as evidenced by the uniform distribution of intensity in the optical transform. To my knowledge no attempt has been made to perform analysis on FIM images before this time. Finally, image (e) shows a sheared and relaxed model structure of a dense random packing of atoms, while image (f) is the optical transform obtained from a limited domain region indicating the tendency towards a hexagonal close packing, i.e. a hexatic structure. This example serves to demonstrate that one is not restricted to actual electron microscope images but the use of models is also possible.

Time Lapse Control

It is often desirable in electron microscopy to have on line comparison of images separated in time to assess microscope performance and observe dynamic changes in materials. For this purpose it was necessary to design a menu driven program which could sample images separated in time by a prescribed interval or obtain updated images for comparison against an original. This program accesses an arithmetic logic unit (ALU) located in the frame store which can add or subtract two images, or more importantly perform Boolean algebra on these images. The Boolean functions are the OR, XOR and AND operations. These have the advantage of avoiding overflow and underflow problems associated with addition and subtraction. Also, they are ideally suited for observing image differences (XOR) or similarities (AND). The computer program has the ability to show only the final result of ALU operation or also show the intermediate digitization steps. A histogram function can also be activated which allows the information content of the processed image to be evaluated, from which decisions can be made as to the amount of change in image features between given time intervals.

Briefly Fig. 13 gives examples of use of the process control program. (a) is an original image of partially amorphized SiO_2, (b) represents the ORed result between two slightly displaced images of (a) and (c) is the XORed result of the two displaced images. While the XORed image looks very similar to a derivative magnitude image, it represents difference features between the two original images. In Fig. 13d to 13f, results of turning on the image analyzer function are also displayed. Here (d) is an original dark-field image of small Au particles. (e) represents the XORed image between two images where no motion has occurred. (f) is the XORed result where the images are shifted. It is apparent by comparing the histograms of (e) and (f) that specimen movement increases the magnitude of the values of the histogram. Hence the XORed result and its histogram represent a means of monitoring image drift or image change. If the image pan function is also used it would be possible to realign an image and therefore minimize the effect of image drift on the TV monitor.

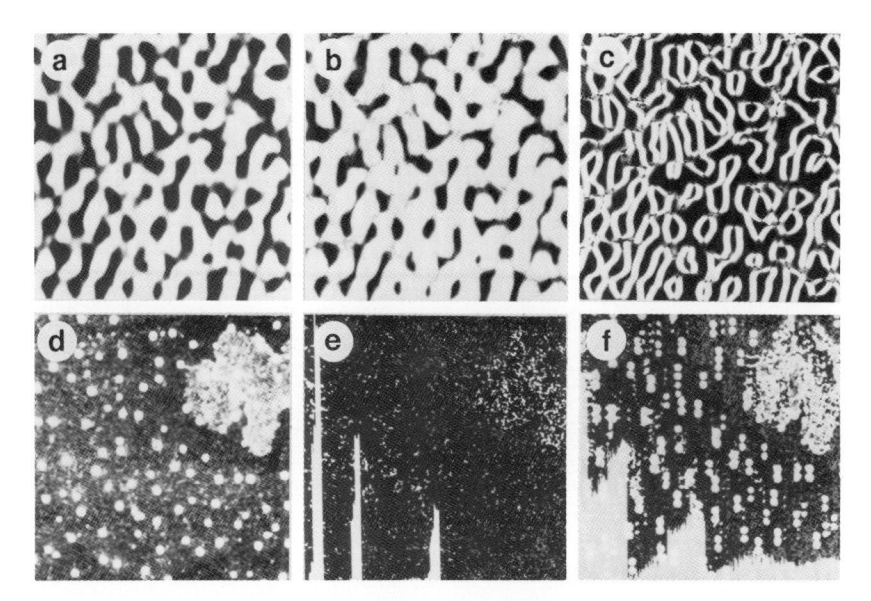

Fig. 13. a) Original bright-field image of partially amorphitized SiO_2, b) ORed image of (a), c) XORed image of (a). Original dark-field image of small Au particles, (b) XORed image and histogram with no motion between two imges, c) XORed image and histogram with motion between two images.

SUMMARY

A time shared digital image processing system has been described in this paper in terms of its hardware components and several software applications areas for electron microscopy. It has been demonstrated that such a system has great flexibility in the types of problems it can address in real time. The use of a mainframe processor with a digital frame store has several advantages for a research environment where large array handling is possible with minimal CPU overhead and maintenance.

ACKNOWLEDGEMENTS

The author would like to thank many of the people who have contributed to the development of the image processing system. In particular, I want to especially thank R.A. Baron who spent many long hours designing the interface and I/O drivers; J. Harry, B. Agule and N. Brenner for their computer consultation. Also, I wish to thank several collaborators who have provided interesting micrographs and models for image processing in specific materials areas.

REFERENCES

1. K.-H. Herrmann, D. Drahl, H.-P. Rust and O. Ulrichs, Optik, *44*, (1976) 393-412.
2. K.C.A. Smith, Inst. Phys. Conf. Ser. No. 61, Proc. EMAG Cong., Cambridge, (1982) 109-113. also Proc. 10th Int. Cong. on Electron Microscopy, Hamburg, West Germany (1982) 123-130.
3. Y. Yokota, M. Tomita, H. Hashimoto and H. Enoh, Ultramicrosc., *6*, (1981) 313-322.
4. E.D. Boyes, B.J. Muggridge, M.J. Goringe, J.L. Hutchinson and G. Catlow, Inst. Phys. Conf. Ser. No. 61, Proc. EMAG. Conf., Cambridge (1982) 119-122. also Proc. 10th Int. Cong. on Electron Microscopy, Hamburg, West Germany, (1982) 523-524.
5. J.C.H. Spence, M. Disko, A. Higgs, J. Wheatley and H. Hashimoto, Proc. 10th Int. Cong. on Electron Microscopy, Hamburg, West Germany, (1982) 519-520.
6. F. Thon, Z. Naturforsch., *21a* (1966) 476.
7. M. Tomita, H. Hashimoto, K. Ikuta, Y. Yokota and H. Endoh, Proc. 10th Int. Cong. on Electron Microscopy, Hamburg, West Germany, (1982) 521-522.
8. S.J. Erasmus and K.C.A. Smith, Inst. Phys. Conf. Ser. No. 61, Proc. EMAG Conf., Cambridge (1982) 115-119.
9. S.L. Erasmus, K.C.A. Smith and D.J. Smith, Proc. 10th Int. Cong. on Electron Microscopy, Hamburg, West Germany (1982) 529-530.
10. W.O. Saxton and D.J. Smith, Proc. 10th Int. Cong. on Electron Microscopy, Hamburg, East Germany, (1982) 527-529.
11. P. Atkin, S.J. Erasmus and K.C.A. Smith, Proc. 10th International Cong. on Electron Microscopy, Hamburg, West Germany, (1982) 525-526.
12. M. Rink, J. of Microsc., *107* (1976) 267-286.

COMBINATION OF CBD AND HIGH RESOLUTION WITH 400 kV

S. Suzuki, T. Honda, Y. Kokubo and Y. Harada
JEOL Ltd., 1418 Nakagami, Akishima, Tokyo 196, Japan

ABSTRACT

A new 400 kV electron microscope has been developed. It can combine high resolution (HR) image observation and convergent beam diffraction (CBD) analysis. The instrument is provided with a specimen height control (Z-control) system to get a wide range of illumination angles. In HR image observation, a resolving power of 0.18 nm is confirmed by ODM. Illumination angle of a converged beam can be changed from 1.0 mrad. to more than 10 mrad. using the Z-control system. Specimen height control is a very useful method to combine HR observation and CBD analysis.

INTRODUCTION

High resolution (HR) observation and diffraction pattern analysis are the basic functions of electron microscope. Recently convergent beam diffraction (CBD) [1∼5] analysis has become popular and it is a powerful technique for crystal structure analysis and for determining the crystal point groups and space groups. But optical conditions to be required for HR image observation and CBD analysis are not consistent with each other. A very good parallel beam with a very small illumination angle (α) is required for HR image observation while highly converged beam with very large α for CBD. To satisfy these two different conditions, the integrated pre-magnetic field (pre-field) of objective lens (OL) should be changed. Changing the specimen position in Z-direction using Z-control system is the best way to change the pre-field.

A new 400 kV electron microscope [6] (Fig. 1) with such a Z-control system and very strongly excited OL ($NI/\sqrt{\Phi^*} \simeq 30$ at 400 kV) is developed to meet those requirements. Its accelerating voltage (HT) is changed by 0.1 kV steps from 100 kV to 400 kV. This paper describes the way to combine HR image observation and CBD analysis in one time with the new 400 kV electron microscope.

CHANGE OF OPTICAL PARAMETERS WITH SPECIMEN POSITION

The objective spherical aberration coefficient (C_S) and chromatic aberration coefficient (C_C), focal length (f_0), focussing AT ($NI/\sqrt{\Phi^*}$) and maximum illumination angle (α_{max}) are very important parameters for electron microscopy. But in a conventional electron microscope, the specimen position is fixed at a certain position to meet specified purpose (e.g. HR observation).

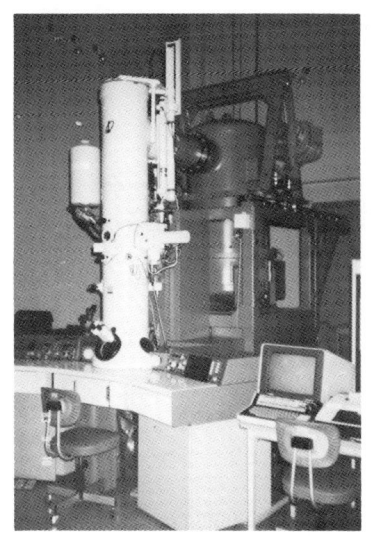

Fig. 1. General view of the new 400 kV electron microscope.

Mat. Res. Soc. Symp. Proc. Vol. 31 (1984) © Elsevier Science Publishing Co., Inc.

And in that case these optical parameters becomes specified values changing with only HT.

Fig. 2 shows calculated values of C_S, C_C and α_{max} depending on specimen position (Z) in the 400 kV electron microscope. It is clearly seen that C_S, C_C and α_{max} change with the specimen position in the Z-direction. In HR image observation C_C, C_S and α_{max} are most important parameters and the specimen position is determined to optimize the optical conditions at maximum HT. In the case of the OL of this electron microscope the specimen position is set at Z = 0.0 mm for HR image observation (C-TEM) mode and the optical parameter are as follows:

C_S = 1.0 mm

C_C = 2.0 mm

α_{max} = 1.2 mrad.
 (OL aperture of 200 µmϕ)

$NI/\sqrt{\phi^*}$ = 22.8

at 400 kV

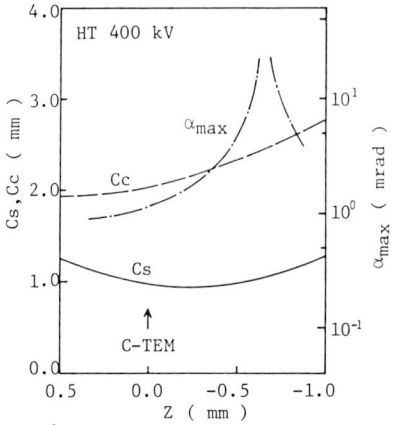

Fig. 2 Relationship between Cs, Cc α_{max} and specimen position.

But this specimen position is not the optimum one at different HT. Fig. 3 shows calculated values of C_S, C_C and $NI/\sqrt{\phi^*}$ at 200 ∿ 400 kV. And Z = -0.2 mm is better specimen position at 300 kV with respect to C_S and C_C. So it is very meaningful to set the specimen position to the optimum position for HR observation.

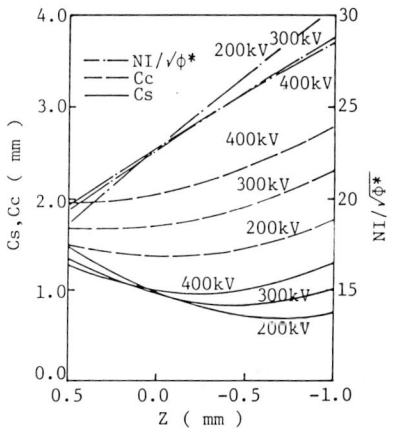

Fig. 3 Relationship between Cs, Cc, $NI/\sqrt{\phi^*}$ and specimen position for various HT.

ILLUMINATION ANGEL AND CBD

Generally, the specimen is set in the OL magnetic field as shown in Fig. 4. Only the magnetic field under the specimen position contributes to image formation. And the pre-field (shaded area) works to converge the electron beam. α_{max} is strongly dependent on the integration of this pre-field, and it is the most important parameter for CBD analysis. The integrated magnetic field necessary for imaging (the area below the specimen in Fig. 4) is almost constant. If the specimen position is moved down from Z1 to Z2 and OL excitation is increased to focus the image, the integrated pre-field is increased and larger α_{max} can be obtained as shown in Fig. 4. Fig. 5 shows the α_{max} and minimum spot size depending on the specimen position, measured at 300 kV and 400 kV. A condenser lens (CL) aperture of 200 µmϕ was used in this measurement. α_{max} was about 1.2 mrad. at C-TEM mode at 400 kV. If CL3 is over focussed to spread the beam on the specimen, α becomes much smaller (<0.5 mrad.). Its illumination angle becomes small enough for HR observation. And very high quality image can be obtained at this specimen position as shown in Fig. 7. Then the specimen position is moved down for CBD analysis. By moving down the specimen position and increasing OL excitation, α_{max} is made larger. Fig. 6 shows the diffraction pattern of Au [001] single crystal at various specimen position at 400 kV. The specimen position of each photo. in Fig. 6 is also shown in Fig. 5. The disks of diffraction pattern overlap each other below the lower specimen position (Z < -0.5 mm). Table 1 shows the relationship between the nearest lattice spacing and illumination angle required for disk-touch (α_t) at 100 kV to 400 kV. Lattice spacing of (200) of Au is 0.204 nm and α_t is about 4 mrad. at 400 kV. Therefore, the nearest disks touch each other at α_{max} (α_t) of 4 mrad., namely, specimen position is Z = -0.48 mm. By moving down the specimen position step by step, α_{max} is increased. Especially below Z = -0.5 mm α_{max} increases sharply and condenser-objective (c/o) condition is achieved at Z = -0.65 mm. After the c/o condition α_{max} becomes smaller again. The specimen position of c/o condition for each HT is listed in Table 2. For CBD analysis, optimum α_{max} depends on lattice spacing of

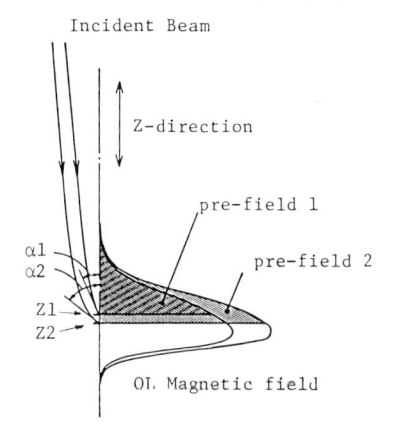

Fig. 4 Schimatic illustration of relationship between specimen position 1 & 2 and α1 & α2 in conection with pre-field 1 & 2.

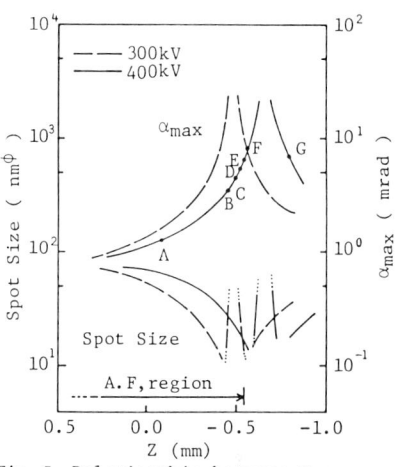

Fig. 5 Relationship between αmax, d_s and specimen position.

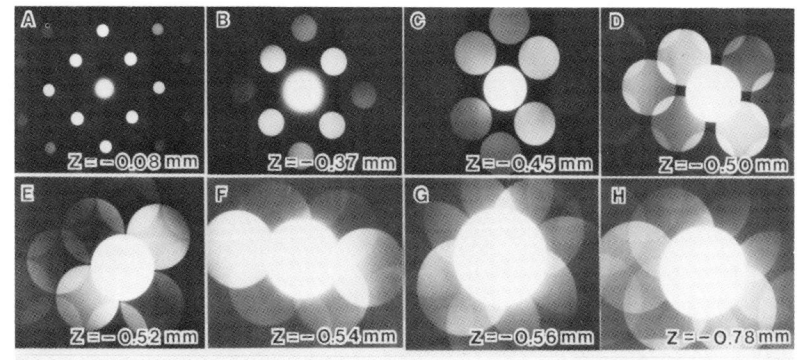

Fig.6 Diffraction pattern of Au [001] single crystal showing the illumination angle at each specimen position at 400kV.

the specimen. Z-control system can move the specimen in minimum 1.5 μm step and very fine specimen position setting is possible. The range of α_{max} covered by this Z-control system using CL aperture of 200 μmφ is 1.0 mrad. to more than 10 mrad. at 400 kV.

Table 1. The relationship between lattice spacing and α (semi-angle) at which disks begin to overlap for each HT.

HT (kV)	0.1 nm	0.15 nm	0.2 nm	0.25 nm	0.3 nm
100 kV	18.5 mrad.	12.3	9.2	7.4	6.2
200 kV	12.5	8.4	6.3	5.0	4.2
300 kV	9.8	6.6	4.9	3.9	3.3
400 kV	8.2	5.5	4.1	3.3	2.7

Table 2. The specimen position of c/o condition and $NI/\sqrt{\Phi^*}$ for each HT.

HT	$NI/\sqrt{\Phi^*}$	Z
200	24.6	−0.35 mm
300	25.5	−0.47 mm
400	26.6	−0.65 mm

SPOT SIZE

The spot size illuminating the specimen (d_s) is also affected by the pre-field. d_s values shown in Fig. 5 are measured under the following conditions:

EM spot size : 7 $NI/\sqrt{\Phi^*}$ (CL1) ≃ 8.0
                    $NI/\sqrt{\Phi^*}$ (CL2) ≃ 6.4

CL aperture   : 200 μmφ

Magnification: x400 K   corrected using lattice image of Au (200).

Spot size   : diameter of 90% intensity of spot recorded on film.

In the range of the specimen position down to Z=−0.55mm, no sherical aberration effects on the spot size (A.F. region) as shown in Fig. 5. The relationship between α_{max} and d_s in this region is expressed as:

$$\alpha_{max} \cdot d_s = \text{const.}$$

By moving down the specimen position and increasing α_{max}, d_s are made samller. But the effect of spherical aberration appears near c/o condition, and a caustic pattern is observed. d_s cannot be decreased under this condition. In the A. F. region, spot size is not affected so much by the size of CL aperture. The minimum spot size which can be obtained in TEM mode is shown in Table 3. So CBD patterns from very small area can be obtained.

Table 3. Minimum spot size at 400 kV

EM spot size	CL aperture (μmφ)	Spot size (nmφ)
7	200	12
7	50	8
8	200	11
8	50	5

Z = -0.55 mm

HR OBSERVATION

HR observation was made under C-TEM condition indicated by an arrow in Fig. 2. Fig. 7 shows the HR image of β-Si₃N₄ taken with [001] illumination at 400 kV. The crystal structure of β-Si₃N₄ is hexagonal, with cell dimensions a = 0.760 nm and C = 0.291 nm. [7] Fig. 8 a and b show optical diffraction (ODM) pattern of carbon film (region A) and structure image (region B), respectively. Theoretical resolution (d_t) is expressed by the following equation.

$$d_t = 0.65 \cdot C_s^{1/4} \cdot \lambda^{3/4}$$

d_t in C-TEM mode at 400 kV with $C_s \simeq 1.0$ mm is 0.17 nm. ODM patterns of Fig. 8 a shows that a spatial resolution of 0.18 nm is already achieved.

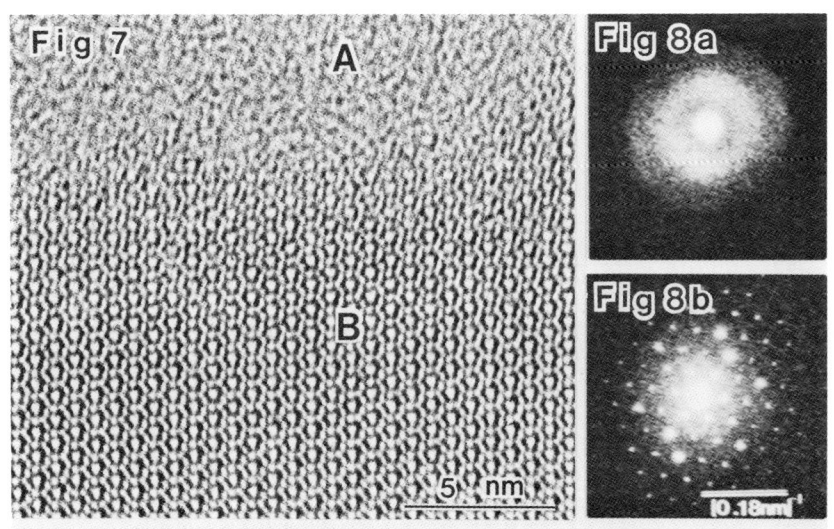

HR image of β-Si₃N₄ with [001] illumination and it's ODM pattern. Fig.8a shows the ODM pattern from area A and Fig.8b from area B.

SUMMARY

It is confirmed that combination of HR image observation with CBD analysis is performed very easily using Z-control system. A typical process is as follows:

1. Set the specimen at C-TEM mode position and make its alignment to perform HR image observation.
2. After HR image observation, move down the specimen to the lower position to get a specified illumination angle over the range of 1.0 mrad. to 10 mrad.
3. Make sure the alignment of illumination system and perform CBD analysis.

REFERENCES

1. P. Goodman, Acta Crystallogr., A31, 804 (1975)
2. B.F. Buxton, J.A. Eades, J.W. Steeds, and G.M. Rackham, Philos. Trans. R. Soc. London, A281, 171 (1976)
3. A. Tinnappel: Ph. D. Thesis, Tech. Univ., Berlin, 1975
4. P. Goodman and R.W. Secomb, Acta Crystallogr., A33, 126 (1977)
5. P. Goodman and A.W. Johnson, Acta Crystallogr., A33, 997 (1977)
6. H. Hashimoto, et al., Proc. of 7th HVEM 15 (1983)
7. R. Grun, Acta Crystallogr. B, 35, 800 (1979)

RECENT DEVELOPMENT IN HITACHI TRANSMISSION ELECTRON MICROSCOPE

SHIGETO ISAKOZAWA,* ISAO MATSUI,* SHOJI KAMIMURA,* AND AKIRA TONOMURA**
*NAKA WORKS, HITACHI, LTD., 882 Ichige, Katsuta, Ibaraki 312, Japan;
**CENTRAL RESEARCH LABORATORY, HITACHI, LTD., Kokubunji, Tokyo 185, Japan

The 200 kV electron microscope has been extensively utilized as a high grade model for diversified applications. This paper reports image resolution available at present with the Hitachi 200 kV Electron Microscope Model H-800 and possible techniques for improving present resolution limit which depends on the aberrations of objective lens.

The Model H-800 consists of 7-stage electron lens system; 2 illumination lenses and 5 imaging lenses. It permits image observation without rotation ($<\pm 5°$) over the entire magnification range of 100x to 1,000,000x. This design allows orientation matching between a selected area image and the diffraction pattern at any magnification available. The vacuum system of either dry or wet pumping is available. The microscope column is vacuum-sealed with liner tube.

Fig. 1 represents performance of objective lens designed for an ultrahigh resolution microscopy. A clearance of 2 mm is provided between the lower polepiece and the specimen in order to ensure stable operation of the goniometer stage which permits specimen tilting by $\pm 10°$. In this setting, the spherical aberration coefficient C_S and the chromatic aberration coefficient C_C become 1.0 and 1.2 mm respectively.

Fig. 2 shows a TEM image and optical diffractogram of a carbon film specimen taken at the Scherzer defocus condition using the above objective lens. Carbon graphite was used for calibration. As is evident from the halo pattern of the diffraction image, theoretical resolution (about 2.3 Å) expected from an accelerating voltage of 200 kV and spherical aberration coefficient C_S was achieved.

Fig. 3 exemplifies recording of a magnetic substance of magnetic tape. Point resolution of 2 to 3 Å is clearly observed.
As explained above, the Model H-800 is capable of providing a theoretical resolution (about 2 Å) which is determined by spherical aberration. In order to improve resolution further, a different technique is required.

The electron beam holography is well known as a promising technique for enhancing the resolution beyond the limit of spherical aberration. The Hitachi Central Research Laboratory has been engaged in research of electron beam holography for more than 10 years. [1], [2]
This report introduces principle and recent results of spherical aberration correction by the electron beam holography utilizing a field emission electron gun which assures excellent beam coherence.

Fig. 4 shows a schematic diagram for making electron beam holograms. Specimen is irradiated with a highly coherent electron beam emitted form the FE electron gun and is brought into focus by the objective lens. On the focused image, the electron beam passing the non-specimen area is superimposed as a reference wave through biprism, thereby forming a hologram. It includes all the effects of spherical aberration. When optically reproducing the hologram, an image corrected for aberration is obtainable by employing an optical lens having the same spherical aberration coefficient value which is opposite to that of the electron lens, as shown in Fig. 5.
Figures 6 (a) and (b) show a hologram of gold particle and a corrected image of the same specimen after optical correction for aberration. Fig. (b) is a lattice image of 2.4 Å and that of 1/2 spacing.

The electron beam holography offers the following advantages.
(1) Point resolution is enhanced (better than 1 Å) because the spherical aberration of electron lens is corrected at the stage of optical reproduction.
(2) Inelastically scattered electrons in a specimen neither interfere with the reference-wave nor contribute to the formation of a reproduced image, so energy filtered images are available.
(3) Direct microscopy of magnetic field is permitted by observing the electron beam whose phase is changed due to the magnetic field of a specimen.

Although the electron beam holography has some problems for practical use, they will be resolved and this technique will provide a point resolution of better than 1 Å in the near future.

Following delivery of high performance ultrahigh voltage electron microscopes (H-1250S, H-1300) to Tokyo Institute of Technology and Hokkaido University in 1981, the Model H-1250 was delivered to National Institute of Physiology for bio-medical science in 1982, and the Model H-1250ST was delivered to Nagoya University in 1983 as a world's first ultrahigh voltage microscope dedicated for STEM. Although these microscopes have functional characteristics in each field of application, the basic configuration and performance are similar except for the Model H-1250ST. The characteristics, basic performance and application of ultrahigh voltage electron microscopes are explained below mainly for the Model H-1250S.

Theoretical resolution at high voltages has been improved by a) stabilization of high voltage due to use of a high voltage metal film resistor, b) reduction of aberration with a new high-excitation objective lens, c) an ultrahigh vacuum system for minimizing specimen contamination.[3]
Fig. 7 shows an external view of the Model H-1250M.
Fig. 8 shows configuration of evacuation system.

(1) Basic performance
The measured aberration coefficients of the objective lens polepiece, which is commonly usable both for the top and side entry systems, are as follows; C_S = 2.5 mm, C_C^u = 3.5 mm and C_C^i = 2.6 mm at 1,000 kV. Point to point resolution on an optical diffractogram was 0.16 nm. Fig. 9 details the results. The mechanical vibration of three turbomolecular pumps is attenuated by anti-vibration bellows, and a acoustic damper/shield has been used for these pumps. Fig. 10 shows a lattice resolution of 0.1 nm.

(2) Structure image
The structure image was observed with a ±30° top-entry specimen goniometer stage.
Fig. 11 shows structure model and structure image of $W_8Nb_{18}O_{69}$. [4)]
Individual strings of metal atoms in the corner or edge sharing metal oxygen octahera are clearly resolved. An X-ray structure analysis showed that a distance between the corner sharing ones in 0.384 nm, while a distance in the projection between the edge sharing ones is 0.237 nm. [5)]

The Model H-1250ST which is the microscope dedicated for STEM is now under installation at Nagoya University. It is provided with an EELS unit as well as TEM function. Its details will be reported at some other occasions. The author would like to express his sincere appreciation to Prof. K. Yagi, Tokyo Institute of Technology, who gave us helpful advice and provided application data.

REFERENCE

1) Tonomura, A., Matsuda, T. and Endo, J.: Jpn, J. Appl. Phys., 18, 9 (1979)
2) Tonomura, A., Matsuda, T. and Endo, J.: Jpn, J. Appl. Phys., 18, 1373 (1979)
3) G. Honjo, K. Yagi, K. Takayanagi, S. Nagakura, S. Katagiri, M. Kubozoe and I. Matsui, Rroc. 6 the Int. Conf. HVEM (1980) p22
4) K. Yagi, K. Takayanagi, K. Kobayashi and S. Nagakura, Rroc. 7 the Int. Conf. HVEM (1983) p11
5) P. S. Rothe and A. D. Wadsley, Acta Crys. 19, 38 (1965)

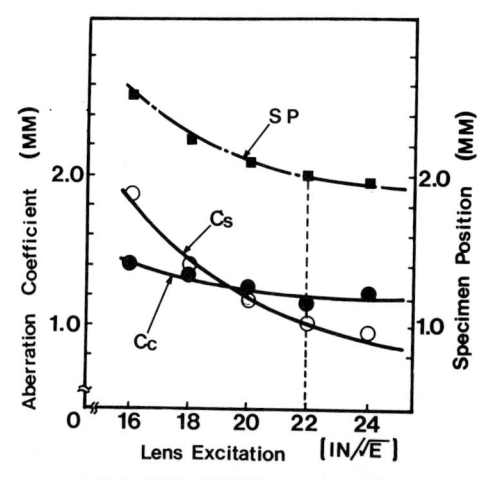

Fig. 1 Characteristics of objective lens

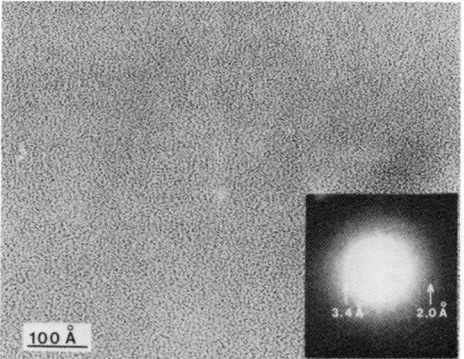

Fig. 2 Optical diffraction of carbon film

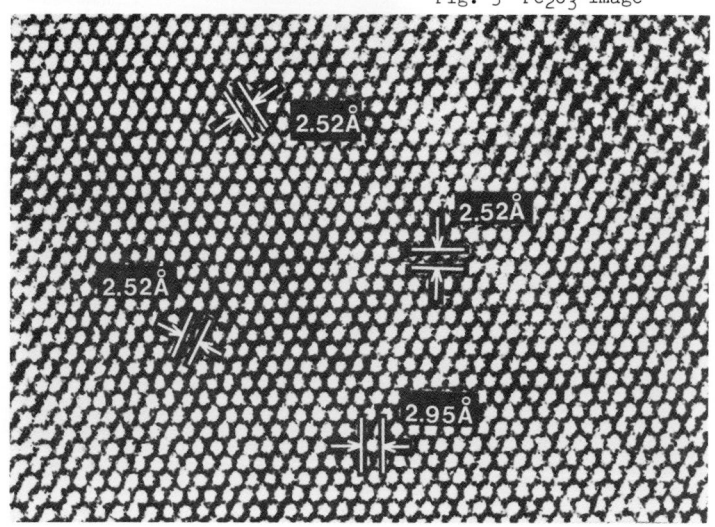

Fig. 3 Fe_2O_3 image

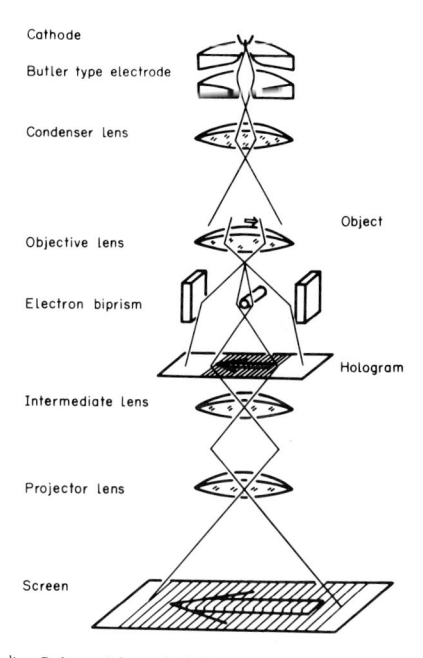

Fig. 4 Schematic of interence electron microscope

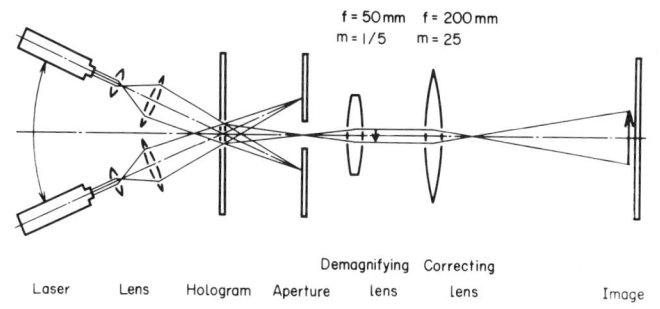

Fig. 5 Arrangement of the correction of spherical aberration

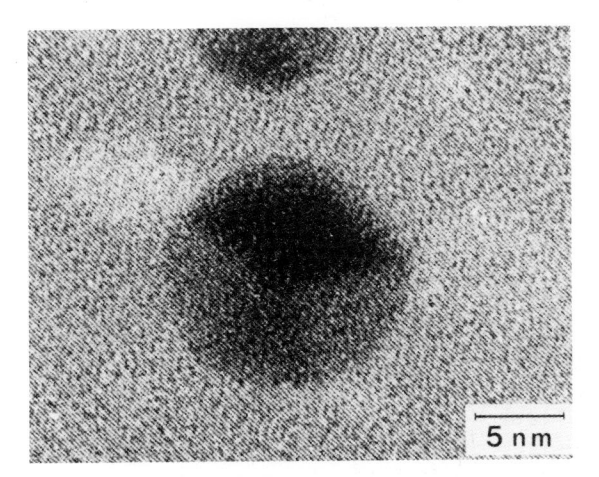

(a) Image hologram of a gold particle

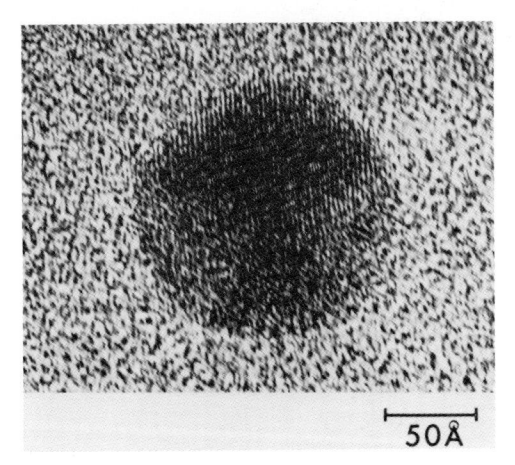

(b) Corrected image of a gold particle

Fig. 6 Hologram and corrected image

69

Fig. 7 Out view of H-1250M

Fig. 8 Pumping line of H-1250S

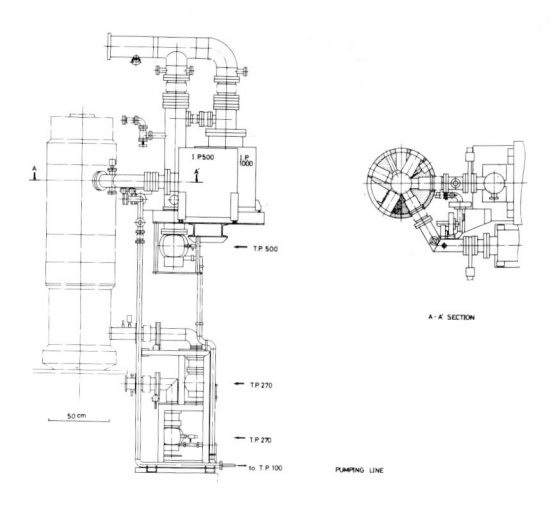

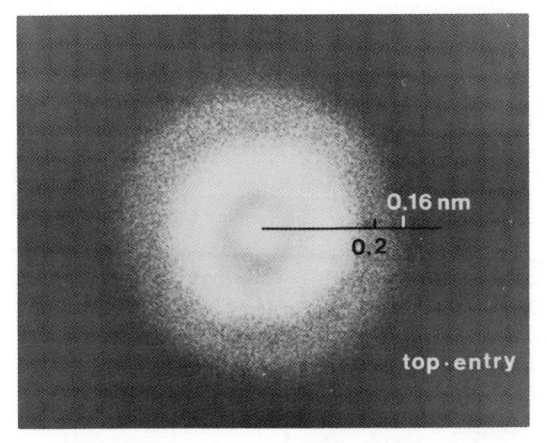

Fig. 9 An optical diffractograpm from an image of an amorphous carbon film taken at 1000 kV

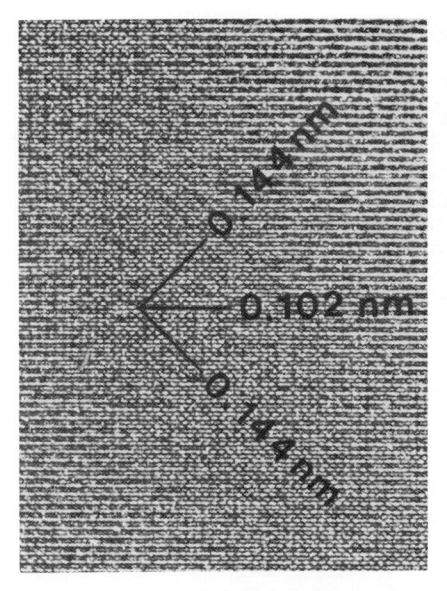

Fig. 10 0.1 nm lattice fringes of Au (001) film taken at 1000 kV (axial illumination)

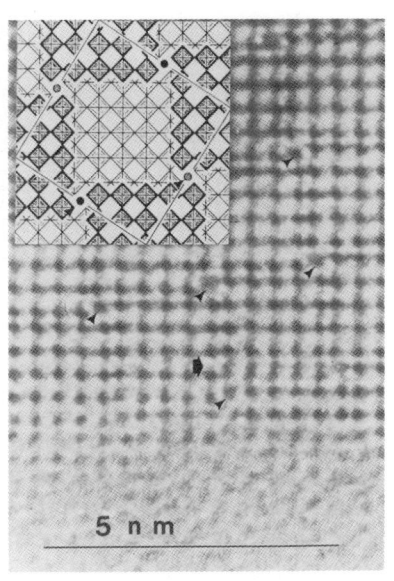

Fig. 11 Structure model and a high resolution structure image of a $W_8Nb_{18}O_{69}$ crystal

Electron Microscopy and Analysis at Higher Voltages: The Philips EM430 300kV Microscope

by Peter N. Hagemann, Electron Optics Laboratory, Philips Scientific and Industrial Equipment Division, N.V. Philips' Gloeilampenfabrieken, Eindhoven, The Netherlands.

Presented by John S. Fahy, Product Manager, Electron Optics, Philips Electronic Instruments, Inc., Mahwah, N.J.

Introduction

The past decade has seen the evolution of the Philips TEM from a "microscope" to an integrated microanalytical system. A current example of this trend is the EM 420*, an instrument whose design concepts are reflected in the most recently offered TEM, the compact, high kV, high resolution EM430. These design concepts include:

- A high resolution objective lens, with analytical capability, i.e., the Twin lens concept.
- A eucentric, large tilt, side-entry goniometer.
- A clean, dry, differentially pumped ultra-high vacuum environment for the specimen, which permits high current densities on the specimen at high magnifications...or in small probes, without instrument-induced contamination.
- An electron gun suitable for high brightness LaB_6 emitters.
- An illumination system allowing large area illumination or small probe forming on the specimen for TEM imaging or localized measurements, respectively.
- An imaging lens system which allows the selection of the imaging or diffraction mode independent of the illumination conditions and which permits, for example, direct imaging of the probe.
- A chromatically corrected magnification system to maintain high image quality in thick specimens at low magnifications.
- Integrated interfacing of peripheral detector systems, i.e., a high collection angle X-ray detector and an electron energy loss spectrometer focussed onto a fixed object plane.
- A microscope system suited to a multi-user/multi-discipline environment.

Higher Voltages: The Advantages for Electron Microscopy
Penetration

The scattering cross section for elastic as well as inelastic scattering decreases with increasing electron energy. With a given specimen thickness, the electron beam is thus less attenuated by elastic large angle scattering as the voltage becomes higher. Inelastic scattering causes an image blurring d_C given by:

$$d_C = C_C \alpha \frac{\Delta E}{E_O}$$

The reduction of chromatic image blurring for a given specimen thickness with increasing electron energy can be threefold:

- the mean energy loss ΔE decreases because of the decrease in the inelastic scattering;
- The relative energy loss $\Delta E/E_O$ decreases because E_O increases;
- the required objective aperture is smaller because of the decreased scattering angles.

The chromatic aberration figure C_C of the objective lens determines, together with the increased voltage, whether full advantage of the reduction of chromatic aberration can be obtained. At medium and low magnifications correction of the chromatic magnification error of the imaging system is essential to faithfully transfer the resolution of the objective lens onto the plate or final screen, Andersen and Fahy (1976).

High Resolution Imaging

The point resolution limit d_o is determined by microscope parameters only, since one assumes a specimen thin enough so that negligible inelastic scattering occurs. d_o is then given by:

$$d_o = 0.65 \, C_s^{1/4} \, \lambda^{3/4}$$

The spherical aberration values for the Twin and Super-Twin lenses are 2.0mm and 1.2mm respectively. On going from 120kV to 300kV, the point resolution drops from 0.3nm to 0.2nm for the Super-Twin lens, while the Twin lens provides a resolution of 0.34 and 0.23nm respectively and a specimen tilt range of ± 60°.

Beam Broadening

The effect of beam broadening is most striking for signals which can only be measured point by point, for example, the X-ray signal. With increasing specimen thickness, the electron beam is broadened due to multiple elastic and inelastic scattering. With the decrease of the scattering cross sections at higher voltages, the beam broadening is reduced for a given specimen thickness. The formula of Goldstein (1979) suggests a $1/E_o$ dependence of the beam broadening:

$$b = 625 \, \frac{Z}{E_o} \, \left(\frac{\rho}{A}\right)^{1/2} t^{3/2}$$

with: b in cm
E_o in keV
ρ = density g/cm^3
A = atomic weight
Z = atomic number
t = specimen thickness in cm

In electron energy loss spectroscopy and STEM imaging, the effective beam broadening can be reduced by selecting appropriate entrance apertures in the detection system.

In X-ray analysis the excited volume contributes fully to the signal and cannot be decreased afterwards. An increase of the acceleration voltage improves the lateral resolution of the X-ray analysis. With respect to the "thin film" criterion, however, there is no change of the foil thickness depending on the electron energy, since this is determined by X-ray absorption in the specimen itself. The limiting thickness t of a thin foil is given by (e.g. Zaluzec 1979): $\mu \rho t < 0.1$

where u = mass absorption coefficient;
ρ = density;
t = thickness of the specimen.

Electron energy loss spectroscopy, on the other hand, gains in usable foil thickness by an increased accelertion voltage since the mean free path of the electron between two inelastic scattering events is larger. The useful thickness for EELS can be estimated as a function of the mean free path for plasmon scattering, which is the main contributor to inelastic scattering. $t < a \lambda_p$ Here is a given value by Joy (1983) of 0.1 and by Zaluzec (1982) of 1. Measurements made by Seveley et al (1974) on Al and C show an increase of λ_p by a factor of two when going from 100 to 300kV. In order to double λ_p yet again would require in excess of 1 MeV electron energy.

EM430 Specific Design Considerations

Specific design considerations, in addition to the general design objectives previously discussed for the EM430 (shown in Figure 1a) may be noted as follows:

Compact installation. The EM430 comprises three major items: the microscope column, the high tension generator and the power supply cabinet. The floor area needed is 4 x 5m. Figure 1b shows the arrangement of the three components in a floor plan. The height required is 2.9m for normal operation, including filament exchange, plus the height of an external lifting device used only during installation.

Six step selectable high voltage... from 50kV to 300kV. All lens currents as well as the deflection coil and alignment currents are coupled to the high tension. Since there are no mechanical alignments, additional realignments are minimal when the voltage is changed.

Optional, variable, high tension control... allows the continuous voltage change between the 50kV steps, and is designed for applications in convergent beam or critical voltage experiments.

Energy dispersive X-ray detector... interfaced to the column with a detector elevation angle of 20° and a solid angle of collection of 0.13 sterad. The detector is permanently positioned. There is no need to retract the detector when not in use, since the silicon diode is shielded mechanically by a shutter (Philips patent) whenever high energy backscattered electrons could reach the detector diode. Because this occurs only at very low magnifications, i.e. when the objective lens is switched off, full operational freedom for mode selection is maintained.

Electron Energy Loss spectrometer... has its fixed object plane at the projector crossover plane, which lies in the differential pumping diaphragm. The projector crossover stays fixed at this plane in order to guarantee the transfer of the image and diffraction pattern through the 200μm aperture without field limitations.

Projection chamber... contains a 16cm diameter large viewing screen and a small observation screen for use in conjunction with the binoculars at 12 X light optical magnification.

The major components of the electron optics and the detector systems are shown in Figure 2. The symmetrical Twin objective lens is equally well suited for imaging in TEM and the immediate generation of small TEM probes. The objective lens is generally held at a fixed excitation. Therefore the auxiliary lens compensates the objective prefield for large area illumination as used for TEM imaging, see Figure 2a. In order to obtain small probes, the objective prefield is used as a third condenser lens, Figure 2b. The auxiliary lens is then optically switched off.

The electron probe can be focussed on the specimen to less than 2nm in diameter. Positioning of the probe can be observed directly from the TEM image. When a STEM system is coupled to the microscope, the beam deflection coils above the specimen are used to provide double deflection scanning.

The objective lens parameters and the associated magnification ranges are tabulated on the following page.

	Objective Lens	EM430 Twin	EM430 Super Twin
TEM	C_S	2.0mm	1.2mm
and	C_C	2.0mm	1.2mm
STEM	f	2.7mm	1.7mm

Total Magnification Ranges

50-200kV	50 to 870,000	50 to 1,350,000
250kV	50 to 750,000	50 to 1,150,000
300kV	50 to 600,000	50 to 850,000

Materials Science

High resolution imaging is applied in basic material science especially when studying lattice defects, grain boundaries and interfaces between different materials. In many cases it is essential that high resolution imaging be compatible with the analytical capabilities of the microscope in order to allow a full characterisation of the problem under investigation. Figure 3 shows twins in (110) silicon. The micrograph was obtained using the Twin lens and the ± 60°, side entry eucentric goniometer.

In thicker specimens, the lattice image is faded out due to the inelastic scattering and loss of coherence caused by multi-beam image can beneficially be applied to the analysis of lattice defects using classical diffraction contrast (Hashimoto, 1974). Figures 4a and 4b show micrographs, stacking faults and line dislocations in a titanium-aluminum foil in bright field and in dark field, respectively. The multi-beam image in Figure 4c shows that the fringe contrast of the stacking faults and the oscillating contrast of the dislocations vanishes towards the bottom surface of the foil.

Micrographs of a molybdenum jetted disc specimen are shown in Figure 5. The micrographs demonstrate a higher penetration and a lower chromatic image blurring at 300kV, Figure 5a, when compared to 100kV, Figure 5b. The higher penetration is indicated by the exposure times which were 2.8 seconds for a and b, respectively.

For materials science, the large usable foil thickness is especially advantageous for the observation of ceramics and minerals, which in general can only be prepared by ion thinning. In the semiconductor technology, high voltage electron microscopy can be applied to the study of integrated circuits. The several micron-thick active layer of such components can be directly imaged after removing the supporting silicon substrate. Figure 6 shows an area of a PROM (programmable read-only memory) which can be programmed selectively by melting minute nickel-chromium fuses. An example of a melted fuse is shown in the enlargement in the inset. The cleanness of the break governs the operation effectiveness of the device.

The following example, Figure 7a, shows a micrograph of a ferritic steel (2-1/4 Cr-a Mo wt%). The specimen is of the disk type and therefore has a large inherent magnetic field. In the EM430 both the objective astigmatism and the condenser astigmatism could be fully corrected even up to 45° of tilt. Correction of the condenser astigmatism is essential to control the electron probe for X-ray and/or microdiffraction analysis. X-ray analysis revealed two types of precipitates: one was chromium rich and the other molybdenum rich. Microdiffraction pattern information, the arrowed precipitate, as shown in the inset, complemented the image and X-ray microanalysis data.

Convergent beam electron diffraction has proved to be a very sensitive microanalytical tool, Steeds (1979). It is applied in lattice parameter determination, measurements of chemical variations and measurements of lattice strain. In crystallography, it is also used for point and space group determination. In order to support quantitative X-ray or energy loss data processing, convergent beam diffraction can be used as a means of thickness measurement. On the EM430, the convergent beam diffraction is fully compatible with high resolution imaging, since electron probes of less than 2nm diameter are available without any change of specimen position or orientation. Figure 8 shows a convergent beam diffraction pattern for TaSe$_2$: a) at room temperture; b) at liquid nitrogen temperature using the cooling holder of the cryo system. Cooling the specimen causes an increase of diffraction intensities at the higher order Laue zones. Figure 9 shows a convergent beam diffraction pattern from (111) diamond. The enlargement of the central disc exhibits fine high-order Laue zone lines at a specific acceleration voltage of 280.65kV, which was selected with the variable high tension control.

References

Anderson, W. H. J. and Fahy, J. S (1976) Imaging thick specimens with the EM400 high Resolution Electron Microscope. Philips Electron Optics Bulletin, EM107, 10-12.

Fahy, J. S., Plomp, F. H., Rakels, C. J. and Thompason, M. N. (1976, 1977) The EM400 Transmission Electron Microscope Design. Philips Electron Optics Bulletin, EM110, 6-13.

Favard, P. (1980) Recent Aspects of Development of High Voltage Electron Microscopy. Electron Microscopy, 1980 (Den Haag), 2, Biology, 112-117.

Goldstein, J. I. (1979) Principles of Thin Film X-ray Microanalysis. Introduction to Analytical Electron Microscopy, ed. J. Hren et al., Plenum Press, New York and London, 83-120.

Hashimoto (1979) High Voltage TEM-contrast Theory. High Voltage Electron Microscopy, ed. P. R. Swann et al., Academic Press, London and New York, 9-21.

Joy, D. C. (1983) Practical Quantification for Energy Loss Spectra. Scanning Electron Microscopy, 1983, 1, 505-515.

Steeds, J. W. Convergent Beam Electron Diffraction. Introduction to Analytical Electron Microscopy, ed. J. Hren et al., Plenum Press, New York and London, 387-422.

Wolosewick, J. J. and Porter, K. R. (1979) Microtrabecular Lattice of the Cytoplastmatic Ground Substance--Artefact or Reality. J. Cell Biology, 82, 114-139.

Zaluzec (1979) Quantiative X-Ray Microanalysis: Instrumental Considerations and Applications to Materials Science. Introduction to Analytical Electron Microscopy, ed. J. Hren et al., Plenum Press, New York and London, 121-167.

Zaluzec (1982) Private communication.

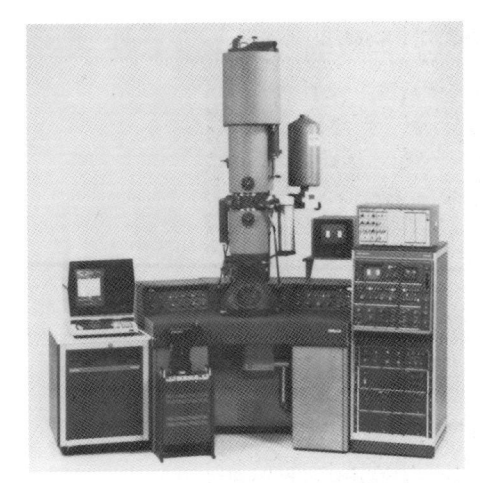

Figure 1a. View of the EM430 300kV transmission electron microscope equipped with STEM system and energy dispersive X-ray analyser.

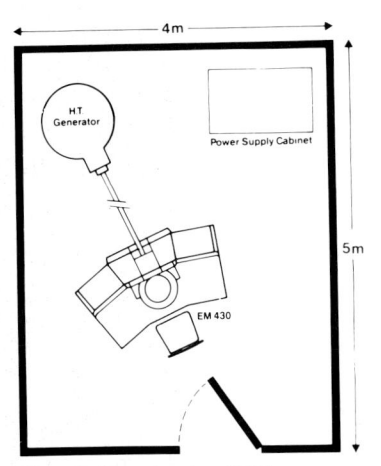

Figure 1b. Floor plan of an EM430 microscope room.

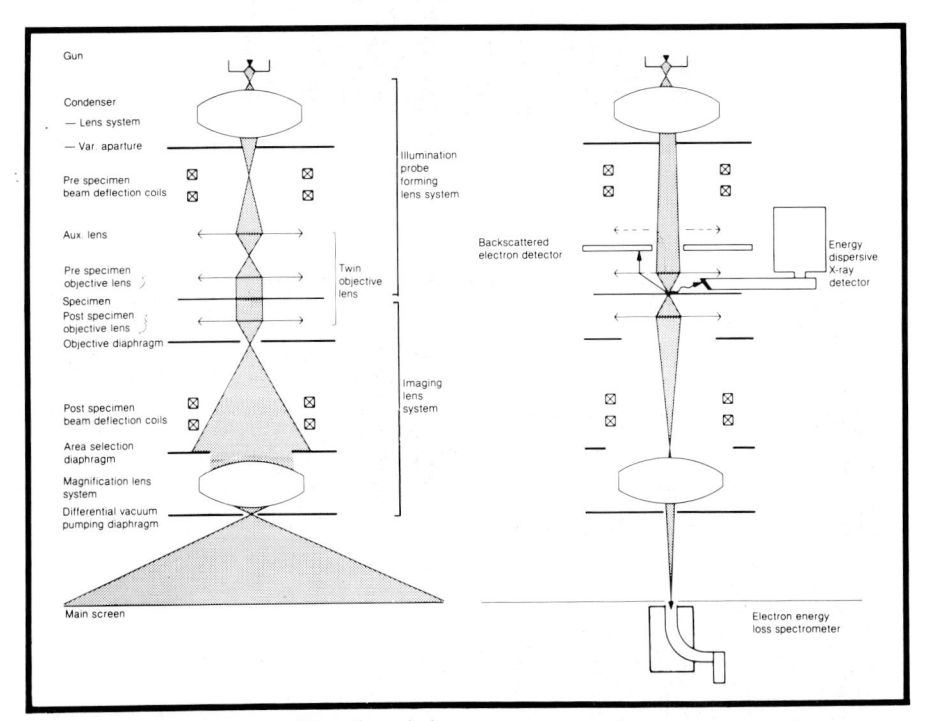

Figure 2. Schematic diagram of the major optical and detector components of the EM430. The ray path shows: a) large area illumination and imaging. b) probe forming and probe imaging for point analyses or microdiffraction.

Figure 3. High resolution image of a twinned area in (110) oriented silicon. The image is taken with the high tilt Twin lens which allows a specimen tilt of ±60°. Specimen courtesy of Dr. S. L. Sass, Cornell University, Ithaca, N.Y., U.S.A.

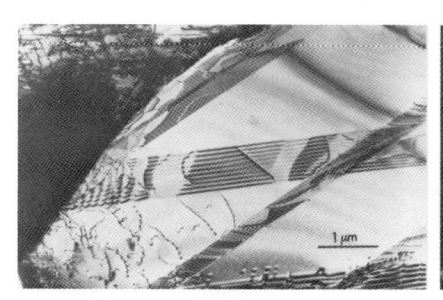

Figure 4a, b, and c. Stacking faults and dislocations in a titanium/aluminium alloy. a) in bright field; b) in dark field; c) in multi-beam imaging condition. The latter shows the cancellation of the fringe contrast where the inclined stacking fault and dislocations intersect the bottom surface of the foil (arrows). Specimen courtesy of Dr. H. Fraser, University of Illinois, Champaign/Urbana, Illinois, U.S.A.

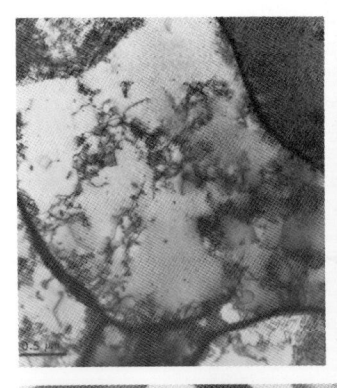

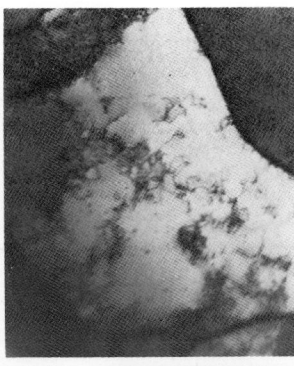

Figure 5a and 5b. Demonstration of the resolution and intensity improvement that can be achieved at 300kV (figure 7a) as compared to 100kV (Figure 7b). The relative intensity of the micrographs can be judged by their exposure times which differed by a factor of approximately 20 times. Specimen courtesy of Dr. Titchmarsh, UKAEA, Harwell, England.

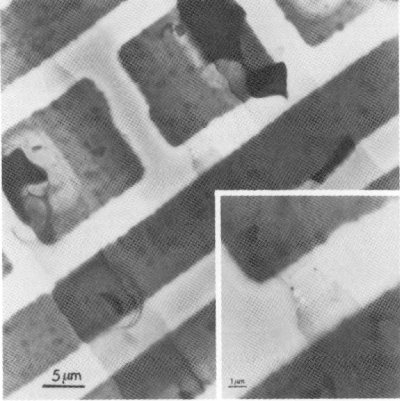

Figure 6. Programmable read-only memory showing the melted fuse at high magnification. The cleanness of the break of the connection strip is essential for the performance of the device. Courtesy of Dr. W. Stacy, Philips Research Laboratories. Signetics Corporation, Sunnyvale, California, U.S.A.

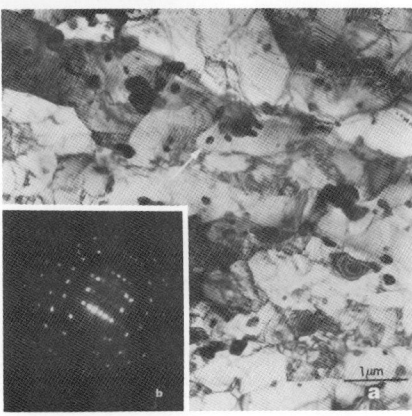

Figure 7. Ferritic steel with 2¼ Cr-1 Mo wt•. a) image; b) microdiffraction pattern. The latter is taken from the precipitate indicated in (a) by the arrow. Specimen courtesy of Dr. J. Titchmarsh, UKAEA, Harwell, England.

Figure 8a. Convergent beam diffraction pattern at room temperature.

Figure 8b. Convergent beam diffraction pattern at liquid nitrogen temperature. Cooling the specimen causes an increase of the high angle diffraction intensities. Courtesy of Dr. J. Steeds, University of Bristol, England.

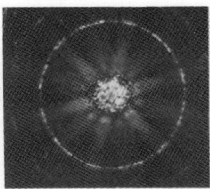

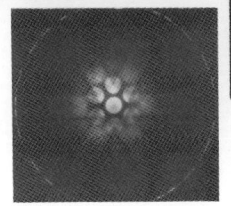

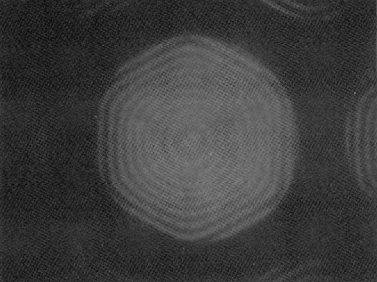

Figure 9a and 9b. Convergent beam diffraction pattern. Figure 9a. from (111) Diamond at a nominal acceleration voltage of 280.65kV using the variable high tension control unit. The enlargement of the central disk, Figure 9b, exhibits fine high-order Laue zone lines.

DETERMINATION OF THE SPECIFIC SITE OCCUPATION OF RARE EARTH ADDITIONS IN $Y_{1.7}Sm_{0.6}Lu_{0.7}Fe_5O_{12}$ THIN FILMS BY THE ORIENTATION DEPENDENCE OF CHARACTERISTIC X-RAY EMISSIONS

KANNAN M. KRISHNAN, PETER REZ[*], RAJA MISHRA AND GARETH THOMAS
Lawrence Berkeley Laboratory, Materials and Molecular Research Division, National Center for Electron Microscopy, University of California, Berkeley, CA 94720

[*] Present address: V. G. Microscopes, Ltd., England.

A
ABSTRACT

The orientation dependence of characteristic x-ray emissions have been used to determine specific site occupations of Rare Earth additions in epitaxially grown films of $Y_{1.7}Sm_{0.6}Lu_{0.7}Fe_5O_{12}$. A theoretical formulation based on the assumption of highly localized inner shell excitations was used not only to predict specific site sensitive orientations, but also to refine experimentally observed data employing a constrained least squares analysis to give probabilities for the occupation of the RE additions in the different crystallographic sites. Thus, it has been shown that in this compound the preference for the RE additions is a predominantly octahedral occupation with a probability $\geq$ 95%. Some of the assumptions and limitations of the technique have also been discussed.

INTRODUCTION

The uniaxial magnetic anisotropy in epitaxially grown garnet films consists of a growth-induced component and a stress-induced component. The latter component arises due to magneto-elastic interactions [1]. The former component, as the name implies, originates from the growth process and is explained in terms of the presently accepted "site preference" model [2,3,4]. According to this model the uniaxial anisotropy is attributed to the preferential occupation of the Rare Earth (RE) elements in the different crystallographic sites of the garnet lattice. But the results obtained by different investigators [5,6] using this model for the RE occupations are controversial, if not contradictory. However, it has been observed that there is an order of magnitude reduction and subsequently, ultimate loss of the value of the anisotropy constant (K_u) on annealing [7]. This observation is an indirect proof for the site preference model of anisotropy, but specifically which sites are involved is unclear [8].

Traditionally, this cation distribution problem has been experimentally investigated by one of the following methods: (1) Magnetization; (2) Magnetic Anisotropy; (3) Optical Spectroscopy; (4) Mössbauer Spectroscopy and (5) Spin Echo (see Ref. [9] for a detailed review). These techniques are limited in their application to bulk samples and hence can resolve average distribution only. Here we present an alternative technique for the evaluation of specific site occupancies of RE substitutions in $Y_3Fe_5O_{12}$ (YIG) based on the orientation dependence of electron-induced characteristic x-ray emissions. The advantage of this technique over the ones mentioned earlier is that a converged electron probe is used and hence local areas of the sample can be studied. Ideally, this should allow us to determine RE distributions as a function of the non-equilibrium growth process, by studying sequential slices of thin films.

THE TECHNIQUE: PHYSICAL DESCRIPTION

As a result of the dynamical scattering of an incident plane wave of high energy electrons, a standing wave pattern is set up within the crystal unit cell [10]. This intensity modulation within the crystal is dependent on the incident beam orientation and under favorable conditions will be a maximum on certain crystallographic sites. Hence, highly localized scattering events, like inner shell excitations, show a strong orientation dependence [11,12]. Further, it has been shown that for certain systematic or "planar channelling" conditions, the characteristic x-ray emissions induced by the channelling or blocking of the incident electrons, can provide information on the occupancy of specific crystallographic sites by different elements, provided that an a priori knowledge of the distribution of some reference elements is available [13,14,15]. This is particularly applicable for crystals with a layered structure (i.e. crystals that in some crystallographic projection can be resolved into alternating layers of parallel non-identical planes [A,B,A,B,...], where the appropriate systematic orientation can be determined by mere inspection. For example, the spinel structure compounds can be resolved in the [001] projection into alternating planes of tetrahedral and octahedral sites and hence a $\bar{g}$ = 400 systematic orientation can be easily seen to be appropriate for this kind of experiment. The garnet structure on the other hand, is more complicated and a number of crystallographic projections calculated using the ORTEP [16] program failed to indicate a systematic orientation that could be specific site sensitive. As a result, the electron induced characteristic x-ray intensities for different orientations and different site occupations have to be calculated in order to determine the appropriate systematic or "planar channelling" condition suitable to study site occupations for this particular crystal structure.

THEORETICAL FORMULATION

Assume that the electron-induced characteristic x-ray emission of an element 'q' at the coordinate $\bar{\rho}$ in the unit cell is proportional to the intensity of the electron standing wave at $\bar{\rho}$ and the probability of the localization of the standing wave at $\bar{\rho}$, $P(\bar{\rho})$. It is reasonable [17] to assume that $P(\rho)$ is highly localized, i.e. a delta function at the mean atomic positions. Hence, for a crystal of thickness 't'

$$N_q = \sum_{RSI} \int_0^t \phi^* \phi \, dz$$

where N_q is the intensity of the characteristic x-ray emission of element q and the summation is over the appropriate crystallographic sites of interest in the unit cell. ϕ is the amplitude of the scattered wave at any depth z expressed as a linear combination of Bloch waves (for an incident plane wave) in the conventional dynamical theory formulation [10].

On carrying out the integration and rearranging the different terms, the characteristic x-ray intensity per unit thickness can be expressed as

$$N_q = \sum_{RSI} \left[\sum_{g,h} \exp((\bar{h} - \bar{g})x) \left\{ \sum_{\substack{j,l \\ j=l}} C_o^{(j)*} C_g^{(j)*} C_o^{(l)} C_h^{(l)} \right. \right.$$

$$\left. \left. + \sum_{\substack{j,l \\ j>l}} C_o^{(j)*} C_g^{(j)*} C_o^{(l)} C_h^{(l)} \frac{\sin[(k_z^{(j)} - k_z^{(l)})t]}{(k_z^{(j)} - k_z^{(l)})t} \right\} \right]$$

where $C_o^{(j)}$ are Bloch wave coefficients and $k_z^{(j)}$ are components of the wave vector for the electrons in the crystal.

CALCULATIONS

The electron induced characteristic x-ray emission intensities were computed using the above expression for $MgAl_2O_4$, a compound with the spinel structure. A 15-beam, $\bar{g}$ = 400 systematic excitation condition was chosen and the calculated intensities at an acceleration voltage of 100kV were found to be in agreement with experimental results [18]. This served as a verification for the validity of the theoretical formulation.

X-ray emission intensities were then calculated for complete RE occupation in any one of the three crystallographic sites of the prototype compound YIG for different systematic excitation conditions (i.e. $\vec{g} = 00\bar{2}$; $\vec{g} = \bar{2}20$; $\vec{g} = 1\bar{2}1$, etc.) to determine an orientation with specific site sensitive characteristic x-ray emission. These 11-beam calculations were performed for a range of crystal thicknesses (250Å - 2000Å) and for a range of specimen tilts ($0 < k_x/g < 2$). The x-ray production was found to be insensitive to crystallographic orientation for excitation of the ($\vec{g}$ = 002) and ($\vec{g} = \bar{2}20$) systematic conditions. On the other hand strong orientation dependence for the ($\vec{g} = 1\bar{2}1$) systematic excitation were observed at the first order Bragg diffraction condition ($k_x/g = 0.5$) and for negative and positive excitation errors (Fig. 1), i.e. channeling for dodecahedral site substitution of RE additions; blocking for tetrahedral site substitutions and an intensitivity to orientation for octahedral site substitutions.

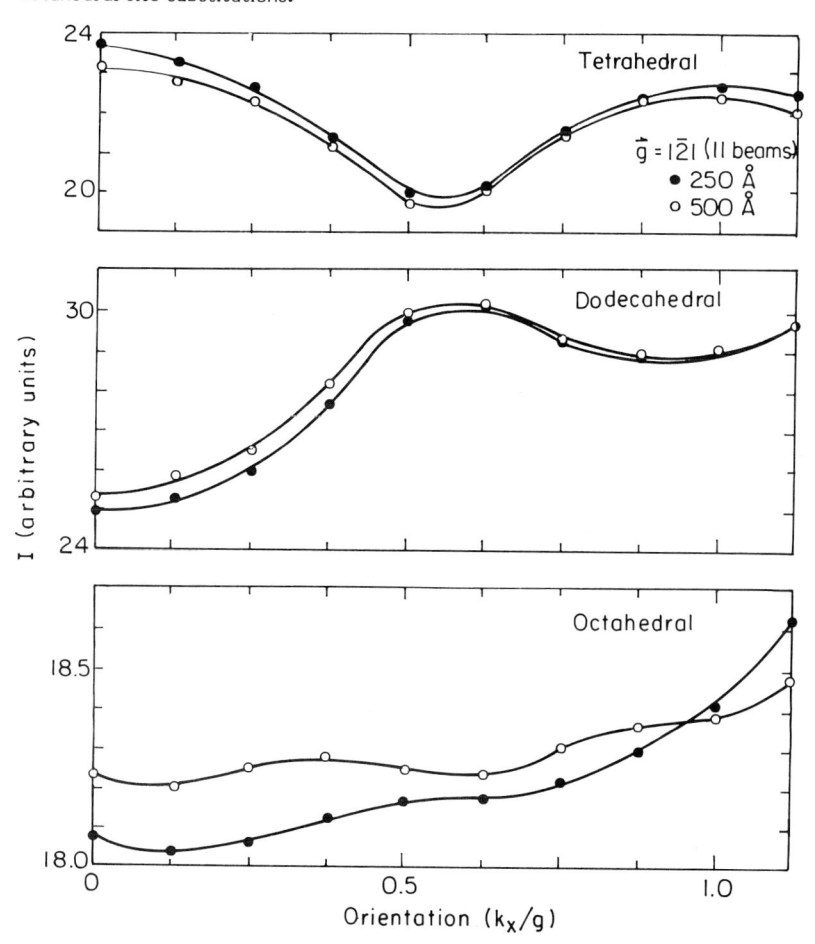

FIG. 1. Variation of electron-induced characteristic x-ray emission with orientation for a garnet structure. These calculations were carried out for an 11-beam, $g = 1\bar{2}1$ systematic row and for a range of incident beam orientations ($0 \le k_x/g < 1.0$).

EXPERIMENTAL DETAILS

Samples of nominal melt composition $Y_{1.7}Sm_{0.6}Cu_{0.7}Fe_5O_{12}$, 0.94μm thick, prepared by liquid phase epitaxy on GGG substrates at a growth temperature of 969°C and a growth rate of 1.76μm/min were studied. Details of the preparation of these samples are given elsewhere [19]. Thin foils of these samples for TEM studies were prepared by the ion-milling technique.

Experiments were performed on a Philips 400 ST analytical TEM fitted with a LaB_6 filament and an energy dispersive x-ray analyzer. A low background double tilt holder to reduce x-ray background intensities was used in all experiments. Stray x-ray generation was further minimized by using gridless self-supporting specimens. The specimen thickness was estimated to be approximately 500Å. The incident beam divergence was a few milliradians and the probe diameter was 500-1000Å. The latter is determined by the level of contamination and the actual counting time. The characteristic x-ray emissions were collected at different orientations of the collimated electron beam based on the calculations described earlier. A strong [1$\bar{2}$1] systematic row was excited and spectra acquired at six different values of the specimen tilt described by the parameter: k_x/g = 0, 0.375, 0.5, 0.875, 1.00 and 1.125 where k_x/g is defined such that k_x/g = 0.5 for the first order Bragg diffraction condition. The specimens were oriented using convergent beam electron diffraction patterns.

In order to ensure proper statistics, spectra were collected (at the appropriate orientations) for a total counting time of 600 seconds and a counting rate of ~3000 counts/sec. A hole count was taken at periodic intervals and subtracted along with the continuous background from the spectra. Because of the large counting time (600 secs) and the susceptibility of the specimen to damage and contamination, the spectra was collected at a different specimen sampling area at each specimen tilt. Care was taken to ensure that the specimen thickness was not significantly different for different acquisitions. However, small differences in counting rates or changes due to the small displacement of the probe have to be incorporated in the analysis. This was done by scaling the integrated elemental intensities using normalized values of the total integrated intensities of the whole spectra.

The scaled integrated intensity for the two RE additions are shown in Table 1. The standard deviations of the observations are approximately 0.5% and 0.65% for $Lu(L_\alpha)$ and $Sm(L_\alpha)$ respectively.

TABLE 1. Characteristic x-ray intensities (normalized)

K_x/g	0	0.375	0.5	0.875	1.0	1.125
$Lu(L_\alpha)$	50807	49747	49130	50142	48985	50437
$Sm(L_\alpha)$	31911	31754	31808	32073	31218	31805

ANALYSIS

The results (Table 1) are statistically significant, but a clear agreement with the theoretical predictions is not obviously evident. This is interpreted as being due to a mixed distribution of the RE elements in all three possible crystallographic sites. The probabilities of the distributions can then be determined by a least squares analysis, based on the algorithm of constrained least squares by Lawson and Hanson [20].

An error term is defined here as the difference between the experimentally observed intensity and an intensity calculated as a summation over all sites of the product of the theoretical value for complete occupation of each site (calculated earlier) and a weight factor representing the probability of occupation of that specific site. For each element of interest, i.e. Sm and Lu, a summation over all orientation of the square of this error term is minimized, subject to the constraint that all the weights are positive [21]. The results of the least squares refinement for the weights of the occupation for each element in a particular site is shown in Table 2.

TABLE 2. Relative weights for the site occupancy of Rare Earth additions

	Octahedral	Tetrahedral	Dodecahedral
Lu	0.2141±0.02	0.0049±0.0005	**
Sm	0.1679±0.0164	**	0.0095±0.0007

DISCUSSION

A significant though reasonable assumption in this experimental technique is that the x-ray emission process is highly localized, i.e. it can be approximated by a delta function at the mean atomic positions. Based on band structure calculations, the inner shell radii have been estimated [22] to be of the order of 0.02Å. Further, Bourdillon et al. [16] have shown complete localizations for all excitation energies within the detectability limits of conventional energy dispersive x-ray detectors. Further, reductions in the orientation-dependent x-ray emission intensities for temperature rises up to 200°C has been shown [11] to be insignificant when compared with reductions due to angular spread, based on computations for a range of Debye-Waller factors. These arguments bear out the assumption.

It is acknowledged that the present approach neglects absorption effects in the crystal and the fact that a convergent probe as opposed to parallel illumination was used in the experiment. A more rigorous formulation would have to include these in the analysis. However, the latter could easily be accommodated by redoing the calculations for a whole range of incident orientations and incorporating these values in the least squares refinement.

Within the limitations of the assumptions of the technique, it can be inferred that the RE substitution in the octahedral sites for this compound are highly probable (~95%). These results are in good agreement with earlier studies [23,24] particularly for the distribution of the small rare-earth ion Lu^{3+} which can be easily accommodated in the smaller octahedral sites. A determination of the contribution of these ions to the magnetic anisotropy demands the exact knowledge of the valence state of the cation in addition to the site occupancy and is beyond the scope of this work.

ACKNOWLEDGEMENTS

This work was supported by the Director, Office of Energy Research, Office of Basic Energy Sciences, Materials Sciences Division of the U. S. Department of Energy under Contract No. DE-AC03-76SF00098. The specimens were obtained from Dr. V. J. Fratello of Bell Labs, N. J. We would also like to thank Drs. R. Sinclair and A. Pelton for the use of the AEM facilities at Stanford and Dr. D. Cockayne for helpful criticism and comments.

REFERENCES

1. P. Hansen, "Physics of Magnetic Garnets", International School of Physics "Enrico Fermi", 125 (1978).
2. H. Callen, Appl. Phys. Lett 18, 311 (1971).
3. A. Rosencwaig, W. J. Tabor and R. D. Pierce, Phys. Rev. Lett. 26, 779 (1971).
4. E. M. Gyorgy, A. Rosencwaig, E. I. Blount, W. J. Tabor and M. E. Lines, Appl. Phys. Lett. 18 479 (1971).
5. E. M. Gyorgy, M. D. Stuge and L. G. Van Uiter, A.I.P. Conf. Proc. 18, 70 (1974).
6. A. Akselrad and H. Callen, Appl. Phys. Lett. 19, 464 (1971).
7. F. B. Hegedorn, W. J. Tabor and L. G. Van Uiter, J. Appl. Phys. 44, 432 (1973).
8. J. W. Nielsen, Ann. Rev. Mater. Sci. 9, 87 (1979).
9. Ref. 1: R. Krishnan, p. 505.
10. P. B. Hirsch, A. Howie, R. B. Nicholson, D. W. Pashley and M. J. Whelan, Electron Microscopy of Thin Crystals, London, Butterworth (1965).
11. P. Duncumb, Phil. Mag. 7, 2101 (1962).
12. D. Cherns, A. Howie and M. H. Jacobs, Z. Naturforsch. 28A, 565 (1973).
13. J. Taftø, J. Appl. Cryst. 15, 452 (1979).
14. J. C. H. Spence and J. Taftø, J. Microscopy 130, 147 (1983).
15. K. M. Krishnan, L. Rabenberg, R. K. Mishra and G. Thomas, J. Appl. Phys. to be published, 1983.
16. C. K. Johnson, "ORTEP--A Fortran Thermal-Ellipsoid Program for Crystal Structure Illustrations", ONRL-3794, Oak Ridge National Lab., June 1965.
17. A. J. Bourdillon, P. G. Self and W. M. Stobbs, Phil. Mag. A 44, 1335 (1981).
18. J. Taftø and J. C. H. Spence, Ultramicroscopy 9, 243 (1982).
19. J. W. Mathews, "Epitaxial Growth", Academic Press, N. Y. (1975).
20. C. L. Lawson and R. J. Hanson, "Solving Least Squares Problems", Chapters 23 and 25, Prentice-Hall, N. J. (1974).
21. Kannan M. Krishnan, Ph.D. Thesis, University of California, Berkeley, to be published.
22. J. Gjonnes and R. Høier, Acta Cryst. A27, 166 (1971).
23. L. Suchow and M. Kakta, J. Sol. Stat. Chem. 5, 329 (1972).
24. S. L. Blank, J. W. Nielsen and W. A. Biolsi, J. Electrochem. Soc. 123, 845 (1976).

SECTION II

Semiconducting Materials

TEM/EBIC INVESTIGATIONS OF STRUCTURAL DEFECTS IN POLYCRYSTALLINE SOLAR CELLS

D.G.AST, B.CUNNINGHAM* AND R.GLEICHMANN.
Material Science and Engineering, Cornell University, Ithaca N.Y., 14853

INTRODUCTION

Most solar silicon is grown in the form of cast ingots, e.g. HEM, Silso and UCP or in the form of thin continous ribbons, e.g. EFG, RTR and Web. HEM and EFG will be considered as representative example of each catagory.

HEM is solidified in rectangular SiO_2 molds and cooled slowly over a period of days. The material is generally single crystalline, except for the upper corners which solidify last. The slow cooling rate tends to promote segregation of impurities to grain boundaries [1] and permits structural defects to arrange themselves in stable patterns minimizing long range stresses. Consequently, HEM is essentially free of residual stresses.

Ribbon materials, on the other hand, invariably contain large residual stresses. An evaluation of the relation between the rate of solidification, the thermal gradients at the solid-liquid interface and the post solidification temperature profile shows that these stresses are a consequence of the high growth speed and can not be avoided without reducing the growth rate [2]. EFG ribbons can be grown from either quartz or graphite crucibles, but a graphite die with a narrow slot is always used to shape the liquid Si prior to solidification [3]. As a result of the contact of liquid Si with graphite, EFG contains C at the solubility limit (about 2.10^{18} cm^{-3}). The amount of oxygen depends on the crucible material and the gas environment in the die top vicinity and can be varied by the intentional introduction of CO or CO_2 [4]. The maximum achievable O concentration is about a factor two below the C concentration. For this reason, the defect chemistry of EFG is always dominated by C and very different from standard CZ where oxygen is the major non-electrically active chemical impurity.

In the following sections, we consider first the crystallography and electrical activity of twin related boundaries, since these boundaries comprise the majority of the boundary structures found in solar silicon. We then proceed to discuss the influence of processing on the grown-in defect structure and it's electrical activity.

ELECTRICAL ACTIVITY OF HIGHER ORDER TWIN BOUNDARIES

Grain boundaries which can be generated by n consecutive twinning operations on non-coplanar {111} planes are termed higher order twins of order n in the notation of Kohn [5], and $\Sigma=3,9,27$ for n=1,2,3 in the notation of Bollmann [6]. Σ is the inverse fraction of common lattice sites formed if the crystals lattices on either side of the boundary were mutually penetrating. This fraction is generally different from the fraction of coincident sites at the actual boundary which depends on the boundary plane orientation. Experimentally one finds that almost all high angle grain boundaries in Si are <110> tilt boundaries with Σ = 3,9 or 27 [7,8]. These twin related orientations are strongly preferred over other low Σ orientations, e.g. Σ=5 which does not appear to occcur in cast Si.

To predict the location of the boundary plane, the CSL theory [6] assumes that the density of coincident sites is reversely related to grain boundary energy. Lattice images have shown that the preferred boundary plane is,

---
*Present address: IBM East Fishkill, Hopewell Junction, N.Y. 12533.

however, not always the plane comprising the highest density of coincident sites. When the boundary plane with the highest density of coincident sites is asymmetric, the crystal prefers a symmetric boundary plane even if the latter has a lower coincident site density. The inability of the CSL theory to correctly predict the boundary plane most likely arises from the lack of a formalism to preserve tetrahedral bonding across interfaces. This requirement can be implemented by the repeating group description first formulated by Hornstra [9] for specific rotation angles corresponding to low Σ <110> tilt boundaries. A recent extension of this concept shows that symmetric <110> tilt boundaries with {110} median lattice plane can be constructed without broken bonds for any arbitrary tilt angle between 0 and $108°$ [10]. Asymmetric boundaries, however, cannot be constructed without broken bonds.

The importance of symmetry considerations is further indicated by high resolution images which show that boundaries with (macroscopically) asymmetric grain boundaries planes are frequently dissociated on an atomistic scale (i.e. nm) in such a way that sections of symmetric grain boundaries are generated. This has previously been demonstrated for a Σ=27 boundary [8] and recently for a Σ=9 boundary , see Fig. 1 [11]. Such a dissociated boundary presents a different atomic structure to adjacent grains and may, therefore, getter impurities differently. Since investigators of electrical properties tend to focus on strongly electrically active, and hence - most likely - asymmetric grain boundaries, asymmetric gettering offers a possible explanation for the large difference in the minority carrier lifetime at either side of a grain boundary [12].

The concept of symmetry can be extended further since the [110] direction contains two mirror planes, (1$\bar{1}$0) and (001). It is therefore possible to form a second group of symmetric tilt grain boundaries with median lattice plane {001}. In melt grown silicon, boundaries of this symmetry type occur much less frequently then those of the {110} type and tend to have a more complicated structure. The best known example of a boundary of the {001} symmetry type is the first order incoherent 112/112 twin boundary commonly observed at the termination of first order coherent twins. Both computer simulation [13] and lattice images [11] of this boundary suggest the existence of (at least) two different configurations.

The electrical activity of both symmetry types has been studied in HEM [14] as well as in other types of solar silicon [15]. Fig. 2 shows optical (a,b) and EBIC (c,d) images of first order coherent twin bands located on different sets of {111} planes. The matrix is oriented $\sim$ <111> and the twin bands $\sim$ <115>. A Sirtl type etch was used so that only features in the matrix are etched. At the intersections of twin bands, second order twin boundaries are formed. Two stages of the formation of a symmetric 221/221 boundary are shown in Figs. 2(a) and (b). The corresponding EBIC image, 2(c) shows that the first order 111/111 coherent twin boundary as well as the second order 221/221 boundaries are not electrically active (the absence of EBIC contrast along the first order coherent twin boundaries indicates that the boundary is free of secondary dislocations [16]). Conversely, strong EBIC contrast is observed along incoherent first order sections, including the very small section in the upper left corner of Fig. 2(a) which in the EBIC image 2(c) appears as a black dot. The different thickness of intersecting twin bands allows a direct evaluation of the role which symmetry plays in the energy and electrical activity of second order twin boundaries. In Fig. 2a, the bands are of equal thickness so that a symmetric grain boundary can be formed along the entire length of the intersection. In 2 (b), the intersecting twin bands are of unequal thickness so that a straight boundary connecting both corners would be asymmetric in character. Such asymmetric corner to corner boundaries are also formed in HEM. These boundaries are straight within the resolution of optical and EBIC microscopy

(~ 0.5μm) and electrically active along their entire length. Electrical activity at an asymmetric second order twin boundary has been documented previously for the 111/115 boundary [17].

Using the reasonable assumption that the defect structure in HEM is close to equilibrium, one can obtain some insight into the relative energy of symmetric and asymmetric second order twin boundaries from the following consideration. The second order twin boundary emerging from the inside corner in Fig. 2(b) is symmetric as indicated by both optical measurements (which show that it accurately bisects the orientation of the adjoining first order twins) and the lack of EBIC contrast. This boundary can not proceed to the outside corner without changing it's symmetry, since twin band of unequal thickness meet in Fig. 2(b). Rather then forming an asymmetric straight boundary from corner to corner, the symmetric boundary terminates midway. The remaining, roughly equally long section, consists of two first order incoherent twin boundaries enclosing matrix material. Since both this arrangement and completed boundaries exist in HEM, the energy of a straight asymmetric second order boundary must be comparable to the energy of an incoherent first order boundary plus one half of the energy of a symmetric second order grain boundary, which is likely to be a small contribution. Similarily, Fig. 1 indicates that the energy of an asymmetric second order twin boundary is comparable to that of an incoherent first order twin plus a coherent twin; the latter is known to be small, ~30 erg/cm^2.

The tentative relation between symmetry and absence of electrical activity appears to extend to the (001) type symmetry case, since the 112/112 boundary in HEM is also found not to be electrically active [14]. This boundary, too, can be constructed without broken bonds [18]. Since it is likely, as pointed out before hand, that the boundary exists in more than one configuration, further experiments are needed to verify the absence of electrical activitiy in all 112/112 type boundaries.

Since short segments of all boundaries described by Kohn [5] exist in HEM, the electrical activity of other higher order twins can be examined. The major obstacle in this endeavor is the difficulty to accurately determine the boundary plane of short sections. However, in all cases where a boundary lacked EBIC contrast, a symmetric orientation could be assigned to the boundary within experimental error [14].

Considering that boundaries in HEM are formed in sections which accumulate high amounts of impurities during solidification, the absence of electrical activity at symmetric first and second order boundaries is quite remarkable. Even if the boundaries are not grown in (some boundaries in HEM may form by deformation twinning) it is easily demonstrated that rapidly diffusing transition metals [19] would have no difficulties in reaching grain boundaries during cool down. It appears therefore that the symmetric boundaries considered above not only lack dangling bonds but also do not act as (efficient) sinks for impurities.

PROCESS INDUCED CHANGES IN THE DEFECT STRUCTURE OF EFG RIBBONS

The Defect Structure of Unprocessed EFG.

EFG invariably develops a structure comprised of first order coherent twins perpendicular to the (110) ribbon surface and parallel to the [1$\bar{1}$2] growth direction which is termed the equilibrium defect structure [3]. Recent doping experiments with transition metals supply convincing evidence that the defect structure reflects an equilibrium of surface tensions at the liquid solid interface. When transition metals (TM's) are added to EFG, TM-C-Si complexes are formed. The large size and spherical shape (see Fig. 3) of these precipitates indicates that they are formed in the melt, most likely in the die, which contains a built up of impurities rejected from the liquid/solid interface. Precipitates reaching the solid liquid interface disturb the equilibrium of surface tensions between liquid, solid and twin. The twin spacing is therefore altered whenever a precipitate is incorporated into the ribbon. An example is shown in Fig. 4. In the vicinity of the particle, the boundary contains numerous facets, whereas facets are never observed in precipitate free material. The process believed to generate these facets is shown schematically in Fig. 5.

Process induced changes in the defect structure.

Diffusion induced defects and their interaction with grown-in defects will be discussed first. Following this section, we describe the structure and electrical activity of bulk defects, changes in these properties during processing and the decrease in electrical activity which can be achieved with a new procesing cycle in which a high temperature anneal of the ribbon preceeds formation of the p-n junction.

Depending on the surface concentration, P diffusion generates misfit dislocations in CZ silicon when the emitter depth exceeds ~0.4 µm. It is remarkable that diffusion induced misfit dislocations have never been observed in EFG cells where junction depths are between 0.3 and 0.6 µm. Although it is possible that the absence of misfit dislocations is caused by a low surface concentration of P, the observations below favor an alternative explanation based on the premise that generation of misfit dislocations is more difficult in EFG than in CZ.

C can surpress formation of dislocations by two mechanisms: (i) by increasing the friction stress (solution hardening) or (ii) by enhancing the concentration and mobility of of point defects, or complexes between C and point defects [20,21]. Since misfit strain is present at the emitter interface as seen from the observations that bulk dislocations glide into the emitter-base interface to relief misfit stresses, see Fig. 6, we believe that the former mechanism is more likely. Besides simply acting as a solution hardening agent, C when present in concentrations comparable to the average dopant concentration, may also surpress the formation of misfit dislocations in a more complex manner similar to dual doping [22].

Helical dislocations are found in the vicinity of the junction (Fig. 7). These helices are formed from grown in dislocations which absorb point defects generated by the P diffusion. An analysis of the helix geometry indicates that the point defect supersaturation is ~ 500. The sense of the helix was not analysed but results in CZ silicon indicate that the diffusion generated point defects are Si self interstitials. The enhanced outdiffusion of C in the diffused region [23] also supports an interstitial interpretation, since C is believed to diffuse predominantly via an interstial mode. The point defect promoted outdiffusion of C near the surface in turn reduces the number of available nucleation sites for oxygen. (The near surface concentration of oxygen remains essentially unchanged, since the high temperature diffusion of oxygen is not influenced by the presence or absence of point defects.) For this reason, one expects that somewhat larger oxygen precipitates are formed near the surface. Such larger, but still TEM invisible [24] donor complexes might be the light enhancement centers [25] observed in EFG.

In the bulk of the material, processing introduces the following changes: (i) the dislocation density between grains decreases (ii) <110> small angle tilt boundaries are formed and (iii) small crystalline precipitates, ~10 nm in size, nucleate at some nodes of these boundaries.

The tilt boundaries are formed by a polygonization like process. In the initial stage, Fig. 8, many of the dislocations contain either periodic kinks or jogs, suggesting the absorption of point defects into the sub-boundary. A fully developed sub-boundary with a tilt angle of ~4° is shown in Fig. 9. Two types of dislocation nodes are formed as the result of dislocation interactions, see Fig. 10(a,b). In Fig. 10 b, the stacking faults are out of contrast and dissociated dislocations and dissociated nodes are visible. Very small (~ 20 nm) crystalline precipitates, the composition of which could not be determined are located on dislocation segments between nodes. Such a boundary is likely to be an efficient gettering center but may also, since the <110> tilt axis is not colinear with the <110> surface direction, penetrate the depleted region. In this case, the cells appear to have high reverse dark currents and inferior junction characteristics.

As one would expect, even more pronounced changes occur in the defect structure when EFG ribbons are processed in an experimental manner, aimed at improving the bulk minority carrier lifetime [23], which involves a high temperature anneal at 1200 °C prior to formation of the p-n junction. In this case, tilt boundaries with misfits up to 30° are formed. Examples of annealing induced boundaries and the interactions of twins and dislocations are shown in Figs. 11 to 13. Fig. 11 shows the emission of a twin boundary from a large angle (~20°) formed by annealing. Fig. 12 shows the pile up of lattice dislocation at a microtwin. Dislocations entering the twin dissociate and, moving in the twin plane, can be seen to accumulate towards the top of the figure. Fig. 13 shows an example of a lattice dislocation which, after having entered a twin, dissociates into two partials, thus reducing the twin thickness by one one atomic layer. Since EFG, because the presence of large residual stresses, is difficult to thin to ~100Å without breaking we illustrate the process with a lattice image taken in a CVD ribbon sample deposited at a high temperature (1100°C), Fig. 14. (This CVD is used as the feedstock for RTR ribbon to ribbon growth). Termination of the microtwin in the lower section involves a Frank partial, suggesting that microtwins may originate at point defect loops.

The influence of a high temperature anneal on the minority carrier recombination at grown-in defects is illustrated in Figs. 15(a,b) which show EBIC images of completed solar cells (the p-n junction was used to collect the EBIC current) fabricated from the same ribbon. Fig. 15 (a) was obtained from a specimen diffused after growth, while Fig. 15(b) is the companion piece annealed at a high temperature prior to formation of the junction. A side by side comparison shows that pre-annealing dramatically lowers the electrical activity of the grown in defect structure.

This interesting and potentially useful effect is not understood, but two working hypothesis have been formulated:

(i) Defects introduced during ribbon growth, when annealed, undergo structural changes, which lower the electrical activity. It is well established that deformation induced dislocations (and most dislocations seem to be introduced during cool down of the ribbon) change properties when annealed at high temperatures. A high temperature anneal eliminates the ESR activity [26] and converts the dislocations into efficient sinks for point defects [27].

(ii) The high temperature anneal (a) disperses transition metal impurities originally gettered at dislocations and grain boundaries and (ii) simultaneously leads to the formation of new gettering sites. These new gettering sites getter the majority of transition metal impurities during subsequent cooling.

Presently, not enough data are available to distinguish between these two models. However, since C is known to be present in supersaturation at 1200°C, the second mechanism appears presently to be more probable. The new gettering sites, in this view, would be carbon or carbon related precipitates. Carbon based gettering does not appear to have been investigated previously. It should be particularely effective in gettering those transition metals which are known strong carbide formers. A carbon based precipitation mechanism is also supported by preliminary observations on the rate of the decay of the EBIC contrast with time. A model which assumes that the process is controlled by diffusion indicates that a slowly moving species, with a diffusion coefficient equal (or similar) to C must be involved.

ACKNOWLEDGEMENTS

The authors would like to thank Dr. J.P.Kalejs for providing specimens of experimentally processed ribbons. This research was sponsored by DOE under Contract 76ER02899 and by DOE/JPL under Contract 956046. The Material Science Center at Cornell provided analytical facilities.

REFERENCES

1. T.Sullivan and D.G.Ast, to be published in Proc. of the 1983 MRS Meeting
2. J.C.Lambropouluous, J.W.Hutchinson, R.O.Bell, B.Chalmers and J.P.Kalejs, J. Cryst. Growth, in press.
3. F.V.Wald, in: Crystals: Growth Properties and Applications 5 (Springer Verlag Berlin, Heidelberg, 1981) p.147
4. J.P.Kalejs, M.C.Cretella, F.Wald and B.Chalmers, Proc. Electrical and Optical Prop. of Polycryst. or Impure Semiconductors and Novel Si Growth Methods, Eds. K.V.Ravi and B.O'Mara, Electrochem. Soc. (Pennington, N.J., 1980) ECS 80-5, page 242.
5. J.A.Kohn, Amer.Minereralogist 41,778 (1958).
6. W.Bollmann, Crystal Defects and Crystalline Interfaces (Springer 1970)
7. C.Fontaine and A.Rocher, J. of Microscopy, 118,105 (1980)
8. B.Cunningham, H.P.Strunk and D.G.Ast, Scripa Met 16,349 (1982).
9. J.Hornstra, J.Phys.Chem.Solids, 5,409 (1959).
10 M.D.Vaudin, B.Cunningham and D.G.Ast, Scripta Met 17,191 (1983).
11.B.Cunningham and D.Ast, to be published.
12.J.Marek, Proc. 16th IEEE Photovoltaic Specialists Conference (September 1982) (0160-8371/82/0000-6272 IEEE), p.627
13.R.C.Pond, D.J.Bacon and A. Bastaweesy, to be published in: Proc. Micros. Semicon. Materials, Oxford 1983 (Inst. Phys. Conf. Ser.).
14.T.Koch and D.G.Ast, An EBIC Study of HEM polycrystalline Si, DOE/JPL Report 956046-2, 1982 (Avail. from LASS Doc.Center, JPL, Pasadena, Ca).
15.C.Dianteill and A.Rocher, to be published in ref. 13.
16.H.Strunk and D.G.Ast, 38th Ann.Proc.EMSA, Ed. G.W.Bailey, p.332 (1980).
17.B.Cunningham, H.Strunk and D.G.Ast, Appl.Phys.Lett.40, 237(1982).
18.E.R.Weber, Appl. Phys. A 30, 1 (1983).
19.C.Fontaine and D.A.Smith, Appl. Phys. Lett. 40, 153 (1980).
20.K.P.O'Donnell, K.M.Lee and G.D.Watkins, Physica 116, 258 (1983).
21.D.D.Watkins and K.L.Brower, Phys. Rev. Lett.36, 1329 (1976).
22.T.Yonezawa, W.Watanabe, Y.Koshino, H.Ishida, H.Muraoka and T.Ajima, in Semiconductor Silicon, Eds. R. Huff and E. Sirtl (Electrochem. Soc., Princeton, NJ, 1977, ESC 77-2) p. 658.
23.J.P.Kalejs, private communication.
24.R.Takizawa, A.Ohasawa, K.Honda and T.Toyokura, Appl.Phys.Lett. 43, 766 (1983).
25.C.T.Ho and J.B.Mathias, J.Appl.Phys. 54, 5993 (1983).
26.M.Suezawa, K.Sumino and M.Iwaizumi, Inst.Phys.Conf.Ser.No 59,407 (1980)
27.A.J.de Kock in : Handbook of Semiconductors, Vol 3. Ed. SP Keller (North-Holland, New York 1980), p.247.

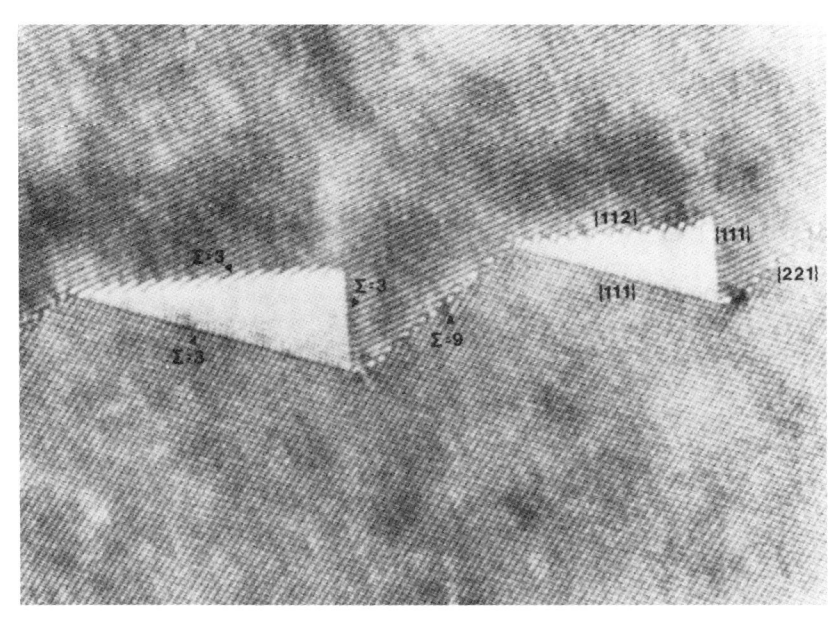

Fig. 1. Dissociation of an asymmetric Σ=9 tilt boundary into a coherent first order,Σ=3, boundary and an incoherent 112/112,Σ=3, first order boundary. (The sample is RTR high temperature CVD polysilicon).

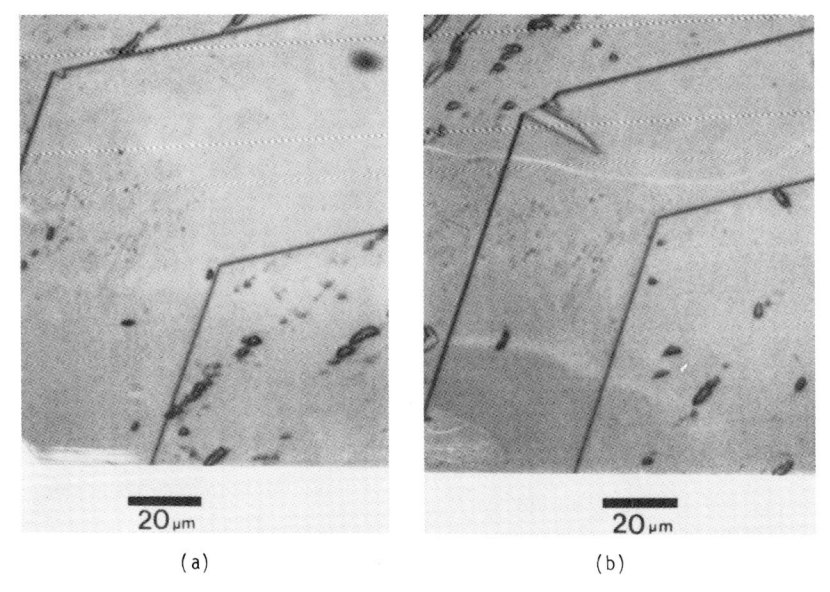

(a) (b)

Figs. 2(a,b). Optical micrographs of intersecting first order,Σ=3, twin bands in HEM. The matrix orientation is ~<111>. The twin orientation is ~<115>. Second order,Σ=9, boundaries are formed at intersections.

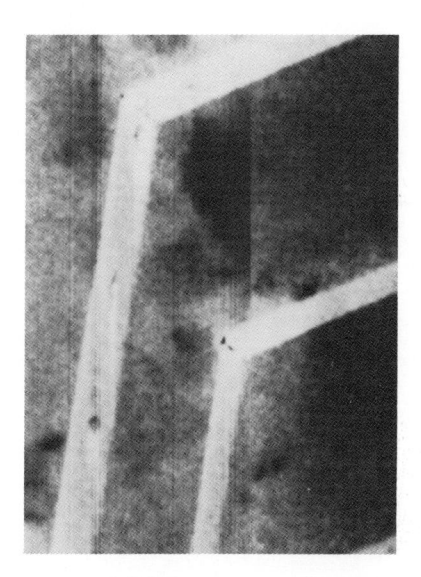

Fig. 2(c). EBIC image of the areas shown in Figs. 2(a,b).

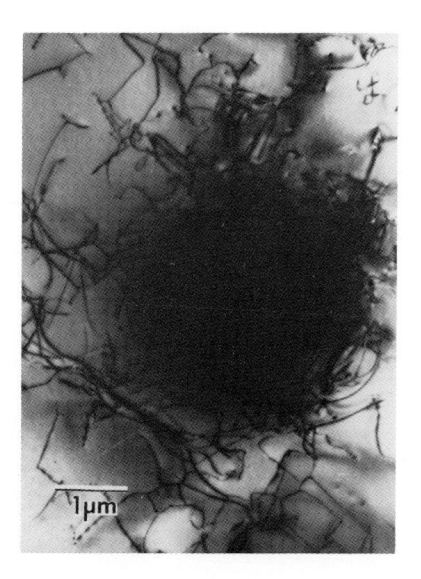

Fig. 3. Precipitate formed in EFG as the result of Fe doping.

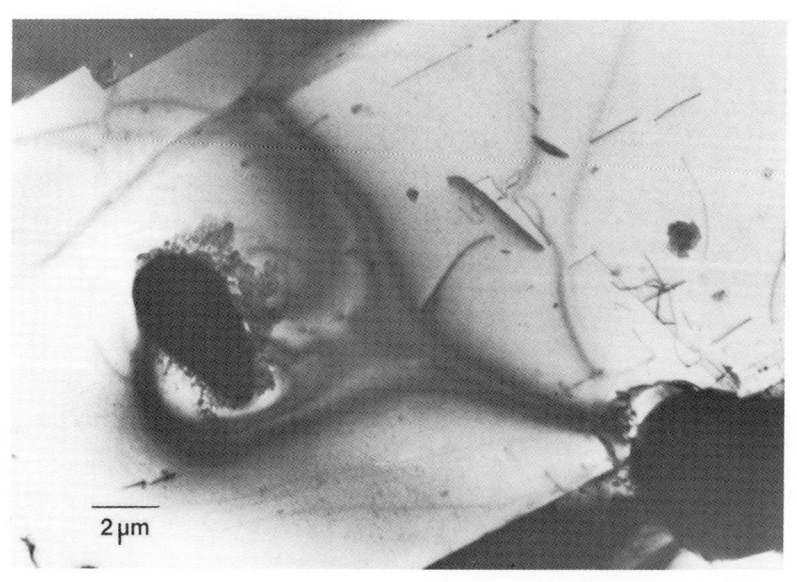

Fig. 4. Example of a facetted equilibrium defect structure in EFG as the result of the incorporation of an Fe related precipitate.

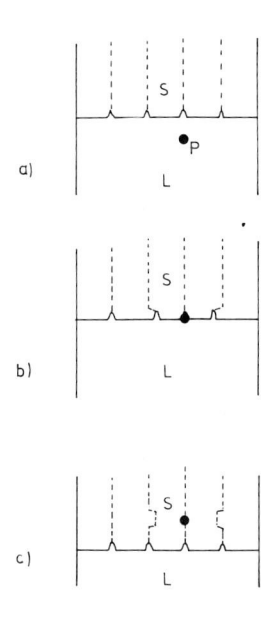

Fig. 5. Schematic diagram depicting succesive stages (a to c) of the incorporation of a particle into a growing EFG ribbon.

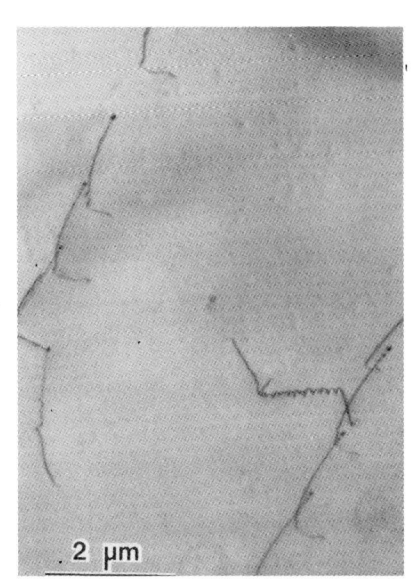

Fig. 6. Example of bulk dislocations gliding into the emitter base interface to relief misfit strain in processed EFG.

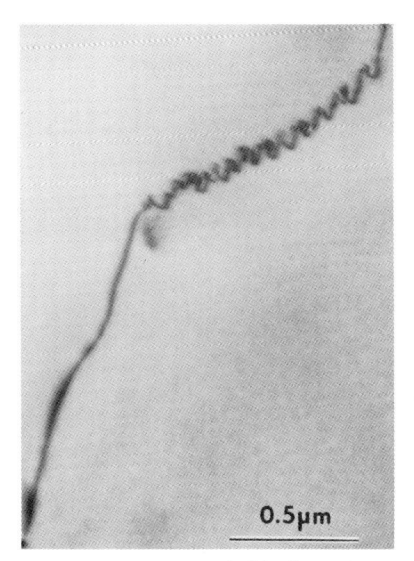

Fig. 7. Example of helical dislocations, formed near the junction of EFG cells.

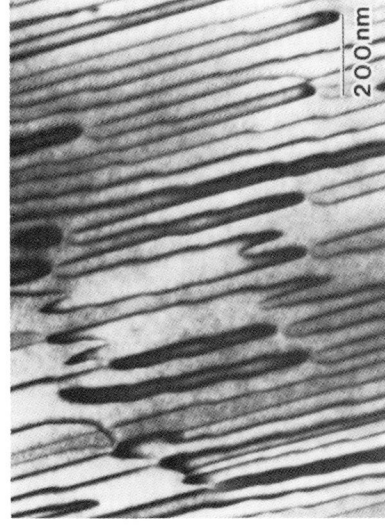

Fig. 8. Example of the early stage of the formation of a processed induced sub-boundary in EFG.

94

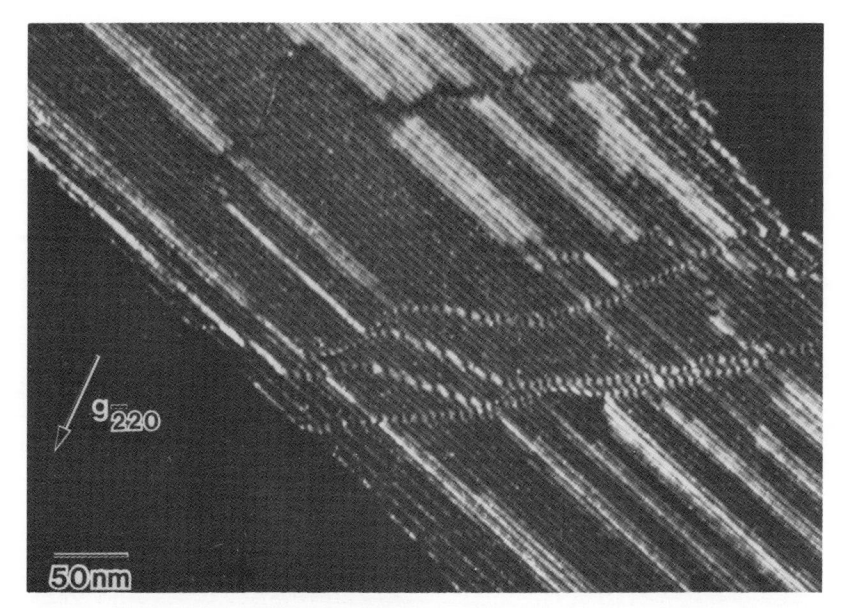

Fig. 9. Example of a fully developed process induced sub-boundary in EFG.

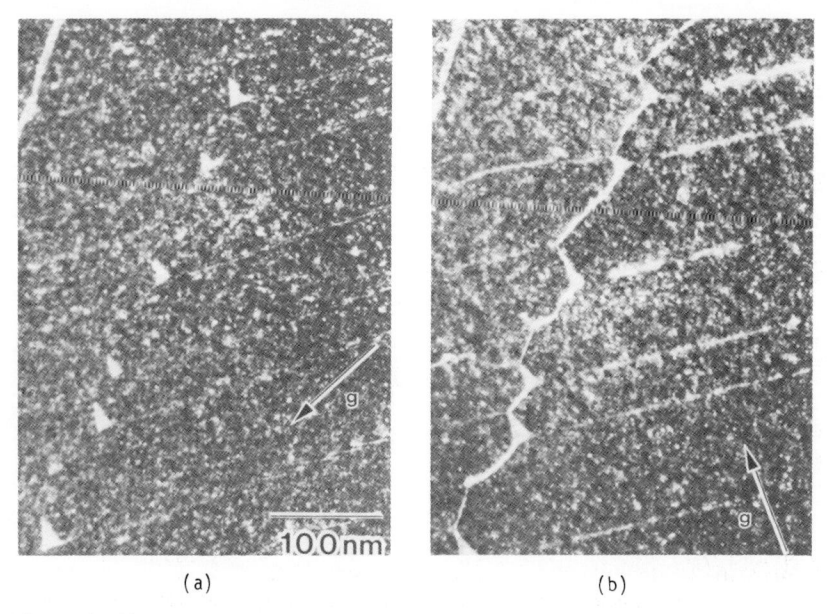

(a)

(b)

Fig. 10(a,b). Weak beam images of dislocation nodes formed in process induced sub-boundaries. Small precipitates are formed at some nodes.

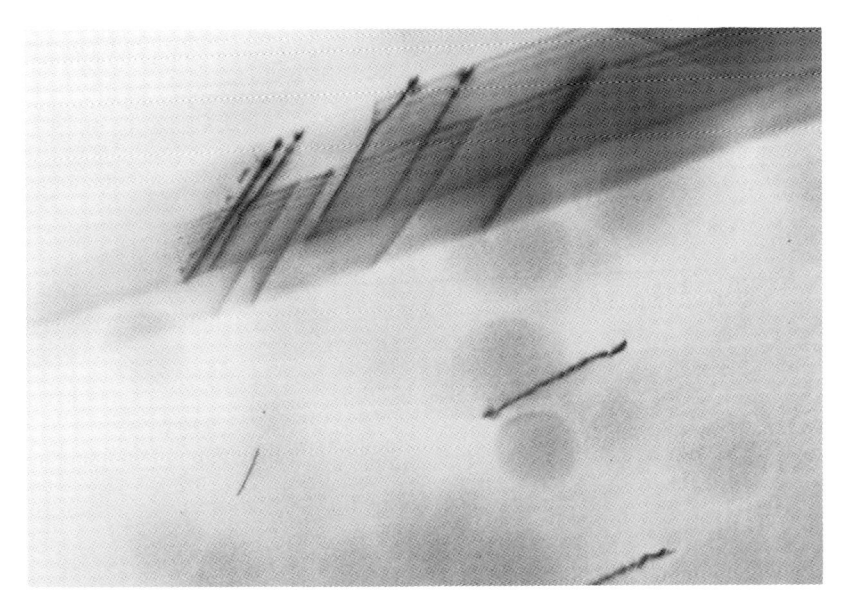

Fig. 12. Lattice dislocations piling up against and entering into a twin boundary. EFG, processed in two steps.

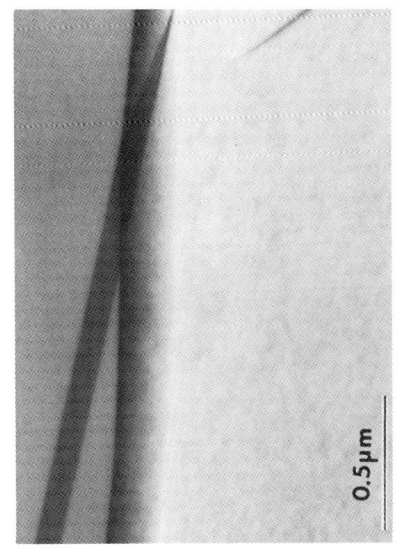

Fig. 11. A sub-boundary emitting a first order, Σ=3, twin. EFG, processed in two steps.

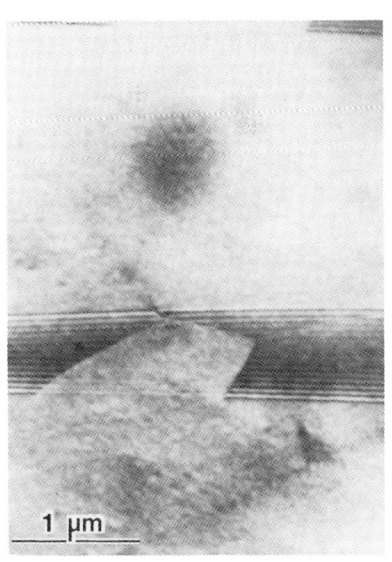

Fig. 13. Example of lattice dislocations dissociating in a twin boundary in two step processed EFG.

Fig. 14. Example of reduction of twin thickness in single layer steps. (This micrograph was taken in a high temperature deposited CVD sample).

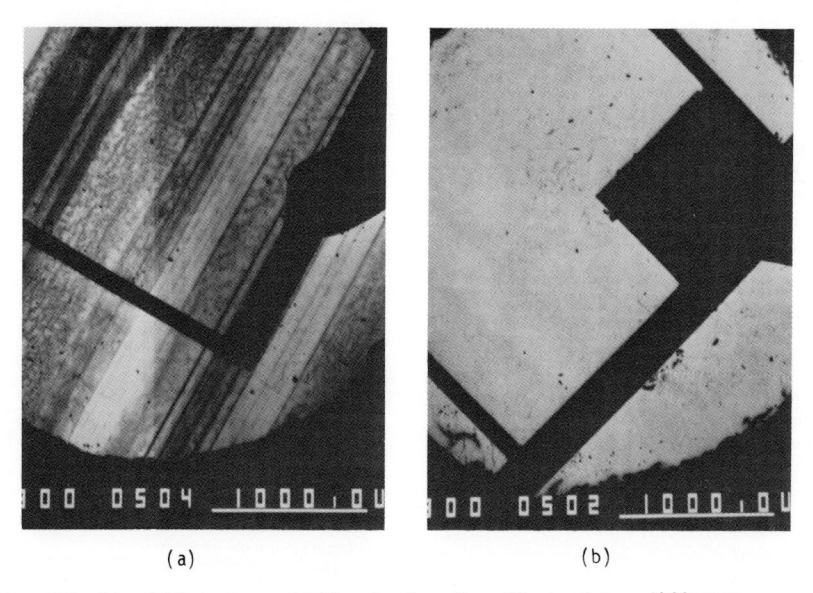

(a) (b)

Fig. 15(a,b). EBIC images of EFG, showing the effect of two different processing (a) one step processing, in which the diffused junction is formed in the as grown material, (b) two step processing, in which a high temperature anneal preceeds junction formation.

DOPANT SITE LOCATION BY ELECTRON CHANNELING IN ION IMPLANTED SILICON.*

S. J. PENNYCOOK, J. NARAYAN AND O. W. HOLLAND
Solid State Division, Oak Ridge National Laboratory, Oak Ridge, TN 37830

ABSTRACT

A simple ratio technique using the phenomenon of electron channeling can be used to measure the substitutional concentrations of dopants in semiconductors on a submicron scale. A comparison was made between electron and ion channeling measurements on Si-Sb alloy samples having a range of nonsubstitutional fractions of Sb. Good agreement was obtained but both measurements indicated considerably more nonsubstitutional dopant than could be accounted for by precipitates observed in electron micrographs. The discrepancy can be explained if the precipitates are coherent in the early stages of growth and have their planes located interstitially with respect to the Si planes. The sensitivity of the electron channeling measurements to the implantation profile was investigated and found to be small. The determination of local dopant profiles in the electron microscope is described.

INTRODUCTION

The lattice location of dopants in semiconductors is critical in controlling their electrical activation. The determination of lattice location must rely on the use of a channeling effect using ion, x-ray or electron beams. Ion channeling analysis has become well established for such measurements,[1] but the minimum area of sample that can be analyzed is of the order of millimeters, or possibly 0.1 mm with a focused or collimated beam. The same limitation applies with x-ray beams. In modern integrated circuit devices active areas may be smaller than a micron and a technique is needed to perform a channeling analysis on a submicron scale. Here we show how the electron channeling effect can be used for lattice location analysis, using a focused electron beam in a transmission electron microscope (TEM). The size of the analyzed region is limited only by the probe forming capabilities of the microscope and may be potentially on a nanometer scale.

The electron channeling effect occurs when a parallel beam of electrons is incident on a crystal close to a Bragg reflecting orientation. The effect is similar to the x-ray channeling effect which has been used for site location studies.[2,3] Instead of propagating through the crystal as a planar wavefront, a standing wave intensity profile is set up having the same periodicity as the channeling planes. The maxima in the profile lie either on the planes or between them for incident beam orientations just inside or just outside the exact Bragg reflecting condition, respectively.[4] A distinct change in the intensity of characteristic x-ray emission is seen between these two channeling orientations. For dilute solutions of dopant, the matrix planes still channel the electron beam, and if matrix and dopant x-ray yields vary identically between the two orientations, then the dopant necessarily lies within the matrix channeling planes (Fig. 1a). If the dopant x-ray yield does not vary, it is randomly located with respect to the matrix planes (Fig. 1b). The exact matrix yield variation depends on the crystal thickness and the exact two orientations chosen. The ratio technique, where the dopant and matrix yield variations are compared, avoids the

*Research sponsored by the Division of Materials Sciences, U. S. Department of Energy under contract W-7405-eng-26 with Union Carbide Corporation.

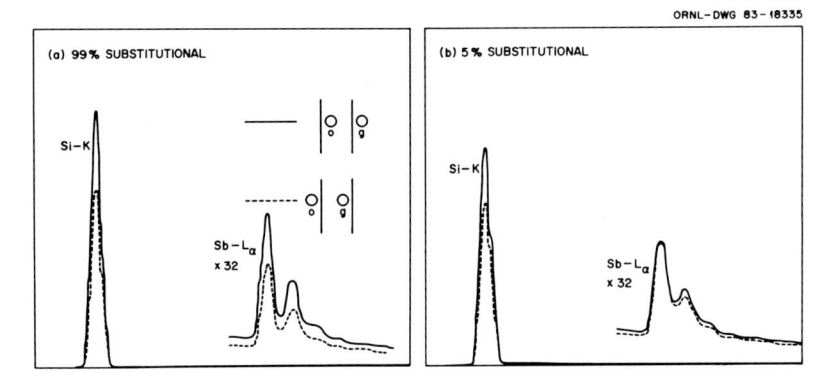

FIG. 1. X-ray spectra from electron channeling analysis of Si-Sb alloys 99% and 5% substitutional. Insert shows schematically the beam tilt conditions showing incident and diffracted beams with respect to the Kikuchi lines (traces of the matrix planes).

need for detailed dynamical theory calculations of the electron intensity profile.[5,6]

SUBSTITUTIONAL ALLOYS

Uniform Sb Profile

For highest accuracy of electron channeling analysis the dopant should be uniformly distributed through the sample thickness, since the standing wave pattern slowly decreases in amplitude and changes in form as it propagates through an increasing thickness of crystal. In these investigations (100)Si specimens were implanted with ^{121}Sb$^+$ ions at different energies and doses chosen so as to produce an approximately uniform dopant concentration of 1.6 at. % in the top 100 nm of the samples. After solid-phase-epitaxial (SPE) regrowth in a furnace at 550°C for 30 min. a supersaturated solid solution is formed free of extended defects with the dopant trapped in substitutional sites.[7,8] An ion channeling analysis of this material is provided in Fig. 2, which shows an Sb profile with uniform concentration and with greater than 99% of the Sb in substitutional sites. This material is ideally suited for an electron channeling study and has been reported in detail previously.[9] It was found that a correction factor was required to take account of the different spatial localization of the Si-K and Sb-L$_{III}$ excitations. The Sb-L$_{III}$ excitation is of a higher energy (4123 eV) than the Si-K excitation (1939 eV) and is more highly localized. It therefore shows a slightly greater x-ray yield variation than the Si-K. Using this correction the substitutional fraction of Sb is given by

$$F_s = \frac{1}{C} (X_{Sb}-1)/(X_{Si}-1) \qquad (1)$$

where X_{Sb}, X_{Si} are the ratios of Sb-L$_\alpha$ and of Si-K x-ray intensities obtained in the two channeling orientations. $C = 1.10 \pm 0.04$ is the experimentally determined delocalization correction factor for Sb in Si using {220} planar channeling.[9]

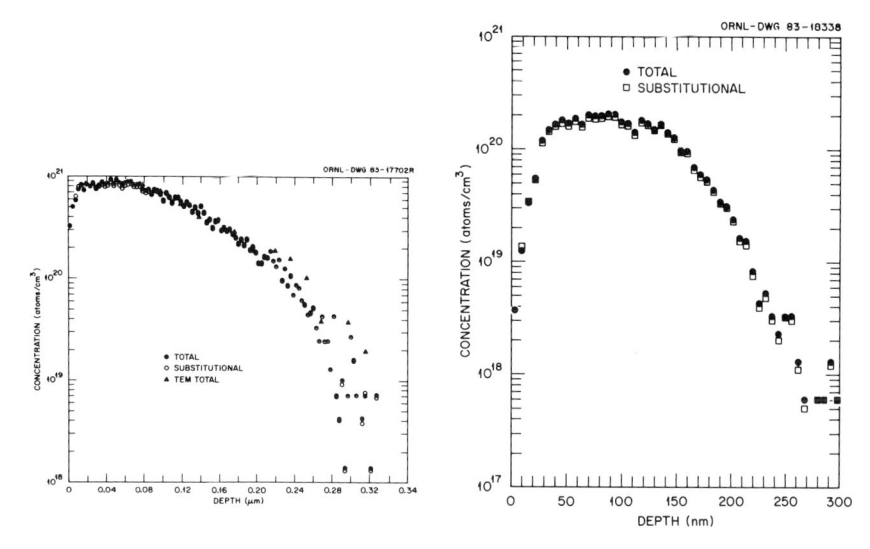

FIG. 2. Ion channeling analysis of graded energy ^{121}Sb$^+$ implanted Si. Also shown is the profile determined from a small region of sample in TEM.

FIG. 3. Ion channeling analysis of graded energy ^{209}Bi$^+$ implanted Si.

Uniform Bi Profile

We have also studied a (100)Si sample containing a uniform concentration of Bi, as shown in Fig. 3. Here we found the delocalization correction factor to be 1.20 ± 0.07. A larger correction is needed because the Bi-L$_{III}$ excitation requires an energy transfer of at least 16,396 eV; and is, therefore, more highly localized than the Sb-L$_{III}$ excitation.

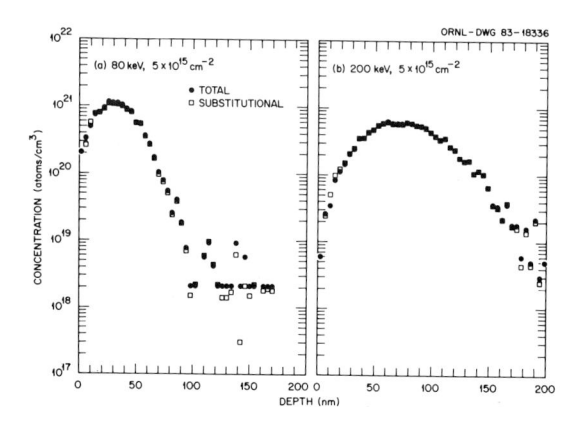

FIG. 4. Ion channeling analysis of Si implanted with ^{121}Sb$^+$ at a) 80 keV, 5 × 10^{15} cm^{-2} and b) 200 keV, 5 × 10^{15} cm^{-2}.

Gaussian Sb Profiles

We have studied two samples with non-uniform dopant profiles to test how accurate the electron channeling technique will be in a general situation. The two samples were implanted with $^{121}Sb^+$ ions at single energies chosen so as to produce the peak Sb concentration at the top or the bottom of a 100 nm thick sample. Specifically, one sample was implanted with Sb at 80 keV, 5×10^{15} cm^{-2} and the other at 200 keV, 5×10^{15} cm^{-2}. Ion channeling analyses of these samples are shown in Fig. 4, showing both samples to be over 99% substitutional. The electron channeling analysis of the 80 keV implanted sample gave a figure of 109% substitutional, while the 200 keV sample gave a figure of 95% substitutional. The inaccuracy produced by the non-uniform dopant distribution is, therefore, less than 10% in sample thicknesses of around 100 nm. The discrepancies are caused by the gradual change in form of the electron intensity profile with depth. When electrons are channeled along the atomic planes they suffer increased scattering, so that by a depth of about 700 nm all the electron current is channeled between the atomic planes, irrespective of the incident beam orientation. In a sample 100 nm thick, the standing wave modulation is larger near the electron-entrance surface than the electron-exit surface, so that dopant distributed close to the entrance surface gives x-ray yield ratios higher than the average matrix ratios, whereas dopant close to the exit surface will show a reduced x-ray yield ratio. If the dopant profile is known then possibly these effects can be allowed for.

Depth Profile Determination

The dopant depth profile in small regions of sample can be determined directly in the electron microscope. The total integrated dopant concentration, c, is measured as a function of sample thickness, t, and the slope of a plot of ct vs t gives the depth profile, c(t). The sample thickness is accurately measured using thickness fringes[4] and convergent beam electron diffraction[10] at the exact Bragg condition, and the electron beam is then tilted so that no Bragg reflections are strongly excited for an accurate x-ray analysis of total dopant concentration. The results for the uniform Sb implantation are shown in Fig. 5 and the depth profile obtained from the slope of this curve is shown on the ion channeling profile of Fig. 2. The ion channeling analysis was used as the quantitative calibration of x-ray detector efficiency, so that the peak concentrations match exactly. The distribution in depth also matches quite well, although errors occur at large depths where the integrated x-ray intensity became approximately constant. The TEM depth profile is determined using approximately a one micron area of sample and offers much higher spatial resolution, although less depth resolution, than the ion channeling analysis. The only requirement is that profile does not change significantly across the region analyzed. Spatial inhomogeneity is probably responsible for the scatter seen on Fig. 5. In principle, by performing an electron channeling analysis as a function of thickness the substitutional dopant profile could be determined.

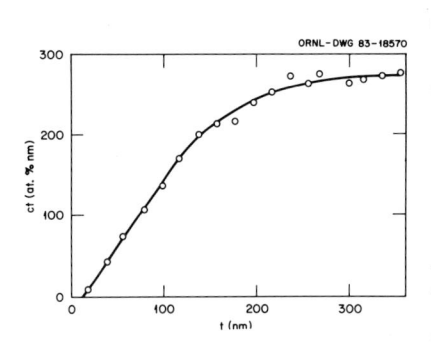

FIG. 5. Plot of ct vs t for TEM determination of dopant profile.

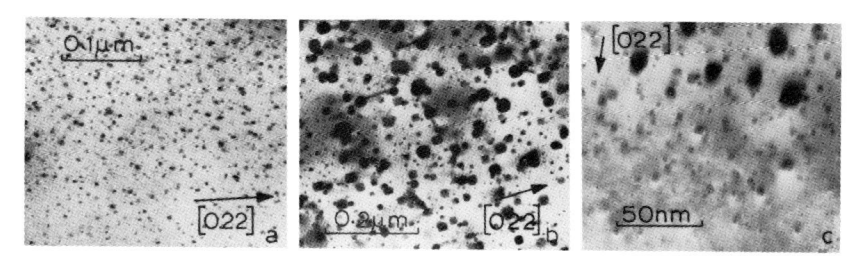

FIG. 6. TEM images of uniform Sb implant during the annealing sequence
a) after 680°C/20 min. anneal; b) after 950°C/20 min. anneal; c) 200 keV
Sb implant after 700°C/20 min. anneal in dynamical diffraction conditions
showing coherency strains.

ANNEALING STUDIES

A thin sample of the uniform Sb implant was annealed sequentially at
eight temperatures from 680°C–950°C to precipitate out increasing amounts of
dopant in excess of the solubility limit (Fig. 6a,b). Electron channeling
analysis was carried out after each anneal. Fully precipitated samples have
been studied previously, and it was shown that the largest precipitates were
partially coherent with Si{111} planes, but not coherent with Si{220}
planes. In this case, the precipitated Sb has a random location with
respect to the {220} planes, and the substitutional fraction of Sb is given
by Eqn. (1). The fraction of Sb in precipitate form was calculated from
the electron micrographs.[9] Figure 7 shows the nonsubstitutional fraction
determined by electron channeling plotted against the fraction of Sb in
precipitate form, measured from the micrographs. After the highest temperature anneals, the two estimates agree closely, and the remaining substitutional dopant is consistent with the solubility limit of Sb in silicon.
In the early stages of precipitation there appears to be over twice as much
nonsubstitutional dopant than can be accounted for in the observed precipitates.

COMPARISON OF ELECTRON AND ION CHANNELING ANALYSES

To confirm the results of electron channeling three bulk samples were
annealed at temperatures of 720, 740 and 780°C and samples prepared for
electron and ion channeling analysis. The results of ion channeling analysis for total and substitutional concentrations are shown in Fig. 8. The
substitutional fraction obtained by integrating over depth from 5 to 110 nm
is shown in Table I, and agrees well with the electron channeling analysis.
The precipitated fraction was again determined from the micrographs and
found to be lower than the nonsubstitutional fraction. These results are
also plotted in Fig. 7. and are in good agreement with the sequentially
annealed thin sample.

TABLE I. Analyses of bulk annealed uniform Sb implants.

Anneal Temp/°C	% Nonsubstitutional electron channeling	ion	% Precip.
720	41 ± 4	42	18
740	49 ± 4	48	26
780	57 ± 5	59	46

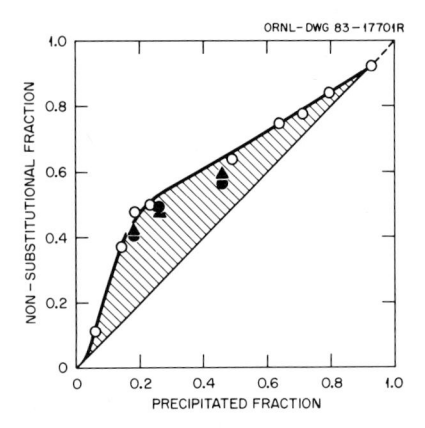

FIG. 7. Nonsubstitutional fraction of Sb vs fraction of Sb in precipitate form. Circles indicate electron channeling analysis, triangles ion channeling analysis. Solid symbols denote samples annealed in bulk form.

DISCUSSION

We now consider possible explanations for the "lost dopant" which is apparently nonsubstitutional but not in precipitate form. One possibility is that this is in the form of precipitates too small to be observed in the electron micrographs. In this case a very high number density would need to be present to account for the amount of lost dopant being of the same order as that in observed precipitates. A further anneal at the same temperature for the same time would cause these precipitates to become visible. In fact only twice as many precipitates were observed after the second anneal,[11] ruling out this possibility.

The second possibility is that the observed precipitates are coherent with the silicon matrix when small. Coherent precipitate formation has been observed in laser annealed Tl implanted Si.[12] If coherent precipitates were located with their planes in an interstitial position between the matrix planes, the x-ray yield variation from the Sb in these precipitates would be reversed, giving the false impression that twice as much Sb was nonsubstitutional. Flux peaking effects in ion channeling can also lead to an enhanced scattering yield from interstitially located atoms by up to a factor of two.[13] Therefore, the electron and ion channeling measurements would agree, but both be in error by roughly a factor of two on the fraction of nonsubstitutional dopant. Evidence for the presence of coherency strains comes from the observation of black-white contrast around precipitates in micrographs taken under dynamical diffraction conditions (Fig. 6c). From the micrographs we cannot determine the relative location of the Sb and Si planes. It is possible that these coherent precipitates were located with their planes aligned with the Si planes so that the Sb in them would appear substitutional. In this case, the channeling results could only be explained if the entire nonsubstitutional dopant was still in solution. It is possible that dopant complexes may form, in which the Sb is pulled off its regular lattice position. These would be below the critical size where a cluster becomes a nucleus capable of further growth, and so could not be detected by further annealing. However, complex formation is thought to be unlikely since the Sb is diffusing rapidly during the early stages of precipitation[11], and it is unlikely that a complex would diffuse rapidly. Another possibility arises since the rapid diffusion seen in the early stages of precipitation is due to a high concentration of interstitials which had been trapped during SPE regrowth.[11] The Sb may perhaps lie in interstitital locations after annealing, although this too is thought to be unlikely. The most likely situation is that the coherent precipitates formed during the early stages of precipitation have their planes interstitially located between the Si planes. The "lost dopant",therefore, simply represents an error in the electron or ion channeling analysis caused by interstitially located atoms. As the precipitates grow, their coherency is lost, and they become randomly located with respect to the {220}Si planes.[14] Electron and ion channeling analyses then correctly give the nonsubstitutional fraction of dopant, in agreement with the precipitated fraction.

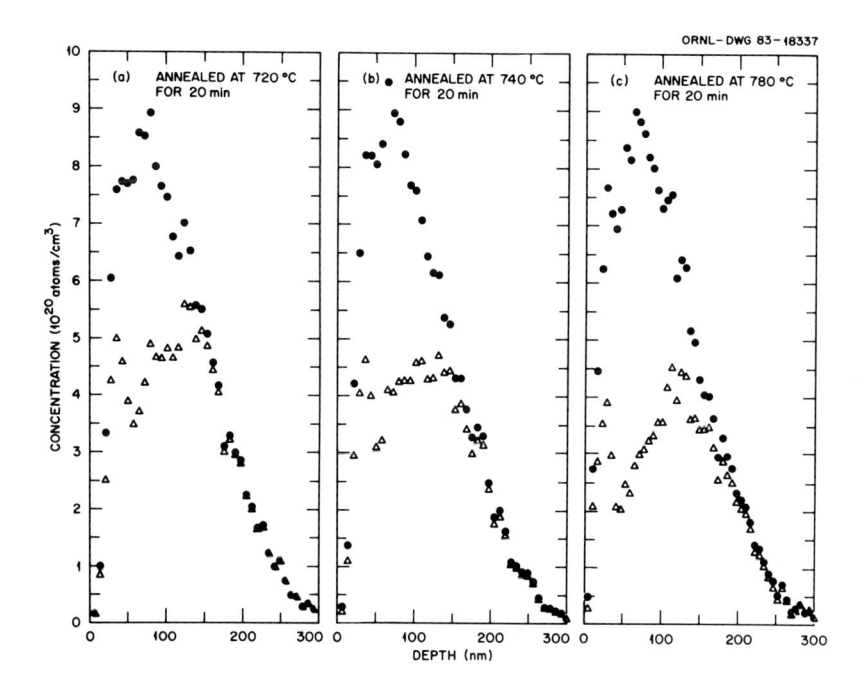

FIG. 8. Ion channeling analyses of uniform Sb implanted Si (shown in Fig. 2) following bulk annealing. Note the linear concentration scale.

SUMMARY

Electron channeling analysis can provide a determination of the substitutional fraction of dopants in semiconductors in quantitative agreement with the results of ion channeling analysis but from submicron-sized regions of samples. In addition the TEM image provides valuable information on the formation and coherency of precipitates. This information combined with a channeling analysis, can determine whether the coherent precipitates have their planes aligned with the matrix planes, or in an interstitial location between them.

REFERENCES

1. B. R. Appleton, p. 97 in Defects in Semiconductors, ed. by J. Narayan and T. Y. Tan, Elsevier, New York (1981).
2. B. W. Batterman, Phys. Rev. Lett. 22, 703 (1969).
3. P. L. Cowan, J. A. Golovchenko and M. F. Robbins, Phys. Rev. Lett. 44, 1680 (1980).
4. P. B. Hirsch, A. Howie, R. B. Nicholson, D. W. Pashley and M. J. Whelan, Electron Microscopy of Thin Crystals, Butterworth, London (1965).
5. J. C. H. Spence and J. Tafto, p. 523 in Scanning Electron Microscopy/ 1982/II, SEM Inc. AMF O'Hare, Ill. (1982).
6. J. Tafto and J. C. H. Spence, Science 218, 49 (1982).
7. J. Narayan and O. W. Holland, Appl. Phys. Lett. 41, 239 (1982).
8. J. Narayan and O. W. Holland, Phys. Stat. Solidi (a) 73, 1225 (1982).
9. S. J. Pennycook, J. Narayan and O. W. Holland, Appl. Phys. Lett., in press (1983).
10. P. M. Kelly, A. Jostons, R. G. Blake and J. G. Napier, Phys. Stat. Sol. (a) 31, 771 (1975).
11. S. J. Pennycook, J. Narayan and O. W. Holland, in Ion Implantation and Ion Beam Processing of Materials, Proc. 1983 MRS Meeting, in press.
12. B. R. Appleton, J. Narayan, C. W. White, J. S. Williams and K. T. Short, Nucl. Instrum. and Methods 209/210, 239 (1982).
13. J. U. Anderson, O. Andreasen, J. A. Davies and E. Uggerhoj, Radiat. Eff. 7, 25 (1971).
14. S. J. Pennycook, J. Narayan and O. W. Holland, J. Appl. Phys. (in press).

HIGH RESOLUTION TEM STUDIES OF DEFECTS NEAR Si-SiO$_2$ INTERFACE

J. H. Mazur and J. Washburn
Materials and Molecular Research Division
Lawrence Berkeley Laboratory
University of California
Berkeley, California 94720

ABSTRACT

Small defects with habit parallel to $\{100\}$ and $\{311\}$ matrix planes were observed using high resolution transmission electron microscopy (HREM) within 100 nm from the Si-SiO$_2$ interfaces after one step oxidation in dry O$_2$ at 900°C, 1000°C and 1150°C of Czochralski (CZ) grown [100] p-type boron doped, 1.5 - 20 Ω cm Si wafers with concentrations of oxygen 1.4 x 10^{18}cm^{-3} and carbon 4. - 10. x 10^{16}cm^{-3}. The defects were less than 10 nm wide and 1 nm thick. The $\{100\}$ and $\{311\}$ defect are interpreted tentatively as thin silica plateletes and $\{311\}$ stacking faults respectively. Distribution of defects near the interface was random although their density appeared to be lower for higher oxidation temperatures. It is not yet clear whether the defects were formed during the oxidation treatments or were present near the surfaces of the as-received wafers.

INTRODUCTION

Silicon crystals grown from quartz crucibles by the Czochralski method typically contain 10^{18}cm^{-3} interstitially dissolved oxygen. This corresponds approximately to the oxygen solid solubility limit at the melting temperature. Thus during processing steps in production of electronic devices involving lower temperature heat treatments, oxygen precipitation from supersaturated solid solution can result [1]. The defects formed are beneficial for gettering metallic impurities if they are away from the active device regions [2,3,4]. However the presence of precipitates within active regions of the devices can be detrimental to performance. For example, higher generation current has been observed in MOS capacitors if these defects are found within the depletion region [5] or near p-n junctions [6]. In this paper, preliminary observations of defects within the depth of the depletion width from the Si-SiO2 interface are reported in wafers after a one-step oxidation at temperatures above 900°C.

The defects were observed during a systematic investigation of the Si-SiO$_2$ interface structure using lattice imaging electron microscopy of cross-sectional specimens.

EXPERIMENTAL PROCEDURES

Commercially supplied (without specification of thermal history) Czochralski-grown Si [100] wafers, p-type boron doped 1.5 - 20 cm were used for oxidation. These wafers contained 1.4 x 10^{18}cm^{-3} oxygen and 4 to 10 x 10^{16}cm^{-3} of carbon atoms as estimated from I.R. measurements.

The wafers were cleaned in a series of rinses including:
(1) 5:1:1 H_2O: NH_4OH:H_2O_2, (2) deionized water (D.I.), (3) 5:1:1 H_2O:HCl:H_2O_2, (4) D.I., (5) 50:1 H_2O:HF, (6) D.I., and after drying in N_2 were transferred into the furnace with an argon atmosphere. After five minutes the argon was replaced by dry oxygen for the desired time. The wafers were removed from the hot zone of the furnace after again changing to argon ambient during a two minute interval. The oxidations were performed at 900°C, 1000°C, and 1150°C, for times necessary to grow oxide layers 20 nm to 100 nm thick.

Cross-sectional TEM specimens were prepared using the standard technique [7]. High resolution TEM images were obtained using a JEM 200 CX electron microscope operating at 200 kV and equipped with high resolution pole piece. The electron beam parallel to the zone axis [011] allowed resolution of two sets of $\{111\}$ planes in the silicon crystal edge on. The distance between $\{111\}$ fringes corresponds to the distance 0.314 nm spacing of silicon $\{111\}$ lattice planes.

EXPERIMENTAL RESULTS AND DISCUSSION

Observations in the vicinity of the Si-SiO_2 interface within a distance of about one micron were performed on the first set of specimens with about 200 nm of oxide grown at 900°C for 120 minutes. Two characteristic classes of defects were observed and are shown in Fig. 1. The defects marked "a" in Fig. 1 are tentatively interpreted as silica precipitate platelets growing on $\{100\}$ matrix planes. Larger defects having similar contrast have been observed in plan view silicon specimens prepared from CZ silicon crystals after anneals at 870°C and 650°C and are commonly known as "black dot" defects [8,9]. Figs. 2 and 3 show higher magnifications of platelet defects marked in Fig. 1 by "a1", and "a2" respectively. The observed dimensions varied from 4 nm to 10 nm. The thickness was estimated to be 0.3 - 0.6 nm. A second type of defects which appeared occasionally near the platelets are indicated by "b" in Figs. 1, 3. These defects are parallel to $\{311\}$ matrix planes. They also have been previously observed in CZ silicon and have been called $\{311\}$ stacking faults [10]. Other authors [8,9] have considered them to be precipitates growing on the $\{311\}$ matrix planes. They may be elongated in the <110> direction parallel to the beam and thus observed as a cross section 3-4 nm in width [11]. The density of the $\{311\}$ defects appeared to be lower than that of the platelets and the distribution of both types of defects appeared to be random within about 100 nm from the Si-SiO_2 interface.

For the wafers oxidized at 1000°C for eight minutes similar defects were also observed. Figure 4 shows two $\{100\}$ precipitates within 15 nm of the Si-SiO_2 interface. The dimensions and contrast are similar to the "a" defects observed at 900°. The $\{311\}$ defects were also observed at a somewhat greater distance from the Si-SiO_2 interface. The density of both defects was lower than in the specimen oxidized at 900°C.

Specimens oxidized at 1150°C for 25 minutes also contained similar defects as shown in Fig. 5. The density of $\{100\}$ defects was considerably smaller than at 900°C and 1000°C.

To begin to establish the origin of these defects, observations were performed on cross sections of as-supplied wafers. The high resolution image of such a specimen is shown in Fig. 6. Within the area studied, a

107

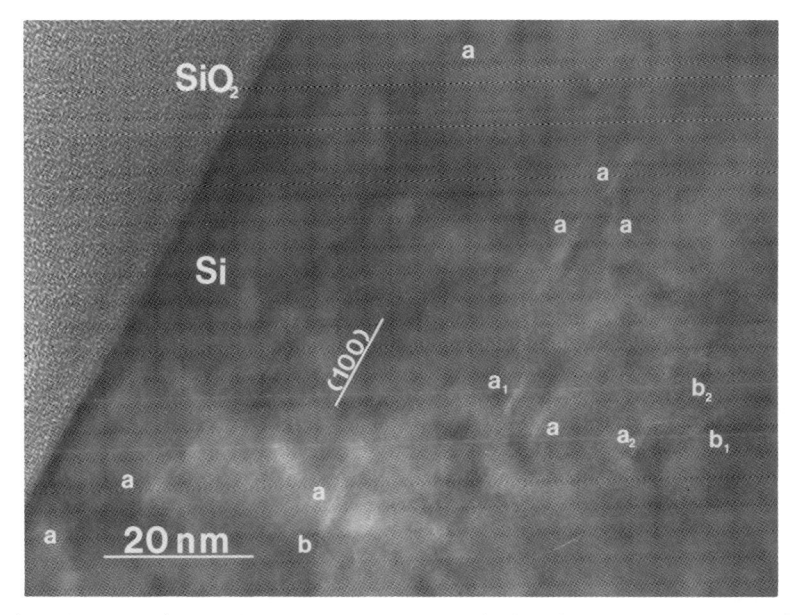

Fig. 1 HREM image of the defects near Si-SiO2 interface after 900°C oxidation.

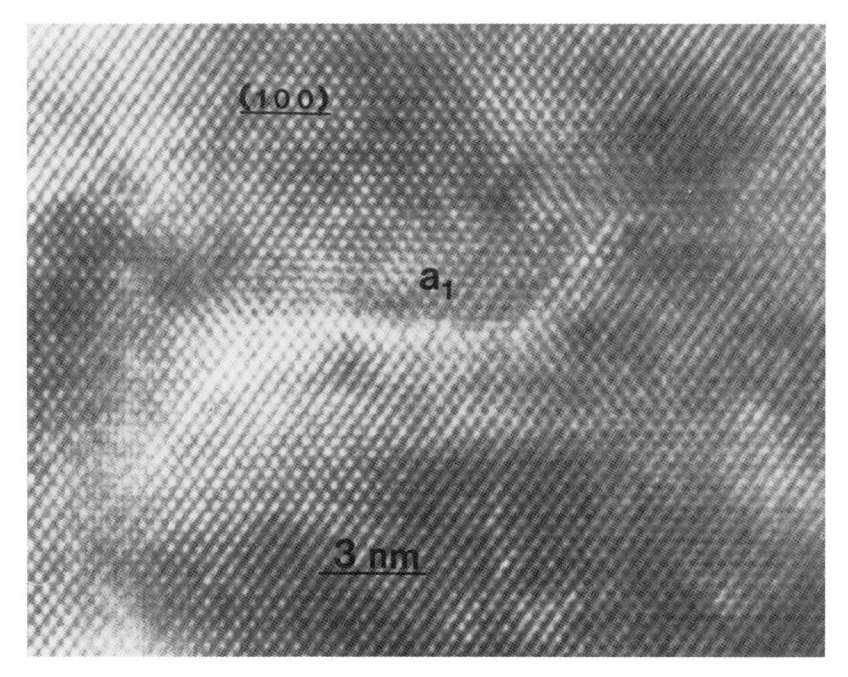

Fig. 2 Higher magnification HREM image of {100} defect marked a_1 in Fig. 1.

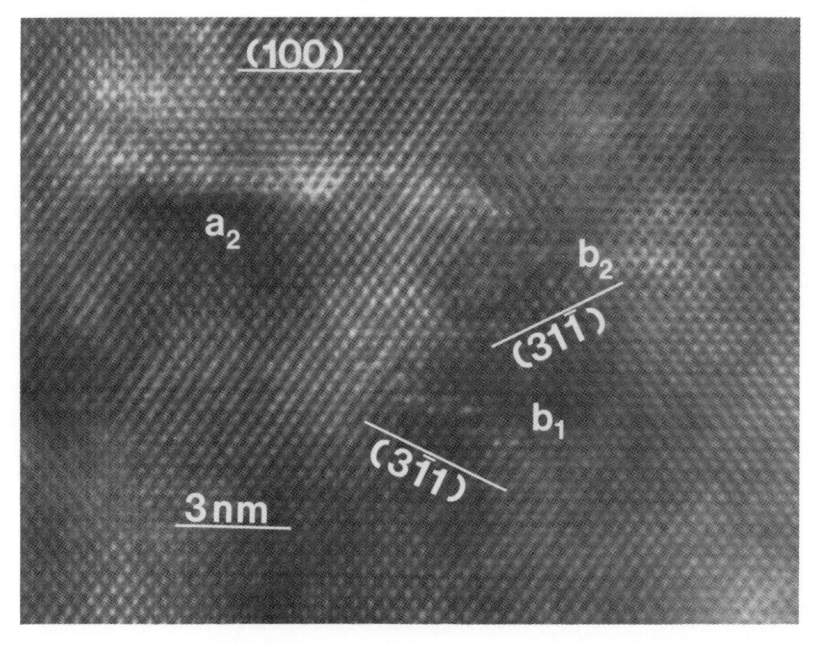

Fig. 3 Higher magnification HREM image of {100} and {311} defects marked a_2 and b_1, b_2 in Fig. 1.

Fig. 4 HREM image of the {100} defects near the Si-SiO$_2$ interface (after 1000°C oxidation).

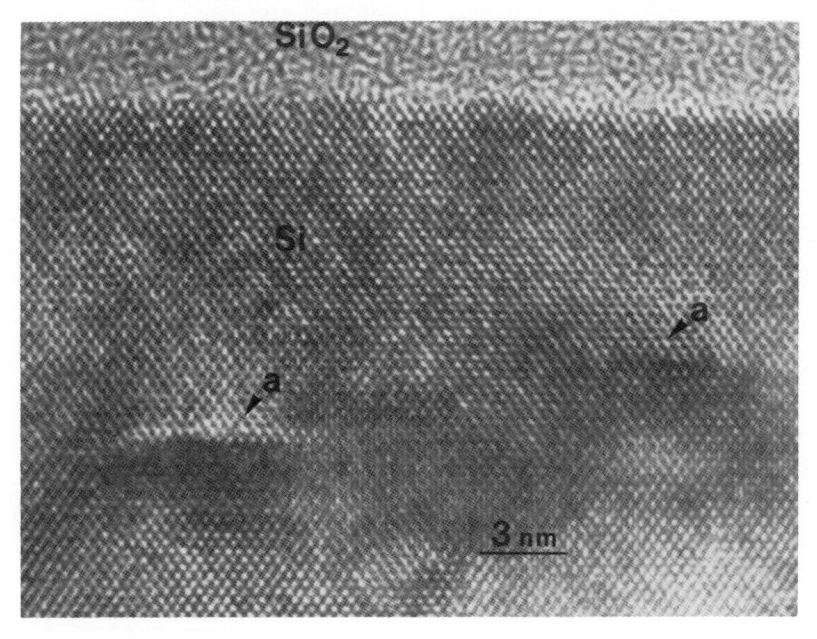

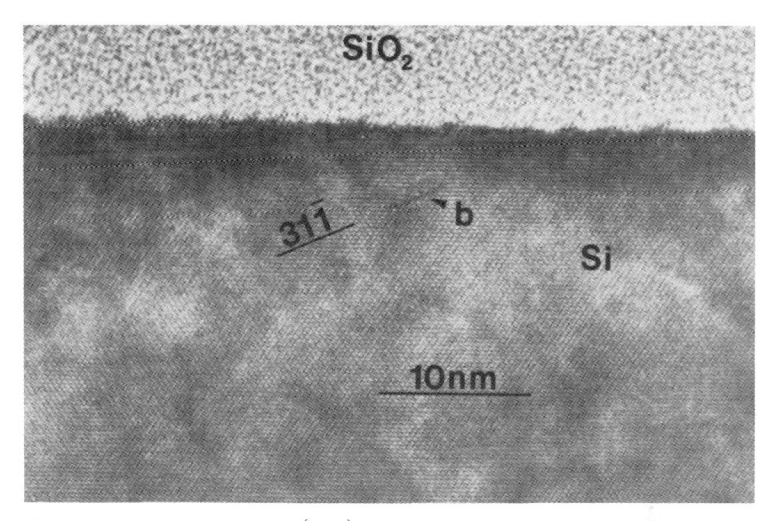

Fig. 5 HREM image of the {311} defect the interface (after 1150°C oxidation).

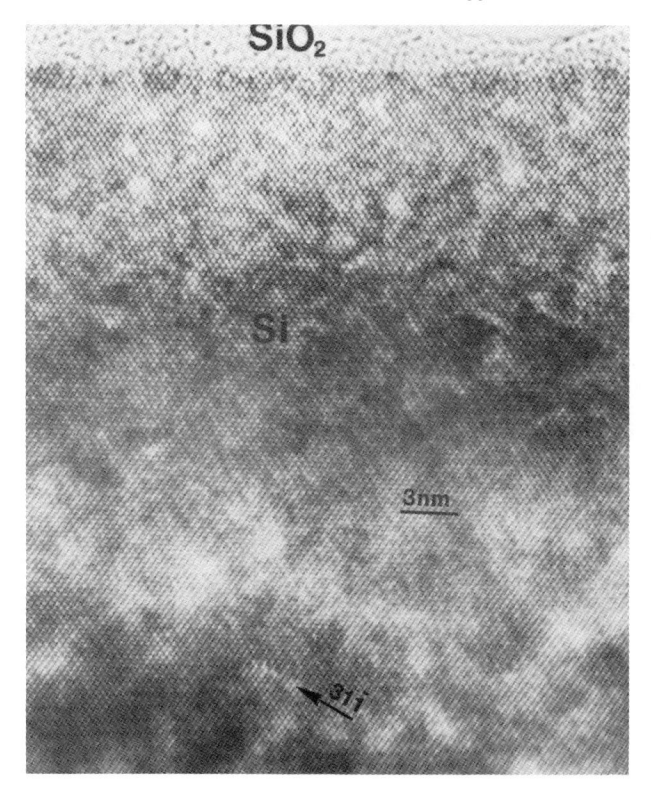

Fig. 6 HREM image of the defect in as-supplied wafer.

few defects were observed with similar contrast to the {311} stacking faults although smaller in size. This observation suggests that some or all of the defects may have been present in the as-received wafers. Because the density of these defects may vary markedly from one wafer to another, or from one part to another in a single wafer, it is not yet possible to conclude that these defects were formed during the oxidation treatment.

However, it is of interest that after an oxidizing anneal even at 1150°, these defects are still present very near the surface.

REFERENCES

1. J. R. Patel in Semiconductor Silicon 1981, H. R. Huff, R. J. Kriegler, Y. Takeishi, Eds (The Electrochemical Society, Pennington, 1981) p. 189.

2. S. M. Hu, J. Vac. Sci. Technol. 14, 17 (1977).

3. T. Y. Tan, E. E. Gardner, W. K. Tice, Appl. Phys. Lettl. 30, 175 (1977).

4. W. K. Tice, T. Y. Tan, Appl. Phys. Lett. 28, 564 (1976).

5. W. J. Patrick, S. M. Hu, W. A. Westdorp, J. Appl. Phys. 50, 1399 (1979).

6. C. J. Varker, J. D. Whitfield, K. V. Rao, J. J. Demer, in Ref. 1, p. 313.

7. J. H. Mazur, R. Gronsky, J. Washburn in Inst. Phys. Conf. Ser. No. 67 (1983).

8. J. Desseaux-Thibault, A. Bourret, J. M. Pennisson, ibid.

9. F. A. Ponce, S. Hahn, T. Yamashita, M. Scot, J. Carruthers, ibid.

10. T. Y. Tan, H. Foell, S. Mader, W. Krakow, Mat. Res. Soc. Symp. Proc. 2, 179-184, (1981).

11. N. Yamamoto, P. M. Petroff, J. R. Patel, J. Appl. Phys., 54, 3475 (1983).

ACKNOWLEDGEMENTS

This work was supported by the Director, Office of Energy Research, Office of Basic Energy Sciences, Materials Sciences Division of the U.S. Department of Energy under Contract No. DE-AC03-76SF00098.

EFFECTS OF MICROSTRUCTURE AND CHEMICAL COMPOSITION ON EBIC CONTRAST IN HEM SOLAR CELL SILICON

T. D. SULLIVAN
Dept. of Materials Science and Engineering
350 Bard Hall, Cornell University, Ithaca, NY 14853

ABSTRACT

EBIC images of an HEM solar cell display extensive spatial innomogenieties in the production of short-circuit current. TEM imaging reveals a variety of morphologies having dimensions on the order of 10 μm embedded in the material. Correlation of a region with high current collection to one of these distinct morphological regions suggests that segregation of B is responsible for the EBIC contrast.

INTRODUCTION

Defect imaging in Si by the electron beam induced current (EBIC) technique [1] has been a popular method over the last ten years for determining the electrical activity of localized defects. Defects such as dislocations, precipitates and grain boundaries have been shown by correlation with transmission electron microscopy (TEM) of the same region to be the sites of recombination centers, i.e. to decrease the magnitude of the current induced by the electron beam [2], [3]. However, formation of the near-surface p-n junction during solar cell fabrication in polycrystalline materials can produce an increased junction depth at dislocation subgrain boundaries [4], which results in enhanced current collection near the boundaries when the junction depth exceeds the minority carrier diffusion length. Similar behavior is observed in an HEM solar cell, but appears to stem from very different factors.

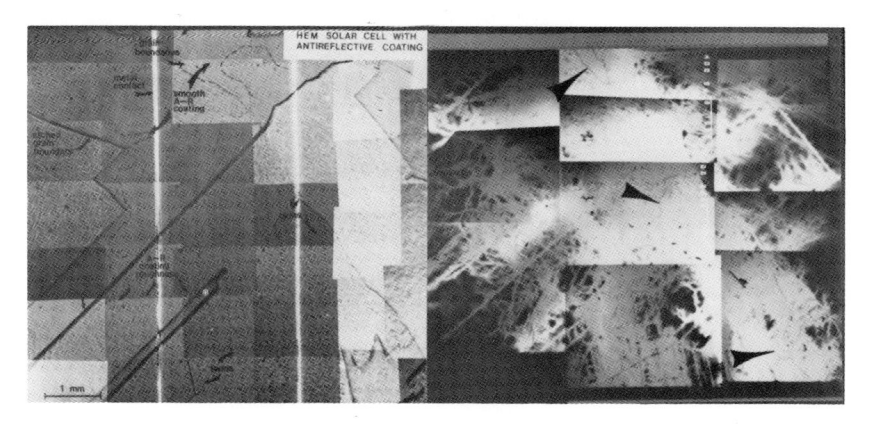

FIG. 1. Left: Optical micrograph of HEM solar cell showing surface defect morphology. Right: EBIC image of same cell after removal of antireflective coating. Arrows indicate grain boundaries appearing at a beam energy of 25 keV.

EXPERIMENTAL

A solar cell fabricated from a section of heavily flawed heat-exchanger-method (HEM) cast silicon displayed a large variation in open-circuit voltage as a function of spot location when excited by a scanned laser spot. Regions of high voltage output, typically about 100 μm wide, often corresponded to grain boundaries, twins and other features identifiable from the surface topography of the cell (fig. 1a). In order to examine the active regions more closely, the antireflective coating (SiO) was dissolved away by immersion in a 10% HF solution for 1 h so the cell could be imaged by EBIC on a scanning electron microscope (SEM). Fig. 1b is the EBIC image of the cell shown in fig.1a obtained by using the built-in p-n junction. High current collection is observed to occur at or within about 30 μm of the features visible in fig. 1a. Lines of darker contrast (enhanced recombination) became visible within the brighter regions when the electron beam energy was increased past 25 keV, which corresponds to an electron penetration depth of about 5 μm (R = 0.0171 $E_B^{1.75}$ [5]). Since the depth of the p-n junction was estimated to be only 1 μm, this suggests that the process of diffusing in the p-n junction may be altering the electrical properties of the near-surface region by a mechanism similar to that cited earlier for the dislocation subgrain boundaries [4].

The unexpected features in fig. 1b are the dead areas exhibiting very low current collection. These areas persist deeper into the material. The EBIC image shown in fig. 2a was obtained, after removal of several microns of surface material by Syton polishing, with a Schottky barrier diode formed by vacuum deposition of 300 Å of Al. Major morphological

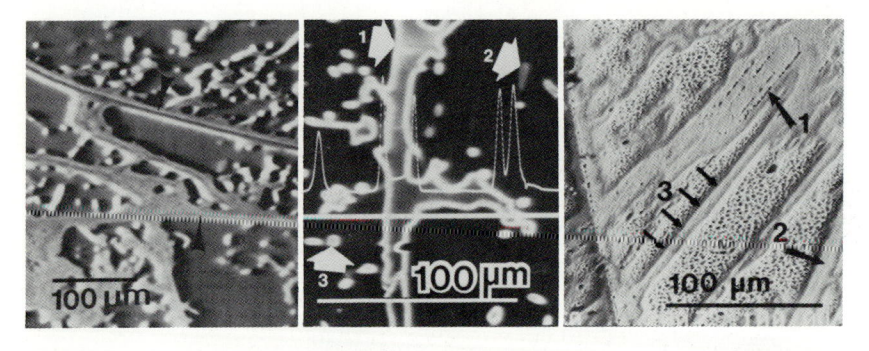

FIG. 2. a) EBIC image of solar cell after removal of more than 5 μm of material. b) EBIC image after removal of 50 μm of material, showing decreased activity within region (1), regions of weaker activity (2), and enhanced activity at borders (3). c) SEM secondary electron image of etched surface with region containing twins(1), grain boundary (2), and raised borders.

features remain, but regions of high current collection are confined more closely to the boundaries and well-defined islands, leaving enlarged dead areas. The boundaries themselves have become clearly visible. Fig. 2b is an EBIC image also obtained with a Schottky barrier after removal of more than 50 μm of the surface material. High collection zones are now restricted primarily to the borders of the islands and borders of the bands about the boundaries. The magnitude of the collected current varies within these zones, and is generally weaker than at the boundary. It is negligably small outside. Fig. 2c is an SEM micrograph of the surface of another region of the same cell after the original surface was etched in a 25% HF solution for 16 h, followed by 4 h in a 50% H_2O_2 solution and an additional HF etch. The more etch-resistant, raised, smoother features correspond to the bright areas in fig. 2a, indicating that these are the high-collection zones. Such a region giving rise to current collection will now be examined.

A 3 mm diameter piece of this material was cut for TEM, and thinned by ion milling. A Schottky barrier was formed by vacuum deposition of 300 Å of Al onto one side of the TEM specimen. A suitable region, previously selected in TEM, was imaged in SEM and EBIC (fig. 3a,b). The specimen was then ion milled for 15 minutes to remove the Al layer, and imaged in scanning TEM (STEM) for chemical microanalysis, and TEM for selected area diffraction (fig. 3 c,d). Three regions are of interest: (1) the relatively clear bulk material, (2) the dark precipitate, and (3) the mottled region. Current collection is low in regions (1) and (2), but high in region (3). Region (2) appears electrically active at first glance, but careful comparison between 3b and 3c reveals that the shapes of the precipitate and the electrically active

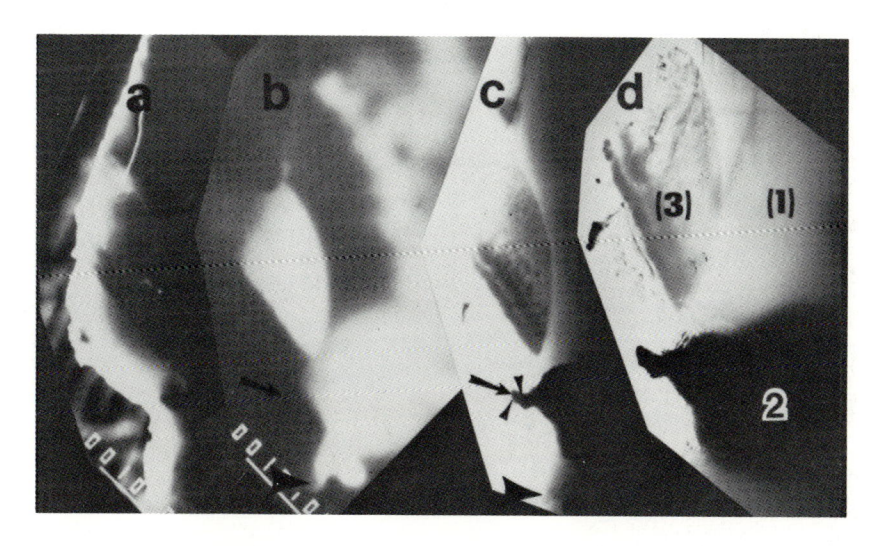

FIG. 3. One location on a TEM specimen imaged in SEM (a), EBIC (b), STEM (c), and TEM (d). In (d) region (1) is single crystal, region (2) is a multi-component polycrystalline precipitate and region (3) is amorphous and contains scattered crystallites. Neither (1) nor (2) is electrically active, but (3) is, apparently due to segregation of B.

area differ. A faint trace of another structure, evident just below the precipitate (see arrows in b,c), is probably a remnant of active material mostly removed along with the Al during the 15 minutes in the ion miller. Current collection in region (3) is obviously high. Energy dispersive X-ray spectra (EDS) indicated regions (1) and (3) contain Si alone. The precipitate, however, was composed of nonuniformly distributed concentrations of Na, Al, Si, S, Cl, K, Ca, Ti, Cr, Mn, Fe, Ni and Zn (fig. 4 Left). The Cu signal appears at about the same intensity in spectra obtained over the entire specimen, as well as in the area of fig. 3, and probably arises principally from the Cu mounting grid used to support the Si. Electron energy loss spectra (EELS) were obtained from both regions (1) and (3) to check for the presence of appreciable concentrations of lighter elements. Only the Si $L_{2,3}$ peak appeared in region (1), but B, C and O peaks also appear in region (3).

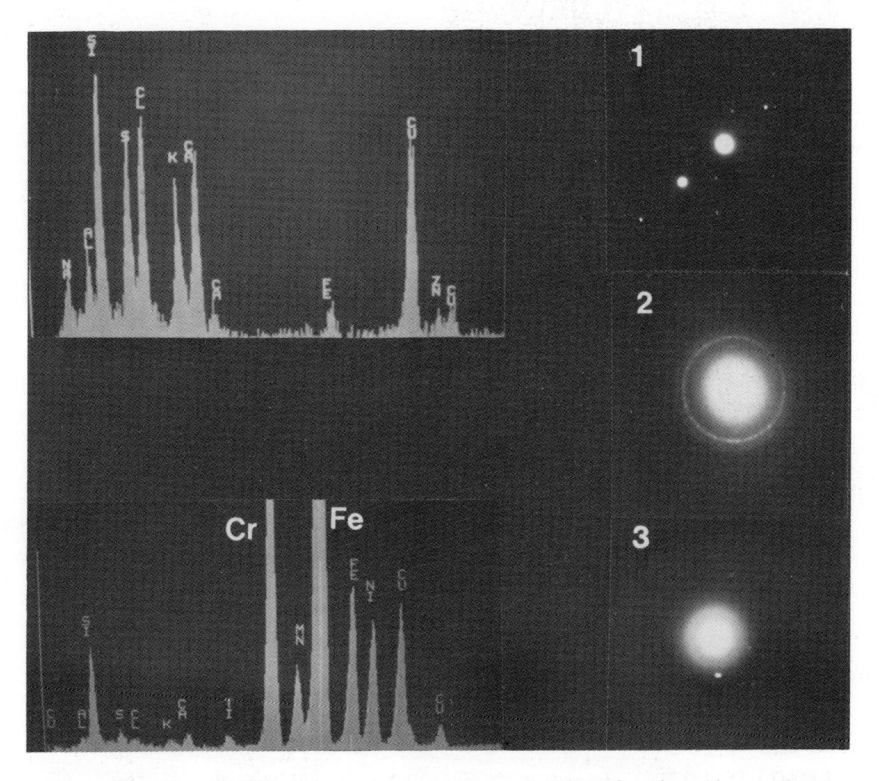

FIG. 4. Left: EDS spectra from dark precipitate in fig. 3, at the two locations indicated by small arrows at the tip in 3c. Right: selected area diffraction patterns from the 3 regions shown in 3d.

Selected area diffraction patterns obtained from the three regions (fig. 4 Right) show that region (1) is single crystal, region (2) is micropolycrystalline, and region (3) is mostly amorphous, but contains microcrystallites 100 - 1000 Å in diameter. Reflection spacings from some of these crystallites match well with Si spacings. Farther back from the edge the amorphous rings are superimposed over the single crystal pattern, indicating that a layer of single crystal material probably lies below the amorphous material. The presence of bend contours a few microns from the edge support this conjecture. Hence region 3 appears to be an amorphous matrix of Si, B, C and some O, containing scattered crystallites of Si, all embedded in single crystal Si. Although other amorphous regions appear in the specimen (fig.5a), a number of other morphologies are also well represented, as shown in the rest of fig. 5. Tendrils of amorphous material sometimes penetrate the crystalline material (b); microcrystals can be dispersed throughout crystalline or polycrystalline material (c);

clouds of impurities can be distributed within single crystal regions (d).

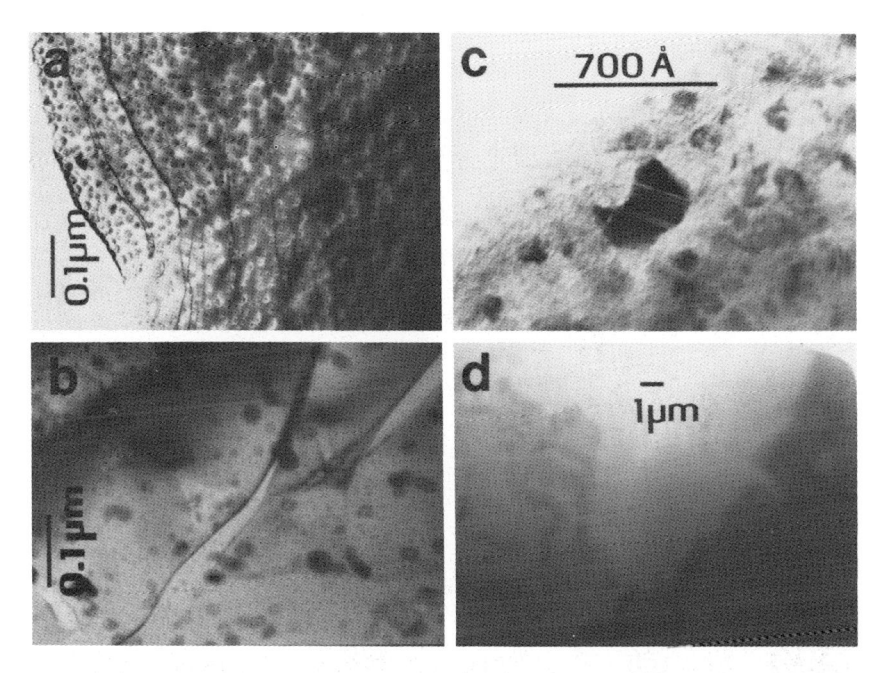

FIG. 5. Four different morphologies observed within the same specimen: (a) neighboring amorphous regions with crystallites, (b) interpenetrating amorphous and crystalline regions, (c) microcrystallites dispersed in single crystal, (d) cloudlike impurities in single crystal material.

DISCUSSION

The presence of detectible concentrations of B in region (3) of fig.3d probably accounts for the observed current collection in EBIC. The amorphous material might ordinally be expected to inhibit current flow and relegate high current collection to the edges of the region. In this case, however, the high doping level coupled with the thinness of the specimen (500 - 1000 Å), probably compensates sufficiently. Since HEM material is cooled slowly over several hours, another possibility to consider is the diffusion of B out of the amorphous region into the neighboring single crystal material. A thin layer of p-type material surrounding the amorphous region could then form a local Schottky barrier and conduct the current induced below the region back around to the surface. Since EELS is generally unable to detect elemental concentrations below 0.1%, the presence of a detectable quantity of B in specific locations suggests that other regions of the material such as region (1) in fig. 3 have been denuded of B. The multi-element precipitate (2) suggests also that appreciable quantities of impurities such as Fe may be distributed throughout the same single crystal regions to act as recombination centers. This could explain the existence of the dead areas observed earlier.

The EBIC contrast observed in fig. 2b can now be understood if defect regions, amorphous or otherwise, are assumed to be surrounded by a layer of single crystal material containing B. A local Schottky diode forms wherever the p-type material intersects the Al-deposited surface and produces current when excited by the electron beam. For defect regions much larger than the electron penetration depth, only these border zones will exhibit bright contrast. However, when the defect region is shallow enough for the beam to intersect the surrounding B-doped layer, current collection can occur to produce the bright contrast within. Other sources of varying EBIC contrast within the regions are variations in B concentration and in carrier lifetime due to material structure and/or concentrations of recombination centers. Although no boundary regions have as yet been analysed, contrast such as that in fig. 2a could be produced by preferential precipitation of elements such as Fe at the boundary itself. Further analysis is intended.

ACKNOWLEDGEMENT

Discussions with P. G. Zielinski, R. Gleichmann, D. Lilienfeld and D. Ast have been most valuable in proceeding with this work.

REFERENCES

1. H.J. Leamy, L.C. Kimerling, S.D. Ferris, SCANNING ELECTRON MICROSCOPY/1976 (Part IV), Chicago, pp. 529-538.
2. J. Heydenreich, H. Blumtritt, R. Gleichmann, H. Johansen, SCANNING ELECTRON MICROSCOPY/1981 (Part I), AMF O'Hare (Chicago) pp. 351-365.
3. H. Strunk, B. Cunningham, D. Ast, in "Defects in Semiconductors," MRS Symposia Proc. 2, Boston, J. Narayan, T. Y. Tan, eds. (North-Holland, New York 1980) pp. 297-302.
4. S.M. Johnson, R.W. Armstrong, R.G. Rosemeier, G.M. Storti, H.C. Lin, W.F. Regnault, in "Grain Boundaries in Semiconductors", MRS Symposia Proc. 5, Boston, H.J. Leamy, G.E. Pike, C.H. Seager, eds. (North-Holland, New York (1981) pp.179-184.
5. T.E. Everhart, P.H. Hoff, J. Appl. Phys., 42, 13 (1971).

ELECTRON MICROSCOPY OF SEMICONDUCTORS RECONSTRUCTED SURFACES

PIERRE M. PETROFF
AT&T Bell Laboratories, Murray Hill, NJ 07974 USA

ABSTRACT

Surface sensitive transmission electron microscopy (SSTEM) and reflection electron microscopy (REM) have been used to analyze the Si (111) 1x1 → 7x7 surface reconstruction. The SSTEM and transmission electron diffraction results for the Si (111) 7x7 surface are interpreted using several possible "surface dislocation" models. The SSTEM and REM techniques have also been applied to the GaAs (100) MBE deposited surfaces. The rough surface topography for the c(4x4) reconstructed surface is attributed to surface steps motions and bunching upon interruption of the MBE deposition.

INTRODUCTION

Recently, new techniques which make use of small dimension probes (<0.1 μm) and direct surface imaging have emerged for the analysis of surfaces. Among these we note: scanning tunneling microscopy (STM) [1], reflection electron microscopy (REM) and surface sensitive electron microscopy (SSTEM). These offer the opportunity of directly assessing the surface topography on an atomic scale and constitute a promising research approach to the field of surface and interface science.

Electron microscopy of surfaces, in fact, was originally attempted fifty years ago by Ruska [2] who first obtained RHEED diffraction patterns from a metal surface. It is only recently that a renewed interest in using the electron microscope for the study of surfaces has emerged [3]. The surface experiments using an electron microscope can be divided into two classes. First, those which deal with air exposed and/or oxidized surfaces or surfaces protected from the ambient by an amorphous layer. Second, those which aim at understanding the topography of *clean* or *reconstructed* surfaces and, consequently, require either a UHV environment around the sample or a surface which does not oxidize or contaminate prior or during the electron microscopy analysis. Both reflection electron microscopy (REM) [3] and surface sensitive TEM have been applied along with their diffraction information, e.g., RHEED and transmission electron diffraction (TED) to the study of these two classes of surfaces.

In the first class of experiments, REM has been applied successfully to retrieve structural information from metal surfaces [3,4], ionic crystals [5,6] and cleaved semiconductor [7] surfaces. Surprisingly, very little REM experiments have been done to further the understanding of the RHEED patterns and establish correlation with the surface topography.

SSTEM has been applied to the study of atomic steps on noble metal surfaces [8-10], e.g., Au, Pt and silicon [11]. The technique uses the diffraction contrast arising from forbidden bulk Bragg reflection which are allowed for surface layers [9,11,12].

The second category of experiments focuses on the electron microscopy of reconstructed surfaces or clean surfaces. This type of experiments was pioneered by Osakabe [13,14], Yagi [15] and Takayanagi [16]. Using a UHV electron microscope operated in the REM imaging and diffraction mode, Osakabe et al. [13,14] have analyzed in detail the Si (111) 1x1 to 7x7 surface reconstruction. The same group has applied the UHV-SSTEM technique to study the surfaces of Au and other metals [16,17]. The uniform and higher spatial resolution and absence of image distortion are distinct advantages of the SSTEM technique over REM. Equally important is the simpler interpretation of the SSTEM-TED patterns through simple kinematical calculations [18]. This application alone will probably prove to be as significant as the development of LEED to the study of surfaces.

In this paper we discuss the results of new SSTEM and REM experiments on the Si (111) and GaAs (100) reconstructed surfaces.

EXPERIMENTAL AND RESULTS

The observation of clean reconstructed surfaces by electron microscopy requires the presence of a UHV environment around the sample. This condition not present in commercially available microscopes has been achieved by adding to a JEM 200B a liquid Helium cryoshield cooled to ≅4°K. The cryoshield surrounds a sample holder specially built to maintain UHV conditions in the sample chamber. The sample holder allows heating of the sample in situ the microscope and permits sample analysis either in the REM mode or in the SSTEM mode. The details and operation conditions of this UHV attachment have previously been published [19].

A second prerequisite needed for reconstructed surface electron microscopy analysis is the ability to clean the surface in the UHV environment. For SSTEM a third prerequisite is that the sample thickness be as small as possible (≤100Å). This usually requires a means of thinning the sample in situ the microscope either by controlled sublimation of the surface layers or by ion thinning and annealing of the sample. For the Si (111) surface reconstruction REM experiments, the sample in the form of a bar (5 x 3 x 0.5 mm^3) is heated resistively by passing a DC current (~4–10A). Prior to heating in the microscope, an MOS line quality cleaning treatment [20]. was applied to the surface. Upon heating from room temperature the formation of SiC crystals is observed on the surface at ≅750°C. The silicon carbides disappear upon further heating at ≅1100°C leaving a surface with flat areas ≅100 μm x 100 μm with (111) faceted large pits in between. After heating of the sample to ≅1200°C, the sublimation of surface atoms and motion of individual surface steps are observed. The corresponding RHEED pattern of this surface shows a hazy (1x1) structure. The dark field RHEED micrograph of the Si (111) 1x1 surface (Fig. 1A) shows straight surface steps with a black-white contrast indicating the existence of both protruding and reentering surface steps. After cooling below ≅750°C, a 7x7 RHEED pattern is obtained (Figure 1B). The step edges became rough during the transition from the (1x1) to the 7x7 reconstructed surface. The roughening observed is on a scale of a few 100Å. The REM results described above are consistent with those reported earlier by Osakabe et al. [13,14].

FIG. 1 (A) REM dark field micro- of a Si (111) 1x1 surface. Dark areas correspond to surface pits. Arrows indicate surface steps of opposite signs. The electron beam incident direction is close to a <110> direction.

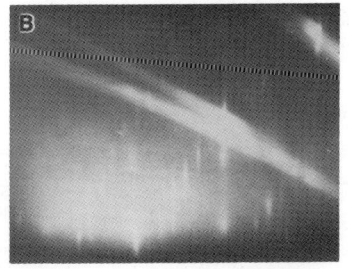

FIG. 1 (B) RHEED pattern of the Si (111) 7x7 reconstructed surface.

The preparation of the Si (111) surface reconstruction for SSTEM experiment follows identical steps to those described from the REM observations with the following differences. Prior to cleaning a small hole (φ ≅ 1mm) is etched chemically through the sample with a mixture of HF,

HNO$_3$ and H$_3$PO$_4$. The sample is then precleaned by use of a previously developed procedure [20]. The sequence of SiC formation and associated pitting of the surface is also present during the surface cleaning for SSTEM. In fact, the thinnest part of the sample are transformed into silicon carbides making them useless for surface microscopy. Careful heating to 1200°C while viewing the sample is again carried out to get rid of the silicon carbides. The sublimation of the surface is observed directly in the bright field image by following the formation of thin areas in the sample. The thinnest areas of the sample are used for carrying out the SSTEM observations. For a thin enough area, imaging of the surface steps is then carried out directly in the bright field or in the phase contrast imaging modes. The step edges for the Si (111) are usually aligned along <110> and <112> directions (Figure 2). The change in contrast from one terrace to another is clearly distinguished and corresponds to that expected from a surface with <111> monoatomic steps [12,21]. SSTEM of the Si (111) 1x1 to 7x7 reconstruction is carried out in the central region (the thinnest part of the sample) far away from the step edges. Upon cooling to T ⩽750°C, triangular defects (50Å to 500Å) in sizes appear in the bright field images of the surface (Figure 3). Contrast analysis of these defects indicates that their strain field is consistent with triangular defects with edges parallel to <110> and a strain field oriented along <111>. The corresponding TED pattern for this surface (Figure 4) shows diffraction superlattice spots characteristic of a 7x7 reconstructed surface. The 7x7 TED pattern and the triangular defects disappear with a loss of the UHV environment in the microscope.

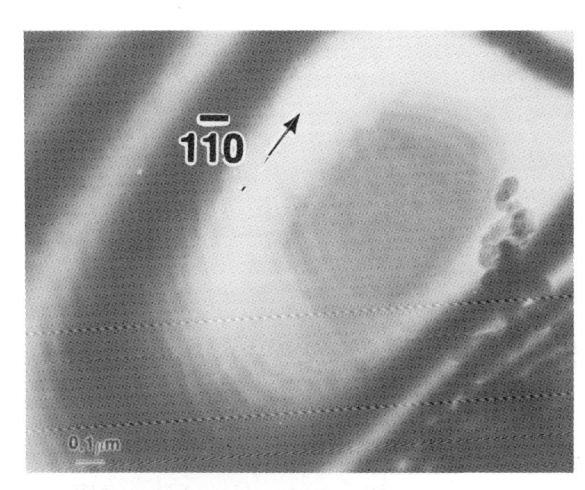

FIG. 2 Bright field micrograph of a Si (111) 1x1 surface in the SSTEM mode of analysis. The contrast changes in the image correspond to the presence of (111) surface monoatomic steps.

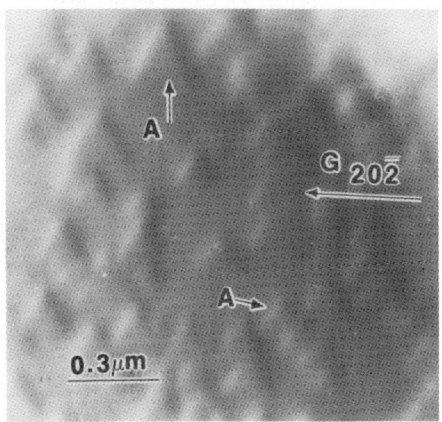

FIG. 3 Bright field micrograph of a Si (111) 7x7 surface in the SSTEM imaging mode.

The GaAs (100) surface are prepared by molecular beam epitaxy (MBE) on (100) GaAs (Si-doped 10^{18} cm^{-3}) substrate. The usual substrate cleaning and MBE deposition conditions have been used for the deposition of a 500Å GaAs epitaxial layer. After the GaAs deposition at 650°C, the sample is brought abruptly to low temperature while an As$_2$ overpressure of 10^{-5} Torr is maintained in the MBE apparatus. With the sample temperature at $\cong$70°C, an amorphous As overlayer $\cong$500Å thick is deposited on the surface. After removal from the MBE apparatus the sample is then cleaved in a small bar (5 x 3 x 0.5 mm^3) prior to insertion in the UHV microscope. The amorphous As layer deposition over the surface has been shown by us and others [22] to efficiently protect the surface from oxygen and carbon contamination during transfer in ambient

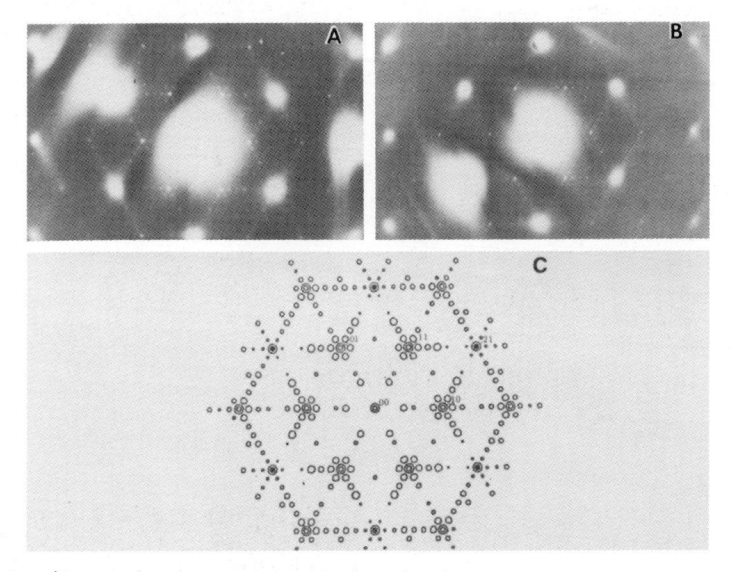

FIG 4. (A and B) TED pattern of the Si (111) 7x7 surface for 2 different samples. (C) Computer rendering of the diffraction spots intensities. Bulk diffraction spots are indicated by filled circles. The size of the open circles gives a measure of the intensities with the largest ones corresponding to the stronger diffraction. Eight intensity levels have been measured.

atmosphere. Resistive heating of the sample to T $\cong$ 250°C in the UHV microscope is then used to sublimate the As layer from the surface. The amorphous As sublimation is followed by direct imaging in the REM mode in the temperature 250°C to 280°C. The As layer leaves the surface in patches and a clean surface free of amorphous As layer is obtained at T $\cong$ 280°C. The first reconstructed surface detected in the RHEED pattern immediately after the As removal is the c(4x4) surface reconstruction (Figure 5A). The streakiness of the superlattice reflection is attributed to the presence of surface steps [23] on the clean surface. Indeed the REM image (Figure 5B) of the c(4x4) surface reveals a high step density with terraces of dimensions $\cong$0.5 x 5 μm^2. The vertical lines in Figure 5B correspond to slip lines introduced by moving dislocations during heating of the sample. The black-white contrast of these lines is a phase contrast due to the surface step produced by the dislocation slip plane intersection with the surface. We note here that the contrast at surface steps is different from that of the slip lines and from that of surface steps on the clean Si (111) surface (Figure 1). This is most probably related to the nature of the strain field at the step edge and to the greater height of the steps on the GaAs (100) MBE surface. Upon heating, a series of surface reconstruction which closely reflect the expected stoichiometry of the GaAs surface [24] are observed. These reconstructions are the c(2x8), c(8x2) and c(4x6) as the surface becomes richer in Ga. The corresponding RHEED patterns in Figure 6 also show streaky superlattice spots which reflect the presence of a large density of terraces with small misorientations from the initial (100) substrate. Finally, we have observed that the dark field images using superlattice reflections clearly show a *nonhomogeneous* surface reconstruction for the GaAs (100) c(4x4) surface.

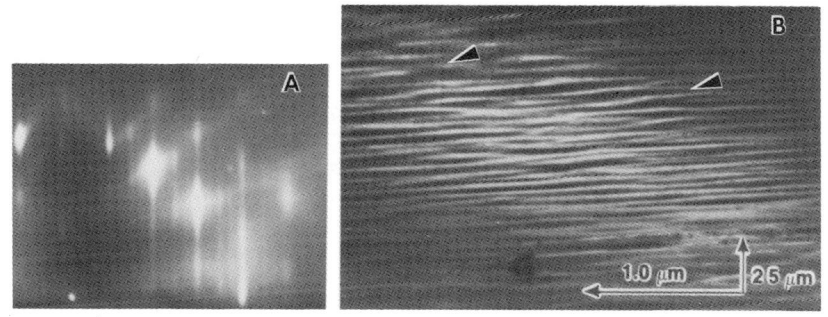

FIG. 5 (A) RHEED pattern of a c(4x4) GaAs (100) reconstructed surface. (B) REM dark field micrograph of a GaAs (100) c(4x4) reconstructed surface deposited by MBE. The electron beam direction is close to a <112> direction.

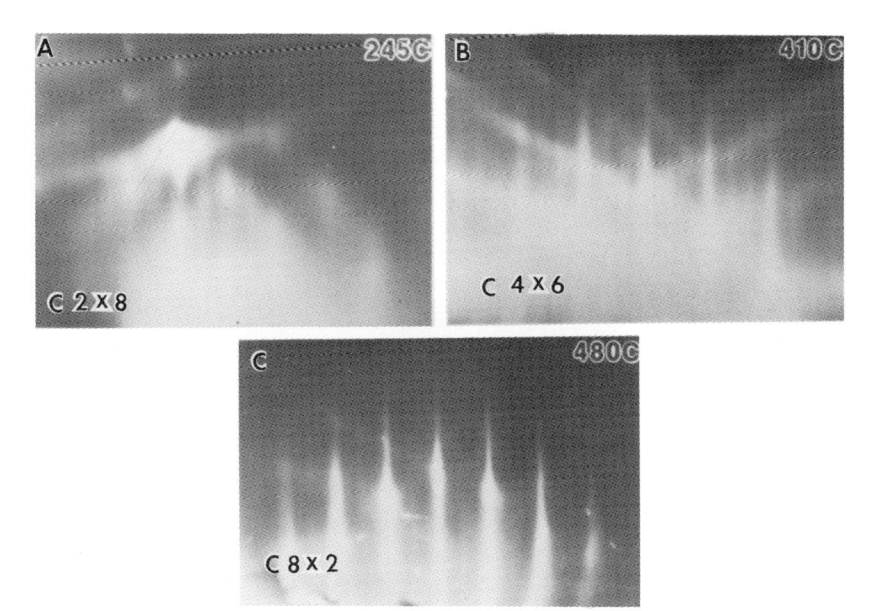

FIG. 6 RHEED patterns corresponding to the various GaAs (100) surface reconstruction (a) c(2x8), (b) c(4x6) and (c) c(8x2).

DISCUSSION

The Si (111) 7x7 Reconstructed Surface

The REM, SSTEM and TED results have led us to propose a new model for the 1x1 → 7x7 surface reconstruction [25]. It is based on the following observations:

a. We have noted as others [13,14] have that important changes in the shape of the step edges occur with the onset of the phase transformation. Osakabe et al. [13] have further reported that the 7x7 phase grows from the step edges towards the step risers. Both observations require atomic transport on the surface over large distances.

b. The backbond induced stresses on the surface layer should produce a misfit strain which will be relaxed during the 1x1 → 7x7 reconstruction [26]. The proposed model for the 7x7 reconstruction compatible with these observation is based on the generation of "surface dislocations" [25]. The "surface dislocation" is the analogue of a misfit dislocation at an interface where a lattice misfit exists. It is thought to originate from the step edge and propagate by glide towards the step riser, thus driving the phase transformation front. It is assumed to have Burger's vectors analogue to those of bulk dislocations in Si, and it is located one or two layers below the surface. In Figure 7, three possible configurations of "surface dislocations" are represented. The initial surface (Figure 7A) will exhibit a step whenever a

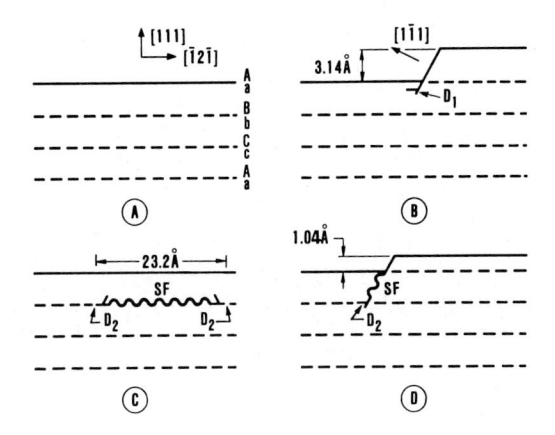

FIG. 7 Schematic of the possible "surface dislocation" configurations leading to a 7x7 surface reconstruction. (A) Initial surface vacuum interface; only the (111) double layers are represented. (B) 60°C surface dislocation with $\underline{b} = \frac{a}{2} <110>$. (C) A dissociated "surface dislocation" consisting of two 30° Shockley partials and a "stacking fault" (SF). (D) A dissociated "surface dislocation" consisting of a 30° Shockley partial and a stacking fault (SF) on a $(1\bar{1}1)$ plane.

perfect 60° type "surface dislocation" with $\underline{b} = \frac{a}{2} <110>$ is introduced by glide on a $(1\bar{1}1)$ plane (Figure 7B). The step height in this case is d = 3.14Å. A screw type "surface dislocation" with $\underline{b} = \frac{a}{2} <110>$ could be dissociated into two Shockley partials with $\underline{b} = \frac{a}{6} <112>$ and a stacking fault (Figure 7C). This configuration yields a flat surface with a wurtzite top layer where the stacking fault resides. A mixed 60° type dislocation gliding in a {111} plane inclined to the surface can dissociate into two Shockley partials (a 30° and a 90° type) with $\underline{b} = \frac{a}{6} <112>$. If the 90° Shockley partial is annealed at the surface, the configuration shown in Figure 7D is present. This "surface dislocation" yields a step of height d = 1.04Å at the surface. To produce the 7x7 reconstructed surface, the "surface dislocations" are introduced every 8 (111) planes. The topography resulting from this operation has been described earlier [25] in detail for the surface dislocation shown in Figure 7B. The surface unit cell consist of two equilateral triangles with edges parallel to <110> directions in the (111) plane. The height of each triangle differ by 3.14Å and their side length is 26.8Å. We further assume that the core of the "surface dislocation" will be reconstructed as in the bulk [27]. Figure 8 shows part of a unit cell with a surface dislocation between the two triangular subunit cells. The dislocation core reconstruction produces a set of 3 dimers along each side of the triangular subunit cell. The "surface dislocation" shown in Figure 7C and 7D when introduced self consistently every (111) plane yield a unit cell consisting of 2 triangles with different topological and crystallographic configurations. The 3 possible configurations are shown in Figure 9. In each case the reconstructed dislocation cores lead to the formation of 3 dimers on each triangle side.

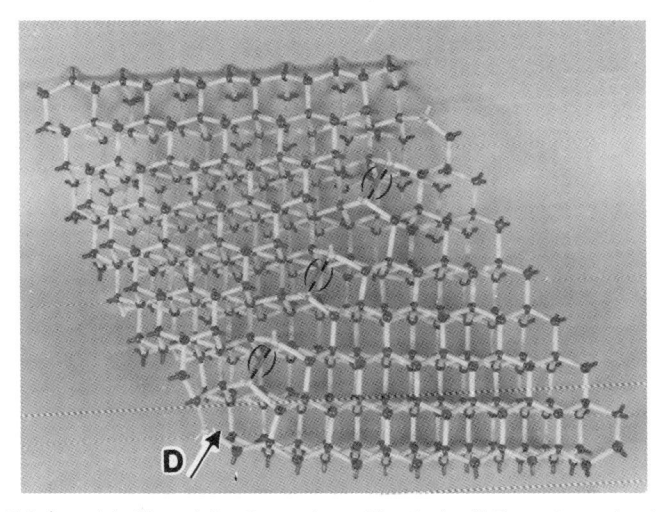

FIG. 8 Stick and ball model of a unit cell of the 7x7 surface showing the reconstructed core of a "surface dislocation" (dotted lines). Three dimers are seen along the dislocation core. The unit cell is bounded on each side by similar "surface dislocations" parallel to <110> surface directions.

This paper will not attempt to discuss in detail the respective merits of the numerous proposed models for the Si (111) 7x7 surface reconstruction. Rather we point out that the various models can be divided in two classes: the adatom models and the triangle-dimer models [28]. We also note that the experimental TED patterns Figure 4 have been compared to computed patterns using a kinematical diffraction calculation for atoms positions corresponding to the two classes of models. The result convincingly allow to rule out the adatom type models [28]. The triangle-dimer models lead to a good agreement between the experimentally and calculated diffraction patterns for a dimer separation equal to $\cong$0.6 the interatomic spacing in the bulk [28].

The GaAs (100) Reconstructed Surface

The various surface reconstruction which are observed as a function of temperature, coincide well to those reported and correlated to the stoichiometry of the surface [24]. The nonhomogeneity in the surface coverage of the reconstructed c(4x4) surface could be indicative of the coexistence of 1x1 domains on the (100) surface.

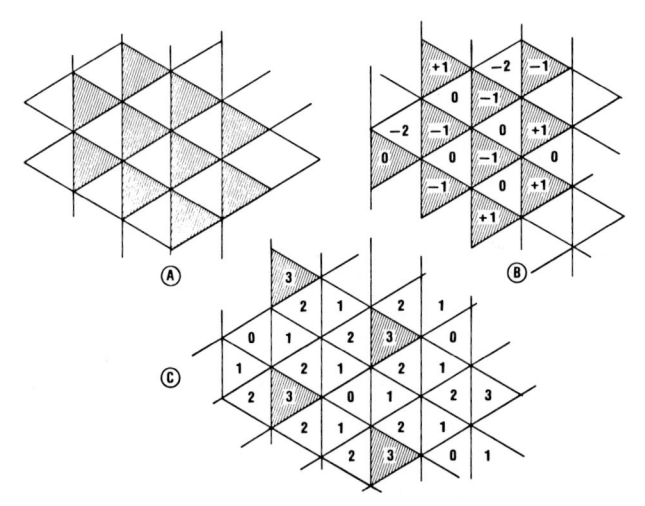

FIG. 9 The three possible topological configurations of the 7x7 surface consistent with the "surface dislocations" discussed in Figure 7. All dislocation lines are parallel to <110> surface directions. (A) For the screw type surface dislocation case. Shaded areas indicate a stacking fault below the surface (i.e., wurtzite layer). (B) For the 60° type perfect surface dislocation. The numbers indicate the height of each triangle in unit of 3.14Å with respect to the initial surface (0). (C) The dissociated 30° surface dislocation. The numbers indicate the height of each triangle in unit of 1.04Å with respect to the initial surface.

A surprising finding was the extent of the surface roughness of the GaAs (100) MBE deposited surface. This roughness is consistent with the streaky RHEED diffraction spots (compare to the Si RHEED pattern in Figure 1) and strongly suggest the existence of a microfaceted surface. From the analysis of the REM images, the average step height is estimated to be between 3 and 5 interatomic distances. This rough surface topography is in sharp contrast to that expected from the observation of atomically smooth MBE interfaces [29] which require an atomically smooth surface (steps with height of ± one atomic spacing) during growth. This apparent contradiction may well be explained by the kinetics of step motions during and upon interruption of the MBE growth [30]. The motion of atomic steps detected by RHEED upon the interruption of the MBE deposition on GaAs (100) surfaces [30] could result in a step bunching and produce the rough surfaces detected in the present experiments. The marked elongation of the terraces along one <110> surface direction detected in the REM images and RHEED spot shapes may reflect an anisotropy of atom mobilities along the two nonequivalent <110> surface directions. These REM observations are consistent with the optical (Nomarsky contrast) observations of GaAs (100) surfaces produced by MBE. The stoichiometry dependence of these various surface reconstruction make it doubtful that a "surface dislocation" model of the type proposed for the Si (111) 7x7 could be applicable to the GaAs (100) surface. This is further supported by our observations that the step and terraces topography did not change through the various surface reconstructions.

CONCLUSIONS

The recent progress in UHV microscopy have opened the possibility of direct imaging of clean reconstructed surfaces for both elemental and compound semiconductors. The diffraction information along with the direct imaging information obtained by REM and SSTEM will undoubtedly help our understanding of these surfaces. As an example, we have described an analysis of the Si (111) 7x7 surface which suggests that a "surface dislocation" model is operative during the reconstruction. In a second example we have shown the rough topography of the clean GaAs (100) c(4x4) reconstructed surface produced by MBE. The surface roughness in this case is thought to originate from step motions and bunching upon interruption of the MBE growth. The stoichiometric dependence of the various reconstructions on the surface do not allow a simple surface dislocation model to be applied to this surface.

ACKNOWLEDGEMENTS

The author wishes to thank R. J. Wilson, A. C. Gossard, E. G. McRae and W. Wiegmann for their collaboration and for valuable discussions during these experiments.

REFERENCES

1. G. Binnig, H. Rohrer, Ch. Gerber and E. Weibel, Phys. Rev. Lett. 50, 120 (1983).
2. E. Ruska, Z. Physik 83, 492 1933.
3. P. E. Hoglund Nielsen and J. W. Cowley, Surface Science 54, 340 (1976).
4. Tung Hsu, Ultramicroscopy 11, 167 (1983); Tung Hsu and J. Cowley, Ultramicroscopy 11, 239 (1983).
5. J. M. Cowley, Surface Science 114, 587 (1982).
6. J. M. Cowley, J. of Microscopy, Vol. 129, 253 (1983).
7. N. Yamamoto and J. C. H. Spence, Thin Solid Films 104, 43 (1983).
8. W. Krakow, Surface Science 111, 503 (1981).
9. D. Cherns, Philosophical Mag. 30, 549 (1974).
10. G. Lehmpfuhl and K. Takayanagi, Ultramicroscopy 6, 195 (1981).
11. S. Iijima, Ultramicroscopy 6, 41 (1981).
12. K. Takayanagi, Japan Journ. Appl. Phys. 22, 1, L4 (1983).
13. N. Osakabe, Y. Tanishiro, K. Yagi and G. Honjo, Surface Science 109, 353 (1981).
14. N. Osakabe, Y. Tanishiro, K. Yagi and J. Honjo, Surface Science 102, 424 (1981).
15. K. Yagi, K. Takayanagi, K. Kobayashi, N. Osakabe, Y. Tanishiro and G. Honjo, Surface Science 86, 174 (1979).
16. K. Takayanagi, Proceedings of 10th International Congress on Electron Microscopy, Hamburg, p. 43 (1982) and Ultramicroscopy 8, 145 (1982).
17. N. Osakabe, Y. Tanishiro, K. Yagi and G. Honjo, Surface Science 97, 393 (1980).
18. J. C. H. Spence and K. Takayanagi, Ultramicroscopy (1983).
19. R. J. Wilson and P. M. Petroff, Rev. Sci. Inst. (1983).
20. R. C. Henderson, J. Electrochem. Soc. 6, 772 (1972).
21. M. Klaua and H. Bethge, Ultramicroscopy 11, 125 (1983).
22. S. P. Kowalczyk, D. L. Miller, J. R. Waldrop, P. G. Newman and R. W. Grant, J. Vac. Sci. Technol. 19, 225 (1981).
23. J. M. van Hove and P. I. Cohen, J. Vac. Sci. Technol. 20, 3, 726 (1982).
24. R. Z. Bachrach, R. S. Bauer, P. Chiaradia and G. V. Hansson, J. Vac. Sci. Technol. 18, 797 (1981).
25. P. M. Petroff and R. J. Wilson, Phys. Rev. Lett. 51, 3, 199 (1983).
26. J. C. Phillips, Phys. Rev. Lett. 45, 11, 905 (1980).
27. P. B. Hirsch, J. de Physique C6, 6, 40, 3 (1979).
28. E. G. McRae and P. M. Petroff, Surface Science, to be published (1984).
29. P. M. Petroff, A. C. Gossard, W. Wiegmann and A. Savage, J. Cryst. Growth 44, 5, (1978).
30. J. M. Van Hove and P. I. Cohen, J. Vac. Sci. Technol. B1, 3, 741 (1983).

INTRINSIC POINT DEFECTS AND DIFFUSION PROCESSES IN SILICON

T. Y. Tan*
IBM Thomas J. Watson Research Center, Yorktown Heights, NY 10598
*Present address: IBM general Technology Division, Hopewell Junction, NY 12533

ABSTRACT

   This paper reviews recent progress in understanding the role of vacancies (V) and self-interstitials (I) in self- and impurity diffusion in Si. Surface oxidation perturbs the thermal equilibrium concentration of point defects and analyses of the resulting effects on dopant diffusion showed that both V and I are present. Developments in experimental and theoretical works on Au diffusion in Si yielded a determination of the I-component and an estimate of the V-component of the Si self-diffusion coefficient. It is hoped that the I and V thermal equilibrium concentrations may be determined in the near future. A number of important physical aspects of the anomalous diffusion of P are now understood but a basically satisfactory model may need further work.

INTRODUCTION

   Intrinsic point defects govern self-diffusion and the diffusion of substitutional impurities in a crystal. By analogy to metals, vacancies have been regarded as the dominant thermal equilibrium point defect species in Si for a long time. However, it was argued by Seeger and Chik [1] in 1968 that Si self-diffusion at temperatures below about 800°C should be mainly due to vacancies (V) while above 800°C mainly due to self-interstitials (I), and by Hu [2] in 1974 that B should be diffusing in Si via an interstitialcy mechanism (I-mechanism) while As via a vacancy mechanism (V-mechanism). These arguments require that V and I coexist in Si at high temperatures under thermal equilibrium conditions. As indicated by the small self-diffusion coefficient, it is likely that the thermal equilibrium concentration of the point defects in Si is extremely smally (presumably $< 10^{-6}$) and hence conventional techniques for studying intrinsic point defects in thermal equilibrium (e.g., quenching and absolute measurement) failed to reveal the nature of Si thermal equilibrium point defects unambiguously. Fortunately, in the last few years progress in two areas showed beyond reasonable doubt that I and V indeed coexist in Si. This makes the earlier assumption that V is exclusively responsible for Si self-diffusion [3-9] no longer tenable. The first area concerns Au diffusion in Si. It was shown that an essential part of Si self-diffusion is carried by I [10,11], but there is little reason to assert that V also are contributing. A decision in favor of the coexistence of I and V came from the second area, namely that of the oxidation-enhanced and -retarded diffusion (OED and ORD) of group-III and -V dopants [12-15]. Subsequently, it was shown that Au diffusion is also consistent with the fact that I and V coexist [16]. In this paper we first discuss some essential points in these two areas and then show that a preliminary quantitative consistency between results obtained on Si self-diffusion coefficients from these two areas can already be obtained. We will also discuss some critical points of the P anomalous diffusion in Si.

BASIC CONSIDERATIONS

   Foreign atoms having no strong bonding interactions with the host crystal atoms and are hence located exclusively in interstices may jump directly between the interstices. This is presumably the mechanism responsible for the diffusion of H, He and some other noble gases in Si. Direct diffusion of substitutional foreign atoms or of host atoms on regular lattice sites involves the exchange of atoms on two or more neighbouring lattice sites (ring-mechanism). So far no example of this kind was found. By contrast, substitutional impurity and self-diffusion of atoms need intrinsic point defects as diffusion vehicles. The V-mechanism is known to control self-diffusion of all metals and Ge, and, as will be discussed, in Si below about 1000°C. In Si, however, the I-mechanism dominates self-diffusion above about 1000°C and plays a prominent role in the diffusion of the substitutional dopants P, B, Al and Ga.

It was shown that a consistent interpretation of the data on OED/ORD phenomena is not possible if either I or V alone were assumed to be present in Si under thermal equilibrium conditions [14]. Therefore, we now assume that both I and V are present and that both contribute to the dopant diffusivities. The Si self-diffusivity, measured with tracer atoms, may be written as

$$D^T = f_I D_I C_I^{eq} + f_V D_V C_V^{eq}, \qquad (1)$$

where D_I, D_V are the diffusivities of I and V respectively, and C_I^{eq}, C_V^{eq} the thermal equilibrium concentrations of I and V respectively in atomic fractions, and f_I and f_V the correlation factors. The diffusivity of a substitutional dopant in thermal equilibrium may be written as

$$D^S = D_I^S + D_V^S = h_I f_I D_I C_I^{eq} + h_V f_V D_V C_V^{eq}, \qquad (2)$$

where the factors h_I and h_V account for the interactions of the dopant atom with I and V respectively. If a perturbation to the thermal equilibrium I or V concentration occurs, it can not be generally assumed that C_I and C_V are independent of each other when a quasi-steady-state condition applies. Instead, we consider that local dynamical equilibrium between I and V is established via the reaction

$$I + V \rightleftarrows 0 \qquad (3)$$

(0 denotes the ideal lattice) which leads to [17,18]

$$C_I C_V = C_I^{eq} C_V^{eq}. \qquad (4)$$

OXIDATION EFFECTS

Oxidation of Si wafer surfaces can also lead to the generation of the oxidation-induced-stacking-faults (OSF) which are interstitial in nature [19]. Considering this together with some OED results, the correct interpretation is that oxidation injects I into Si [2,20]. By now reliable OSF size data were obtained [21-23] and, though still somewhat phenomenological, satisfactory analyses were also carried out [24,25]. Define the superation ratios of I and V respectively as $S_I = C_I/C_I^{eq} - 1$ and $S_V = C_V/C_V^{eq} - 1$ during the oxidation, we obtain [25]

$$\frac{\Omega}{A\alpha_{eff}} \left(\frac{dr_{SF}}{dt} \right) = -(D_I C_I^{eq} + D_V C_V^{eq}) \frac{\gamma}{kT} + D_I C_I^{eq} S_I - D_V C_V^{eq} S_V \qquad (5)$$

for the OSF with a radius r_{SF}. Here A is the area per atom in the fault (6.38×10^{-16} cm^2), Ω the atomic volume (2×10^{-13} cm^3) and γ the stacking fault energy (0.026eV/atom). The dimensionless quantity α_{eff} has a value of ≈ 4 and contains all factors related to interaction potentials between I and the Frank partial dislocation bounding the OSF. Empirically, for (100) Si wafers oxidized in dry O_2, the OSF sizes are fitted satisfactorily by [24,25]

$$r_{SF} = 1640 \exp(-\frac{2.5eV}{kT})t^{3/4} - \frac{4.86 \times 10^9}{kT} \exp(-\frac{5.02eV}{kT})t \quad cm. \quad (6)$$

We use Eq. (3) to relate S_I and S_V by $S_V = -S_I/(1+S_I)$. A comparison of Eqs. (4) and (5) then yields [13,15]

$$S_I = 6.6 \times 10^{-9} \exp(2.52eV/kT) t^{-1/4}. \quad (7)$$

Eq. (7) is a most important piece of information obtained from OSF studies: it is used to obtain quantitative fittings of OED/ORD data. The time averaged value of S_I, $\bar{S}_I$, is obtained by replacing the pre-exponential value with 8.8×10^{-9} in Eq. (7).

The diffusivity of a dopant under oxidation is changed to

$$D_{ox}^S = D_V^S(C_V/C_V^{eq}) + D_I^S(C_I/C_I^{eq}). \quad (8)$$

In term of the normalized diffusivity enhancement defined as $\Delta_{ox}^S = D_{ox}^S/D^S - 1$, the use of Eqs. (4) and (8) leads to [13,14]

$$\Delta_{ox}^S = (2G_I^S + G_I^S S_I - 1) S_I/(1+S_I), \quad (9)$$

where G_I^S is the fractional I-component of the dopant diffusivity under thermal equilibrium conditions defined as $G_I^S = D_I^S/D^S$. Eq. (9) applies to an instantaneous diffusivity enhancement. In experiments the time averaged value $\bar{\Delta}_{ox}^S$ is measured, but Eq. (9) still holds to a good approximation provided the time averaged S_I value is used. A plot of Eq. (9) for three G_I values is shown in Fig.1. The value of $\bar{\Delta}_{ox}^S$ is either positive or nagative (OED or ORD) depending on G_I and $\bar{S}_I$. In Fig.2 we show a fitting of the available Sb ORD data to Eq. (9). It is seen that the fitting is quite satisfactory on a quantitative basis with $G_I \approx 0.02$. This kind of good fitting indicates that the present model of I-V coexisting and attained a local equilibrium is correct, particularly in light of the fact that no other model can be equally satisfactory [14]. The use of Eqs. (7) and (9) allows to determine G_I^S for a dopant at a given temperature. In table 1 we compiled the G_I^S values for a few dopants at 1100°C [12,15, 26-29]. Using the data of Ishikawa et al. [29], Matsumoto et al. [30] found that $G_I^{AS} \approx 42 \exp(-0.54eV/kT)$, $G_I^P \approx 156 \exp(-0.67eV/kT)$ and $G_I^B \approx 860 \exp(-0.83eV/kT)$. Their results agree with those we listed in table 1.

In most oxidation experiments an I supersaturation ($S_I > 0$) is realized, but for very high temperature and thick SiO_2, especially for (111) wafers and for oxidation in O_2 containing a Cl compound, a V supersaturation ($S_I < 0$) may be realized [13,22,25,31]. This leads to ORD of dopants with $G_I > 0.5$ as has been observed for P [31], B [32], Al [28] and Ga [27]. On the other hand, a dopant with $G_I < 0.5$ (e.g., Sb with $G_I \approx 0.02$) should show OED, and this is indeed the case for Sb for an oxidation of (111) wafers at 1160°C for 18hrs [26].

DIFFUSION OF GOLD IN DISLOCATION-FREE SILICON

Gold atoms, which under thermal equilibrium may be incorporated either as substitutional (Au_s) or interstitial (Au_i) atoms, may diffuse via either the kick-out mechanism [10] or the Frank-Turnbull mechanism [33]. The two mechanisms have in common that long range transport of Au atoms occur via migration of Au_i atoms which may either jump directly from interstice to interstice positions or jump from interstice positions to lattice positions to become Au_s atoms. This is true because though the thermal equilibrium concentration of Au_s is much larger than that of Au_i the mobility of Au_i is much much larger than that of Au_s and hence long range transport of Au atoms due to Au_s may be neglected. However, the two mechanisms differ in the ways Au_i snd Au_s atoms interchange. In the kick-out mechanism the interchange involves I according to

$$Au_i \rightleftarrows Au_s + I. \tag{10}$$

This mechanism creates an I supersaturation in the crystal which is balanced by I out-diffusion. This means if the mechanism is operative I is involved in Si self-diffusion but it does not mean a priori that V is not involved in Si self-diffusion. In the Frank-Turnbull mechanism the interchange involves V according to

$$Au_i + V \rightleftarrows Au_s. \tag{11}$$

If operative, this mechanism creates a V undersaturation in the crystal interior which is balanced out by V in-diffusion. This would mean that V is involved in Si self-diffusion. We now assume that local dynamical equilibrium between I, V, Au_i and Au_s is established. This requires that at least two of the three reactions (3,10,11) are sufficiently fast. Under these conditions the nomalized Au_s concentration $C=C_s/C_s^{eq}$ to be measured in an experiment on the diffusion of Au into dislocation-free Si may be approximately described by

$$\partial C/\partial t = \frac{\partial}{\partial x}(D_{eff} \partial C/\partial x) \tag{12}$$

with the effective diffusion coefficient [16, 34-36]

$$D_{eff} = (C^{-2} + D_V C_V^{eq}/D_I C_I^{eq}) D_I C_I^{eq}/C_s^{eq}. \tag{13}$$

In Eq. (13) contributions to the Au_i and Au_s interchange process due to both I and V are accounted for. In deriving Eq. (13), the following approximations are used:

$$C_i = C_i^{eq}, \quad (C_i: Au_i \text{ concentration}) \tag{14a}$$

$$C_s^{eq} \geq C_V^{eq} \tag{14b}$$

and $$C_s \gg (C_s^{eq} C_I^{eq})^{1/2}. \tag{14c}$$

If only V and no I are present Eq. (13) yields a constant effective Au_s diffusivity [10,33]

$$D_{eff}^V = D_V C_V^{eq}/C_s^{eq}. \tag{15}$$

The same result holds if $D_I C_I^{eq}$ is not zero but among the reactions (3,10,11) only (11) operates sufficiently fast. If only I and no V are present Eq. (13) yields a strong concentration dependent diffusivity [10,33]

$$D_{eff}^I = C^{-2}(D_I C_I^{eq}/C_s^{eq}). \tag{16}$$

The same result holds if $D_V C_V^{eq}$ is not zero but among the reactions (3,10,11) only (10) operates sufficiently fast. As observed by Stolwijk et al. [37], Au concentration profiles after diffusion into dislocation-free Si at and above 800°C can not be described by the constant diffusity given by Eq. (15) but are satisfactorily fitted by the use of Eq. (16). Fig.3 shows a typical experimental profile due to Stolwijk et al.. It is clear that an erfc-function type profile as expected for D_V^{eff} failed to fit the experimental data while the use of D_I^{eff} is quite satisfactory. This shows that I is contributing to Si self-diffusion but does not mean V is not contributing for the reasons that: (i) it can not be certain that this is not only due to the fact that among reactions (3,10,11) only (10) operates efficiently; and (ii) the strong dependence of D_I^{eff} on C^{-2} effectively reduces the effect due to $D_V C_V^{eq}$ for most part of the experimental result by this factor (C<1 and hence $C^{-2} \gg 1$). Indeed, after it was decided that I and V coexist via OED/ORD studies [13,14], the contribution of V to the Au_s effective diffusivity was searched for. It was found that while almost perfect fitting were obtained for short time

diffusion data, small but observable deviations occured for the long time
diffusion data if Eq. (16) alone was used for the analysis. Morehead et al.
[16] have now used Eq. (13) to analyse the long time diffusion data. Not only
is a better fitting obtained, this also allowed them to determine that
$D_V C_V^{eq}/D_I C_I^{eq} \approx 1$ at $1000°C$ and ≈ 0.5 at $1100°C$. The quantity $D_I C_I^{eq}$ has been
determined by Stolwijk et al. [37]:

$$D_I C_I^{eq} = 914 \exp(-4.84eV/kT) \quad cm^2 sec^{-1}. \tag{17a}$$

We expect that at $T<900°C$ reaction (3) becomes ineffective and for still
lower temperatures reaction (10) also becomes ineffective. In that case only
reaction (11) will be operating and an erfc-function type profile of C_s
should show. Wilcox et al. [39] found such an Au diffusion behavior at $700°C$.
With their data we obtained that $D_V C_V^{eq} \approx 8.8 \times 10^{-22}$ $cm^2 sec^{-1}$ at $700°C$ [35]. This
together with the data of Morehead et al. [16] yields the rough estimate

$$D_V C_V^{eq} \approx 0.6 \exp(-4.03eV/kT) \quad cm^2 sec^{-1}. \tag{17b}$$

Eqs. (17a), (17b) are ploted in Fig.4 in which a data point on $D_V C_V^{eq}$
obtained by analysing the Ni precipitation behavior [35] of Kitagawa et al.
[40] is also shown. An important point is, as rough estimates, the formation
and migration entropies are now obtained from Eqs. (17a) and (17b) as
$\Delta S_V^f + \Delta S_V^m \approx 3k$ for V and $\Delta S_I^f + \Delta S_I^m \approx 10k$ for I. That is, while I are extended [1] V
seem are fairly point-like. Furthermore, Eqs. (17a), (17b) also roughly
bore out the D^{SD} values obtained from higher temperature experiments vs. that
obtained from lower temperature experiments [41-45]: above about $1000°C$ I
dominate while below $1000°C$ V dominate.

A PRELIMINARY EFFORT TO INTERPRET DATA ON OXIDATION EFFECTS AND GOLD DIFFUSION CONSISTENTLY

Eqs. (12) and (13) fit the Au diffusion experimental data well, which
means the conditions given in Eq. (14) have been fulfilled. Eq. (14c) is
very stringent and has to be satisfied since if it is not then Eqs. (12) and
(13) can not be obtained. Eq. (14b) is less restrictive in that Eqs. (12)
and (13) can always be obtained but for cases that $C_V^{eq}/C_S^{eq} \ll 1$ are not satisfied, the RHS of Eq.(13) needs to be multiplied by the factor $(1+C_V^{eq}/C_S^{eq})^{-1}$.
A practical tolerable limit is that $C_V^{eq} \approx C_S^{eq}$, since if $C_V^{eq} \gg C_S^{eq}$ is satisfied
then the mass action law (11) is violated [10] and the analysis will not be
valid. Stolwijk et al. [37] measured the total equilibrium concentrations of
Au from about $900°C$ to $1300°C$. This is practically C_S^{eq} since the equilibrium
concentration of Au_i is negligible. The expression

$$C_S^{eq} \approx 0.6 \exp(-1.6eV/kT) \tag{18}$$

describes their finding well. The use of Eqs. (14) and (18) yields the <u>upper
bounds</u> of C_I^{eq} and C_V^{eq} as

$$C_V^{eq} < 0.6 \exp(-1.6eV/kT) \tag{19a}$$

and $$C_I^{eq} < 0.006 \exp(-1.6eV/kT) \tag{19b}$$

from about $900°C$ to $1300°C$. The use of Eqs. (16), (17) and (19) yields now

$$D_V > 1 \exp(-2.43eV/kT) \quad cm^2 sec^{-1} \tag{20a}$$

and $$D_I > 1.53 \times 10^5 \exp(-3.24eV/kT) \quad cm^2 sec^{-1} \tag{20b}$$

for the same temperature range. One consequence of Eqs. (19) and (20) is that
$D_I \gg D_V$ and $C_V^{eq} \gg C_I^{eq}$ now hold.

On the other hand, attempts were also made with oxidation studies for
determining the quantity D_I and hence C_I^{eq}. Mizuo and Higuchi [46] performed

an experiment involving adjacent oxidized/unoxidized portions of (100) Si. They examined the horizontal spreading of dopant diffusion profiles from the oxidized into the unoxidized portions and obtained that $D_I(1100°C) \approx 1 \times 10^{-9}$ cm^{2}sec^{-1}. Mizuo and Higuchi have also performed experiments measuring the vertical range of the OED/ORD effects for Sb, P and B at 1100°C by oxidizing the wafer backsides while the dopants were situated at the wafer frontsides that were protected by pre-deposited Si_3N_4 films from oxidation [47]. We have fitted their data using some D values from 1×10^{-9} to 5×10^{-9} cm^{2}sec^{-1} with an apparent success [48]. More recently, Taniguchi et al. [49] performed a vertical range experiment similar to those of Mizuo and Higuchi except they examined the OSF sizes instead of dopant OED/ORD. They obtained

$$D_I \approx 8.6 \times 10^5 \exp(-4eV/kT) \quad cm^2 sec^{-1} \tag{21a}$$

and
$$C_I \approx 4.8 \times 10^{-4} \exp(-0.7eV/kT). \tag{21b}$$

Eq. (21a) yields at 1100°C that $D_I \approx 2 \times 10^{-9}$ cm^{2}sec^{-1}. Thus, the result of Taniguchi et al. [49] seems to agree with that of Mizuo and Higuchi [46] from lateral diffusion. The result of Taniguchi et al. also agrees with the result we obtained [48] from analysing the experimental results of Mizuo and Higuchi on the OED/ORD vertical range [47] if we choose to interpret those D values as D_I. These interpretations, however, contradict Eq. (20b) which requires that at 1100°C $D_I > 2 \times 10^{-7}$ cm^{2}sec^{-1}. In analysing their data, Mizuo and Higuchi [46] as well as Taniguchi et al. [49] assumed that I were the only diffusing species and consequently the D values obtained can only be interpreted as D_I. We need to point out that this is not the only way to interpret the data from those experiments. In the following we show that the alternative possibility that these values may be more properly interpreted as D_V offers a better consistency between the results derived from Au diffusion and oxidation studies. Ignoring the various absorption and generation terms for I and V other than that due to I-V recombination and Frenkel pair creation, the diffusion equations for I and V in the vertical range problem may be written as (the lateral range problem may be treated similarly)

$$\partial C_I/\partial t = D_I \partial^2 C_I/\partial x^2 - k(C_I C_V - C_I^{eq} C_V^{eq}), \tag{22a}$$

$$\partial C_V/\partial t = D_V \partial^2 C_V/\partial x^2 - k(C_I C_V - C_I^{eq} C_V^{eq}). \tag{22b}$$

The use of Eq. (4) in Eq. (22) yields as an approximate description either

$$\partial C_I/\partial t = D_{I,V}^{eff} \partial^2 C_I/\partial x^2 \tag{23a}$$

or
$$\partial C_V/\partial t = D_{I,V}^{eff} \partial^2 C_I/\partial x^2, \tag{23b}$$

with
$$D_{I,V}^{eff} = (D_I C_I^{eq} + D_V C_V^{eq})/(C_I^{eq} + C_V^{eq}) \tag{24}$$

holds. It is more proper to first interpret the D values obtained from oxidation studies [46,48,49] as $D_{I,V}^{eff}$ given in Eq. (24), and, there is no a priori reason to regard it either as D_I or D_V before a further assumption is made. The assigning of the measured D values by Mizuo and Higuchi [46] and Taniguchi et al. [49] as D_I assumes that $D_I C_I^{eq} >> D_V C_V^{eq}$ and $C_I^{eq} >> C_V^{eq}$ in our present treatment which is identical to their simple assumption that $C_V^{eq} \approx 0$. As has already been mentioned, this interpretation contradicts results obtained from Au diffusion studies. A second assumption, that $C_I^{eq} \approx C_V^{eq}$, leads to a similar consequence. A third possibility assumes that $C_V^{eq} >> C_I^{eq}$ holds and we obtain from Eq. (24) that

$$D_V = \alpha D_{I,V}^{eff}, \tag{25}$$

where $\alpha = (1 + D_I C_I^{eq}/D_V C_V^{eq})^{-1}$ which can be computed from Eq. (17). At 1100°C, taking $D_{I,V}^{eff} \simeq 3\times 10^{-9}$ cm^{2}sec^{-1} and $D_I C_I^{eq} \simeq 2 D_V C_V^{eq}$, we have $D_V(1100°C) \simeq 1\times 10^{-9}$ cm^2 sec^{-1} and $C_V^{eq} \simeq 7.5\times 10^{-7}$. No information on D_I and C_I^{eq} is obtained via the use of the assumption that $C_V^{eq} \gg C_I^{eq}$. The solubility of Au as given by Eq. (18) is $C_S^{eq}(1100°C) \simeq 8.1\times 10^{-7}$. Thus, as an order of magnitude estimate, the requirement that $C_S^{eq} \geq C_V^{eq}$, Eq. (14b), has been fulfilled. The present author favors the use of this last assumption for interpreting the oxidation effect results for an obvious reason: an elementary consistency then exist between the oxidation effect and Au diffusion results, while with the other assumptions this is not the case. Oxidation effects and Au diffusion are totally different kind of experiments and only upon the emergence of a consistent interpretation can we begin to believe that our knowledge of point defects in Si has become reliable. We further note that the results of Taniguchi et al. [49] contains the anomalies that Eq. (21) shows, as pointed out to them by the present author, that the entropy associated with I formation is -8k while that with migration is 18k. This unusual result raises some doubts about the accuracy of their experiment.

ON ANOMALOUS DIFFUSION PHENOMENA OF DOPANTS

Diffusion of group-III and -V dopants in Si shows several anomalous features. These features were quite thoroughly discussed in the literature [50-53]. To summarize, Willoughby [51] and Göesele and Strunk [54] mentioned that the most prominent features are: (i) The emitter-push effect. This is the extremely rapid diffusion of base dopant (e.g., B or Ga) of the doubly diffused npn transistor structures which results in an enhanced movement of the base-collector junction; (ii) The kink-tail structure of P diffusion profile. When in-diffusion of P is carried out using a high concentration source, a kink-tail structure of the diffusion profile results, which is conveniently designated as surface and tail regions, Fig.5a. In the tail region the P diffusion rate is much higher than expected for iso-concentration studies; (iii) Movement of buried layers. Buried dopant layers diffuse anomalously rapidly if the P concentration at the surface is high. As will be discussed later, there exists ample evidences that SiP precipitates form during P in-diffusion [55-62], we therefore need to amend the above list by (iv) During P in-diffusion, P concentration in the Si surface region exceeds its normal solubility limit. We consider that the following facts have been established: (i) The dopant diffusivity enhancement are caused by supersaturations of intrinsic point defects. The enhancement are much larger than that due to oxidations; (ii) The point defect supersaturations are related to high surface concentrations of P, and, in some cases of B [63,64] or As [65]; and (iii) Mechanisms involving dislocations in an essential manner, e.g., dislocation climb, are not the origin of the point defect supersaturations.

Since until recently it was widely believed that V are the dominant thermal equilibrium point defect species in Si at high temperatures, elaborate models [66-68] were put forth to explain the dopant anomalous diffusion based on calculating V supersaturations. The model of Fair and Tsai [68] was apparently successful in explaining quantitatively a number of observations, including the kink-tail P diffusion profile. A close examination of the theory would show that the same quantitative description could be arrived at by empirically fitting the diffusion profile by the use of a number of adjustable parameters among which the most important ones are two diffusivity values (one for the surface and one for the tail region). The previous assignment of the diffusion vehicle as V may be regarded as spurious for the reason that in a number of recent experiments it has been determined beyond reasonable doubt that there is a net supersaturation of I and not V below the P diffused layer [62,69-71]. The experiment of Hashimoto et al. [72], however, showed an enhanced OSF shrinkage rate in the high concentration layers of P, As and B. This provided an indication for the possibility that a V supersaturation exists in the diffused layers.

An extensive critique of the existing P diffusion models has been recently put forth by Hu et al. [73] who also proposed their own model to remedy the situation. They have written down some basic equations but not yet performed any calculations. In the following we discuss a few points that may be appended to the discussion of Hu et al.

The Reason of P In-Diffusion With A Concentration Exceeding Solubility Limit

There should be little doubt that P diffuses into Si at a concentration exceeding the 'solubility limit'. It is well known that a large portion of the surface region P is electrically inactive. The inactive P atoms are in the form of SiP precipitated out during the diffusion [61,62]. The precipitation phenomenon should be at least partially responsible for the generation of excess I below the surface region. Citing the fact that an explosion can produce gas pressure far exceeding the equilibrium pressure of a system as an analogous example, Hu et al. [73] argued that the 'nascent P atoms' with a 'fugacity' (chemical potential?) far exceeding that of a saturated solution are produced at the oxide-Si interface, and hence they enter Si in supersaturation. 'Nascent P elements' are produced, they argue , because the usual diffusion souce is a P doped oxide layer on the Si surface and P diffusion usually proceeds under oxidizing conditions, i.e, there are chemical reactions. However, their discussions were not detailed and clear enough to allow a distinction to be made between the possibilities: (i) By the use of the analogy to an explosion and the word 'nascent' they mean the P atoms gained the excess chemical potential due to the dynamical reason that these atoms are newly produced by a chemical reaction; or (ii) The use of the words 'nascent P element' implies the creation of a new phase of material, the elemental P, at the oxide-Si interface, which serves as the true diffusion source. If (i) is true then what we will discuss in the following amounts to proposing a reason different from that of Hu et al. for explaining the in-diffusion of P to a concentration exceeding the normal P solubility in Si. If (ii) is true then Hu et al. have envisaged the same mechanism we will discuss into some detail (this interpretation of their statements would, however, make their explosion analogy a kind of point-less).

We propose to examine the fact that P can diffuse into Si in excess of the normal solubility limit from the point of view of a phase equilibrium type of mechanism. For the two component system with constituents Si and P the Si rich α phase (P exist in solution in Si) can only coexist with the SiP phase in equilibrium for the practical temperature ranges we are concerned with. The normally known solubility limit of P in Si (in the α phase) is defined with respect to the equilibrium coexistence of the α phase and the SiP phase.   This usual definition of the P solubility in Si bore out the observations that when it is exceeded SiP precipitates form. Because of the use of a P doped oxide layer source in usual diffusion experiments, the Si material near the oxide-Si interface is not in equilibrium with SiP but rather with the P rich oxide for which the equilibrium does not naturally exist in the two component phase diagram of P and Si. The material in this artificial equilibrium with Si may be, e.g., regarded as just the P-rich oxide itself or a pure P phase (on atomic scale) created at the interface. For our present purpose it is not yet important to know what exactly this material is, so we simply call it the phase XP. Since XP is not SiP, its coexistence with Si requires that now the P solubility limit in Si near the interface be redefined with respect to the equilibrium coexistence of XP and Si which shoud be generally different from that obtained for the equilibrium coexistence of SiP and Si. Here we postulate that the solubility limit of P in Si derived from the equilibrium coexistence of XP and Si exceeds that for the equilibrium coexistence of SiP and Si. This requires that the relative free energy situation shown schematically in Fig. 6 hold between the different materials. Experiments can be carried out to check whether the present proposal has some merit: (i) The use of SiP as a surface source material. This experiment should show that P diffuses into Si at a concentration equal to the normal solubility limit; (ii) The use of a gaseous phase P as a source material, e.g., experiments with an encapsulatin consisting of side-by side P powders and Si wafers. If positive, then this is an indication that our proposal may indeed be correct. This is so since the gaseous P material is not SiP and we expect the P atoms enter Si at a concentration differing from the normal solubility limit. On the other hand, if negative, it does not immediately mean that our proposal is incorrect, since the solubility limit derived from the Si and gaseous P coexistence may happen to be equal to or very close to the normal solubility limit. In such an experiment care must be taken so as not to have either too high or too low a P gas presure since then either SiP will form on the Si wafer surface or there will not be a large enough source concentration which negates the experimental results. A solid P source deposited on top of the Si surface is not a good experimental arrangement since then SiP will form.

Descriptions of The P Concentration Profile

The kink-tail structure in the P diffusion profiles may be described by invoking two different types of mechanisms: (i) Two stream diffusion [50]; and (ii) Concentration dependent diffusivity. These are examined in the following.

(i) Two stream diffusion. Hu [50] proposed the use of this type of description. Not yet discussing the physical meaning, we can write

$$\partial C_A/\partial t = \partial(D_A \partial C_A/\partial x)/\partial x + k_2 C_B - k_1 C_A \quad (26a)$$

and

$$\partial C_B/\partial t = \partial(D_B \partial C_B/\partial x)/\partial x + k_1 C_A - k_2 C_B. \quad (26b)$$

Eq. (26) describes a hypothetical situation of having two physically distinguishable species A and B codiffusing into Si with the possibility that the physical identities of A and B are interchangeable via the kinetic reaction

$$A \underset{k_2}{\overset{k_1}{\rightleftarrows}} B. \quad (27)$$

The best way to see how the kink-tail type diffusion profile developes is to assume that A and B are totally different chemical species so that $k_1=k_2=0$ in reaction (27) and we are looking for $C_{AB}=C_A+C_B$. If now D_A and D_B have different enough values, the situation shown in Fig. 5b for the A and B profiles results and they add up to the kink-tail structure in Fig. 5a for C_{AB}. To invoke this kind of mechanism, we note that, though D_A and D_B may still be concentration dependent, the a priori reason for obtaining the kink-tail profile in C_{AB} is the fact that A and B are distinguishable entities and they are diffusing independently of each other, i.e., Fig. 5a and 5b are well arrived at by assuming that D_A and D_B are constants.

To apply this mechanism to P diffusion, of course, it can not be assumed that there exist two different kinds of P elements. Instead, one searches for two possibly different physical states that P atoms may exist in the Si lattice. As Hu et al. [73] have proposed, a natural first choice is to assume that (a) a portion of the P atoms occupies the substitutional sites (P_s) while the rest the interstitial sites (P_i). It needs to be further assumed that (b) conversion between P_s and P_i is very slow so that the use of $k_1 \simeq k_2 \simeq 0$ is a good approximation in using Eq. (26). This means that there is little P_i-P_s mutual conversion for the diffusion times involved. The necessity of this last assumption may be understood by examining the consequence if the P_i-P_s conversion is rapid. In that case the concentration ratio of P_i and P_s (A and B in reaction (27)) quickly reaches a constant value [74]

$$C_{Pi}/C_{Ps} = k_2/k_1 = k, \quad (28)$$

and Eq. (26) reduces to

$$\partial C_P/\partial t = \partial(D_P^{eff} \partial C_P/\partial x)/\partial x. \quad (29)$$

In Eq. (29) $C_P = C_{Pi} + C_{Ps}$ is the total P concentration and the effective diffusivity D_P^{eff} is given by

$$D_P^{eff} = (D_{Ps}+kD_{Pi})/(1+k), \qquad (30)$$

with D_{Ps} and D_{Pi} being the diffusivities of P_s and P_i respectively. Eqs. (29) and (30) do not yield a kink-tail structure in the diffusion profile of P without further assuming that D_P^{eff} is strongly concentration dependent. We further note that if the two fractions of P atoms are not physically distinguishable, then even if two or more diffusion mechanisms operate simultaneously, a kink-tail profile can not be obtained. Common wisdom indicates that atomically disolved P atoms in Si are electrically active, thus Hu et al. needed to further assume that (c) both P_s and P_i are shallow donors.

The model of Hu et al is supposed to be consistent with all known physical phenomena [73]. However, a consideration of the above discussed assumptions (a) to (c) raised some concerns of the basic soundness of the model: (a) Common wisdom also has it that there is but one shallow donor associated with P doping. While it is legitimate at this stage to postulate that P_i are also donors, it seems to be too coincidental that the donor is also a shallow one which happened to be at an energy level identical to that due to P_s; (b) High concentration P diffusion can induce the formation of a dislocation network to accomodate the Si lattice parameter change in the diffused layer due to the incorporation of P atoms to a large concentration. Analyses of such dislocation networks indicated that the P diffused layer has a smaller lattice parameter which is consistent with the knowledge that P atoms are on substitutional sites (the ratio of the covalent radii of P atoms to Si atoms is 0.94)[75]. In order to explain all the published diffusion profiles by the two stream mechanism, it needs to be also assumed that C_{Pi} in some cases can be equal to C_{Ps}. If this is true, it seems that by now there should exist some indication that the oppsite, i.e., the P diffused layer has a lattice parameter larger than that of the undiffused Si, is true; and (c) The requirement that P_i and P_s are converting into each other very slowly is a very stringent condition. Considering the diffusion times involved in a typical experiment, a simple estimate would show that the energy barrier against the $P_i \to P_s$ conversion has to be higher than 3eV and quite likely of the magnitude of 3.5 to 4eV. This is the same energy barrier preventing the interstitialcy pair P_i-Si or P_s-I from dissociation, and hence the pair needs to be regarded as very stable. If so, it is not easy to understand why P_i can diffuse very fast which is also required in the model.

Whether the model of Hu et al.[73] is basically sound and whether our present concerns about the model are needed can be checked experimentally: Diffusion of P can be carried out with reduced source strength so that the P concentration in the surface region is within the solubility limit but with the kink-tail structrure still produces. Such samples can be used to measure: (a) Electrically active P concentration vs. the total P concentration via, e.g., SIMS; (b) Donor levels and their concemtrations; and (c) Whether a substantial portion of P is not in substitutional sites via, e.g., nuclear reaction analysis (not performed for P up to now).

(ii) <u>Concentration dependent diffusivity</u>. All available models [66-68] other than that of Hu et al. belong to this category. The P diffusion is basically described by

$$\partial C_{Ps}/\partial t = \partial[D(C_{Ps})\partial C_{Ps}/\partial x]/\partial x, \qquad (31)$$

where $D(C_{Ps})$ has a strong dependence on C_{Ps} which is also the total P concentration. These models were not satisfactory for reasons briefly discussed here and more extensively by Hu et al. [73].

Fair and Tsai [68] were already successful in obtaining a quantitative fitting to some available data and the model of Hu et al. [73] appears also as capable of achieving a quantitative fit. These efforts may be viewed as yielding satisfactory phenomenological descriptions. However, most quantities that were supposed to be meaningful should now be regarded as fitting parameters with their physical meaning yet to be defined.

CONCLUDING REMARKS

As reviewed in this article, experimental and theoretical progresses in the last few years in the field of diffusion in Si yielded the following outstanding features:

(i) It is shown beyond reasonable doubt that I and V coexist in Si in thermal equilibrium and under oxidizing conditions at high temperatures.

(ii) Au diffusion phenomena in Si is satisfactorily explained and is consistent with the conclusion that I and V coexist. Au diffusion data allowed a determination of the quantity $D_I C_I^{eq}$ and an estimate of the quantity $D_V C_V^{eq}$ which demonstrated that I in Si is quite extended while V seems to be fairly point-like.

(iii) It appears to be hopeful to determine or estimate the quantities D_I, D_V, C_I^{eq} and C_V^{eq} separately.

(iv) Many physical aspects associated with P anomalous diffusion are now understood. A truely satisfactory quantitative model, however, is awaiting further work.

TABLE I. Fractional diffusivity G_I^s via self-interstitials for some group-III and -V dopants at 1100°C. The ratio of the radii r_s of the substitutional dopant and r_{Si} of a Si atom is also given.

	group-III dopants			group-V dopants		
	B	Ga	Al	P	As	Sb
r_s/r_{Si}	0.75	1.08	1.08	0.94	1.01	1.16
G_I^s	0.8-1.0	0.6-0.7	0.6-0.7	0.5-1.0	0.2-0.5	0.02
ref.	12	27	28	12,29	12,15,29	12,26

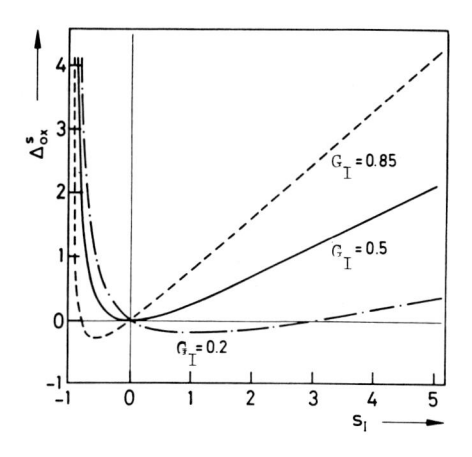

Fig. 1. Prediction of dopant diffusivity changes due to an oxidation. The model assumes that I and V coexist in Si in thermal equilibrium and in oxidation at high temperatures and that during the oxidation I and V attained local equilibrium.

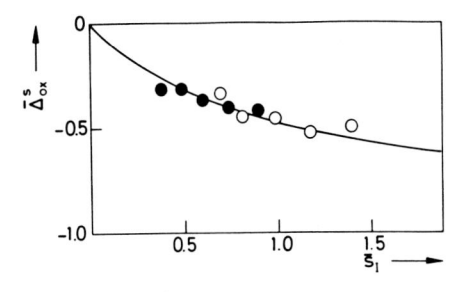

Fig. 2. Quantitative fitting of available Sb ORD data from (100) wafers to the calculated $G_I=0.02$ curve per Eq. (9). Unfilled circle: Mizuo and Higuchi [12]; filled circle: Tan and Ginsberg [26].

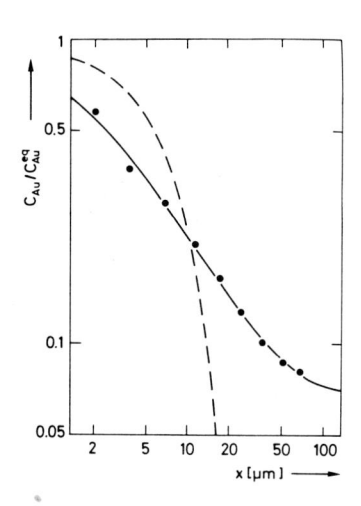

Fig. 3. Gold concentration profile after a 1hr diffusion into dislocation-free Si at $900°C$ [37]. Dash line: an erfc-function fitting per Eq. (15); solid line: fitting per Eq. (16).

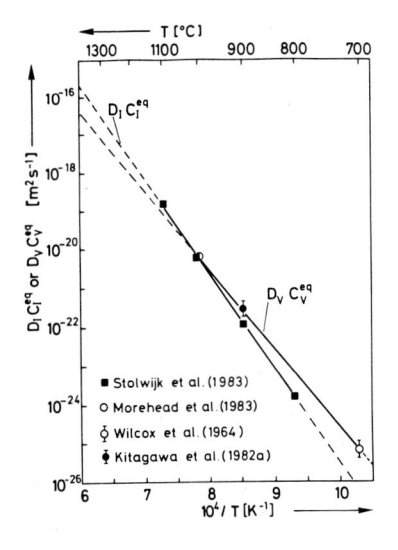

Fig. 4. $D_I C_I^{eq}$ and $D_V C_V^{eq}$ vs. reciprocal temperature. The quantity $D_I C_I^{eq}$ is due to Stolwijk et al. [37], and $D_V C_V^{eq}$ due to the data of Morehead et al. [16] and the estimate of Gosele and Tan [35].

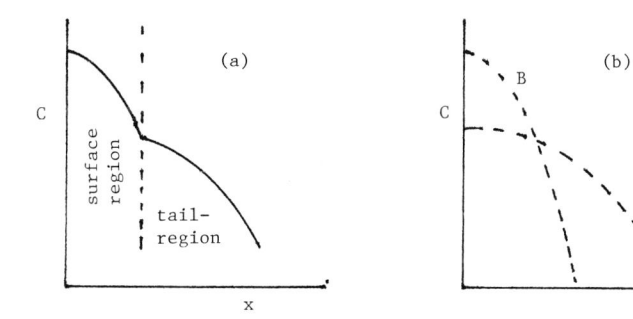

Fig. 5. (a) Schematic drawing of the kink-tail structure in P diffusion; and (b) Concentration profiles of physically distinguishable A and B atoms resulting from independent diffusion processes. The kink-tail profile in (a) may be obtained by adding the concentrations of A and B togather-- if there is a reason to do so.

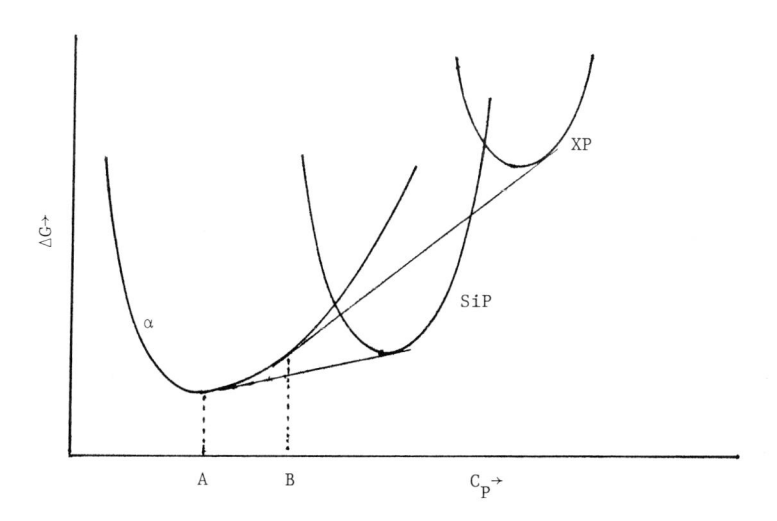

Fig. 6. A schematic diagram showing the possible relative free energy vs. composition situation between three phases: α-, SiP and an unknow XP phase. There is no natural equilibrium coexistence between α- and XP phases in the two component system of Si and P. The free energy minimum of XP is postualated to be higher than that of SiP so that when an artificial coexistence of XP and Si is created the P solubility limit in Si (α-) near the XP-Si interface exceeds that due to the equilibrium coexistence of SiP and Si. This is postulated as a reason for P diffusing into Si in excess of the normally known P solubility in Si which is defined with respect to the coexistence of SiP and Si (α-). A: P solubility limit in Si per coexistence of SiP and Si (α-) phases. B: P solubility limit in Si per coexistence of XP and Si (α-) phases.

REFERENCES

1. A. Seeger and K. P. Chik, Phys. Stat. Solidi 29, 455 (1968).
2. S. M. Hu, J. Appl. Phys. 45, 1567 (1974).
3. B. J. Masters, Sol. State Commun. 9, 283 (1971).
4. P. M. Petroff and A. J. R. de Kock, J. Cryst. Growth 30, 117 (1975).
5. D. Shaw, Phys. Stat. Solidi B72, 11 (1975).
6. R. B. Fair, in Semiconductor Silicon 1977, H. R. Huff and E. Sirtl eds. (Electrochem. Soc., Princeton, 1977) p. 968.
7. J. A. van Vechten, Phys. Rev. B17, 3197 (1978).
8. J. C. Burgoin and M. Lanoo, Rad. Effects 46, 157 (1980).
9. H. Kitagawa, K. Hashimoto and M. Yoshida, Jpn. J. Appl. Phys. 23, 2033 (1981).
10. U. Gösele, W. Frank and A. Seeger, Appl. Phys. 23, 361 (1980)
11. U. Gösele, F. Morehead, W. Frank and A. Seeger, Appl. Phys. Lett. 38, 157 (1981).
12. S. Mizuo and H. Higuchi, Jpn. J. Appl. Phys. 20, 739 (1981).
13. T. Y. Tan and U. Gösele, Appl. Phys. Lett. 40, 616 (1982).
14. T. Y. Tan, U. Gösele and F. Morehead, Appl. Phys. A31, 97 (1982).
15. D. A. Antoniadis and I. Moskowitz, J. Appl. Phys. 53, 9214 (1982).
16. F. Morehead, N. Stolwijk, W. Meyberg and U. Gösele, Appl. Phys. Lett. 42, 690 (1983).
17. S. Prussin, J. Appl. Phys. 43, 2850 (1972).
18. E. Sirtl, in Semiconductor Silicon 1977, H. R. Huff and E. Sirtl eds. (Electrochem. Soc., Princeton, 1977) p. 4.
19. G. R. Booker and W. J. Tunstall, Phil. Mag. 13, 71 (1966).
20. P. S. Dobson, Phil. Mag. 24, 567 (1971); 26, 1301 (1972).
21. S. M. Hu, Appl. Phys. Lett. 27, 165 (1975).
22. H. Shiraki, Jpn. J. Appl. Phys. 15,1 (1976).
23. S. P. Murarka, J. Appl. Phys. 48, 5020 (1977).
24. B. Leroy, J. Appl. Phys. 50, 7996 (1979).
25. T. Y. Tan and U. Gosele, J. Appl. Phys. 53, 4767 (1982).
26. T. Y. Tan and B. J. Ginsberg, Appl. Phys. Lett. 42, 448 (1983).
27. S. Mizuo and H. Higuchi, Denki Kagaku (J. Jpn. Electrochem. Soc.) 50, #4 (1982).
28. S. Mizuo and H. Higuchi, Jpn. J. Appl. Phys. 21, 56 (1982).
29. Y. Ishikawa, Y. Sakino, H. Tanaka, S. Matsumoto and T. Niimi, J. Electrochem. Soc. 129, 644 (1982).
30. S. Matsumoto, Y. Ishikawa and T. Niimi, J. Appl. Phys. 54, 5049 (1983).
31. R. Franscis and P. S. Dobson, J. Appl. Phys. 50, 280 (1979).
32. C. Hill, in Semiconductor Silicon 1981, H. R. Huff, R. J, Kriegler and Y. Takeishi eds. (Electrochem. Soc., Pennington, 1981) p. 988.
33. F. C. Frank and D. Turnbull, Phys. Rev. 104, 617 (1956).
34. U. Gösele and T. Y. Tan, in Aggregation Phenomena of Point Defects in Silicon, E. Sirtl and J. Gorissen eds. (Electrochem. Soc., Pennington, 1983) p. 17.
35. U. Gösele and T. Y. Tan, in Defects in Semiconductors II, S. Mahajan and J. W. Corbett eds. (North-Holland, NY, 1983) p. 45.
36. T. Y. Tan, F. Morehead and U. Sösele, in Defects in Silicon, L. C. Kimerling and W. M. Bullis eds. (Electrochem. Soc., Pennington, 1983) p. 325.
37. N. A. Stolwijk, B. Schuster, J. Hölzl, H. Mehrer and W. Frank, Physica 116B, 335 (1983).
38. W. Frank, A. Seeger and U. Gosele, in Defects in Semiconductors, J. Narayan and T. Y. Tan eds., (North-Holland, NY, 1981) p. 37.
39. W. R. Wilcox, T. J. LaChapelle and D. H. Forbes, J. Electrochem. Soc. 111, 1377 (1964).
40. H. Kitagawa, K. Hashimoto and M. Yoshida, Jpn. J. Appl. Phys. 21, 276 (1982).
41. J. M. Fairfield and B. J. Masters, J. Appl. Phys. 38, 3148 (1967).

42. H. J. Mayer, H. Mehrer and K. Maier, in Lattice Defects in Semiconductors 1976 (Inst. Phys. Conf. Series 31, London, 1977) p. 186.
43. L. Kalingnowski and R. Sequin, Appl. Phys. Lett. 35, 211 (1979); 36, 141 (1980).
44. I. R. Sanders and P. S. Dobson, J. Mat. Sci. 9,1987 (1974).
45. A. Hirvonen and J. Anttila, Appl. Phys. Lett. 35, 703 (1979).
46. S. Mizuo and H. Higuchi, Jpn. J. Appl. Phys. 21, 272 (1982).
47. S. Mizuo and H. Higuchi, J. Electrochem. Soc. 129, 2292 (1982); Jpn. J. Appl. Phys. 22, 12 (1983).
48. T. Y. Tan, U. Gösele and F. Morehead, (1982) unpublished.
49. K. Taniguchi, D. A. Antoniadis and Y. Matsushita, Appl. Phys. Lett. 42, 96 (1983).
50. S. M. Hu. in Atomic Diffusion in Semeconductors, D. Shaw ed. (Plenum, NY, 1973) p. 217.
51. A. F. W. Willoughby, J. Phys. D10, 455 (1977); Rep. Prog. Phys. 41, 1665 (1978).
52. A. F. W. Willoughby, in Impurity Doping Process in Silicon, F. F. Y. Wang ed. (North-Holland, NY 1981) p.1
53. A. Seeger, W. Frank and U. Gosele, in Defects and Radiation Effects in Semiconductors 1978 (Inst. Phys. Conf. Series 46, London, 1979) p.148.
54. U. Gosele and H. Strunk, Appl. Phys. 20, 265 (1979).
55. M. C. Duffy, F. Barson, J. M. Fairfield and G. H. Schwuttke, J. Electrochem. Soc. 115, 84 (1968).
56. P. F. Schmidt and R. Stickler, J. Electrochem Soc. 111, 1188 (1964).
57. T. W. O'Keefe, P. F. Schmidt and R. Stickler, J. Electrochem. Soc. 112, 878 (1965).
58. E. Kooi, J. Electrochem. Soc. 111, 1383 (1964).
59. R. J. Jaccodine, J. Appl. Phys. 39, 3105 (1968).
60. A. Armigliato, D. Nobili, M. Servidori and S. Solmi, J. Appl. Phys. 47, 5489 (1976).
61. D. Nobili, A. Armigliato, M. Finnetti and S. Solmi, J. Appl. Phys. 53, 1484 (1982).
62. R. J. Jaccodine, in Defects in Semiconductors II, S. Mahajan and J. W. Corbett eds. (North-Holland, NY, 1983) p.101.
63. J. E. Lawrence, J. Appl. Phys. 37, 4106 (1966).
64. C. L. Claeys, G. J. Declerck and R. J. van Overstraeten, Revue de Physique Appliquée 13, 797 (1978).
65. H. Shibayama, H. Masaki, H. Ishikawa and H. Hashimoto, J. Electrochem. Soc. 123, 743 (1976).
66. F. N. Schwettmann and D. L. Kendall, Appl. Phys. Lett. 19, 218 (1971); Appl. Phys. Lett. 20, 2 (1972).
67. M. Yoshida, Jpn. J. Appl. Phys. 19, 2427 (1980).
68. R. B. Fair and J. C. C. Tsai, J. Electrochem. Soc. 124, 1107 (1977).
69. C. L. Claeys, G. J. Declerck and R. J, van Overstraeten, in Semiconductor Characterization Technique, P. A. Barnes and G. A. Rozgonyi eds. (Electrochem. Soc., Princeton, 1978) p. 336.
70. A. Armigliato, M. Servidori, S. Solmi and I. Vechi, J. Appl. Phys. 48, 1806 (1977).
71. H. Strunk, U. Gösele and B. O. Kolbesen, Appl. Phys. Lett. 34, 530 (1979).
72. H. Hashimoto, H. Shibayama, H. Masaki and H. Ishikawa, J. Electrochem. Soc. 123, 1899 (1976).
73. S. M. Hu, P. Fahey and R. W. Dutton, J. Appl. Phys. Nov, 1983.
74. Actually the reactions involved in P_i-P_s convertion are (a) $P_i \rightleftarrows P_s + I$, and (b) $P_i + V \rightleftarrows P_s$, instead of the form of reaction (27). The use of reaction (27) in the discussion amounts to employing the assumption that values of C_I and C_V (not equal to C_I^{eq} and C_V^{eq}) resulting from reactions (a) and (b) are regarded as constants. This allows to elucidate in a simple and understandable manner of (i) what is a two stream diffusion, and (ii) why P_i-P_s convertion needs to be very slow.
75. W. F. Tseng, S. S. Law and J. W. Mayer, Phys. Lett. 68A, 93 (1978).

STRUCTURAL ANALYSES OF METAL/GaAs CONTACTS AND Ge/GaAs AND AlAs/GaAs HETEROJUNCTIONS

T. S. KUAN
IBM Thomas J. Watson Research Center
Yorktown Heights, New York 10598

ABSTRACT

Cross-sectional observations of interface structures at high magnification by TEM and STEM offer excellent lateral and depth resolution which are not available by other analytical techniques. This technique has recently been applied by the author to study metal/GaAs contacts and MBE-grown semiconductor heterojunctions. Specific observations on alloyed Au-Ni-Ge/GaAs and Pd/GaAs interfaces and on epitaxially grown Ge-GaAs and AlAs-GaAs superlattice structures are reviewed in this paper. The high spatial resolution of the energy dispersive x-ray analysis performed in a STEM is most useful in the studies of complicated metal/GaAs interfacial reactions. However, to resolve on an atomic scale the compositional profile across a heterojunction, one has to rely on diffraction technique. The detailed analyses of these interfacial reactions and structures are shown to be essential to the understanding and control of the electronic behavior associated with these interfaces.

INTRODUCTION

Recently, increasing attention has been drawn to the digital applications of GaAs FET devices. Two of these devices, the self-aligned metal-semiconductor field effect transistor (MESFET) [1] and the high electron mobility transistor (HEMT) [2] are shown in Fig. 1. To fabricate these devices in large-scale-integration, major improvement and tight control of material and process parameters are required. For instance, high quality, stable, and reproducible ohmic contacts at the source and drain are urgently needed for the MESFET. The ability of the gate contact to withstand high temperature annealing following the n+ implantation is also important. The HEMT contains a selectively doped AlGaAs/GaAs heterojunction fabricated by molecular beam epitaxy (MBE). To achieve high electron mobility and conductivity this device requires a sharp doping profile and an atomically abrupt AlGaAs/GaAs interface.

The cross-sectional TEM is most suitable for examining these metal/GaAs interfaces and semiconductor heterojunctions. To analyze the phases formed at metal/GaAs interfaces, special care is taken to optimize the sensitivity and spatial resolution of the energy dispersive x-ray analyses (EDX), which are performed in a VG HB5 STEM using an electron probe size of about 0.5 nm. Since probe spreading in a thin sample (with thickness about 20 nm) is small, chemical information can be collected from an area much less than the grain size of the reacted phase. The high-resolution, multibeam lattice images obtained by the JEOL JEM-200CX are useful for detecting lattice defects such as misfit dislocations at the interface. However, this technique is not sensitive to defects whose nature does not involve perturbations of the lattice (e.g., an antiphase boundary or an AlGaAs/GaAs interface). The most sensitive imaging mode for a particular defect structure can generally be chosen beforehand by comparing the defect images simulated for different imaging modes. To determine the compositional abruptness across a heterojunction is a difficult task. In this case the satellite reflections from a periodic superlattice structure can offer direct and reliable observations if the dynamical nature of the electron diffraction in the crystal is properly included in the data analysis.

In the following sections the analyses by EDX and various imaging and diffraction techniques of the alloying reactions at Au-Ni-Ge/GaAs and Pd/GaAs contacts and the structures of Ge/GaAs and AlAs/GaAs heterojunctions will be discussed.

144

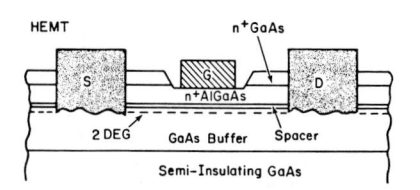

Fig. 1. Cross-sectional view of a self-aligned GaAs MESFET and a HEMT.

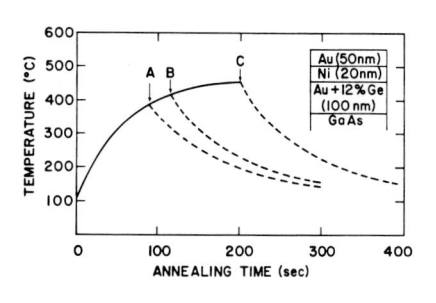

Fig. 2. Temperature vs. time curves of the three samples during and after the annealing.

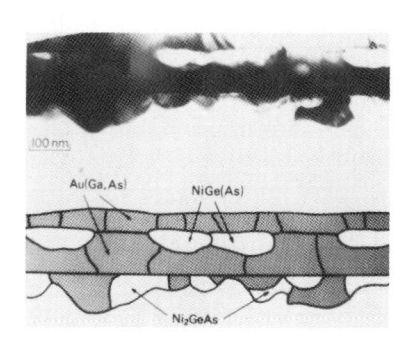

Fig. 3. Bright-field image of sample B.

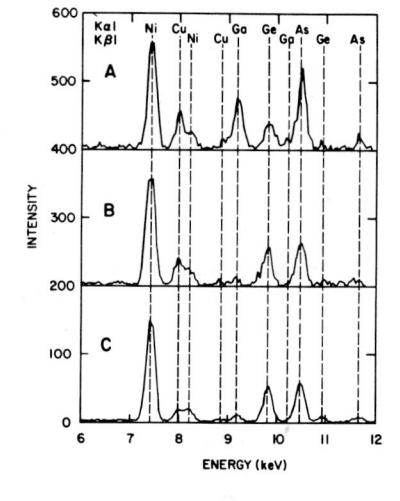

Fig. 4. Typical x-ray spectra from the NiAs phase in samples A, B, and C.

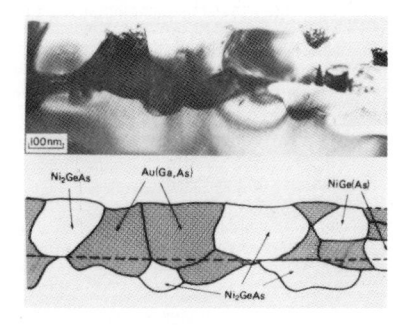

Fig. 5. Bright-field image of sample C.

Au-Ni-Ge OHMIC CONTACT TO GaAs

The resistance of the Ni-Au-Ge/GaAs ohmic contact is very sensitive to the annealing history of the contact and therefore its reliability and reproducibility are difficult to control. The alloying behavior of this contact has been observed previously by Auger electron spectroscopy, x-ray diffraction, and Rutherfold backscattering [3,4,5]. However, its details were not well understood. Cross-sectional TEM was thus used to examine closely the interface structures which resulted at different stages of alloying in hopes of finding the structural elements that govern the resistivity of the contact [6].

Observations were made on Au/Ni/Au-Ge/GaAs contacts before and after three different heat treatments in a forming gas (Ar plus H_2) ambient as shown in Fig. 2. In the (100) GaAs substrate an ion-implanted conducting channel was formed using 150 keV Si^{28} with a dose of $3 \times 10^{12}/cm^2$, resulting in a peak doping of $2 \times 10^{17}/cm^3$ at a depth of about 160 nm. As measured by a standard transmission line method, these contacts exhibit vastly different resistances after annealing. The resistivity of sample A is very high ($>10^{-4}$ ohm-cm^2) and some measurements find it to be non-ohmic. The resistivity of sample B, which has been annealed for a period of only 25 sec longer than sample A, is very low (about 1×10^{-6} ohm-cm^2), and the resistivity of sample C is about a factor of five higher than that of sample B.

At the as-deposited stage before any heat treatment, selected-area electron diffraction and x-ray spectroscopy detect no interactions, i.e., no compound phase formation, among the metal layers and GaAs. In sample A, however, some major reactions similar to those observed previously by AES and x-ray diffraction are observed, e.g., the formation of NiAs phase in the substrate and NiGe phase between the two Au layers, the penetration of Au into the substrate and out diffusion of mostly Ga and some As into the Au, NiGe, and NiAs phases. The contact remains very similar in overall structure after slightly longer annealing (sample B) except for slight increases in grain size and subtle changes in chemical compositions in the NiAs and NiGe phases. Fig. 3 shows a bright-field image of sample B with a schematic map underneath depicting all the phases identified by electron diffraction and x-ray analyses in the same sample area. Almost all the Ga dissolved in NiAs and NiGe is expelled from these phases and more Ge has diffused from the NiGe layer to the lower NiAs phase. This can be seen by comparing the Ga and Ge peak intensities in the x-ray spectra obtained from these two phases in samples A and B. The spectrum from the NiAs phase (Fig. 4) in sample B indicates that this phase contains about equal amounts of Ge and As, and the phase is in fact a ternary phase with composition close to Ni_2GeAs. Electron diffraction patterns indicate that the Ni_2GeAs phase has a hexagonal structure with lattice parameters very close to that of the NiAs phase (a = 0.360 nm, c = 0.501 nm) [7]. It is possible that the atomic structure of the ternary phase is identical to that of NiAs except that the As sites are evenly occupied by Ge and As atoms. Most of the Ni_2GeAs grains are epitaxially oriented with $(101)Ni_2GeAs||(001)GaAs$ and $[001]Ni_2GeAs||[\bar{1}11]GaAs$.

It is generally believed that a low resistance ohmic contact to n-type GaAs is achieved by the presence of a heavily doped n^+ layer at the metal/GaAs interface where the carrier transport occurs mainly through tunneling. Following this model, the high resistivity exhibited in sample A should be because most of the doping element Ge is still trapped in the NiGe layer, and no Ge is found in the Au phase. In sample B, the high concentration of Ge (about 25 wt%) in the Ni_2GeAs phase may facilitate the formation of a Ge-doped n^+ layer at the $Ni_2GeAs/GaAs$ interface and the measured high conductivity can be attributed to the good contact achieved at the Ni_2GeAs areas.

The contact after 200 sec of annealing is quite different in structure from the previous samples as shown by a typical image in Fig. 5 obtained from sample C. The grains have increased in size and the layered structures are no longer present. As a result more Au grains are in contact with the GaAs substrate. The reduction in the conductive $Ni_2GeAs/GaAs$ areas during this stage of alloying may account for the increase in contact resistivity. The penetration depth of the reacted phases is found to be less than 100 nm.

Since the Ni_2GeAs phase is agglomerated at the Au/GaAs interface, the measured contact resistance is therefore sensitive to the size and distribution of this phase. As discussed recently by Braslau [8], the measured resistance from a nonhomogeneous contact can be dominated by the spreading resistance due to the small size of the conductive areas. Based on our TEM observa-

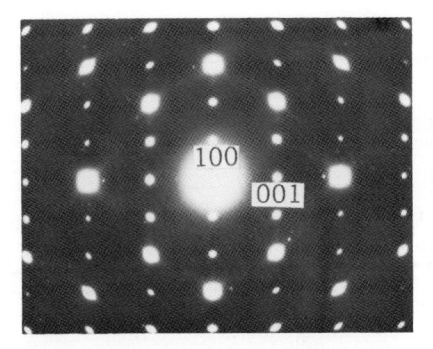

Fig. 6. Diffraction pattern from the Pd$_2$GaAs phase formed at room temperature on an atomically clean (100) GaAs surface.

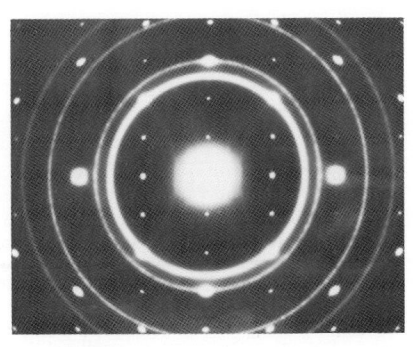

Fig. 7. Diffraction pattern from the unreacted Pd and the Pd$_2$GaAs phase formed at room temperature on an air-exposed (100) GaAs surface.

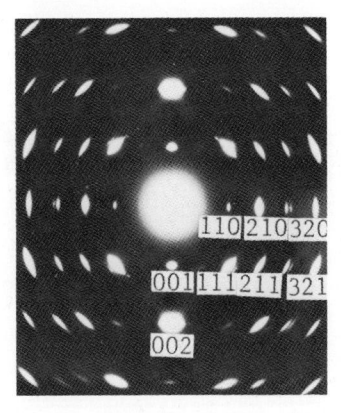

Fig. 8. Diffraction pattern from the Pd$_2$GaAs phase formed at 250°C on an atomically clean (110) GaAs surface.

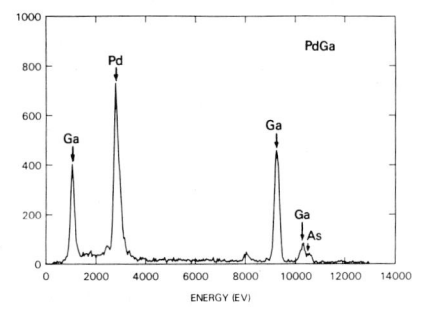

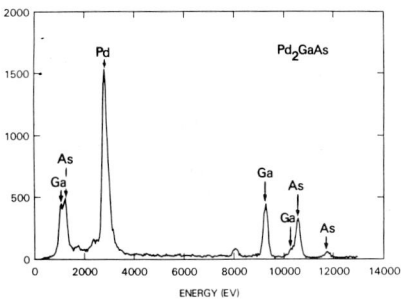

Fig. 9. X-ray spectra from the PdGa and Pd$_2$GaAs phases formed at 250°C on a (110) GaAs surface.

Fig. 10. Diffraction pattern from the PdGa phase formed epitaxially on an atomically clean (100) GaAs surface at 500°C.

tions, the deterioration of the Ni-Au-Ge contacts upon aging is related to the penetration and spreading of Au and the decrease of low-resistance Ni_2GeAs areas. Thus, to attain an ohmic contact with better reliability, one should investigate non-Au-based metallization schemes or non-alloyed contacts.

Pd/GaAs INTERFACIAL REACTIONS

Several binary phases resulting from Pd/GaAs reactions have been reported in the literature. Olowolafe et al. [9] observed by x-ray diffraction the formation of $PdAs_2$ and PdGa at 250°C, $PdAs_2$, Pd_2Ga, and PdGa at 350°C, and PdGa at 500°C in He or forming gas ($N_2 + H_2$) ambient. Oustry et al. [10] found by RHEED only $PdAs_2$ at 300°C and PdGa at temperatures higher than 400°C in 10^{-10} Torr vacuum. Zeng and Chung [11], however, reported x-ray results showing the formation of Pd_2Ga and PdGa at 250°C, $PdAs_2$, Pd_2Ga, and PdGa at 350°C and PdGa at 500°C in Ar ambient. They also reported that no $PdAs_2$ was found in 10^{-6} Torr vacuum at all temperatures. These somewhat inconsistent results might be due to the different GaAs surface condition, ambient, and Pd layer thickness used in each study. The purpose of our TEM study is to determine if the GaAs surface condition and the ambient can actually affect the reactions [12,13].

Our TEM observations indicate that a 15-nm-thick Pd film reacts completely on an atomically clean (100) GaAs surface at room temperature and forms a new ternary phase in 10^{-10} Torr vacuum (Fig. 6). This ternary phase is epitaxially oriented on (100) substrate and is stable up to 250°C. If the same Pd layer is deposited onto an air-exposed (100) GaAs surface (and thus covered with a thin native oxide layer), then only part of Pd is reacted to form the same ternary phase as shown in Fig. 7. Further annealing in vacuum, forming gas (Ar plus H_2), or pure Ar at 250°C would complete the reaction and a uniform ternary phase layer results as in Fig. 6. The same ternary phase also forms on atomically clean (110) GaAs surfaces and in this case the ternary phase is not epitaxial to the substrate but maintains a textured orientation (Fig. 8). On the (110) surface, a small amount of PdGa phase embedded in the ternary phase is also found at 250°C. The EDX analyses on the ternary phase and the PdGa phase in the same film indicate that the composition of the ternary phase is close to Pd_2GaAs (Fig. 9). The diffraction patterns from the ternary Pd_2GaAs phase can be indexed as from a hexagonal unit cell with $a_o = b_o = 0.672$ nm and $c_o = 0.340$ nm. The epitaxial relationship with the (100) GaAs substrate is: $(1\bar{2}0)Pd_2GaAs$ || $(100)GaAs$; $[001]Pd_2GaAs$ || $[011]GaAs$.

At 350°C the 15-nm-thick Pd layer reacts with GaAs in vacuum to form PdGa phase. In Ar or forming gas ambient some Pd_2Ga and $PdAs_2$ phases are also formed. At 500°C only the PdGa phase is formed. This phase is randomly oriented on (110) substrate and epitaxially oriented on (100) substrate. In the later case the relationship $(110)PdGa$ || $(100)GaAs$; $[\bar{1}11]PdGa$ || $[011]GaAs$ is found as shown in Fig. 10.

In the Pd/Si system the crystal orientation of Pd_2Si formed is sensitive to the Si surface condition [14]. The reaction of Pd on GaAs is more complicated in that the kinetics of various Pd-Ga and Pd-As reactions occurring concurrently during the annealing can be affected not only by the substrate orientation and surface oxide but also the ambient and the thickness of the Pd layer.

Ge-GaAs INTERFACE

Among all semiconductor heterojunctions the Ge/GaAs system is particularly interesting because of its nearly perfect lattice match and different band gaps. A periodic superlattice consisting of alternating Ge and GaAs layers is ideal for studying the interface structure and its dependence on growth parameters [15,16].

Fig. 11 shows a typical cross-sectional, bright-field image of a (100) oriented Ge/GaAs superlattice grown at 400°C, taken with the incident electron beam exactly parallel to the <110> direction normal to the growth axis. The electron diffraction pattern from the same superlattice indicates that the superlattice has a single crystal structure. The high contrast observed between the Ge and the GaAs layers in Fig. 11 is mainly due to the variations in sample thicknesses

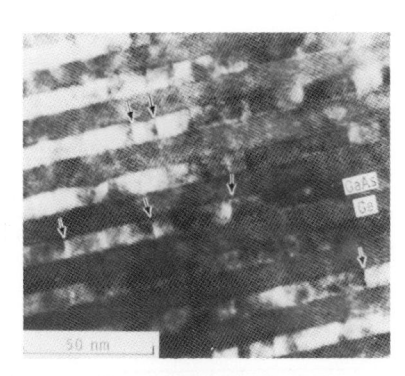

Fig. 11. Cross-sectional, bright-field image of a Ge(10 nm)-GaAs(8 nm) superlattice.

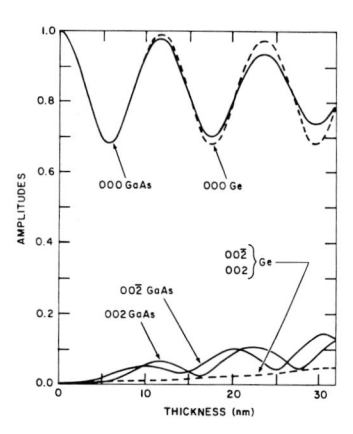

Fig. 12. Calculated thickness dependence of the amplitudes of the 000, 002, and 00$\bar{2}$ reflections from GaAs and Ge crystals.

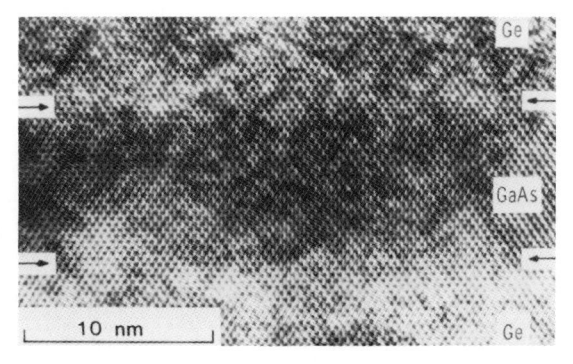

Fig. 13. 13-beam lattice image showing the perfect lattice match at the Ge/GaAs interfaces and no lattice irregularities in both GaAs and Ge layers. The approximate positions of the interfaces are indicated by arrows.

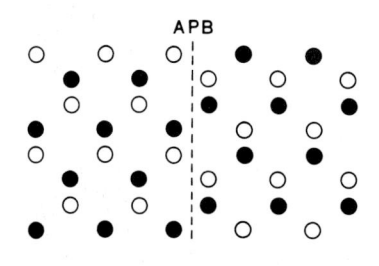

Fig. 14. Schematic diagram of a GaAs crystal containing an antiphase boundary.

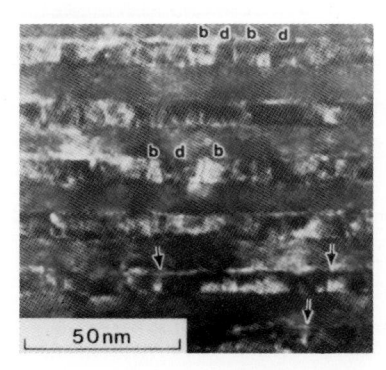

Fig. 15. Cross-sectional, 002 dark-field image of the same superlattice as in Fig. 11.

resulting from preferential etching during the ion-milling of the sample. As can be seen from the calculated transmitted beam amplitude in the (110) pattern from GaAs or Ge (Fig. 12), a slightly thinner GaAs (e.g., 5 nm thinner) can give rise to as much as a factor of two difference in bright-field image intensities between Ge and GaAs.

In the bright-field image (Fig. 11) contrast features indicating the presence of planar defects lying roughly parallel to the growth axis are always observed in the GaAs layers but not in the Ge layers. These planar defects are most likely antiphase boundaries (APBs), since in a 13-beam lattice image taken from the same superlattice no lattice irregularities in the GaAs layers are observed (Fig. 13). As shown schematically in Fig. 14, these APBs can result from the coalescence of GaAs domains independently grown on the Ge surface. The crystal on both sides of an APB are related to each other by a 180° rotation along any of the six equivalent <110> directions, and not by a lattice displacement. Therefore, they cannot be created or annihilated by movement of dislocations. Due to the lack of two-fold axes along the <110> direction in the GaAs structure, the intensities of the 002 and the $00\bar{2}$ reflections from GaAs are not the same, as calculated in Fig. 12. Since the 002 reflection from one side of a APB is equivalent to the $00\bar{2}$ from the other side of the APB according to the rotational symmetry, a change in the (002) dark-field image intensity is expected across an APB. This intensity change is actually observed in a 002 dark-field image taken from the same superlattice (Fig. 15). Here the alternating brighter and darker image intensities observed from neighboring domains in the GaAs layers are indicated by letters "b" (for brighter) and "d" (for darker) above some domains.

A high density of APBs in the GaAs layers can be associated with the GaAs surface roughness during the growth of the layer. The strain field associated with an APB in a thin GaAs layer constrained by the neighboring Ge layers can also enhance the Ge-GaAs interdiffusion. When the layers of the superlattice is only a few monolayers thick, the APBs in the GaAs layers can significantly affect the interface morphology and the growth pattern of the superlattice.

AlAs/GaAs INTERFACE

Interfaces in AlAs/GaAs superlattices have previously been analyzed using mainly bright-field or dark-field imaging modes [17]. In this work it is shown that more reliable analysis of the abruptness of the interface can be achieved by diffraction mode [18]. A cross-sectional, bright-field image of an (100) oriented AlAs(2 nm)/GaAs(2 nm) superlattice grown by MBE at 700°C is shown in Fig. 16. This image is taken with the electron beam exactly parallel to the <110> direction normal to the growth axis. Unlike the Ge/GaAs superlattice in Fig. 11, no APBs are observed in either GaAs or AlAs layers. The ion-milled sample surface is smoother than that of Ge/GaAs, and the image contrast observed between the AlAs and GaAs layers is predominantly diffraction contrast. The electron scattering behavior of AlAs along the <110> direction is quite different from that of GaAs as calculated in Fig. 17. When the crystal thickness is smaller than about 5 nm, the bright-field contrast between AlAs and GaAs is small. As the thickness increases, the bright-field image intensity from AlAs becomes smaller than GaAs until the thickness exceeds 30 nm, where the intensity from AlAs becomes higher than GaAs. This contrast reversion between the AlAs and GaAs layers is observed in Fig. 16. Even though the interfaces in the superlattice appear to be fairly abrupt in bright-field and dark-field images, details about the compositional profile of the interface cannot be derived from the images, since they are generally very sensitive to the imaging condition, sample thickness, surface smoothness and oxide contamination, etc. Lattice imaging can offer even less information about the chemistry at the interface.

Since the layer thickness in this superlattice is small, the satellite reflections from the superlattice can clearly be observed far away from the Bragg reflections (Fig. 18). The intensities of these satellites are very sensitive to the compositional profile of the superlattice. To investigate the effect of the interface abruptness on the distribution of satellite intensities, multislice calculations are carried out for different interface profiles shown in Fig. 19. The dynamical interactions between the Bragg reflections and the satellites are included in the calculation by using 16,384 beams and periodic extension of a large unit cell with two AlAs/GaAs interfaces in each cell. The chemical profile of the interface used in the calculation (Fig. 19) is derived from [19]:

$$c(x,t) = \frac{1}{2} \sum_{n=0}^{\infty} (-1)^n \{ erf[\frac{(2n+1)l - x}{2\sqrt{Dt}}] + erf[\frac{(2n+1)l + x}{2\sqrt{Dt}}]\},$$

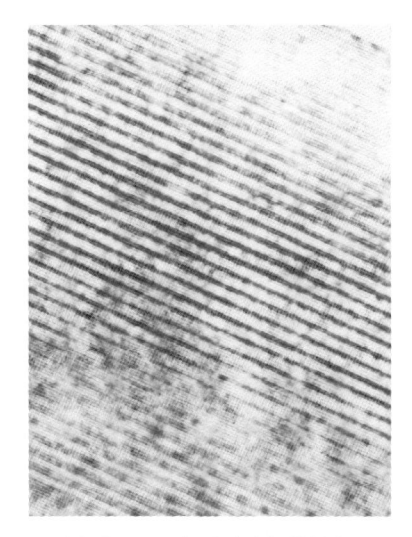

Fig. 16. Cross-sectional, bright-field image of an AlAs(2 nm)/GaAs(2 nm) superlattice.

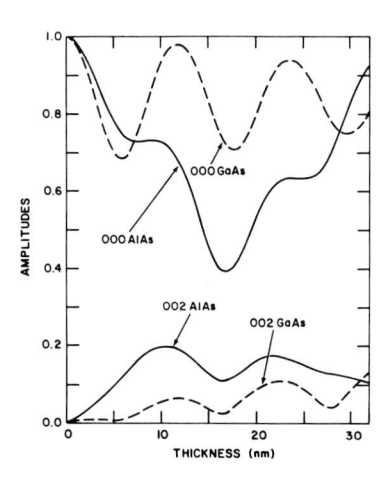

Fig. 17. Calculated thickness dependence of the amplitudes of the 000 and 002 reflections from AlAs and GaAs.

Fig. 18. Diffraction pattern with the satellite reflections from the same superlattice as in Fig. 16.

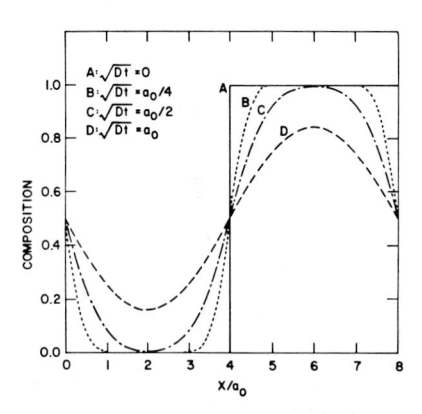

Fig. 19. Compositional profiles of the AlAs/GaAs interfaces in the superlattice used in satellite intensity calculations.

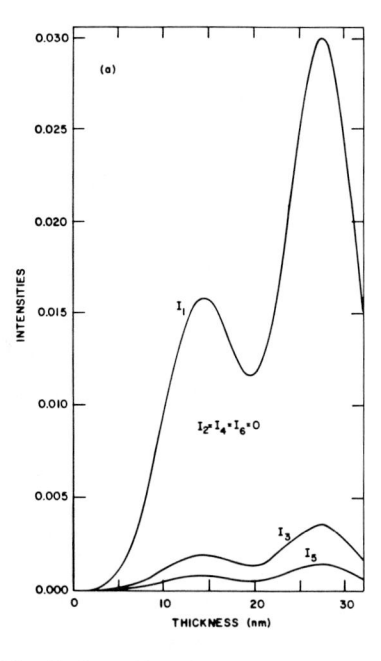

Fig. 20. Intensities of satellite reflections from the superlattice with different interface profiles shown in Fig. 19.

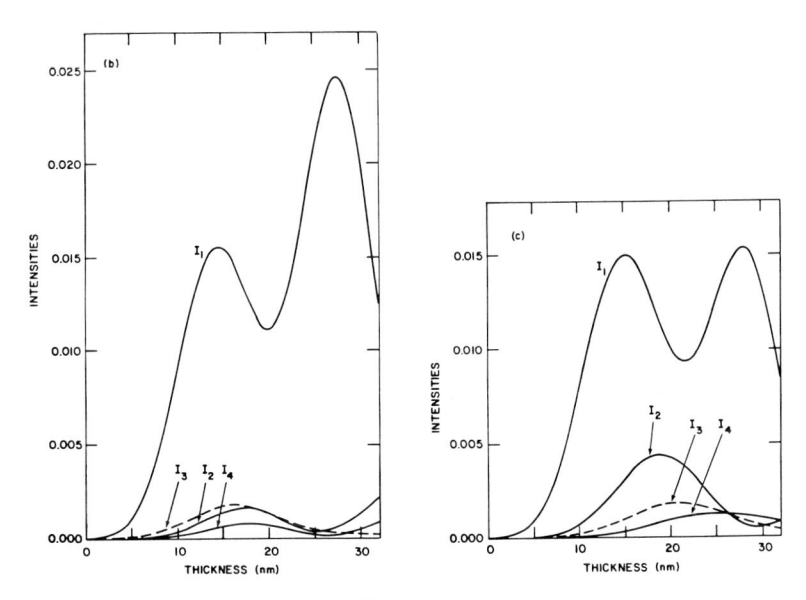

Fig. 20. (continued)

where l is a half of the layer thickness and D the diffusion coefficient. The intensities of the satellites from the superlattices with interface profiles A, B, and C of Fig. 19 are plotted in Fig. 20(a), (b), and (c), respectively. For atomically abrupt interfaces (i.e., atomically flat interfaces without any interdiffusion as described by profile A), the even-order satellites (i.e., second-order, fourth-order, . . . etc.) all have zero intensities (Fig. 20(a)). As the interfaces diffuse from profile A to B, the intensity of the second-order satellite increases from zero to close to that of the third-order (Fig. 20(b)). More spreading of the interfaces will raise the second-order and lower the third and higher order intensities further as shown in Fig. 20(c). If the interfaces are more diffused than profile D, then the calculated third and higher order satellites will become very weak and difficult to detect from a thin sample. By comparing the observed satellite intensities in Fig. 18 to the calculated ones in Fig. 20, we can conclude that the abruptness of the AlAs/GaAs interfaces grown at 700°C is closest to that described by profile C.

The observed spread of the AlAs/GaAs interfaces in this case is mostly due to the slight variations in layer thicknesses and to interface roughness since the surface of the layers (especially AlAs) is probably not atomically smooth during the growth. The interdiffusion between AlAs and GaAs during the short growth time (about 20 min.) at 700°C is believed negligible.

CONCLUSIONS

TEM observations on Au/Ni/Au-Ge and Pd contacts to GaAs, and on Ge/GaAs and AlAs/GaAs superlattices are discussed. It is shown that the reactions occur at the metal/GaAs interfaces are rather complicated and the resultant structures are sensitive to the thermal history, the substrate surface condition, and reaction ambient. In the case of Au/Ni/Au-Ge contact, a correlation between the contact resistance and the interface structure is found. Cross-sectional observation in conjunction with EDX analysis using fine probe in a STEM offer the high spatial resolution necessary for these investigations.

The MBE grown Ge/GaAs and AlAs/GaAs superlattices are ideal structures for interfacial studies. The defect structures and the growth mechanism behind them in the Ge/GaAs system is examined by a combination of bright-field, dark-field, and lattice imaging techniques. The intensities of the satellite reflections from the AlAs/GaAs superlattice are found to be extremely sensitive to the abruptness of the AlAs/GaAs interfaces. From these satellite reflections, the chemical profile at the AlAs/GaAs heterojunction can be derived with atomic level resolution.

ACKNOWLEDGMENTS

The author would like to thank P. E. Batson, C. A. Chang, W. I. Wang, J. L. Freeouf, T. N. Jackson, H. Rupprecht, and E. L. Wilkie for cooperations in these projects and N. Braslau, R. Ghez, D. A. Smith, and T. N. Theis for helpful discussions.

REFERENCES

1. N. Yokoyama, T. Mimura, M. Fukuta, and H. Ishikawa, 1981 IEEE International Solid-State Circuits Conference Digest of Technical Papers, pp. 218.
2. T. Mimura, K. Joshin, S. Hiyamizu, K. Hikosaka, and M. Abe, Jpn. J. Appl. Phys. **20**, L598 (1981).
3. G. Y. Robinson, Thin Solid Films **72**, 129 (1980).
4. M. Ogawa, J. Appl. Phys. **51**, 406 (1980).
5. M. Wittmer, R. Pretorius, J. W. Mayer, and M-A. Nicolet, Solid-State Electron. **20**, 433 (1977).
6. T. S. Kuan, P. E. Batson, T. N. Jackson, H. Rupprecht, and E. L. Wilkie, J. Appl. Phys. **54**, 6952 (1983).
7. R. W. G. Wyckoff, *Crystal Structures*, Second Edition, Vol. 1, (Interscience Publishers, Inc., 1963) pp .122.
8. N. Braslau, J. Vac. Sci. Technol. **19**, 803 (1981).
9. J. O. Olowolafe, P. S. Ho, H. J. Hovel, J. E. Lewis, and J. M. Woodall, J. Appl. Phys. **50**, 955 (1979).
10. A. Oustry, M. Caumont, A. Escaut, A. Martinez, and B. Toprasertpong, Thin Solid Films **79**, 251 (1981).
11. X. Zeng and D. D. L. Chung, J. Vac. Sci. Technol. **21**, 611 (1982).
12. P. Oelhafen, J. L. Freeouf, T. S. Kuan, T. N. Jackson, and P. E. Batson, J. Vac. Sci. Technol. **B1**, 588 (1983).
13. T. S. Kuan, J. L. Freeouf, P. E. Batson, and E. L. Wilkie (to be published).
14. T. S. Kuan and J. L. Freeouf, Proceedings of the 37th Electron Microscopy Society of America Annual Meeting, pp. 696 (1979).
15. P. M. Petroff, A. C. Gossard, A. Savage, and W. Wiegmann, J. Cryst. Growth **46**, 172 (1979).
16. T. S. Kuan and C. A. Chang, J. Appl. Phys. **54**, 4408 (1983).
17. P. M. Petroff, J. Vac. Sci. Technol. **14**, 973 (1977).
18. T. S. Kuan, W. I. Wang, and E. L. Wilkie (to be published).
19. H. S. Carslaw and J. C. Jaeger, *Conduction of Heat in Solids*, (Oxford University Press, Fair Lawn, N. J., 1959) pp. 97.

STRUCTURE OF THERMALLY-INDUCED MICRODEFECTS IN CZOCHRALSKI SILICON

F. A. PONCE*)
Hewlett-Packard Laboratories, Palo Alto, California 94304.

S. HAHN
Siltec Corporation, Mountain View, California 94043

ABSTRACT

The process of oxygen precipitation in Czochralski silicon materials has been studied using high resolution transmission electron microscopy. The resulting structure depends strongly on the thermal history of the material. The initial stages of precipitation involve the formation of clusters exhibiting strain fields which are coherent and isotropic at intermediate temperatures (~700°C). Incoherent defects are formed when the interstitial oxygen precipitates into substitutional sites in the silicon lattice. For long-time anneals, the quasi-equilibrium defect structure ranges from needle-like coesite (450-600°C), silica platelets (600-1000°C) to polyhedral silica precipitates (900-1200°C).

INTRODUCTION

The Czochralski (CZ) technique is widely used to grow silicon crystals for various integrated circuit technologies. Immediately following the growth process, these materials exhibit very few defects, if any, when examined using X-ray topography or chemical etching techniques. During subsequent thermal anneals, microdefects are generated, as shown in Figure 1. Due to their characteristic spiral (swirl-like) distribution in planes perpendicular to the growth direction, these defects are known as *swirl defects* in analogy with similar patterns observed in float zone (FZ) materials [1,2]. The origin of these defects, however, need not be the same in CZ as in FZ materials. The structure associated with the precipitation of oxygen in silicon has been studied using conventional transmission electron microscopy (TEM) [3-5]. Recently, high resolution TEM has been used to directly image the lattice structure of these microdefects [6-8]. In this paper, we report on the structure of thermally induced microdefects associated with swirl-like patterns in p-type, (100) grown CZ silicon materials which has been thermally annealed at various temperatures and for various time intervals.

EXPERIMENTAL PROCEDURE

The material under study was grown by the Czochralski method in a ⟨100⟩ direction, doped with boron (2.2 to 3.6×10^{14} atoms/cm^3, 38-63 Ω·cm). The top portion of a 10cm-diameter ingot was selected for our studies. The thermal history was closely monitored and no post-growth heat treatments were performed before the treatments reported here. {100} wafers and {110} slugs with thicknesses of 0.5 and 1.0 mm, respectively, were used. The interstitial oxygen and carbon contents were measured by Fourier transform infrared spectroscopy. The oxygen content of the as-grown material was 1.7×10^{18} cm^{-3} or 34 parts per million (ppm) in the axial center of the crystal. The substitutional carbon content was below the detection limit of the technique (<0.5 ppm). Further experimental details are given elsewhere [6].

---

*) Present address: Xerox Palo Alto Research Center, Palo Alto, CA 94304, USA.

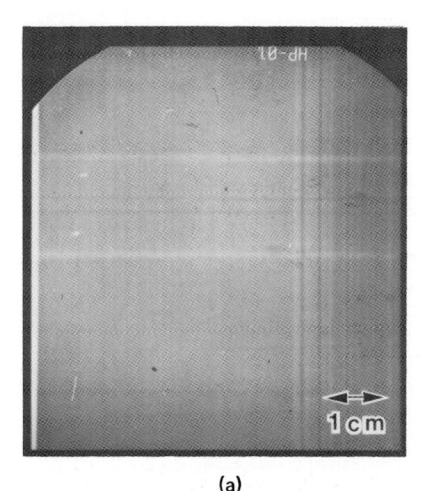

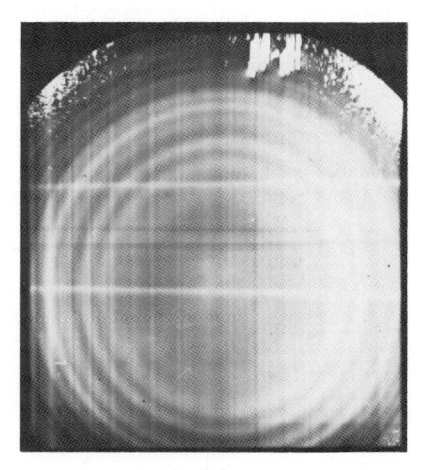

Fig. 1. X-ray transmission topographs of CZ-silicon wafers, using a {220} reflection: (a) as-grown, (b) after heat treatment at 750°C for 100 hrs.

INITIAL AND INTERMEDIATE STAGES OF OXYGEN PRECIPITATION

The microstructure associated with oxygen precipitation depends strongly on the thermal history of the material. For an isothermal process, oxygen precipitation occurs in various stages. We have studied thermal anneals in a O_2 ambient at about 700°C. The initial stages involve the formation of complexes with spherical shape. These are isotropic and coherent clusters which, when observed under two-beam conditions in the transmission electron microscope, exhibit Ashby-Brown contrast [9]. One such microdefect is shown in Fig. 2. This particular material had undergone an anneal at 700°C for 16 hours. Due to the isotropic nature of the defect, the shape of the image rotates with the diffracted beam used in the two-beam image.

After longer anneals at 700°C (~30 hrs), these clusters are no longer isotropic and they exhibit incoherent defects such as small platelets and stacking faults as seen in Fig. 3. A high magnification picture of a typical fault is shown in Fig. 4, which corresponds to the stage of formation of precipitates and interstitials which condense along {111} planes. After still longer anneals, the appearance of {100} precipitates together with {111} extrinsic stacking faults is observed. This is shown in Fig. 5 which corresponds to a heat treatment at 750°C for 100 hours with a drop in interstitial oxygen content from 34 down to 26 ppm [6]. This produces a defect density of about 5×10^{12} cm^{-3} (Figure 6).

These observations indicate that in the initial stages of microdefect formation, oxygen condenses into interstitial clusters which produce isotropic strain fields, manifested in the Ashby-Brown type contrast observed in Figure 2. At subsequent annealing times, oxygen begins to precipitate substitutionally with the formation of {100} platelets and the emmission of Frenkel defects (silicon self-interstitials) which condense into small extrinsic stacking faults [7].

STRUCTURE AFTER LONG TIME ANNEALS

CZ silicon materials are typically far away from equilibrium. Therefore, most of the defect structure that has been reported in the literature ought to depend on the trajectory that the material has followed towards thermodynamic equilibrium (i.e. its

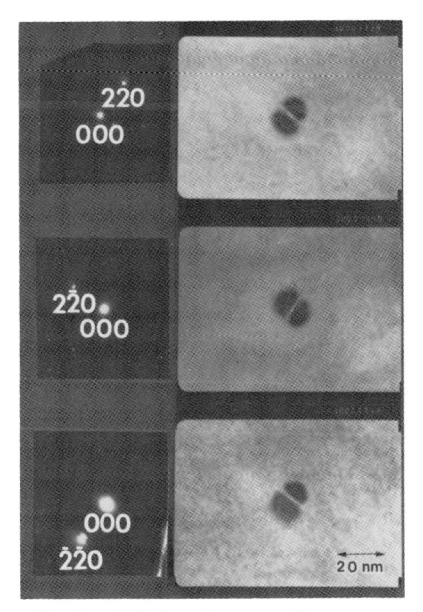

Fig. 2. Initial stages of oxygen precipitation: Isotropic, coherent microdefect exhibiting Ashby-Brown diffraction contrast. (after heat treatment at 700°C for 16 hrs).

Fig. 3. Intermediate stages of precipitation: Cluster with incoherent microdefects. Noticeable distortion in the lattice as well as anisotropic contrast. is observed (after heat treatment at 700°C for 32 hrs).

thermal history). In order to understand the meaning of *equilibrium configuration*, we submitted our materials for prolonged isothermal heat treatments. We have observed three regimes, which we report in this section.

a) Structure at Low Temperatures

We have explored the temperature range from 450 to 600°C and have observed the existence of rod-like defects. These defects have been previously reported in the literature [10,11]. Figure 7 shows a high resolution picture of a rod-like defect corresponding to material annealed at 450°C for 210 hours with a corresponding drop in interstitial oxygen content of 5 ppm. The precipitate consists of a long, needle-like precipitate oriented along a ⟨110⟩ direction and appears to have a crystalline structure. Similar observations were first reported by Bourret et al. [11], who identified this structure as that of coesite, the high pressure form of SiO_2. In addition to rod-like defects, we have observed other more complex defects which we have not been able to identify yet.

b) Structure at Intermediate Temperatures

In the temperature range between 600 and 1000°C, oxygen precipitates into the silicon lattice in the form of {100} amorphous silica platelets. Figure 8 shows a typical platelet corresponding to a thermal anneal at 850°C for 60 hours. These platelets are from 1.0 to 4.0 nm in thickness, and 30 to 50 nm in diameter. At various stages of precipitation they are associated with small extrinsic stacking faults, as in Fig. 6, which grow and coalesce with time.

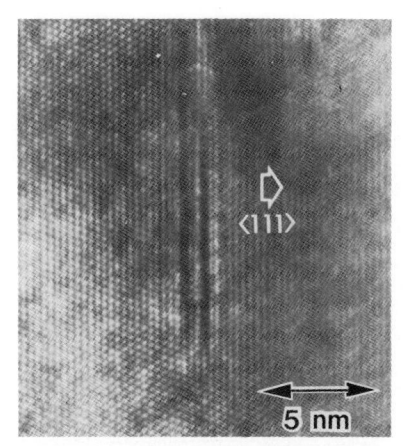

Fig. 4. Small microdefect along {111} silicon plane, occurring in cluster shown in Fig. 3. This is a precursor to the formation of an interstitial loop.

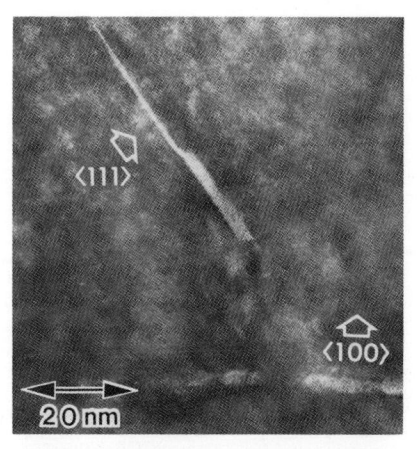

Fig 5. Lattice image of microdefect showing a {100} oxide platelet and a {111} interstitial loop (extrinsic stacking fault).

c) Structure at High Temperatures

The rate of oxygen precipitation in the temperature range from 900-1300°C is very low, and typically few, if any, precipitates would be observed at temperatures above 1200°C. The precipitation of oxygen can be maximized by the introduction of a *nucleation step* at an intermediate temperature (~700°C), followed by a high temperature anneal. The microstructure after a high temperature anneal consists of small polyhedra precipitates [7] formed of amorphous silica and bound by {111} and {100} silicon planes, as shown in Fig. 9. These polyhedra particles have diameters of about 10 nm and, because their density is determined by the nucleation step, they are believed to originate in the {100} platelets produced during the nucleation process. For heat treatments around 1200°C, these microdefects exhibit little or no strain contrast in the surrounding silicon lattice, which makes them difficult to detect using diffraction contrast techniques.

d) Generation of Dislocations

It has been believed that dislocation loops are to be associated with the precipitation of oxygen in silicon. In the study reported here, we have found that this is not the case. We did not observe a *single* dislocation loop in the cases where slow heating and cooling rates were used. We did simulate a swirl enhancement treatment where fast heating and cooling rates were used. The result was the decoration of defects with dislocation tangles and punched-out dislocation loops, a sample of which is shown in Fig. 10.

CONCLUSIONS

From our observations, oxygen precipitation in CZ silicon is significantly dependent on the thermal history of the material. For isothermal processes, there are indications that in the early stages, interstitial oxygen groups together to form spherical clusters which are isotropic and coherent. This would be a condensation process similar to the formation of clouds. With time the oxygen in these clusters precipitate

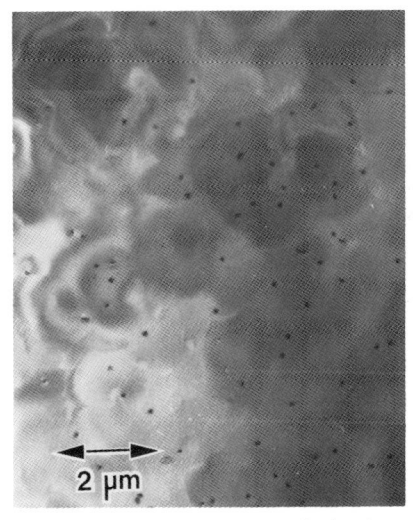

Fig. 6. Bright field micrograph showing defect distribution following 750°C anneal for 100 hrs.

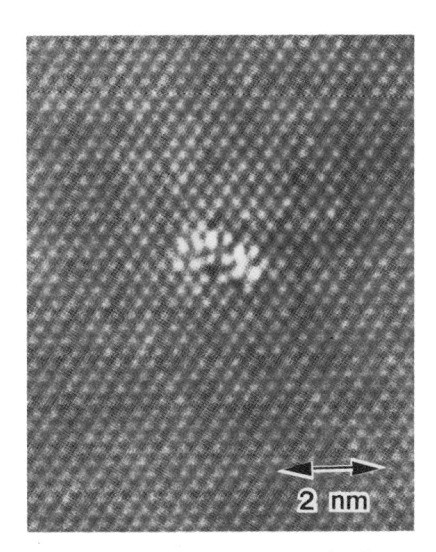

Fig. 7. Needle-like precipitate after heat treatment at 450°C for 210 hrs.

substitutionally with the formation of oxide particles.

Depending on the temperature in question, after prolonged heat treatments, the microstructure will range from needle-like coesite precipitates (between 450-600°C), amorphous silica platelets (between 600 to 1000°C), to polyhedral silica particles (above 900°C). The variation of defect morphology with temperature (from needle-like at low temperatures to sphere-like polyhedra at high temperatures) can be explained by taking into consideration the free energies associated with strain in the lattice, interface area, and generation of Frenkel defects [8].

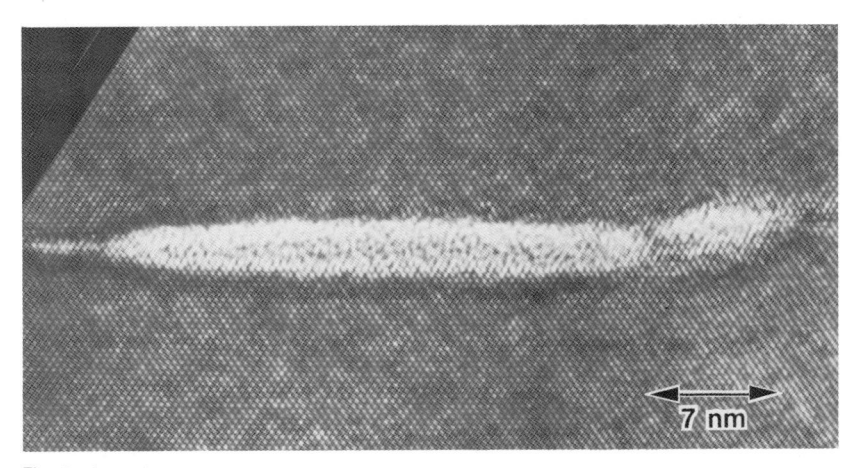

Fig. 8. Amorphous silica platelet, parallel to {100} silicon planes, following heat treatment at 850°C for 60 hrs.

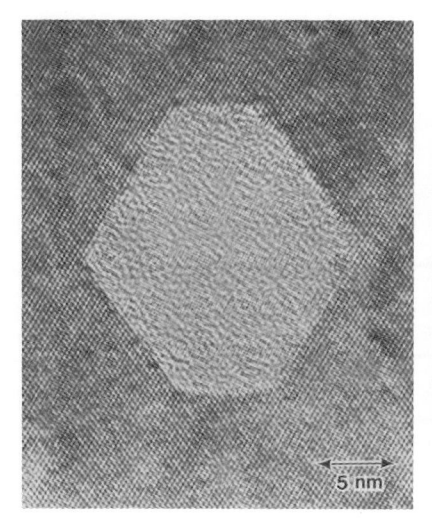

Fig. 9. Polyhedral silica precipitate after heat treatment around 1200°C for 64 hrs, following a *nucleation step* at 750°C for 100 hrs.

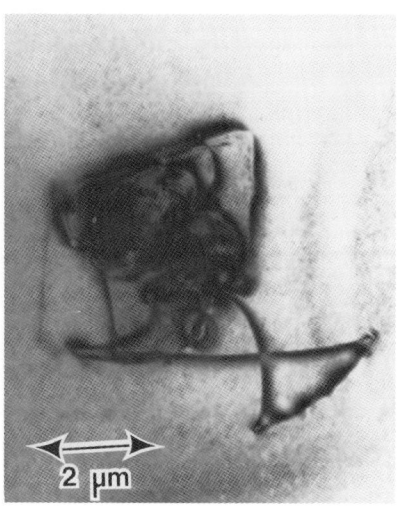

Fig. 10. Dislocation tangle around defect similar to the one shown in Fig. 2, following a *swirl-enhancement treatment* involving rapid heating and cooling rates.

ACKNOWLEDGMENTS

The authors are grateful to G. Anderson, G. Reid and R. Smith for technical assistance. Helpful discussions with T. Yamashita, J. R. Carruthers and M. Scott are also gratefully acknowledged.

REFERENCES

1. A. J. R. de Kock, *1976 Crystal Growth and Materials*, E. Kaldis and H. J. Scheel, eds. (North Holland, Amsterdam, 1977) p. 662.
2. S. M. Hu, J. Vacuum Sci. Technol. **14**, 17 (1977).
3. T. Y. Tan and W. K. Tice, Philos. Mag. **34**, 615 (1076).
4. D. M. Maher, A. Staudinger, and J. R. Patel, J. Appl. Phys. **47**, 3813 (1976).
5. K. Tempelhoff, F. Spiegelber, R. Gleichmann, and D. Wruck, Phys. Status Solidi **A 26**, 213 (1979).
6. F. A. Ponce, T. Yamashita, and S. Hahn, *Defects in Silicon*, W. M. Bullis and L. C. Kimmerling, eds. (Electrochemical Soc., Pennington, NJ, 1983), pp. 105-114.
7. F. A. Ponce, T. Yamashita, and S. Hahn, Appl. Phys. Lett. **43**, 1051 (1983)..
8. W. A. Tiller, S. Hahn and F. A. Ponce, J. Electrochem. Soc. (to be published).
9. M. F. Ashby and M. Brown, Phil. Mag. **8**, 1083 and 1649 (1963).
10. K. Tempelhoff and F. Spiegelberg, *Semiconductor Silicon*, H. Huff and E. Sirtl, eds. (Electrochemical Soc., Pennington, NJ, 1977), pp. 585-595.
11. A. Bourret, J. Thibault-Desseaux and D. N. Seidman, J. Appl. Phys. **55**, 825 (1984).

CRYSTALLINITY, MORPHOLOGY, AND CONDUCTIVITY OF BORON-DOPED MICROCRYSTALLINE SILICON

G. RAJESWARAN, J. TAFTO, R. L. SABATINI, AND P. E. VANIER
Division of Metallurgy and Materials Science
Brookhaven National Laboratory, Upton, New York 11973

ABSTRACT

Boron-doped microcrystalline (μc) silicon films produced by rf glow discharge from dilute (1%) mixtures of SiH_4 in H_2 show a critical dependence of conductivity on deposition conditions. The dark conductivity was related to the microscopic features using electron microscopy. The μc-Si:H films contain clusters of crystallites embedded in an amorphous matrix. The size of the crystalline clusters is typically 0.2 μm in diameter, and the size of the individual crystallites is about 2.5 nm. Electron micrographs of samples prepared at substrate temperatures T_s=135°C, 150°C, 165°C, and 180°C show that the number of crystalline clusters increases with T_s up to 165°C. At T_s=180°C, the crystallites completely disappear. When the concentration of SiH_4 in H_2 is decreased to 0.25%, the microstructure shows a high density of crystallites with no apparent clustering.

INTRODUCTION

Highly doped microcrystalline silicon, useful as contact layers in solar cells [1], can be deposited from the rf glow discharge deposition of SiH_4, H_2 and either PH_3 or B_2H_6 [2,3]. The dark conductivity (σ_d) of μc-Si:H depends strongly on the ratio x of SiH_4 flow rate to the total flow rate ($SiH_4+H_2+PH_3$ or B_2H_6), the substrate temperature T_s, PH_3 or B_2H_6 concentration, and the rf power. Typical values of the above parameters are x=1%, T_s=150-250°C, $[PH_3]$ or $[B_2H_6]/[SiH_4]$=1-5% and rf power=200 mW/cm^2. This paper describes the microstructural features of boron-doped μc-Si:H films obtained under conditions similar to those mentioned above. The changes induced in the microstructure when deposition conditions, such as x and T_s, are varied have been studied using a scanning electron microscope (SEM) and a transmission electron microscope (TEM).

EXPERIMENTAL

All the depositions reported in this study were performed in an rf capacitively-coupled system with 25 cm diameter electrodes, 3.5 cm apart, with radially inward gas flow, at a pressure of 1 Torr and a total gas flow rate of 100 sccm. A sketch of the deposition system is shown in Fig. 1. Gas flows were controlled with modified Veeco PV-10 piezoelectric valves and measured with Tylan FM-360 flowmeters, using MKS 257 flow controllers. Total pressure was measured with a MKS 222 capacitive pressure transducer. The discharge was excited by a Plasma-Therm HFS-500E 13.56 MHz generator.

The substrates used were Corning 7059 glass. The substrates were mounted on the upper electrode and the temperature was regulated to ±1°C

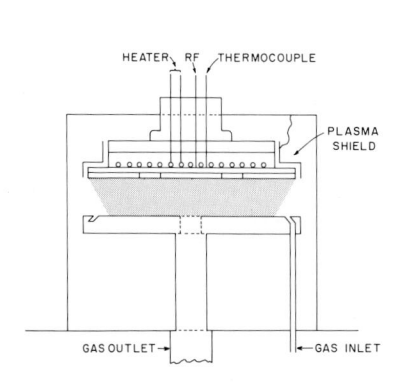

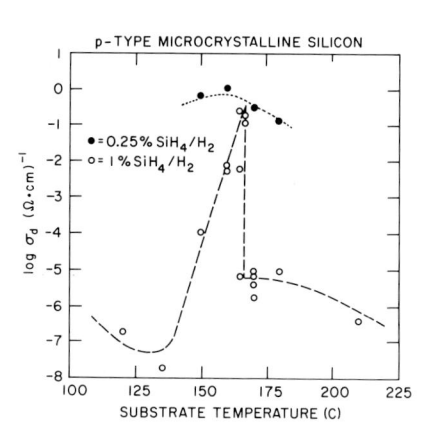

FIG. 1. The rf glow discharge system used to deposit µc-Si:H.

FIG. 2. Dark conductivities of p-type µc-Si:H films prepared at various T_s.

by an Eurotherm proportional/integral controller. Samples of µc-Si:H for measurements of conductivity were deposited on glass substrates on which molybdenum pads were predeposited in a gap cell configuration. The thicknesses of the deposited films ranged from 0.2-0.5 µm. Other samples prepared at the same time were studied using a JEOL JXA-35 scanning electron microscope (SEM) and a JEOL 100-CX transmission electron microscope (TEM). In order to perform TEM, the glass substrates with the µc-Si:H films were submerged in dilute HF to dissolve the glass. The Si films could then be floated onto standard electron microscope grids. The SEM experiments were performed with 10 keV electrons and the TEM with 120 keV electrons.

RESULTS AND DISCUSSION

Figure 2 shows the variation of σ_d with T_s for x=1% and x=0.25%. For x=1%, a critical dependence of σ_d on T_s is seen. Four films (x=1%) with T_s=135°C, 150°C, 165°C, and 180°C, respectively were selected for analysis. Two other films from x=0.25% with T_s=150°C, and 180°C, respectively were also selected for a comparison between x=1% and x=0.25% deposition conditions. The µc-Si:H films from x=1% conditions are interesting samples. Their conductivity increases about 5 orders in magnitude when T_s increases from 135°C to 165°C and then drops four orders in magnitude within a span of 5°C. The µc-Si:H films from x=0.25% conditions do not exhibit this critical T_s dependence. In addition, the conductivities of these films are higher than those for x=1% films in the range of T_s between 100-200°C.

Bright field TEM images of µc-films prepared with x=1% are shown at low magnification (5000x) in Fig. 3. The pictures indicate black spots in an otherwise featureless matrix. The number of these spots increases with increasing T_s from 135°C to 165°C (Fig. 3a, b, and c). For T_s=180°C, no spots are observed (Fig. 3d). The µc-Si:H film with the largest concentration of black spots (T_s=165°C) is also the most conductive among x=1% films. The size of each spot is ~0.2 µm independent of T_s. TEM diffraction patterns reveal that the spots correspond to polycrystalline silicon while the featureless medium is amorphous. Each spot must have originated from an isolated nucleation site on the substrate. Figure 4 shows a comparison between the diffraction patterns arising from µc-Si:H and

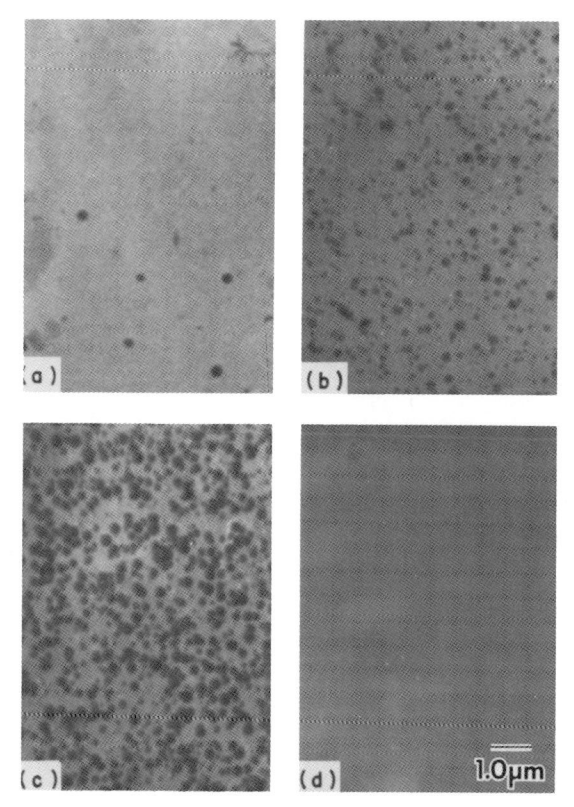

FIG. 3. Bright field TEM images at (a) T_s=135°C, (b) 150°C, (c) 165°C, and (d) 180°C; x=1%.

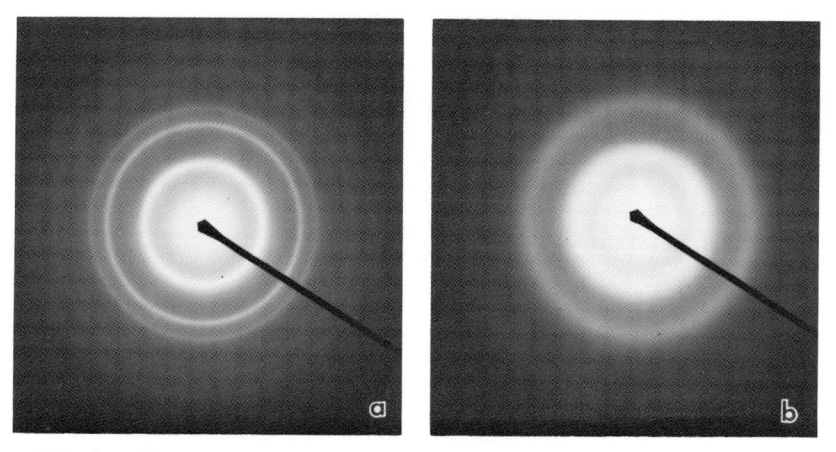

FIG. 4. Electron diffraction patterns of Si (a) x=0.25%, T_s=150°C, microcrystalline (b) x=1%, T_s=135°C, amorphous.

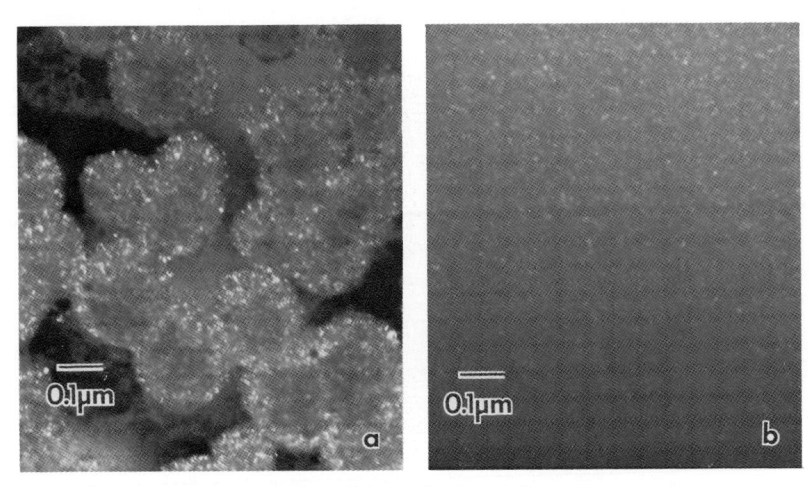

FIG. 5 Dark field images of (a) x=1%, T_s=165°C, microcrystalline and (b) x=0.25%, T_s=150°C, microcrystalline films.

amorphous-Si:H films. In Fig. 4a, the diffraction pattern of a highly conducting film grown with x=0.25%, T_s=150°C conditions shows sharp and pronounced rings indicating that the film is crystalline. The rings broaden into halos in Fig. 4b which corresponds to a film (x=1%, T_s=135°C) that is mostly amorphous (see Fig. 3a). The diffraction patterns of the amorphous films agree with those reported in the literature [4]. For T_s= 180°C, the diffraction patterns show no indication of crystallinity. The absence of μc regions for T_s>165°C is remarkable, and is not understood in detail but could be due to changes in local strains, H concentration or B concentration in the film [5,6].

A dark field image of the most conductive film (x=1%, T_s=165°C) at 50000x magnification is shown in Fig. 5a. This micrograph clearly reveals that the spots are clusters of small crystallites. The individual crystallites, seen as white spots, have typical dimensions of 2-3 nm. The clusters are randomly connected to each other and thus contribute to high conductivity. Figure 5b is a similar micrograph (dark field, 50000x) of a x=0.25%, T_s=150°C μc-Si:H film. This film appears homogeneous and is polycrystalline, as shown in Fig. 4a. The size of each crystallite is ~2.5 nm.

Figure 6 shows the bright field image (5000x) corresponding to x=1%, T_s=150°C sample of Fig. 3b after the film has been attacked by HF for a long time. The area around the black spots (clusters) which appears white in this figure has been etched more rapidly than the rest of the film indicating that the amorphous material in the immediate neighborhood of the clusters is chemically more reactive than the bulk of the amorphous phase. This may result from either an increase in H or B concentration, or from a more porous or highly strained microstructure surrounding the clusters. The continued growth of the cluster depends on renucleation of crystallites within this region. In the case of Fig. 5b corresponding to x= 0.25%, T_s=150°C μc-Si:H, the density of cluster nucleation sites may be so large that the growing clusters actually coalesce and appear homogeneous.

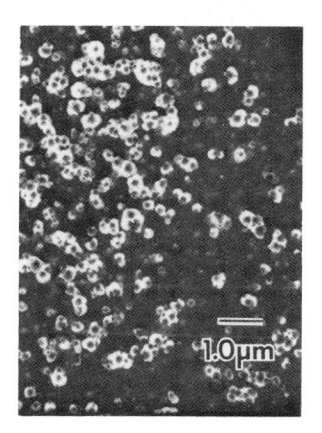

FIG. 6. Bright field image corresponding to Fig. 3b after the film was etched in HF for a long time.

The surfaces of the as deposited films were studied by SEM. These studies again indicate the existence of 0.2 μm clusters. Figure 7 indicates the surface topography of two conducting films (x=1%, T_s=150°C and x=0.25%, T_s=180°C) at 60° tilt. The surface features are influenced by the clusters, with increased film thickness in the vicinity of the clusters. The concentration of clusters in Fig. 7b is higher than that in Fig. 7a, consistent with TEM observations and as expected from the values of σ_d for the corresponding films (see Fig. 2).

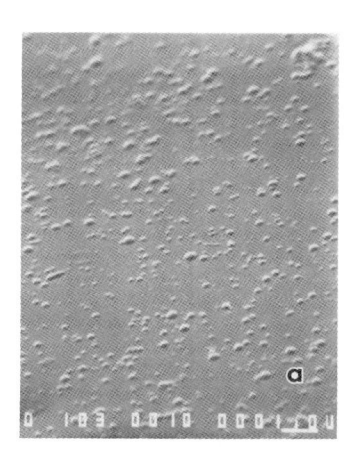

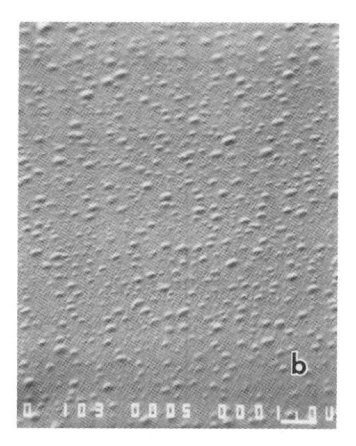

FIG. 7. Surface topographs of (a) x=1%, T_s=150°C, microcrystalline and (b) x=0.25%, T_s=180°C microcrystalline films showing the existence of growing clusters.

CONCLUSIONS

The macroscopic properties (such as σ_d) of boron-doped µc-Si:H are governed by their microscopic features. Microcrystalline silicon consists of clusters of crystallites ~2.5 nm in size. The size of each cluster is ~0.2 µm independent of substrate temperature during growth. When SiH_4 is more highly diluted in H_2 (x=0.25% instead of x=1%) the film growth rate decreases and the density of clusters increases, thereby increasing the conductivity. Each cluster is believed to originate from a single nucleation site. More nucleation occurs at existing nucleation sites, resulting in a mass of small grains growing in size as the film grows. The resulting µc regions would be expected to be cone shaped with each cone standing on end at its original nucleation site. Surface topography is consistent with this picture.

ACKNOWLEDGMENTS

This research was conducted under the auspices of the U.S. Department of Energy, Division of Materials Sciences, Office of Basic Energy Sciences under Contract No. DE-AC02-76CH00016.

REFERENCES

1. Y. Hamakawa, K. Fujimoto, K. Okuda, Y. Kashima, S. Nonomura, and H. Okamoto, Appl. Phys. Lett. 43 (7), 644-646 (1983).
2. Y. Mishima, S. Miyazaki, M. Hirose, and Y. Osaka, Phil. Mag. B46, 1 (1982).
3. G. Willeke, W. E. Spear, D. I. Jones, and P. G. LeComber, Phil. Mag. B46, 177 (1982).
4. S. C. Moss and J. F. Graczyk, Phys. Rev. Lett. 23, 1167 (1969).
5. F. J. Kampas, J. Appl. Phys. 53, 6408 (1982).
6. G. Rajeswaran, F. J. Kampas, P. E. Vanier, R. L. Sabatini, and J. Tafto, Appl. Phys. Lett (in press).

INTERFACIAL REACTIONS OF NICKEL FILMS ON GaAs

L.J. CHEN, AND Y.F. HSIEH*
Department of Materials Science and Engineering, National
Tsing Hua University, Hsinchu, Taiwan, ROC.

ABSTRACT

Transmission electron microscopy (TEM) and X-ray diffraction (XRD) were performed to study the interfacial reactions of Ni/GaAs contact system as a result of isothermal annealing and two step annealing. Ni_2GaAs was observed to exhibit preferred orientation relationships with respect to GaAs substrate after 300-350°C annealing. The compound decomposed to NiAs and Ga-compounds after 400°C annealing. NiAs was also found to grow preferentially on GaAs. The step annealing was found to be ineffective in varying the morphological structure of interface.

I. INTRODUCTION

Silicon has been in dominant position as the base material for IC industry for many years. Recently, scientists and technologists have made tremendous inroad in finding other materials which possess superior properties to silicon and yet technically and economically feasible for mass production. GaAs has appeared to be the most promising candidate material for such an endeavor. The main advantages of GaAs over Si are: considerably higher mobility, larger and direct band gap, simpler technology, possibly better performance and reliability [1]. All devices based on semiconductors need metal contacts either to form an active Schottky barrier or an ohmic contact. Until recently, very little work has been done on metallizing GaAs integrated circuits. Extensive researches were carried out in the area of doped metal contacts, particularly fundamental physics and chemistry of contact formation mechanisms and new techniques for improving and controlling the properties of contacts during the past few years [2]. Although several investigations have been made on the most popular system, Au-Ge-Ni, few of them elucidate the role of Ni, whereas Au is known to act as a base metal, Ge acts as a doping element. As far as the effect of the addition of Ni in the alloyed system is concerned, previous study revealed that the important reaction is likely that between Ni and GaAs but not that between Ni and other deposited film constituents. Since the number of constituents involved in this reaction is so great that it is difficult to isolate the reaction in which Ni participates. This leads to an intensive study of the simpler system, Ni/GaAs, instead of the practical contact system, Ni-Au-Ge/GaAs. In this paper, we report results of the investigation of interfacial reactions of nickel thin films on GaAs.

II. EXPERIMENTAL PROCEDURES

N-type, Si doped, GaAs wafers were grown by gradient-freeze method with

----

* Present address: Materials Research Laboratory, Industrial Technology and Research Institute, Hsinchu, Taiwan, ROC.

etch pit density less than 1x10⁵ cm⁻². The surface normal was deviated from the [001] direction toward the [110] direction by an angle of 1-3°. Nickel thin films were e-gun deposited onto GaAs wafers at a vacuum better than 8x10⁻⁷ torr. The deposition rate was controlled in a range between 0.9 and 1.2 Å/sec. Thermal treatments were performed in 3-zone diffusion furnace with dry nitrogen ambient which was purified by passing through a Ti-getter tube.

Transmission electron microscope examinations were performed with JEOL-100B and JEOL-200CX electron microscopes. X-ray diffraction (XRD) study was carried out in a Shimadzu ZD-5 diffractormeter. The angular scans were from 20-60° (2θ) with a scanning speed 4° min⁻¹.

III. RESULTS

The alloying behaviors of Ni/(001) GaAs were studied for various processing temperature and time. The samples were annealed at 200-800°C for 5-180 mins. Table 1 lists all the phases identified through TEM examinations.

Time (mins) (°C) Temp.	15	60	180
200	Ni		
300	Ni+Ni₂GaAs		
350	Ni₂GaAs		
450	NiAs+Ni₂GaAs		
500	NiAs+Ni₂GaAs+β-Ga₂O₃		
600			
700	β-Ga₂O₃		
800			

Table 1 Schematic diagram of phases identified under different annealing conditions.

The as deposited Ni film on (001)GaAs is polycrystalline, as shown in Fig. 1, with randomly oriented grains 150 Å in average size. Microstructural changes were found as a result of interfacial reactions

Fig. 1 Bright field micrograph (B.F.) of an as deposited film with corresponding diffraction pattern (D.P.)

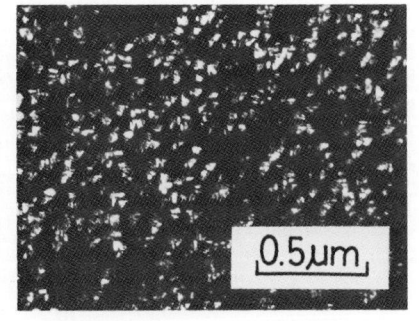

Fig. 2 Dark field micrograph (D.F.), 300°C, 10 mins.

between Ni thin film and GaAs following heat treatments. The main results are summarized as follows,

(i) 200°C annealing — no reaction between Ni and GaAs was detected even after 3 hrs annealing. Grain growth was also not evident.

(ii) 300°C annealing — Ni₂GaAs islands, 600 Å in average size, were detected along with Ni for samples annealed at 300°C for 5-10 mins. Fig. 2 shows Ni₂GaAs island structure which was found at the interfacial region between Ni and GaAs. The diffraction pattern near [001]GaAs direction, Fig. 3, of a sample annealed for 60 mins exhibited a symmetric pattern. The spot pattern instead of ring pattern indicated that the ternary phase was epitaxially grown. The morphology of this reaction product was noticed to change with the diffraction conditions. From the analysis of diffraction patterns, the orientation relationships:

(0001)Ni₂GaAs // (111)GaAs,   (11$\bar{2}$0)Ni₂GaAs // (110)GaAs

were established. Once the annealing time was prolonged to 180 mins, part of Ni₂GaAs was decomposed to NiAs and Ni-Ga compounds with granular structure.

(iii) 350°C annealing — all Ni grains disappeared to from Ni₂GaAs after 5 mins annealing. The morphology and orientation relationships were similar to those at 300°C annealing for 30-60 mins. Highly symmetric diffraction pattern near (001)GaAs was also observed at this temperature.

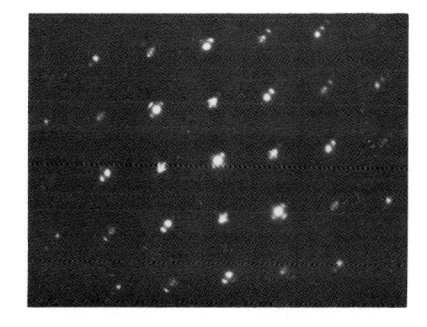

Fig. 3   D.P., 300°C, 60 mins.   Fig. 4   D.P., 500°C, 30 mins.

(iv) 400-600°C annealing — the metastable phase, Ni₂GaAs, was partially decomposed to NiAs and Ni-Ga compounds after 5 mins annealing at 400-600°C. These phases appeared as fine grains with strongly preferred orientation relationships with respect to GaAs matrix. From the analysis of the diffraction pattern, see Fig. 4, the orientation relationships of the newly formed phase, NiAs (Fig. 5), with respect to GaAs were identified to be:

(11$\bar{2}$0)NiAs//(220)GaAs,   (10$\bar{1}$1)NiAs//(200)GaAs   with 2.32° deviation.

Inspection of the diffraction pattern indicates a pseudo four fold symmetry of the product phase. It was composed of two [1$\bar{1}$0$\bar{1}$]NiAs standard patterns, 90° rotation relative to each other, and a [001]GaAs matrix pattern. As the matrix was tilted to [111] orientation, the orientation relationship:

[0001]NiAs // [111]GaAs

was also established. The results are consistent with what was found from [001] diffraction pattern analysis.

The decomposed Ni-Ga compounds with small lattice spacing difference

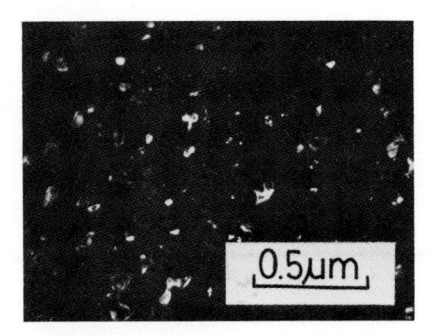

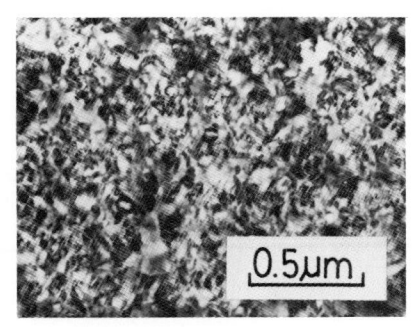

Fig. 5 D.P., 500°C, 30 mins.   Fig. 6 B.F., 500°C, 30 mins.

with GaAs matrix (<0.5%) can hardly be distinguished. Whether Ni-Ga compounds were formed or not will be discussed in detail in the discussion section. The morphological structure of the samples annealed at this temperature range showed similar layered structures, see Fig. 6. The stacking sequence of this layered structure was analyzed to be NiAs/ Ni_2GaAs/GaAs. Because Ga atoms, resulted from GaAs dissociation or Ni_2GaAs decomposition, have the tendency to diffuse out to the surface to react with the electronegative oxygen atoms, an oxide film was generally observed to cover the layered structure for annealing time exceeding 30 mins. NiAs grains were found to coexist with the oxide grains in some area of the sample, indicating they were geometrically connected. The oxide was identified to be fine grained β-Ga_2O_3. The oxidation commonly occurred after long time annealing at 400-500°C or following short time annealing at 500-600°C. Similar microstructures, with exception of slight grain size changes, were found after different time and temperature annealing.

(v) 700-800°C annealing — this temperature range exceeds the dissociation temperature of GaAs (630°C). Considerable amount of gallium readily reacted with oxygen to form β-Ga_2O_3 and arsenic was evaporated. Grain growth of β-Ga_2O_3 was evident with annealing temperature increased.

Two step annealing was found to be ineffective, as far as the phase formation and microstructural changes were concerned, compared to single step annealing. There was little sign of grain growth for prolonged annealing at the 1st or 2nd step annealing temperature.

IV. DISCUSSION

Ohmic contact systems containing Ni, such as Au-Ge-Ni and Au-Zn-Ni, have been widely used in various III-V compound semiconductor devices. The participation of Ni in these contact systems leads to the great improvement in the surface uniformity following heat treatment. Previous study [3] of the simple contact system, Ni/GaAs, indicated that there was a tendency for diode characteristics to have an abrupt change when the annealing temperature exceeded 450°C and the instability of the diode characteristics at low annealing temperatures (250-350°C) may be related to the microstructural changes. Ogawa [4] has also investigated this contact structures with the aid of Auger electron spectroscopy (AES) reflection high energy electron diffraction (RHEED) and transmission electron diffraction (TED). The contact system was annealed with a thin Cr film deposited to protect the sample surface. Experimental results of Ogawa revealed that the alloying reactionwas found to occur from 200°C, producing polycrystalline Ni_2GaAs at substrate interface. As annealing time was increased at 300°C, the reaction product Ni_2GaAs was transformed to a superstructure with preferred orientation relationships as described before.

For annealing at 400°C, Ni$_2$GaAs commenced to decompose to NiAs and β-NiGa. A layered structure, NiGa/NiAs/GaAs was detected by AES after 500°C annealing. The result is different from the observation of the present study that the resultant structure was NiAs/Ni$_2$GaAs/GaAs. There has not been any direct and obvious evidence to assure that Ni-Ga compounds did form, although the possibility exists that the decomposition reaction: Ni$_2$GaAs → NiAs + Ni-Ga (β-phase, β'-phase and γ-phase) occurs. On the assumption that Ni-Ga compounds did form, the phases were probably: β-phase (Ni$_3$Ga$_2$), β'-NiGa (Ni$_2$Ga$_3$) and γ-phase (NiGa) due to small lattice mismatch (<0.5%) with respect to GaAs. In the cases that preferential orientation relationships, as those between NiAs and GaAs, do exist among β-phase, β'-phase, γ-phase and GaAs, the diffraction spots of these phases can hardly be distinguished from those of GaAs matrix. The formation of Ni-Ga can not be precluded. If the Ni-Ga compounds were not present, the question concerning where was the decomposed Ga remained. A clue may be provided by the electronegativity principle. Since there was no metallic film deposited on Ni/GaAs for protection, the tendency of Ga to outdiffuse through Ni$_2$GaAs and NiAs and form Ga$_2$O$_3$ on NiAs surface correlated with an increase in electronegativity in the sequence:

Ga(1.6) → Ni(1.8) → As(2.0) → O(3.5)

This suggests that the electropositive Ga atoms are more prone to react with the most electronegative element, oxygen, instead of nickel. Surface oxidation reaction may suppress the formation of Ni-Ga compounds. The following reactions are proposed to occur:

Ni$_2$GaAs → NiAs + Ni + Ga   and

2Ga + 3/2 O$_2$ → β-Ga$_2$O$_3$,   2Ni + GaAs → Ni$_2$GaAs

The reactions resulted in the layered structure: β-Ga$_2$O$_3$/NiAs/Ni$_2$GaAs/GaAs. All the resultant compounds were positively identified for samples annealed at 400-600°C.

Ni$_2$GaAs was formed at the early stage of alloying reaction, e.g. after annealing at 300°C for 5 mins. The phase was of rod-shape and resulted in highly symmetric diffraction pattern near [001]GaAs direction. Scarce data on crystal structure, was available for Ni$_2$GaAs in the literature, except the measured lattice spacing proposed by Ogawa [4]. The phase was detected by careful analysis of the off-[001] diffraction patterns. The size of the fine precipitates did not change significantly after annealing at 300-350°C. The entrenchment of Ni$_2$GaAs is probably related to the uniform alloying behavior of interfacial reaction of Au-Ge-Ni/GaAs system. The superlattice of diffraction spots near [001]GaAs direction of samples annealed at 400-600°C, see Fig. 3, was then probably resulted from the Ni$_2$GaAs phase. The existence of Ni$_2$GaAs was not evident in the diffraction pattern owing to the small lattice spacing difference between Ni$_2$GaAs and GaAs.

V. SUMMARY AND CONCLUSIONS

The metal contact system, Ni/GaAs, has been investigated by transmission electron microscopy and X-ray diffraction to study the interfacial reactions between metal films and GaAs. The main conclusions are summarized as follows:
   (1) No alloying reaction was found below 300°C annealing.
   (2) Ni$_2$GaAs islands exhibiting epitaxial orientation relationships with respect to GaAs were observed at the interfacial layer after annealing at 300°C for 5 mins. The low reaction temperature is attributed to the high reactivity of Ni with GaAs.
   (3) Part of Ni$_2$GaAs decomposed to NiAs and promoted the formation of

β-Ga$_2$O$_3$ in samples after 400-600°C annealing. A layered structure, β-Ga$_2$O$_3$/NiAs/Ni$_2$GaAs/GaAs, was found.

(4) The symmetric diffraction pattern (Fig. 4) of samples annealed at 400-600°C was composed of two superimposed [1$\bar{1}$0$\bar{1}$] diffraction patterns of NiAs. The preferred orientation relationships were found to be:

(11$\bar{2}$0)NiAs // (220)GaAs,   [0001]NiAs // [111]GaAs,

(10$\bar{1}$1)NiAs // (200)GaAs   (2.32° deviation)

[1$\bar{1}$01]NiAs // [001]GaAs

(5) Ni$_2$GaAs and NiAs were found to form in samples annealed at temperatures higher than 300°C and 400°C, respectively. Both phases show preferred orientation relationships with respect to GaAs.

(6) Only β-Ga$_2$O$_3$ oxide films were observed in samples heated to the temperature above the dissociation temperature of GaAs (630°C).

(7) Two step annealing was found to be ineffective in Ni/GaAs system as far as the phase formation and the microstructural changes are concerned.

ACKNOWLEDGEMENTS

The research was supported in part by Materials Research Laboratory, ITRI, Hsinchu, Taiwan, ROC.

REFERENCES

1. L. Hollan, J.P. Hallais, and J.C. Brice in: Current Topics in Material Science, Volume 5, E. Kaldis, ed. (North-Holland, Amesterdam, 1980).
2. J.M. Woodall and J.L. Freeouf. JVST 19(3), 794 (1981).
3. I.D. Romanova, N.K. Maksimova, E.N. Pekarskii and M.P. Yakubenya, Translation from Izrestiya Vysshikh Uchebnykh Zavedenii, Fizika, No. 4, 151 (1976).
4. M. Ogawa, Thin Solid Films, 70, 181 (1980).

TEM OBSERVATION OF DEFECTS IN InGaAsP AND InGaP CRYSTALS ON GaAs SUBSTRATES GROWN BY LIQUID PHASE EPITAXY

O. UEDA, S. ISOZUMI, S. KOMIYA, T. KUSUNOKI, AND I. UMEBU
FUJITSU Laboratories Ltd., 1677 Ono, Atsugi 243-01, Japan

ABSTRACT

Defects in InGaAsP and InGaP crystals lattice-matched to (001)-oriented GaAs substrate successfully grown by liquid phase epitaxy, have been investigated by TEM and STEM/EDX. Typical defects observed by TEM are composition-modulated structures, dislocation loops, non-structural microdefects, and stacking faults.

INTRODUCTION

InGaAsP/InGaP double-heterostructures (DH's) lattice-matched to GaAs substrates are expected to have high luminescence efficiency and band gaps varying in a wide range from 1.4 eV for GaAs and to 1.9 eV for $In_{0.48}Ga_{0.52}P$. In recent years, much effort has been devoted to the growth of these materials [1-3] and cw operation of the lasers fabricated from these materials at room temperature have been reported elsewhere [4]. Elimination of various defects in the crystal is one of the most important problems for reliable visible lasers. In this paper, we describe for the first time detailed TEM and STEM/EDX investigations of defects in InGaAsP and InGaP crystals grown on (001)-oriented GaAs substrates by liquid phase epitaxy.

EXPERIMENTAL

InGaAsP and InGaP epitaxial layers were grown on (001)-oriented GaAs substrates by liquid phase epitaxy at 780°C. The lattice mismatch was about $1 \times 10^{-4} - 1 \times 10^{-3}$. The thicknesses of the InGaAsP and InGaP layers were 0.1 - 0.4 μm and 2.0 - 2.5 μm, respectively. Three types of wafers were examined as follows: InGaAsP/GaAs, InGaP/GaAs, and InGaAsP/InGaP/GaAs. The photoluminescence peak wavelength, λ_{PL}, of the InGaAsP layers was in the range 710 - 825 nm. Thin specimens for TEM were prepared by chemical etching. TEM observation was carried out in a Hitachi H-800H electron microscope with an accelerating voltage of 200 kV. STEM/EDX analysis was also carried out in this microscope. A Kevex system 7000 for quantitative analysis was used for the EDX measurement.

RESULTS AND DISCUSSION

Composition-Modulated Structures

In all crystals of InGaAsP and InGaP, periodic diffraction contrasts in two equivalent directions of the $\langle 100 \rangle$ and the $\langle 010 \rangle$ were observed. Typical TEM images of the modulated structures are shown in Figs. 1(a)-1(c), obtained by 220 reflection. The photoluminescence peak wavelength, λ_{PL}, of these layers were 721, 730, and 800 nm, respectively. The modulation wavelength is in the range 10 - 50 nm. These images are quite similar to those observed in spinodally decomposed metal-alloys such as, Cu-Ni-Fe [5], Au-Ni [6], Al-Zn [7] etc. Contrast experiment on these modulated structures was carried out by using different reflections of 220, $2\bar{2}0$, 400, and 040. When $\bar{g}=220$ and $2\bar{2}0$, both the [100]- and the [010]-oriented modulated structures are in contrast. On the other hand, under the $\bar{4}00$ reflection, the [100]-oriented modulated structures is out of contrast, and under the 040 reflection, [010]-oriented one is out of contrast. Therefore, these modulated structures involve periodic strain field in two equivalent directions of the ⟨100⟩ and the ⟨010⟩. However, there were not structural defects in relation to these modulated structures. It is possible that these modulated structures are due to compositional variation in the crystal.

In order to investigate this variation, STEM/EDX analysis was performed. A typical STEM image under the 220 reflection is shown in Fig. 2(a). An EDX measurement was performed along the line A-B in this figure with a spatial resolution of about 10 nm. The results are shown in Fig. 2(b), such as the intensity ratios of the In Lα line to the Ga Kα line, the As Lα line to the Ga Kα line, and the P Kα line to the Ga Kα line. Compositional variation is clear and the variation of the intensity ratios is estimated to be $\pm$ 4 - 5%.

Next, the influence of the hetero-structure on the modulated structure was investigated. In both the region near the top surface of the InGaP layer, and the region near the interface between the InGaP layer and the substrate, composition-modulated structures are clearly observed. Therefore, the composition modulation is a bulk phenomenon, and not limited to the interface region.

According to the calculation of the spinodal isotherms in InGaAsP quarternary system by de Cremoux et al. [8], the composition of the InGaAsP layers examined in the present work are within the spinodal region, in which the solid solution is thermodynamically unstable, and tends to separate into two quasi-stable phases, during the growth and/or the cooling process. Therefore, the composition modulation could be associated with the spinodal decomposition in the crystal. In the spinodally decomposed alloys which have assymmetry in elastic coefficients, the modulated waves are considered to be promoted so as to minimize the strain contribution to free energy in the direction of the [hkl] which minimizes the elastic coefficient $Y_{[hkl]}$ as follows [9,10]:

$$Y_{[hkl]} = \frac{c_{11} + c_{12}}{2} \left[3 - \frac{c_{11} + 2c_{12}}{c_{11} + 2(2c_{44} - c_{11} + c_{12})} \times \frac{1}{l^2m^2 + m^2n^2 + n^2l^2} \right]$$

Using this equation, we estimated the $Y_{[hkl]}$ for GaAs, GaP, InAs, and InP, and it was found that in all cases, $Y_{[100]}$ gives the minimum value. Therefore, it is well explained that the modulated structures were generated in three equivalent directions of [100], [010], and [001].

Dislocation Loops

Dislocation loops were often observed in both the undoped InGaAsP and the InGaP layers. A typical TEM image of them in undoped InGaAsP layer (λ_{PL}=790 nm) under the 220 reflection with a positive s is shown in Fig. 3. These loops are 10 - 100 nm in diameter, and their long axes are in the ⟨100⟩ or the ⟨110⟩ direction. In order to determine the characters of these loops, inside-outside contrast analyses were carried out on six loops denoted by L_1 - L_6 in Fig. 3. The results are summarized in Table I. In all cases, the deviation parameter s was positive. The "x" indicates that the loop is out of contrast. From these results, the Burgers vectors are determined to be a/2[10$\bar{1}$] for loops L_1 and L_2, a/2[$\bar{1}$0$\bar{1}$] for loops L_3 and L_4, and a/3[$\bar{1}\bar{1}$1] for loops L_5 and L_6. Therefore, the characters of the loops are determined to be intrinsic for loops L_1 - L_4, and extrinsic for loops L_5 and L_6. These results indicate that both vacancy type loops and interstitial type ones were generated in the same crystal. In III-V compound semiconductors and alloy semiconductors, extrinsic dislocation loops are induced only by heavy doping. They are thought to be generated by the condensation of excess interstitials due to segregation of the dopant atoms. Therefore, the generation of the both types of loops indicates that the composition modulation can induce excess vacancies and interstitials.

Concerning the defect formation in III-V group compound semiconductors and alloy semiconductors, we have already reported that there is clear correlation between the formation of dislocation loops in heavily doped as-grown crystals and recombination enhanced defect motion in actual optical devices, as shown in Table II [11]. According to this correlation, it is possible to predict that in InGaP and InGaAsP crystals lattice-matched to (001)-oriented GaAs substrates, recombination enhanced defect motion occur easily.

Non-structural Microdefects

Two types of non-structural microdefects were observed. Typical TEM images of one type of defects observed in InGaAsP layers are shown in Figs. 4(a) and 4(b). Figure 4(a) shows a bright field image with the 220 reflection with a positive s, and Fig. 4(b) shows a weak beam image with the weak 220 reflection with a positive s_{660}. These defects were observed only in InGaAsP layers with intensely modulated structures grown directly on GaAs substrates. In the bright field image, the defects were observed as strong dark contrast regions with diameter of 0.1 - 0.2 μm (denoted by arrows). While, in the weak beam image, clear strong bright contrast regions (including dislocation-like contrasts) were observed corresponding to these defects. Therefore, these defects are considered to be induced by local large stress concentrated in the crystal due to strong composition modulation, and are peculiar to the InGaAsP layers grown directly on GaAs substrates.

Figure 5 shows a TEM image of precipitate-like microdefects (the other type of defects) observed in InGaAsP layer. These defects were non-structural and 100 - 200 nm in diameter. Some of these defects contain small dislocations. These defects were not out of contrast as a whole under four different reflections of 220, 2$\bar{2}$0, 400, and 040. Thus, there is a strain field in equivalent directions around them. Since the epitaxial layers were not intentionally doped, these defects were expected to be generated by the precipitation of the residual impurities.

Stacking Faults

Stacking faults were also observed very rarely, and were generated in the case, where the liquid-solid interfacial reaction occurred at the early stage of growth. Typical images of the stacking faults are shown in Fig. 6, taken by the 400 reflection. Irregularly shaped faults were observed and the size of the faults was different from each other. By contrast experiment, these faults were proved to be intrinsic in character. In this image, intensely modulated structures and high density stress-induced microdefects were observed. Therefore, these irregularly shaped faults were considered to be generated by the relaxation of local stress involved in these crystals. Multiple stacking faults were also observed.

CONCLUSION

Defects in InGaAsP and InGaP crystals successfully grown on (001)-oriented GaAs substrates have been investigated for the first time, and the following results were obtained.
(I) Composition-modulated structures were observed in all of the crystals. It is possible that these modulated structures are associated with spinodal decomposition in the crystal during the crystal growth.
(II) Two types of dislocation loops were often observed even in undoped crystals.
(III) It is expected that, in InGaAsP and InGaP crystals lattice-matched to GaAs, the recombination enhanced defect motions, i. e., climb motion and glide motion, can occur easily.
(IV) Non-structural microdefects and irregularly shaped stacking faults were observed very rarely. These defects can be eliminated by improving the purity of the crystal and/or controling the growth condition.

ACKNOWLEDGMENT

The authors would like to express their thanks to T. Misugi, O. Ryuzan, T. Kotani, and M. Takusagawa for their valuable discussion. They also thank A. Yamaguchi for friutful discussion throughout this work.

REFERENCES

1. B.W. Hakki, J. Electrochem. Soc. 118, 1469 (1971).
2. G.B. Stringfellow, J. Appl. Phys. 43, 3455 (1972).
3. S. Mukai, M. Matsuzaki, and J. Shimada, Jpn. J. Appl. Phys. 20, 321 (1981).
4. S. Mukai, H. Yajima, Y. Mitsuhashi, and J. Shimada, Appl. Phys. Lett. 43, 24 (1983).
5. R.J. Livak and G. Thomas, Acta Met. 22, 589 (1974).
6. B. Golding and S.C. Moss, Acta Met. 15, 1239 (1967).
7. K.B. Rundoman and J.E. Hilliard, Acta Met. 15, 1025 (1967).
8. B. de Cremoux, P. Hirth, and J. Ricciardi, Gallium Arsenide and Related Compounds (Vienna) 1980, Inst. Phys. Conf. Ser. 56, 115 (1980).
9. J.W. Cahn, Acta Met. 10, 179 (1962).
10. L.D. Landau and E.M. Lifshitz, Theory of Elasticity (Pergamon Press 1959) pp. 16.
11. O. Ueda, I. Umebu, and T. Kotani, J. Crystal Growth 62, 329 (1983).

Table I. Summaries of $(\bar{g}\,\bar{b})$s analyses of dislocation loops in InGaAsP layer (λ_{PL}=790 nm).

$\bar{g}$	s	L_1 and L_2	L_3 and L_4	L_5 and L_6
2$\bar{2}$0	+	Outside	Inside	Inside
2$\bar{2}$0	+	Outside	Inside	x
$\bar{4}$00	+	Outside	Inside	Inside
$\bar{4}$00	+	Inside	Outside	Outside
040	+	x	x	Inside
2$\bar{4}$2	+	x	Inside	Outside
$\bar{b}$		a/2[10$\bar{1}$]	a/2[$\bar{1}$0$\bar{1}$]	a/3[$\bar{1}\bar{1}$1]
Character		Intrinsic	Intrinsic	Extrinsic

x: out of contrast.

Table II. Comparison of defect formation in III-V group compound semiconductors and alloy semiconductors.

Material	Formation of dislocation loops in as-grown crystals	Recombination enhanced defect motion
GaAs	Yes	Easy
GaAlAs	Yes	Easy
GaP	Yes	Easy
GaAsP	Yes	Easy
InP	No	Difficult
InGaAsP on InP	No	Difficult
InGaP (this work)	Yes	Easy?
InGaAsP on GaAs (this work)	Yes	Easy?

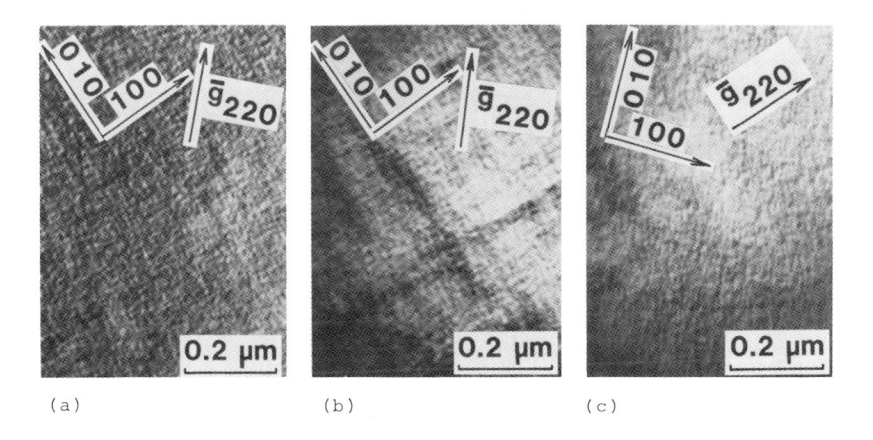

Fig. 1. Bright field TEM images of modulated structures in InGaAsP layers. (a) λ_{PL}=721 nm; (b) λ_{PL}=730 nm; (c) λ_{PL}=800 nm.

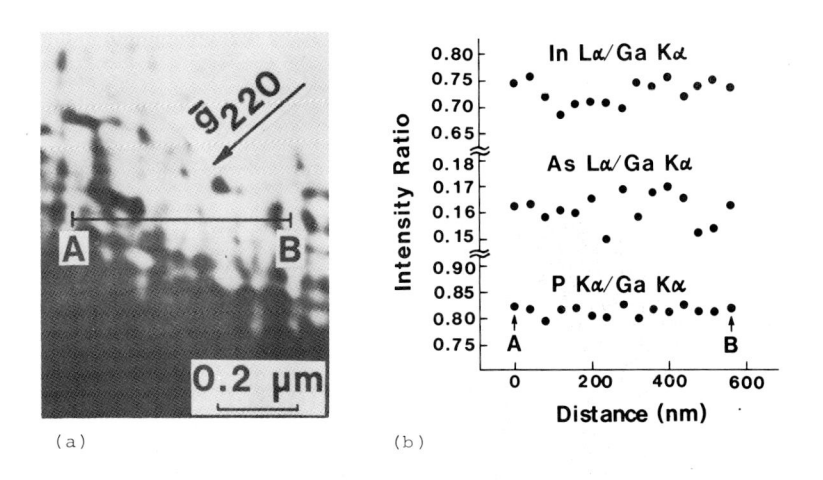

Fig. 2. STEM/EDX analysis of the modulated structure in InGaAsP layer (λ_{PL}=790 nm). (a) A STEM image; (b) intensity ratios along the line A-B in Fig. 3(a).

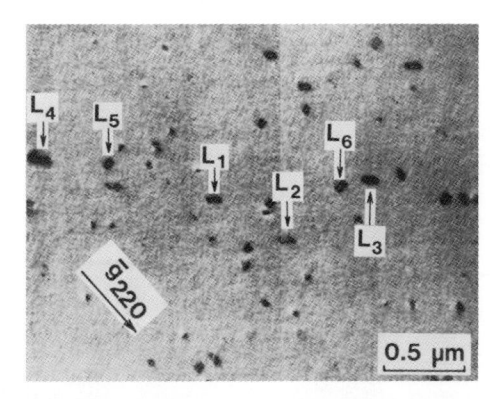

Fig. 3. A TEM image of dislocation loops in InGaAsP layer.

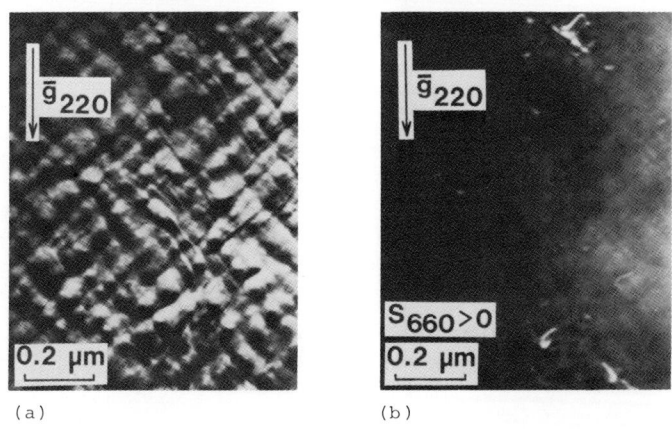

Fig. 4. TEM images of stress induced microdefects in InGaAsP layer on GaAs substrate. (a) A bright field image; (b) a weak beam image.

Fig. 5. A TEM image of precipitate-like microdefects in InGaAsP layer.

Fig. 6. A TEM image of stacking faults in InGaP layer on GaAs.

SECTION III
Surfaces and Interfaces

REFLECTION ELECTRON MICROSCOPY AND DIFFRACTION FROM CRYSTAL SURFACES

J.M. COWLEY
Department of Physics, Arizona State University
Tempe, Arizona 85287, USA

ABSTRACT

The recent revival of techniques for the imaging of crystal surfaces, using electrons forward-scattered in the RHEED mode and employing modern electron microscopes, has lead to the introduction of valuable new methods for the study of surface structure. Either fixed beam or scanning transmission electron microscopy (STEM) instruments may be used and in each case a lateral resolution of 10Å or better is possible. Simple theoretical treatments suggest that the contrast from surface steps may be attributed to a combination of phase-contrast, diffraction contrast and geometric effects. With a STEM instrument the image information can be combined with information on the local composition and crystal structure by use of microanalysis and microdiffraction techniques. Examples of applications include studies of the surface structure of metals, semiconductors and oxides, and the surface reactions.

INTRODUCTION

The early attempts by Ruska in the 1930's and others in the 1950's to image the surfaces of bulk specimens by using electron microscopy showed some success but were forgotten after scanning electron microscopy, using secondary electrons, proved to be a much more flexible and convenient method for examining surface morphology. The motivation for the revival of reflection electron microscopy (REM) in the 1970's [1] was the hope that images obtained using diffracted beams from crystals would provide useful information concerning structural variations on their surfaces and play the same role for surfaces as dark field imaging of thin films does in TEM. The method showed some promise although limited by the severe forshortening of the images resulting from the small diffraction angles of fast electrons and by the lack of clean conditions in conventional electron microscopes.

The full power of the technique was adequately demonstrated only when it was applied by Osakabe et al [2] to near-perfect surfaces of silicon crystals in a special microscope having high-vacuum capabilities and facilities for in-situ specimen surface preparation. At about the same time the special technique of Low Loss Imaging, using SEM instruments in a mode reminiscent of the earlier REM work, was introduced by Wells [3]. A forward scattering geometry was used to generate images of surfaces with quite high resolution. An energy filter was introduced to select for detection only those electrons which have lost no more than a small fraction of their incident energy. With a scanning system a variety of signals can be obtained in addition to the higher intensity signals, favored for imaging, which include the electrons either transmitted through thin specimens or reflected from surfaces, either with or without energy losses, or the secondary electrons. Local chemical analysis or images showing the distributions of chemical elements are obtained by detecting the characteristic X-rays emitted or the characteristic energy losses of electrons in solids. The STEM instrument, having the very bright and small electron source of a field emission gun allows all of these signals to be used with high spatial resolution, comparable with the resolution of a modern TEM image.

We will proceed by first summarizing and illustrating the current capabilities of the various techniques with some discussion of the image contrast in particular and then provide a few examples from recent applications.

HIGH ENERGY ELECTRON TECHNIQUES

Reflection High Energy Electron Diffraction (RHEED)

As in the case of TEM, consideration of the possibilities for the imaging of crystals must start with the diffraction pattern. RHEED patterns such as figure 1, obtained from relatively flat faces of large single crystals have been familiar since the early days of electron diffraction, fifty years ago. Almost without exception, the features of such patterns can be described in general terms but no complete theoretical analysis can be applied to the understanding of all details of the intensity distributions.

The inner pattern of almost regularly spaced spots is attributed to elastic diffraction by the crystal lattice plus some small angle inelastic scattering. It is recognized that strong dynamical effects must contribute to the intensity distributions of these spots and some limited measure of agreement between observed and calculated intensities has been obtained for very simple structures and chosen orientations [4]. More recently it has been argued that in the case of weak superstructure reflections due to surface reconstructions [5] and in a few other favorable cases the weak scattering assumptions of the kinematical diffraction theory may be applied, at least as a first approximation. While such assumptions may sometimes appear reasonable, a full theoretical justification for their use is not easily found.

If most of the diffraction intensity comes from transmission through small projections from the surface which are steep-sided and more than about 50Å high, the spot patterns will be regularly spaced for single crystals and will, in any case, resemble those given by transmission samples. For very flat single crystal faces for which steps, projections or other forms of roughness do not "shadow" more than a small part of the surface (see fig.2.), the spots will be displaced by refraction of electron beams at the surface.

Dynamical diffraction theory suggests that the displaced spots will remain sharp. The common observation that spots are extended into streaks has been given various explanations as arising from distortions of the surface, or surface steps or the presence of surface layers of spacing or composition different from that of the bulk crystal, but no adequate theoretical treatments of these effects have been given.

The available dynamical diffraction theories for RHEED apply to perfect crystals terminated at a planar surface [6,7,8] or else make a column approximation in which the same assumptions apply, independently, for all local regions of the surface [9].

By analogy with the development of diffraction theory for transmission electron diffracttion it may be anticipated that a more complete understanding of RHEED patterns may be achieved when more experience has been obtained with diffraction from very small perfect crystal regions or regions containing individual, well characterized crystal defects, using the methods mentioned below in connection with the discussion of STEM and microdiffraction.

FIG. 1 RHEED pattern of Pt

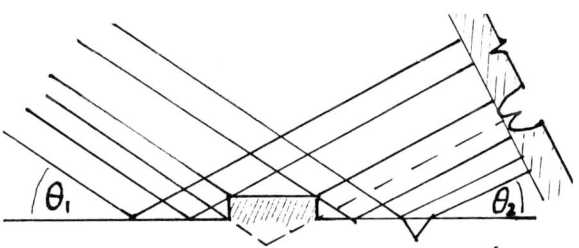

FIG. 2 Diagram: surface diffraction and REM intensity

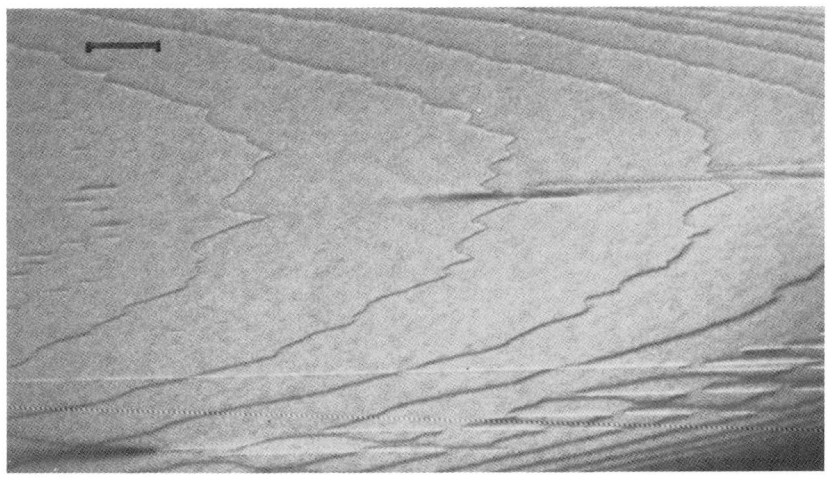

FIG. 3 REM image of surface of Pt crystal showing growth steps, slip traces and dislocations. Marker: 0.1µm

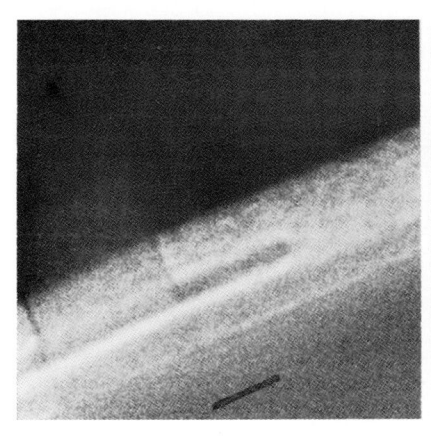

FIG. 4 SREM image of surface of MgO smoke crystal showing steps, etch pits. Marker 100Å

FIG. 5 SREM image of surface of Au with grain boundaries and corrugations. Marker: 0.1µm

Reflection electron microscopy (REM)

To obtain an image such as figure 3 of a relatively flat crystal face one uses the objective aperture of a TEM instrument to select one of the diffraction spots of a pattern such as figure 1 and uses the selected electrons to form the image with a magnification of usually 10^4 to 10^5. To avoid off-axis aberration effects, the selected diffracted beam should lie on the axis of the objective lens so that, as with transmission dark-field images, the crystal face is tilted by the Bragg angle θ_B (corrected for refraction) and the incident beam is tilted by an angle $2\theta_B$. Simple adaptions of TEM specimen holders can be made to allow thick specimens to be mounted in an appropriate configuration [10].

The wavey lines of figure 3 are surface steps, one atom high. Several dislocations intersecting the surface are indicated by the characteritic black-white streak contrast due to the associated strain fields. Straight lines due to slip traces are also visible.

The evidence for the single-atom height of the growth steps in such images is indirect. For example, the steps are observed to be associated with screw dislocations [10] or the movements of steps across a surface is correlated with the rate of evaporation of atoms from the surface [12]. The origin of the contrast given by such steps will be discussed later. In view of the limited understanding of RHEED pattern intensities, as outlined above, it is difficult to provide an adequate theoretical discription of all features of REM intensity distributions but simple approximations make it possible for useful descriptions to be given of the variations of relative intensities associated with localized features in images of moderate resolution.

Estimates of the resolution attainable in REM are strongly dependent on the assumptions made regarding the imaging conditions. For very small particles on a surface illuminated by a diffracted wave emerging from the surface one might expect that the resolution in REM would approach that for TEM since the imaging geometries are comparable. If one considers, however, that in such a case the electrons emerging from the surface may have an energy spread of about 100eV due to multiple inelastic scattering processes in the crystal, one would predict a resolution limited to about 10Å by the chromatic aberration of the objective lens. Such an energy limitation could presumably be reduced by suitable energy filtering. The best resolution observed in REM to date is probably that shown by Hsu [13] in resolving arrays of steps on a gold surface with an almost regular spacing of 9Å.

The high resolution possibilities for REM imaging refer only to lateral resolution i.e. to the resolution of detail in directions perpendicular to the incident beam. For the angles of incidence used, corresponding to Bragg angles between 1 and 4 degrees the image is usually foreshortened by a factor of from 20 to 40 and the resolution is correspondingly limited in directions almost parallel to the incident beam.

Small steep projections from a flat surface will, on the other hand, not be foreshortened. They will appear with full magnification (and resolution). Furthermore they will cast long shadows along the surface since they will intercept both the beam indicent from the source and the diffracted beam leaving the surface (figure 2). Thus the REM image, like the RHEED pattern will be very strongly affected by surface roughness.

Images such as figure 3 showing surfaces steps and other features with high contrast and good resolution have been obtained for samples of noble metals, semiconductors and other materials by the use of standard, commercial TEM instruments having no provision for the preparation or maintenance of clean surface conditions. There was ample evidence for the accumulation of amorphous contamination layers on the surface. Fortunately the influence of contamination layers on the images may be minimized if a small objective aperture is used to select a diffraction spot from the crystal and exclude most of the diffuse scattering from the amorphous material.

However it is evident that the range of applications for REM is severely limited unless cleaner specimen conditions can be achieved. For all but the least reactive materials the surface structures may be modified by exposure to air and the study of reactions at surfaces can be severely hindered or modified by a presence of amorphous carbon layers or other absorbed material.

A major step towards the REM study of clean surfaces and their reactions has been made by Yagi et al [14] in their use of a special high vacuum instrument having means for the in situ heating, ion bombardment and evaporation of materials on the surface. This has permitted them to make studies of clean silicon surfaces showing the details of the processes of formation and transformation of the reconstruction superstructures on the surface and of the surface superstructures formed with the addition of small amounts of metal [2,15]. The current move towards the building of new TEM instruments having comparable or better vacuum and specimen treatment facilities should extend the power of the REM method considerably.

Scanning reflection electron microscopy (SREM)

For reflection, as in the case of transmission imaging, it is possible in principle to reproduce all the imaging modes of the fixed beam instruments by use of equivalent scanning modes and vice versa [16]. In practise the quality of the STEM and SREM images may be limited by signal-to-noise considerations and other factors, but the scanning modes may, on the other hand provide a variety of new possibilities for which the fixed beam instruments are not appropriate. These possibilities include the use of special detector configurations for the imaging mode and the microanalysis and microdiffraction studies of small selected features of the image.

SREM images should have resolution comparable with that of REM and in images such as figure 4 the resolution is clearly 10Å or so. As in this figure, single-atom surface steps have been seen in SREM but never with the same high contrast and clarity as in REM.

In SREM, as in STEM, it is customary to reduce the electron noise as much as possible by using the largest possible sizes for the objective (illuminating) and detector apertures consistent with the required resolution. However, as will become evident from the subsequent analysis of factors affecting image contrast, such an approach may reduce the contrast given by small surface steps and some other features of the specimen.

The loss of contrast may come from two sources. Firstly, the use of large apertures involves the integration over a wide range of angles of incidence or collection so that contrast features which are dependent on coherent interference effects, including phase contrast, will be reduced. Secondly, when large objective or collector apertures are imployed a large proportion of the image signal may come from the diffuse background usually present in reflection diffraction patterns rather than from the sharp Bragg reflections. The contribution to the image from the diffuse scattering is of relatively poor resolution and shows little dependence on the diffraction effects. When the detector aperture collects large regions of the diffraction pattern containing only weak diffraction spots, the image shows mostly the surface morphology as in figure 5, with little evidence of even the gross changes in diffraction conditions associated with a high-angle grain boundary.

Low Loss Imaging

As a variant of the SEM mode using high energy backscattered electrons, Wells [3] found that for incident beams making angles of about $30°$ with the specimen surface, good resolution could be obtained by detecting forward-scattered electrons, especially if the counter-field energy filter is used to prevent contributions to the image signal from electrons which have lost more than a few hundred eV in inelastic scattering processes.

The geometry of the scattering process has been studied by Wells [17] who showed that both the image resolution and the penetration into a surface of an Al specimen may be about $100 Å$ for 15keV electrons. The images should show some dependence on atomic number and some dependence on crystal orientation may appear as a result of channelling effects.

In low-loss imaging experiments on the surfaces of thin metal foils, Smith and Treacy [18] have used a STEM instrument and achieved reasonable resolutions. The experimental arrangements and imaging conditions for these low-loss imaging experiments are very close to those of the present author using STEM to obtain images such as figure 5. This suggests that a continuous range of imaging conditions, from the LLI modes to the diffraction- based STEM modes, is readily available to provide a variety of image signals emphasizing various aspects of the specimen surface structure.

Microanalysis of surfaces

When a strong diffracted beam is produced at the flat surface of a perfect crystal the penetration of the electrons into the crystal may be very small. The excitations of atoms giving rise to characteristic X-ray emission or characteristic energy losses may occur mostly in surface layers having the thickness of only a few planes of atoms. In principle, then, the methods of microanalysis used in conjunction with SREM imaging should provide very sensitive probes of the composition variations of the surface.

The few experiments made to date to test these possibilities have not been very encouraging. Krivanek et al [19] showed that, while some inner shell absorption edges could be detected in energy loss spectra obtained from diffraction spots in RHEED patterns from silicon , the background to the spectra tended to be high, especially for the stronger low angle reflections.

The situation requires further investigation. Simple considerations suggest that the effective depth of penetration of electrons into a surface and the contributions to the EELS spectrum backgound from multiple inelastic scattering could be very strongly influenced by the presence of steps or other perturbations of the surface. Electrons incident in, or scattered into, directions deviating from the Bragg angle can penetrate much deeper into the crystal and will be free to undergo multiple inelastic scattering processes before leaving the surface. It may well be, however, that the predominant contribution to the background of ELS spectra comes from multiple surface plasmon losses [20] which are difficult to avoid.

Microdiffraction from surface structures

As in the case of STEM, an important advantage of the scanning mode for the imaging of surfaces is that the beam may be stopped at any selected point within the image and the diffraction pattern from the corresponding part of the specimen may be recorded. The region giving the diffraction pattern may have a diameter comarable with the resolution limit of the microscope which may now be 3-5Å. Some applications for diffraction with these very small beam diameters are now being explored for the STEM case [21] but for SREM microdiffraction has, as yet, been used only with beam diameters of 10Å or more.

As in the STEM case, the microdiffraction patters are essentially convergent beam diffraction patterns. The beam diameter at the specimen is inversely proportional to the objective aperture angle if a coherent electron source, such as a field emission gun is used. The objective aperture size determines the diameters of the individual diffration spots in the diffraction pattern. Hence patterns from regions of small diameter inevitably have spots which are of large size. Also the diffraction spots may have irregular shapes and intensity distributions [22]. While these details may in principle be interpreted in terms of specimen structure, such interpretations are currently not feasible in most cases. As a consequence it becomes difficult to use the patterns for accurate determination of lattice constants or the analysis of complicated structures. However, the microdiffraction patterns may be extremely useful for many purposes.

The insert in figure 7, for example, is the diffraction pattern from one of the small gold crystals, about 20Å in diameter, aligned on the face of a large crystal of MgO. Such patterns show very clearly the variations of the structure or alignment of the small gold crystals corresponding to any of the features visible in the image.

CONTRAST IN REM AND SREM IMAGES

Examination of the details of REM images such as figure 3 shows that the contrast of the images of surface steps varies strongly form one part of the image to another. Close to the in-focus line across the middle of the image, in a direction perpendicular to the beam, the contrast disappears for "down" steps although not for "up" steps which appear as thin dark lines. Away from this region the contrast of both up and down steps increases with a strong anti-symmetric (black-white) component of the intensity distribution.

The contrast variation with defocus is strongly reminiscent of that for the TEM imaging of thin phase objects. This phase contrast component of the imaging of steps was recognized by Osakabe et al [2], who proposed that strain contrast, or contrast due to variation of diffraction conditions associated with local lattice strains, should also be present for steps, as for dislocations.

The diffraction contrast due to the strain fields of dislocations was treated theoretically by Shuman [9] using a column approximation in which it was assumed that the intensity for any point of the surface is equal to that given by a perfect crystal having the orientation of the small local region within the strain field. Shuman predicted the asymmetric, black-white, double streak form of the contrast which appears, for example, in figure 3, indicating the intersection of a dislocation with the surface.

For large steps or other features on the surface which have dimensions comparable with, or larger than, the resolution limit of the microscope used, there will be other contributions to the image contrast. As suggested by the diagram of figure 2, shadowing effects will give an image of the object plus an apparent reflection of the object in the surface plane. High steps will thus appear in the image as bands of width equal to twice the height of the step if viewed in one direction (looking up the step from the detector direction in the REM case) or of zero width if viewed in the opposite direction. For images of such steps obtained far out of focus, fringe patterns may appear along the steps as a result of interference between those electron waves deflected by refraction effects at the surface and those not affected by refraction because they emerge from the faces of he steps [23].

Phase contrast imaging of steps

We consider a step on a perfect crystal surface of height very much less than the resolution limit of the microscope. We neglect possible effects of a strain field around the step or the dipole field which may be associated with the step, and consider only the difference in phase between waves reflected from the top and bottom of the step (figure 2). Bragg's Law tells us that if the angles of incidence and reflection, Θ_1 and Θ_2, are both equal to the Bragg angles, Θ_B, for the nth order reflection from the lattice planes of spacing d, equal to the step height, the phase difference will be $2 n\pi$. If the angles Θ_1 and Θ_2 differ for Θ_B by small amounts Δ_1 and Δ_2, respectively, the phase difference depends on Δ_1 only and will be given by

$$E = 2\pi n (1 + \Delta_1 / \Theta_B).$$

Since a phase differece of $2\pi n$ is equivalent to zero phase difference, the important factor will be the deviation from the Bragg angle of the incident beam. For even simple structures with small step heights (1-2Å) a deviation of 10^{-3} radians in the incident beam direction can give large phase differences, comparable, for example, with those of transmission specimens of amorphous carbon several hundred Å thick. Hence the phase contrast effects for step images must be expected to be strong.

An expression for the image intensity is obtained readily by considering the equivalent transmission object having a transmission function

$$q(x) = \begin{cases} 1 & \text{if } x < 0 \\ \exp(i\varepsilon) & \text{if } x > 0 \end{cases}$$

The phase factor ε will be of opposite sign for positive and negative deviations from the Bragg angle and also for steps up and steps down.

The amplitude in the back focal plane, given by Fourier transform of $q(x)$ is multiplied by an aperture function $A(u-B)$, corresponding to an aperture displaced off-axis by a distance B, and the phase factor due to the defocus f and the lens aberrations. Fourier transforming again to give the image amplitude and intensity provides an expression made up of a constant term, two first order terms of weight $\sin \varepsilon$ and two second order terms of weight $(1-\cos \varepsilon)$.

For small phase changes, ε, and $B = 0$, the intensity shows only an anti-symmetric (black-white) component which reverses with the defocus, Δf, as well as with the sign of ε. For a finite objective aperture displacement, B, a symmetrical term is added and this term will reverse sign with B as well as with ε, but not with Δf.

For larger phase changes, ε, the second order terms which are independent of the sign of ε are added. One such term is given by the square of the first order terms and so disappears for $\Delta f = 0$ or $B = 0$. The other second order term is always present, giving a narrow dark line which may be attributed to an "absorption" due to the loss of energy in the parts of the diffraction pattern cut off by the objective aperture.

As a result of this analysis we may deduce that in REM images small surface steps may appear with high contrast but the intensity distributions across the steps may vary strongly with the imaging conditions. The steps may appear bright or dark with antisymmetrical black-white contrast of either sense added to either black or white symmetrical components. This is in agreement with experimental observations.

Useful deductions can be made from the theory regarding the effects of experimental conditions on the observed contrast. It is seen, for example, that use of a large detector aperture in SREM, which is equivalent to the use of a large convergence of the incident beam in REM may reduce the antisymmetrical components of the contrast, reducing the visibility of the steps. This also is in agreement with observatons. One prediction of this theoretical treatment is that for a plane wave incident at the exact Bragg angle, ($\varepsilon = 2\pi n$), no contrast will be visible for any defocus. The presence of a strain field in the crystal or of a dipole field at the step edge could give contrast even under these conditions.

Before making tests for these possibilities, however, it is necessary to test a more fundamental assumption of the theory, namely the use of column approximation. It is not at all clear that for the reflection geometry the column approximation will have even that limited range of validity which it has for the transmission case. The only way in which the validity of the column approximation may be adequately tested is by comparison with complete many-beam dynamical diffraction calculations made for the reflection case with surface defects present. While various proposals have been made for methods by which such calculations could be made, no case is known of such calculations being carried out in practise. The task is not trivial.

Images of dislocations

The contrast generated by the strain field surrounding dislocations intersecting a surface has been calculated by Shuman [9] for the case of screw and edge dislocatons by use of a multibeam dynamical diffraction theory. Making use of a column approximation, Shuman found the variations of diffracted beam intensity for the appropriate deviations from the Bragg condition associated with the lattice deformations. Similar calculations were made by Osakabe et al [12] who showed contrast features, similar to the predicted ones, in REM images of silicon surfaces.

The variation of the image with defocus, not predicted by the Shuman treatment, is one result of an effect which he omitted. For a dislocation having a Burgers vector not parallel to the surface, it is expected that the surface plane of the crystal will be perturbed. The surface will rise or fall along any line passing near the dislocation core. When viewed at a small glancing angle, this will have much the same appearance as a surface step. This will give rise to two contrast-producing effects. The change in height of the surface will give phase changes of the diffracted waves as in the case of a localized step and the same phase contrast effects will be produced. Also there will be purely geometrical effects arising because the varying angle of inclination of the surface will modify the energy density of incident and reflected beams on the surface [1].

This latter type of effect will give contrast variations rather similar to those given by the variations of diffraction intensity but may be of the same or opposite sign. The phase contrast contributions may add anti-symmetric components in the beam direction. Thus any attempts to interpret the image intensities to provide information on the nature of the dislocations involved must be made with careful control of the imaging conditions.

APPLICATIONS OF REM

Surface steps

Individual atom-high steps have been observed on single crystal faces of an increasing number of materials. The original work of Osakabe et al [2, 12] on silicon, showed the favored configuration of steps, the movements of steps at high temperatures associated with vaporization from the surface, and the correlation of surface steps with the emergent screw dislocations.

Observations of steps on (111) surfaces of face centered cubic metals by Hsu et al [11,13] revealed several interesting features. It was possible to distinguish growth steps, which usually meandered across the surface, and steps associated with slip traces occurring on close-packed planes and so intersecting the surface in straight lines in [110] directions. Since intersections of growth and slip steps represent points of high local energy, there is a tendency for atoms to diffuse away from these points, giving characteristic configurations of non-intersecting steps and the appearance of cusps and bows in the reflection images such as figure 3.

It is known from diffraction evidence that vicinal faces, making small angles with principle planes, may contain regularly spaced arrays of surface steps. This has been graphically demonstrated by Hsu [13] in images showing that vicinal faces form on bulk gold specimens with a strong preference for particular step periodicities. Much of the surface of a small gold sphere formed in air may be made up of faces having steps 57, 32 or 9Å apart, giving a characteristic pattern of wrinkles around the flat (111) surface facets.

On crystals having somewhat more complicated structures, such as GaAs, [10], the heights of steps formed by cleavage are more variable and the correlations with dislocations having different possible Burgers vectors becomes more complicated. Steps of height equal to a fraction of the basic unit cell dimension may leave exposed different atom configurations, giving rise to contrast variations across the steps [24].

Surface reconstructions

The periodic arrays of steps on vicinal surfaces may be regarded as surface superstructures having two dimensional unit cell dimensions much greater than the periodicities of the surfaces on the principal planes. Other surface superlattices are formed by rearrangements of the atoms on the principal plane surfaces.

In the most important work on surface reconstructions to date observations of the transition from the 7x7 supertructure to the 1x1 structure and vice versa on silicon (111) surfaces at about $830°C$ [15] have provided important new insights into the nature of this much-studied transformation. Observations on the formation of surface superstructures produced with fractional monolayers of gold or silver on silicon are equally graphic.

Epitaxial growth and surface reactions

One general area where the use of the fast electron surface research techniques should have important advantages is the study of small additions of surface layers to crystal surfaces, either by the use of physical processes such as vacuum evaporation or sputtering or else by chemical reactions of the surface with gases (possibly in situ) or liquids or solids (in absentia). For many specimens prepared in such ways the added structures are not spatially homogeneous. Nucleation and growth at surface imperfections is common. A knowledge of the detailed morphology is essential for the proper understanding of the processes involved and frequently knowledge of local structural and compositional variations is also important.

The number of applications of the fast electron techniques in this area is still very limited. In ultra-high vacuum systems studies have been made with limited spatial resolution of some nucleation and growth processes [25, 26]. Some preliminary studies have been made with STEM of interaction of nobel metals with oxide surfaces [27].

One result of interest is represented by figure 6. Thin layers of gold evaporated indirectly on MgO smoke crystals take the form of small isolated particles of about 20Å diameter, separated by distances of the order of 100Å. Microdiffraction patterns of individual gold particles show them to be untwinned single crystals, very well aligned in epitaxial orientation with the MgO surface. The extraordinary feature of such images is that the gold crystals appear to be aligned along straight lines of length 1000Å or more. Presumable these lines correspond to straight line steps on the MgO crystal surfaces. In the images it is seen that these lines appear to be inclined at angles of about 60 to 120° to each other, but when the foreshortening factor of the SREM image is removed, it is clear that the angles between these lines are 2 to 4°. Measurement of many such angles suggests a distribution of inclinations with a broad maximum at about 3°. It is not easy to understand why such angles between surface steps should be preferred.

Acknowledgment

The examples used to illustrate this report have been drawn from the research programs supported by NSF grants DMR-7926400, DMR-8015785 and INT 81-14363 and by DOE Contract DE-AC02-76 ER02995, making use of the resources of the Facility for High Resolution Electron Microscopy funded by NSF grant CHE-7916098 within the Center for Solid State Science, ASU.

REFERENCES
1. P.E. Højlund Nielsen and J.M. Cowley, Surface Science 54, 340-354 (1976).
2. N. Osakabe, Y. Tanishiro, K. Yagi and G. Honjo, Surface Science 97, 393-408 (1980).
3. O.C. Wells, Appl. Phys. Letters 19, 232-235 (1971).
4. R. Colella and J.F. Menadue, Acta Crystallogr. A28, 16-22 (1972).
5. S. Ino, Japan. J. Appl. Phys. 19, 1277-1290 (1980).
6. R. Colella, Acta Crystallogr. A28, 11-16 (1972).
7. A.R. Moon, Z.f. Naturforsch. 27a, 390-395 (1972).
8. P.A. Maksym and J.L. Beeby, Surface Science 110, 423-438 (1981).
9. H. Shuman, Ultramicroscopy 2, 361-369 (1977).
10. T. Hsu, Sumio Iijima and J.M. Cowley, Surface Science. In press.
11. T. Hsu and J.M. Cowley, Ultramicroscopy. In press.

12. N. Osakabe, Y. Tanishiro, K. Yagi and G. Honjo. Surface Science 102, 424-442 (1981).
13. T. Hsu, Ultramicroscopy 11, 167-172 (1983).
14. K. Yagi, in Scanning Electron Microscopy/1982 Part IV, Om Johari, Ed., (SEM Inc., Chicago 1982) pp. 1421-1428.
15. K. Yagi, K. Takanayagi and G. Honjo in Crystals, Properties and Applications 7, H.C. Freyhardt, Ed., (Springer Verlag, Berlin, 198:") pp. 48-74.
16. J.M. Cowley, Appl. Phy. Letters 15, 58-59 (1969).
17. O.C. Wells in Scanning Electron Microscopy/1972 (Part I) Om Johari and Irene Corvin, Eds. (IIT Research Institute, Chicago, 1972) pp. 169-176.
18. D.A. Smith and M.M.J. Treacy, Applic. of Surface Sci. 11/12, 131-142 (1982).
19. O.L. Krivanek, Y. Tanishiro, K. Takanayagi and K. Yagi, Ultramicroscopy 11, 215-222 (1983).
20. J. Schilling, Z.f. Physik B25, 61-67 (1976).
21. J.M. Cowley, in Scanning Electron Microscopy/1982 Vol 1, Om Johari, Ed., (SEM Inc., Chicago, 1982) pp. 51-60.
22. J.M. Cowley and J.C.H. Spence, Ultramicroscopy 6, 359-366 (1981).
23. P.S. Turner and J.M. Cowley, Ultramicroscopy 6, 125-138 (1981).
24. S. Iijima and T. Hsu in Electron Microscopy 1982, Vol 2, The Congress Organizing Committee, Eds. (Deutsche Gesellschaft f. Elektronenmikroskopie e.V., Frankfurt, 1982) pp. 293-294.
25. J.A. Venables, G.D.T. Speller, D.J. Fathers, C.J. Harlund and M. Hanbuchen, Ultramicroscopy 11, 149-155 (1983).
26. T. Ichinokawa in Proceedings of Ninth International Conference on Solid Surfaces, Madrid, October 1983.
27. J.M. Cowley and Z.-C. Kang, Ultramicroscopy 11, 131-140 (1983).

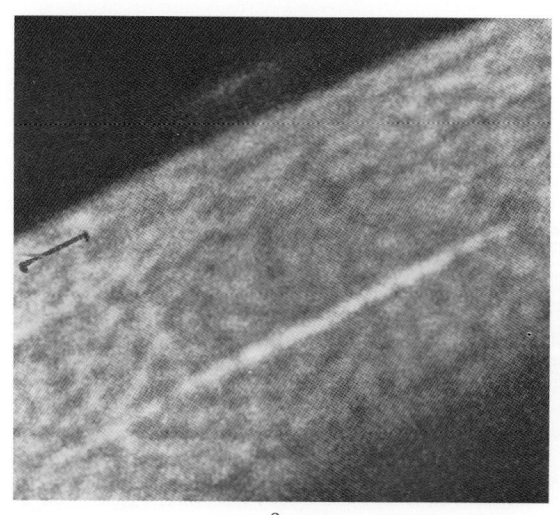

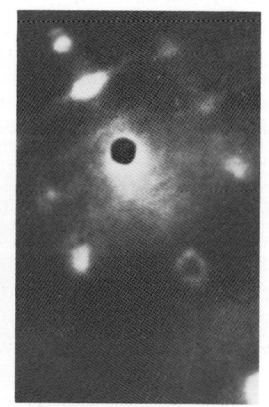

FIG. 6 SREM image of 20Å Au particles on MgO crystal surface. Marker: 100Å. Diffraction pattern from one Au particle and MgO support.

INTERPRETATION OF THE ATOMIC SURFACE STRUCTURE OF Ag$_2$O ON (111) Au THIN FILMS

William Krakow
IBM T.J. Watson Research Center, P.O. Box 218 Yorktown Heights, N.Y. 10598, U.S.A.

ABSTRACT

A high resolution electron microscope investigation of the residual oxidized silver of a few monolayers thickness on the surface of (111) Au films has shown that a reconstructed (2x1) surface structure occurs for (110) oriented Ag$_2$O and can be observed at atomic resolution levels. Image enhancement via a digital frame store processor has revealed improved images which have then been compared to computer simulated diffraction patterns and images of the Ag$_2$O surface. Several iterations of surface structure models and image simulations reveal that the (2x1) reconstruction is consistent with a missing row model. The atomic arrangements of these rows often undergo a translation along the direction of the row to produce cusp like image features. It has also been possible to observe the effect of contraction of the underlying layer which can produce diagonal contrast lines in the images. These features often vary rapidly over lateral distances of a few tens of angstroms and give an indication of the surface topography and the degree ordering of the surface.

INTRODUCTION

In this paper I shall give the results of additional work done to understand the structure of (110) surfaces of Ag$_2$O on (111) Au support films. The first paper on this subject [1] basically dealt with the observation of atomic level structure from high resolution micrographs and the (2x1) surface reconstruction, indexing selected area diffraction patterns and detecting the (2x1) structure in computer generated optical diffraction patterns from original images. Also, general statements were made about the atomic positions of the uppermost and underlying layers of Ag$_2$O based upon the contrast features observed. However, these statements were made with caution since computer modeling and image simulations were not performed at that time. The present paper, therefore, deals with extensive computer experiments for different amounts of Ag$_2$O layer coverages and modification of these layers by shifts of rows and contractions between pairs of Ag atoms. These simulations are then matched as closely as possible to high resolution images which have undergone computer enhancement techniques for noise removal and elimination of long range contrast variations.

Briefly, the imaging of surface structure in a conventional transmission electron microscope (CTEM) has been achieved by dark-field techniques [2] and more recently in an ultrahigh vacuum (UHV) environment [3-6] using bright-field techniques. All of these techniques rely on diffraction contrast mechanisms and generally are limited to perhaps 10 to 20 angstroms resolution. Although the latter work has reported several types of surface reconstructions they may be complicated by an underlying support substrate producing moire' fringes and perhaps interfacial boundaries. Generally the resolution level achieved is approximately one order of magnitude worse than atomic resolution levels possible in high resolution electron microscopes and hence precludes identification of many surface reconstructions.

Atomic level resolution has been achieved by the present author for (001) Au film surfaces [7-9] and more recently with (111) Au films [10] using bright-field phase contrast imaging. The point resolutions required to visualize these structures are 2.86Å and 2.48Å respectively, where this latter number is near the resolution limit of a modern 200kV electron microscope. In these studies very thin, free standing, films of 30 to 40Å thickness were employed. Both top and bottom surfaces were visualized with no evidence of surface reconstructions. This may be due to the fact that the (111) Au specimens were exposed to air, however, a lower resolution study by Heyraud and Metois [11] reported a 23x1 superstructure from the observation of diffraction data and ~63Å fringes in the images, even upon air exposure. The important point to emphasize is atomic resolution can be achieved for imaging surface structure and surface atom positions in Au. Also, as demonstrated in reference [1] it is possible to image surface atoms as light as Ag and hence the Ag_2O surface atomic structure which has reconstructed to a (2x1) superstructure. This result is consistent with several controlled surface studies of the adsorption and interaction of oxygen with the (110) surface of silver investigated by LEED and Auger electron spectroscopy (AES) techniques[12-14]. At saturation coverage these studies all report a (2x1) superstructure with lesser coverages producing (nx1) structures with high values of n. This leads to the conclusion that normal exposure to air has produced the saturated (2x1) structure observed in the high resolution electron microscope.

A recent study of small Au particles by high resolution electron microscopy has reported viewing particle surfaces edge-on [15]. Here a (2x1) surface reconstruction is claimed with a double periodicity being observed by looking down projections of columns of atoms. It must be pointed out that these results do not represent actual atom imaging but rather a bulk type of atomic scale lattice imaging. Therefore, limited information is obtained about the arrangement of the surface or its reconstructions since a projected average image is observed which is complicated by dynamical diffraction effects. As will soon be shown the correct way or more revealing method for imaging surfaces is flat-on in order to reveal a 2-dimensional representation of the surface. By computer modeling it is then possible to generate various different surface and underlying layer modifications to gain insight into the three-dimensional structure of a surface and the contributions of underlying layers to image contrast.

EXPERIMENTAL RESULTS

The experimental procedures for preparing single crystal (111) Au films on thick (~5000Å) Ag have been described in reference [1]. The end result after removal of most of the Ag is a residual amount of Ag_2O supported by an ~30Å thick single crystal (111)Au film suitable for electron microscopy. Electron microprobe analysis has revealed that the average coverage of Ag_2O is approximately 5 to 10Å. A typical selected area electron diffraction pattern of Ag_2O and (111) Au is given in Fig. 1a. Here the six brightest higher order {220} reflections are due to the bulk lattice of Au with spacings of 1.43Å. It is possible to index the other reflections as being due to Ag_2O. No evidence of any sulfides or nitrogen silver compounds were found. The structure of Ag_2O is identical to that of the Cuprit structure with a space group symmetry of Pn3m. The diffraction pattern therefore consists of three different orientations of (110) Ag_2O where each orientation or variant is rotated by 120°. Each variant bears an exact crystallographic relationship to the underlaying (111) Au film. The identification of the three different variants in the diffraction pattern is given in Fig. 1b. Here the variants are represented by open circles, crosses and triangles; while the Au reflections are given by the large and small closed circles. It is also possible to identify what atoms are contributing to each Ag_2O variant by using the space group selection rules of the International Tables for X-ray Crystallography. Fig. 1c shows an example of indexing one variant and assigning atom types to different reflections. It should be noted that a very close correspondence of intensities exists between the experimental diffraction pattern intensities and those listed in the JCPDS tables. Only the fainter {001} reflections with spacings of ~4.7Å are not predicted but can be attributed to monolayer converages of (110) Ag_2O or extra surface layers producing a fractional unit cell occupancy and hence forbidden reflections.

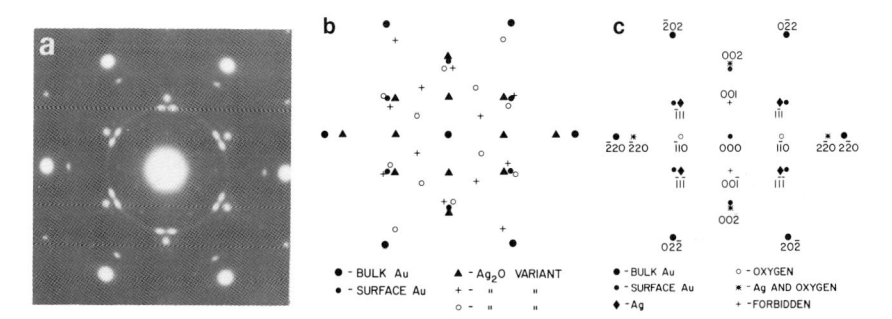

Fig. 1. (a) Electron diffraction pattern of Ag_2O on a (111) Au film. (b) Schematic diagram showing the three variants of (110) oriented Ag_2O. (c) Indexed pattern of one variant of Ag_2O in a (110) orientation.

Now turning to high resolution electron microscope imaging mode, Fig. 2 demonstrates the direct imaging of a Ag_2O patch several hundred angstroms in extent. Here the image was formed with a 200kV incident beam energy in the axial illumination mode at a direct magnification of 680,000X. The image clearly shows a 6.7Å periodicity characteristic of a (2x1) reconstruction which is not allowed in the bulk lattice structure. Orthogonal sets of fringes with spacings of 4.7Å are visible, which are also not allowed. The 2.7Å diagonal fringes can be attributed to {110} reflections of Ag_2O. The images generally are noisy since only a few monolayers of Ag_2O are being visualized. For specific image features in Fig. 2. it can be seen that the local details vary rapidly over a few tens of angstroms. Three regions are particularly evident in this micrograph where there

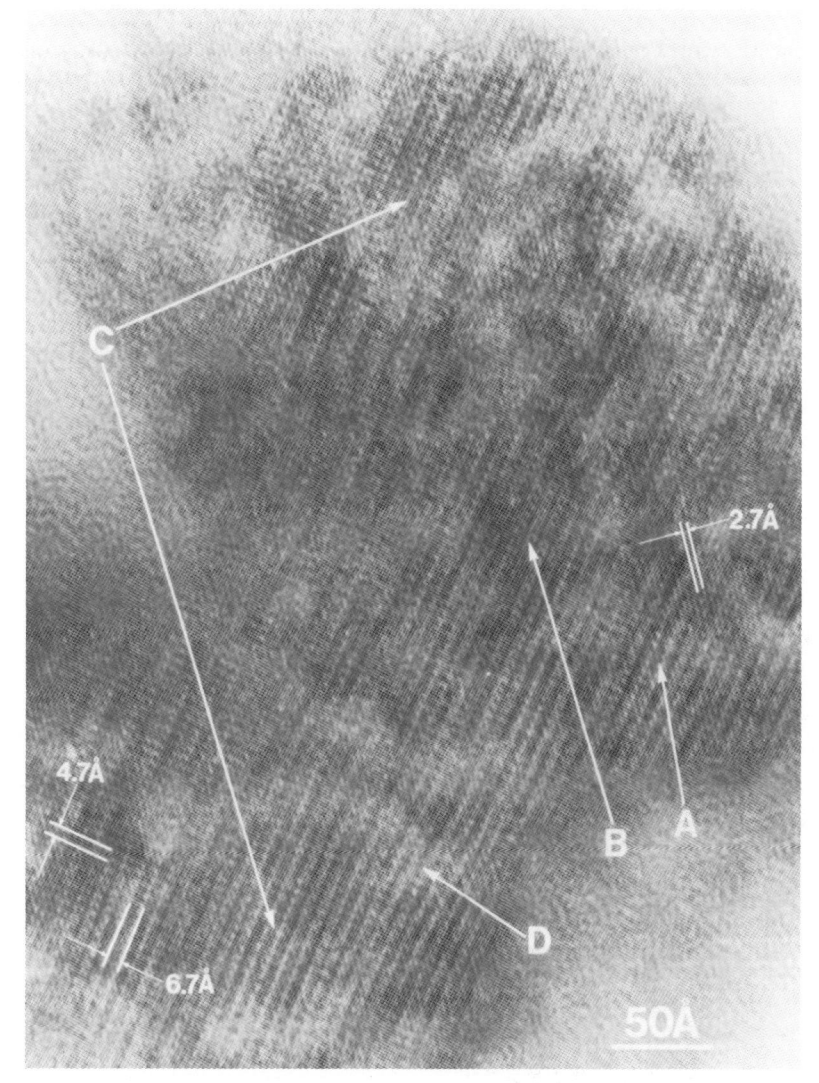

Fig. 2. High resolution bright-field image of an Ag_2O patch.

are: white diagonal fringes (region A), cusplike features (region B), black dots separated by fine white lines (region C). It can be seen that most of these regions are ill-formed and often change from one structure to another. Also there are often lateral shifts of the large 6.7Å spacings as evidenced in region D. This latter effect can best be seen by looking along the length of these fringes. This indicates that the surface which is reconstructed has missing rows of Ag atoms and the position of the missing row is arbitrary across the surface. In other words the uppermost layer can occupy either one row or a row translated by one half the

6.7Å periodicity. This ambiguity accounts for the lack of diffraction spots corresponding to 6.7Å since equal contributions from each of these area types would yield a null result in a diffraction pattern. It is only when very small regions of one type are sampled in computer generated optical diffraction patterns, as demonstrated in ref. [1], that the 6.7Å periods are visible.

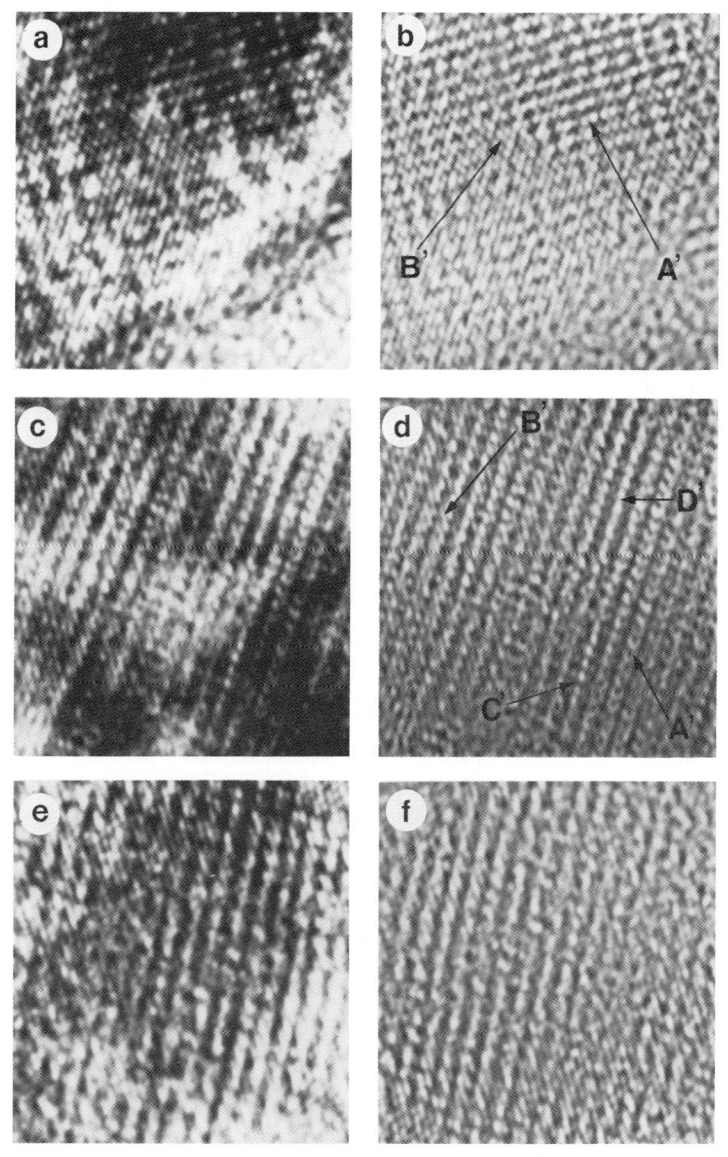

Fig. 3. (a)(c) and (e) Original micrographs of Ag_2O on (111) Au. (b), (d) and (f) are the corresponding digitally filtered images.

In order to compare the real experimental results with computer generated images of the Ag$_2$O structure a reduction in both noise and long range contrast variations of the original images was desirable. This was accomplished by use of digital Fourier filtering program which could simultaneously both high and low pass Fourier filter a digitally acquired image of 256x256 pixels in extent[16]. The original images and results of filtering are given in Fig. 3 for a selection of regions. Specifically, Figs. 3a and 3b show an edge region of Ag$_2$O where both a rectangular mesh is present characteristic of one monolayer coverage and a diamond or centered rectangular structure characteristic of two layers. Image 3c and 3d show a reconstructed region with cusp like bright features and separate diagonally connecting features. Images 3e and 3f show a shift of the 6.7Å fringes of a reconstructed region as well as the local variations in image features. With these few images it is possible to show the correspondence with models and calculated electron microscope images.

Model Building and Computer Experiments

In order to perform image simulations on the Ag$_2$O structure it was first necessary to construct layer models of the (110) surface. Fig. 4 shows two views of the Ag$_2$O structure. In 4a one can see the stacking of lattice planes along the [110] corresponding to an ABA'B stacking sequence. The sequence alternates between Ag-oxygen containing planes designated A and A' and those containing only Ag designated B. The difference between the A and A' layers is a translation vector of 3.35Å along a [110], otherwise, the structures are identical. The orientation used for microscope imaging is given in Fig. 4b where we are looking down the [110] direction. The rectangular meshes generated from single layers are designated at the bottom left in Fig. 4b, where each mesh for the unrelaxed structure would measure 3.35Åx4.74Å. The effect of the Ag atoms on electron scattering is approximately five times greater than that of oxygen, as can be seen by inspecting the values of their respective atomic scattering factors. Therefore, one is basically only visualizing the contrast from Ag atoms in the bright field phase contrast images. The main role of the oxygen atoms would therefore be to produce shifts or rearrangements of th Ag surface atoms into relaxed structures not realized in a bulk lattice.

Also displayed in Fig. 4b are schemes for surface reconstructions. Here every other row of atoms which is designated by the dashed circle around the + symbol would be removed. The cross hatched regions indicate various contrast features that would possibly arise if atom movement of the top layer or next layer below the surface occurred. For the diagonal features at the left and middle the underlying layer would be contracted towards the + symbol of the next layer as indicated by the arrows. The cusp like features would arise by the rows of atoms of the uppermost layer translating along the direction of the filled rows as indicated at the right side of Fig. 4b. While these features can be understood by the proximity of the Ag atoms as seen in projection and the resolution criteria of the microscope employed, it is at best only intended to provide some qualitative agreement with the real images of Figs. 2 and 3. The more exacting assessment of contrast features in the images must be made by constructing computer models which will now be discussed.

In order to perform image simulations for the (110) Ag_2O surface structure it was necessary to first set up various models of the surface structure infered from the contrast features and periodicities observed. This task was somewhat simplified by excluding the scattering of the oxygen atoms hence only A and B layers need be considered as designated by the rectangles in Fig. 4b. Schematically the types of contractions are given in Fig. 5. where the solid circles represent the underlying layer which has arbitrarily been designated as an A layer and the + represents the uppermost layer which is designated as a B layer. Here Fig. 5a corresponds to the case where atoms have been translated towards the + symbol by ~6% which makes the overall diagonal contraction ~12%. The layer designation here is then given as, CA12. Fig. 5b refers to a diagonal contraction of 10% for each atom hence the designation, CA20. Fig. 5c is the case of an x-axis contraction of 10% or a total of 20% and the designation, CAX. If the B layer has every other row of atoms missing, its designation is MB. If the missing B layer is translated along the direction of the row as depicted in Fig. 5d its designation is MBS. In this case the shift was taken as ~1.0Å. Also it is possible for the missing row B layer to shift by half a lattice period of 3.35Å along [110] and is then designated as, MBH. Of course, completely full unshifted and uncontracted layers are simply designated as A and B. Two examples of full models of layers used in the diffraction and image calculation are given in Fig. 6. Here 6a corresponds to a CA20 layer containing 485 atoms while 6b is a MB layer with 245 atoms. From these arrangements kinematical scattering distributions were generated which could then be used to compute various different stacking arrangements of layers by multislice dynamical diffraction. [17] The different layer sequence cases considered are summarized in Table I below.

TABLE I. Layer Sequences Used for Multislice Diffraction Computations

B	A,B	
CA12,B	CA12,B,A	
CA20,B	CA20,B,A	
MB, A*	MB,A,B*	MB,A,B,A*
MB,CA12*	MB,CA12,B*	MB,CA12,B,A*
MB,CA20*	MB,CA20,B*	MB,CA20,B,A*
MB,CAX*	MB,CAX,B*	MB,CAX,B,A*

*All repeated for MB being replaced by MBS and MBH.

A total of 42 different multislice computations were necessary to cover all the cases in Table I and it was usually the case that ten image computations were performed for each of these cases to assess the effect of microscope defocus. Therefore, a total number of images and diffraction patterns approaching nearly 500 cases were generated which demanded several days of computational time on a large mainframe computer. (Details of the computational procedures can be found in reference 17). Here, only a small amount of the generated data can be presented which is most relevant to understanding the Ag_2O structure.

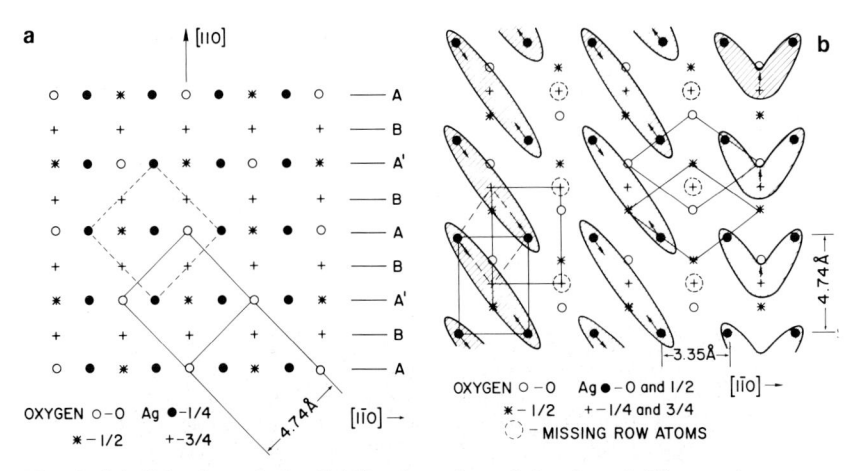

Fig. 4. (a) Side view of the (110) orientation of Ag$_2$O and (b) top view of this orientation which is the viewing direction in the electron microscope.

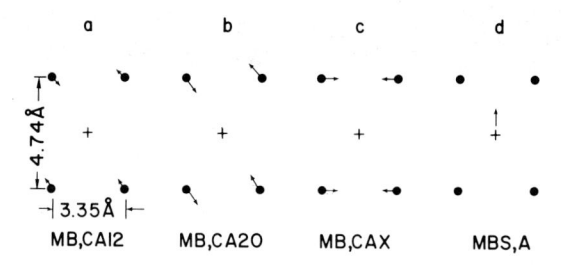

Fig. 5. Possible schemes for reconstructions of various layers of Ag$_2$O viewed along the [110].

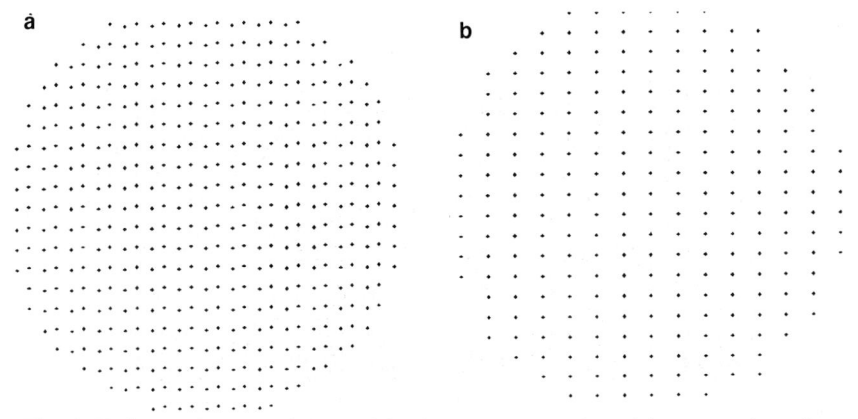

Fig. 6. Full monolayer models used for image computations (a) contraction of an A layer by 12%, (b) missing row model for B layer.

Fig. 7 shows computer generated dynamical scattering diffraction patterns for two different cases of layer stacking where the intensities are displayed logarithmically at 10.0 or 11.0 decades. Image 7a represents an A,B layer stacking producing the typical (110) type pattern. Image 7b is also two layers with the designation MBS, CA12. Here in addition to the four strong {111} reflections there are additional spacings corresponding to the 6.7Å period of the Ag_2O surface and the 4.7Å orthogonal reflections. Also note there are several missing diffraction spots. This pattern shows good agreement with previously presented optical diffraction patterns of ref. [1] and confirms that the amount of contraction for this case is approximately correct based upon the location of the missing reflections.

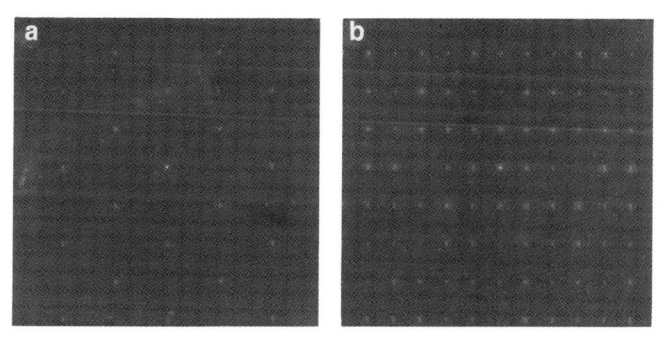

Fig. 7. Computer generated diffraction patterns for: (a) an A, B sequence and (b) for a MBS, CA12 sequence.

Next turning to the imaging computations, the results for one A monolayer are given in Fig. 8a and A, B layer stacking in Fig. 8b. These results are similar to the processed real micrograph of Fig. 3b, where one region designated A' shows a rectangular mesh of white dots and the adjacent region designated B' shows a diamond centered rectangular mesh structure of again white dots. In the real micrograph there is some apparent disorder but similarities with the image computations are still apparent.

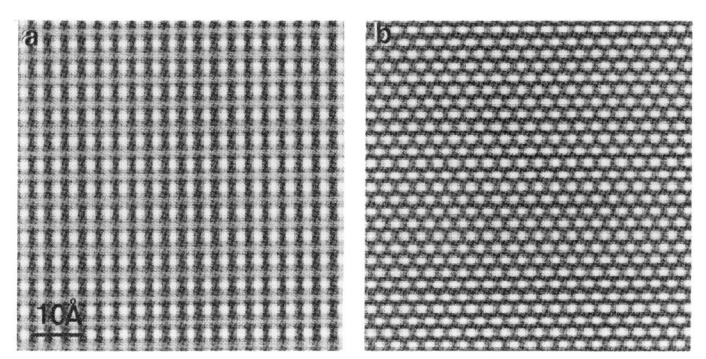

Fig. 8. Computer generated bright field images of: (a) a B layer and (b) an A, B layer sequence.

A demonstration of a focal series is given in Fig. 9 for four layers where the layer sequence is MBS, CAX, B, A. The defocus values in going from left to right are: 750, 850, 900, 950, 1000, 1025, 1050, 1075 and 1100Å. For most of the different experimental images recorded this range was sufficient to reasonably match them to the simulations.

Finally Fig. 10 has selected examples from the image computations which show several features observed in the processed micrographs of Fig. 3d. Fig. 10a with a sequence MBS, CAX, B,A has the cusp like features of Fig. 3d region A$'$ and these features can only be produced if the uppermost layer is a missing row type and the atoms are translated along the row direction. Fig. 10b with a MB, CA12, B, A sequence shows diagonal features which correspond to region B$'$ in Fig. 3d. These features only arise if the underlying layer of atoms is contracted according to the diagonal scheme depicted in Fig. 5a or 5b. If there are both a translation of the upper layer and a diagonal contraction of the underlying layer, skewed cusp type features are formed which are intermediate between images 10a and 10b. If the number of layers is reduced by one, i.e. 3 layers, the general type of contrast observed is similar to that of Fig. 10c which has a three layer sequence MBS, CA12, B. This is very similar to the features of regions C$'$ and D$'$ in Fig. 3d. The most noticeable feature of all three layer regions irrespective of layer structure is narrow white connected features separated by a blackened region somewhat greater in width than the white regions. It must be pointed out that several computations of two layer regions have shown both cusp and diagonal features. However, this is inconsistent with some of the results for one and two layer regions of Fig. 3b which appear to maintain almost a bulk lattice structure arrangement. It is therefore believed that surface reconstructions occur for 3 or more layer coverages. It is obvious that the surface topography varies over monolayer heights across the Ag_2O samples and to investigate these anomalies controlled sample preparation methods must be used. Ideally one would endeavor to have reasonable sample uniformity at least over regions several hundreds of angstroms in extent. For most of the Ag_2O specimens investigated here the modulations of structure appear to be over a few tens of angstroms or less which makes exacting identification difficult.

SUMMARY

This present study has shown that it is possible to observe flat-on a (2x1) surface reconstruction of the (110) Ag_2O surface at the atomic resolution level. It is also possible to match a variety of features observed in real micrographs with computer generated micrographs. From these experiments it is now apparent that the missing row model must undergo translations along the directions of the rows of atoms of the upper most layer. Also the next underlying full layer contracts (at least laterally) towards the atom positions in the missing row. It is then apparent that the surface topography of (110) Ag_2O varies rapidly as a function of spatial position over a few tens of angstroms or less. In other words, the surface is atomically rough. It may well be the case that many surfaces which yield good diffraction data are, in fact, poorly ordered when subjected to atomic resolution imaging. Therefore many structures investigated by surface techniques should also be scrutinized in a high resolution electron microscope preferrably equipped with an ultra high vacuum environment.

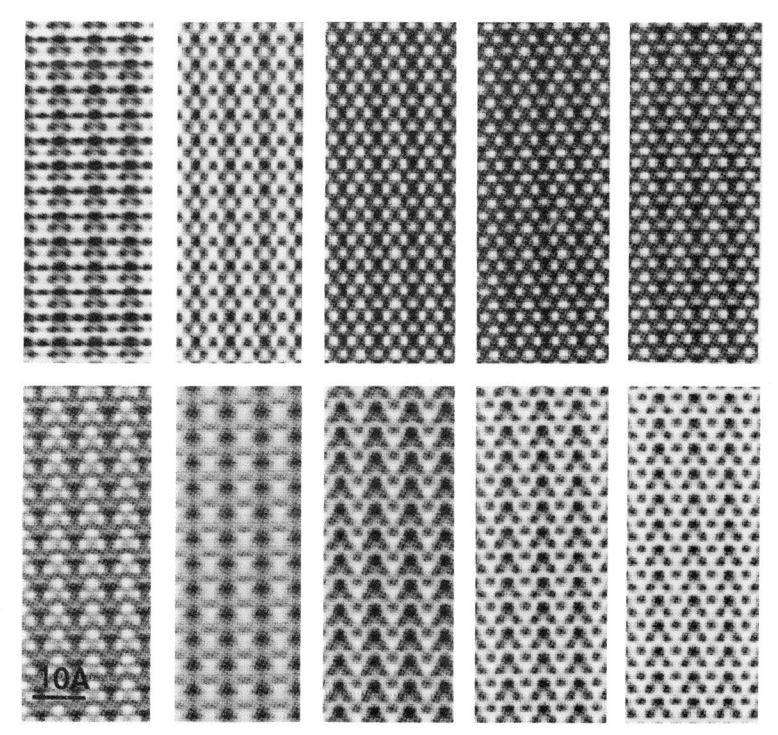

Fig. 9. Computer generated focal series of a MBS, CAX, B, A four layer structure. The defocus values going from left to right are: 750, 800, 850, 900, 950, 1000, 1025, 1050, 1075 and 1100Å.

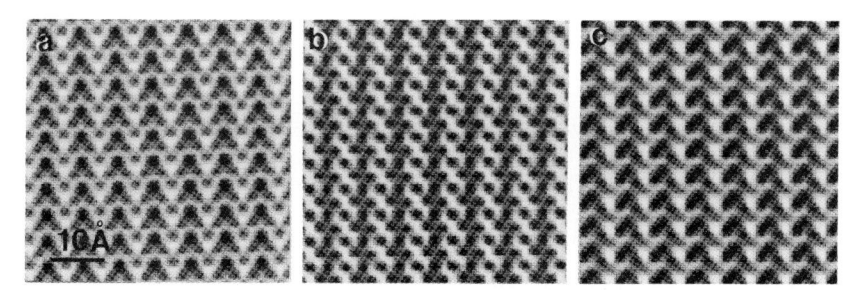

Fig. 10. Examples of computer generated bright-field images observed in real micrographs. (a) MBS, CAX, B,A, (b) MB, CA12, B, A, and (c) MBS, CA12, B sequences.

REFERENCES

1. W. Krakow, "Ultrahigh Resolution Electron Microscopy of the Surface Structure of (110) Ag$_2$O", Submitted to Surface Science (1983).
2. D. Cherns, Phil. Mag. *30*, (1974) 549-556.
3. K. Yagi, K. Takayanagi, K. Kobayashi, N. Osakabe, Y. Tanishiro and G. Honjo, Sur. Sci. *86* (1979) 174-181.
4. Y. Tanishiro, H. Kanoamori, K. Takayanagi, K. Yagi and G. Honjo, Sur. Sci. *111* (1981) 395-413.
5. K. Takayanagi, Sur. Sci., *104* (1981) 527-548.
6. K. Takayanagi, Ultramicrosc., *8* (1982) 145-162.
7. W. Krakow and D.G. Ast, Sur. Sci., *58* (1976) 485-496.
8. W. Krakow, Ultramicrosc., *4*, (1979) 55-76.
9. W. Krakow, Sur. Sci., *111* (1981) 503-518.
10. W. Krakow, Thin Solid Films, *93*, (1982) 235-253.
11. J.C. Heyraud and J.J. Metois, Sur. Sci., (1980) 519-528.
12. G. Rovida and F. Pratesi, Sur. Sci., *52*, (1975) 542-555.
13. H.A. Engelhardt and D. Menzel, Sur. Sci., *57*, (1976) 591-618.
14. H. Albers, W.J.J. Van Der Wal, O.L.J. Gyzeman, and G.A. Bootsma, Sur. Sci., *77*, (1978) 1-13.
15. L.D. Marks and D.J. Smith, Nature, *303*, (1983) 316-317.
16. W. Krakow, Proc. 41st Annual Meeting Electron Microscopy Society of America (Claitors, Baton Rouge, 1983) 398-399.
17. W. Krakow, IBM, J. Res. and Devel. *25*, (1981) 58-70.

FORMATION AND TRANSFORMATION OF AMORPHOUS SILICIDE ALLOYS

K.N. Tu, T. Tien and S.R. Herd
IBM T.J. Watson Research Center, P.O. Box 218 Yorktown Heights, N.Y. 10598, U.S.A.

ABSTRACT

Amorphous silicide films can be formed by rapid quenching using techniques of vapor deposition and ion beam mixing and also by slow heating using solid state interdiffusion and reaction. For example, amorphous $TaSi_2$ films can be formed by sputtering or dual electron guns co-deposition. Amorphous Pt_2Si_3 films have been produced by mixing PtSi and Si at room temperature with an ion beam at about 100 to 300keV. Recently, an amorphous Rh-Si alloy phase has been made by slowly heating to 300°C a very thin crystalline Rh films (~50Å) on amorphous Si. The formation and crystallization behavior of these amorphous silicide alloys has been studied by transmission electron microscopy and electrical conductivity measurement.

I. INTRODUCTION

Amorphous binary alloys consisting of a transition metal and a metalloid element are well known; an example is the amorphous $Pd_{80}Si_{20}$. These amorphous alloys can be formed typically by a rapid quenching from the molten state.[1] With the advance of thin film processing methods, amorphous alloys can now be obtained by coevaporation,[2] by energetic beam mixing[3,4] and even by solid state interdiffusion.[5,6] The first two methods have greatly expanded the composition range in which amorphous alloys can be made. The last method of interdiffusion has shown that rapid quenching is not essential in achieving the amorphous state.

In this paper, we shall illustrate each of these methods by an example of amorphous silicide formation. The as-prepared alloy and its crystallization behavior observed by transmission electron microscopy (TEM) and electrical conductivity measurement will be discussed.

II. CO-EVAPORATED AMORPHOUS $TaSi_2$ FILMS[7]

Thin films of $TaSi_2$ alloys were prepared by simultaneous evaporation of the components in a double electron gun evaporation system. The background pressure during deposition was maintained at or below 2×10^{-7} Torr. Typical rates were 10Å/sec for Ta and 20Å/sec for Si. Thicknesses of 330 and 660Å were projected for Ta and Si, respectively, to obtain a $TaSi_2$ film 1000Å thick. The thin films were deposited at room temperature on several types of substrates;

thermally oxidized Si and sapphire substrates were used for resistivity measurements, and Si substrates for transmission electron microscopic studies.

The as-prepared TaSi$_2$ alloy films were analyzed by x-ray diffraction (XRD), Rutherford backscattering spectroscopy, Auger electron spectroscopy TEM and resistivity measurements. It was amorphous with a resistivity of 275$\mu\Omega$-cm. The composition has a 1:2 ratio of Ta to Si with negligible impurity concentration such as oxygen and carbon. The typical resistivity annealing curve of the TaSi$_2$ alloy film was obtained by recording the *in situ* resistivity at 4°C/min heating rate. The solid line in Fig. 1 shows the resistivities normalized to the as-deposited room-temperature value, 275$\mu\Omega$cm, as a function of temperature. The resistivity decreases slightly in the early stage of annealing. It has a negative temperature coefficient of ~2.5x10^{-4}/°C. This annealing behavior continues to about 300°C where the resistivity suddenly drops. XRD measurements showed a set of weak TaSi$_2$ crystalline phase reflections superimposed on the amorphous background after this annealing step. The diffraction peaks correspond to all the allowed reflections of the TaSi$_2$ crystalline phase.

The resistivity decreases gradually upon further annealing. The decrease is slow at first (300-500°C), fast in the middle region (500-600°C), and slow again. Above 700°C, the resistivity measurement becomes inaccurate because of the displacement of the tungsten probes. The part of the annealing curve with dot and dash lines shown in the figure is the most reasonable prediction we have made from experimental results.

If the heating is arrested along the annealing curve at any point T between 300 and 700°C, and followed by a cooling at the same rate to room temperature, the resistivity increases, reaches a maximum, then decreases (dashed lines in Fig. 1). When the temperature is raised again at 4°C/min the resistivity will trace back along the same curve until it reaches the point T. By increasing the annealing temperature above T, the resisitivity follows the same annealing curve of a film whose heat treatment was not interrupted. For T above 850°C, only the positive slope is obtained (the cooling curve with the dot and dash line in Fig. 1).

The corresponding transmission electron diffraction and micrographs at points A, B and C as shown in Fig. 1 are given in Fig. 2a, 2b and 2c, respectively. Lenticular shape crystallites in an amorphous matrix were observed at the early stage of crystallization [Fig. 2(a)]. Their sizes range from 50Å by 200Å to 200Å by 1000Å. Further annealing of the film makes the crystallites grow to become similar in lengths and widths [Fig. 2(b)]. Their dimensions range from 50 to 1000Å. Because the selected area diffraction pattern shows a spotty ring pattern instead of a diffused ring pattern, these crystallites are on the average larger. For the fully crystallized film annealed at 1000°C for 1 h, it consists of 1000-10000Å TaSi$_2$ grains [Fig. 2(c)].

In relating resistivity measurements to microscopic observation, we note that the micrograph in Fig. 2(c) shows a microstructure having a clear contrast of grains is critical for a low room temperature resistivity of ~50$\mu\Omega$-cm. A clear contrast means a homogeneous composition and low crystalline irregularities, which can only be obtained by annealing TaSi$_2$ above 850°C. However, a high temperature annealing will result in a high thermal stress in the film. Below 850°C, the contrast as shown in Figs. 2(a) and 2(b) is cloudy and the room-temperature resistivity is high. From Fig. 1, we see that the amorphous $TaSi_2$ film crystallizes around 300°C, yet the growth soon becomes sluggish and the resistivity curve bulges with the annealing temperature. Therefore, the slow transformation prevents the resistivity from a precipitous drop. The behavior is quite different from the crystallization of amorphous Pt_2Si_3 alloy to be given in the next section.

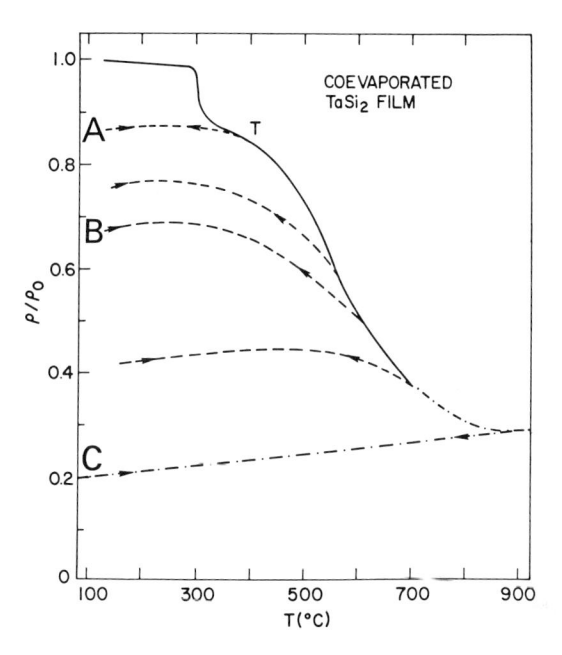

Fig. 1. In-situ resistivity curve as a function of annealing temperature at 4°⁻C/min heating rate for the electron-beam co-evaporated $TaSi_2$ films on thermally oxidized silicon wafer.

II. ION BEAM MIXED AMORPHOUS Pt_2Si_3 FILMS[8-10]

The amorphous and the crystalline Pt_2Si_3 phases were prepared as follows. A thin film of PtSi was first obtained by depositing and reacting thermally at 400°C a 150Å Pt film on Si single-crystal substrates of <100> orientation. Atomic mixing between the PtSi and Si was then initiated by ion bombardment with 300keV Xe or Si ions through the interface. Formation of Si-rich Pt-Si mixed layers on Si was obtained as indicated by Rutherford backscattering measurements. The composition of these layers became progressively more Si-rich with increasing ion dose; a mixed layer with a thickness of 360Å and an average composition of Pt_2Si_3 was formed at a dose of $1x10^{15}$ Xe/cm^2 or $6x10^{15}$ Si/cm^2. The mixed layers exhibited a slight concentration gradient in which the Pt concentration decreased in depth from the surface to the substrate. Examination of this alloy layer by glancing incidence X-ray and transmission electron diffraction

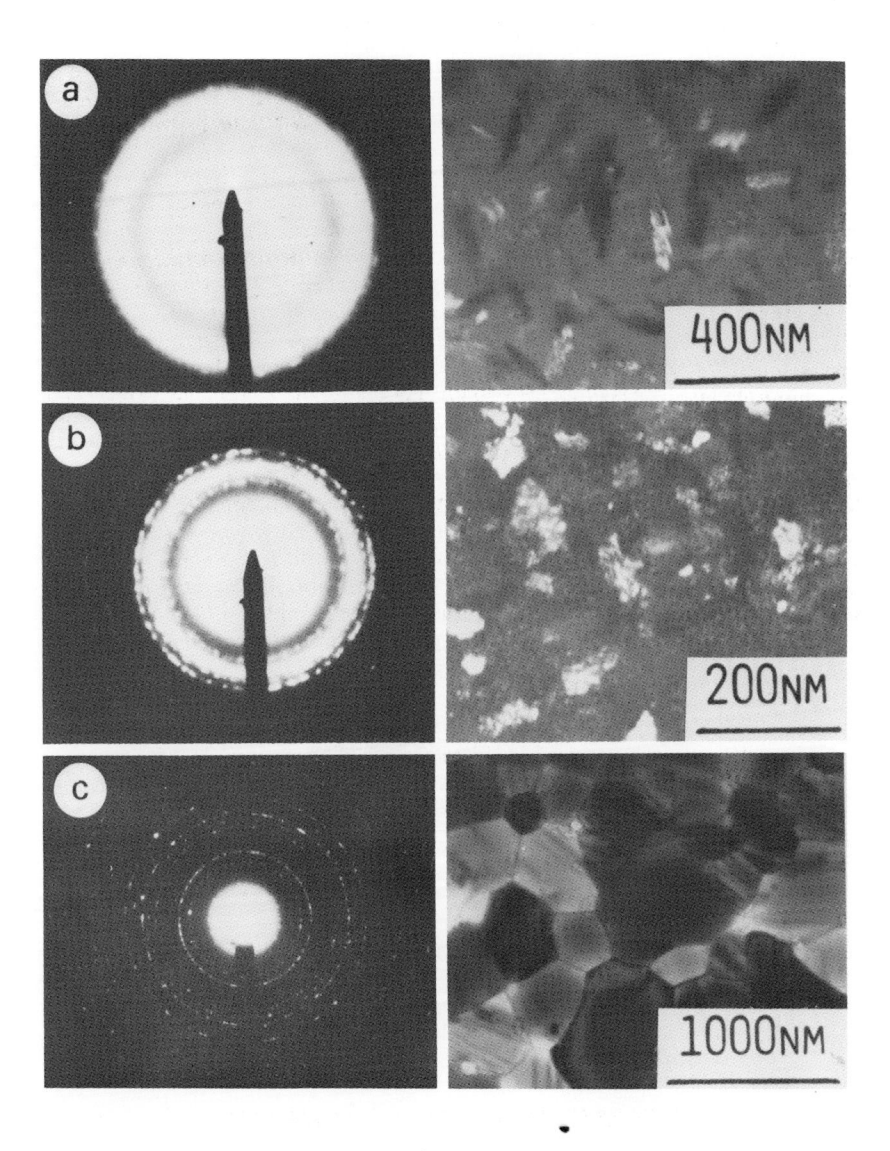

Fig. 2. Bright-field transmission electron micrographs and selected area diffraction patterns of annealed electron-beam co-evaporated TaSi$_2$ films on silicon substrates (a) an early stage of crystallization after 5 hours at 255°C, (b) the crystallized film after 1 hour at 400°C and (c) the fully crystallized film after 1 hour at 980°C.

revealed amorphous-like diffraction patterns without any of the PtSi reflections. The non-crystalline nature of the Si-rich Pt-Si mixed layer was further confirmed by dark-field TEM images with a resolution of ~10Å.

Amorphous Pt_2Si_3 alloys have also been produced using a dual e-gun system by co-deposition of films with the composition of $Pt_{40}Si_{60}$ onto sapphire substrates maintained at liquid nitrogen temperature. Sapphire substrates were chosen so that electrical resistivity measurements could be made. X-ray diffraction measurements showed that the as-deposited films (after warm up to R.T.) were amorphous. The samples exhibited a small in-depth compositional variation as revealed by glancing incident backscattering measurements. Ion-beam-mixing with 250keV Xe^+ ions at a dose of $1 \times 10^{15} cm^{-2}$ were performed to homogenize the film. The crystallization behavior of the amorphous Pt_2Si_3 film was studied by electrical resistivity measurements under a constant temperature rise of 2-3°C/min using the standard four-point probe technique. Figure 3 shows the typical resistivity annealing curves for the co-deposited amorphous Pt_2Si_3 films with ion-beam homogenization. At room temperature, the amorphous Pt_2Si_3 films showed a resistivity of ~370$\mu\Omega$-cm. The resistivity decreases slightly at first with isochronal annealing, then it increases slowly with a small positive temperature coefficient of $\sim 1 \times 10^{14}$ °C^{-1} between 100 and 350°C. After a slight increase at 350°C, the resistivity drops precipitously to ~110$\mu\Omega$-cm around 400°C manifesting the amorphous-to-crystalline transition. A reversible behavior in resistivity with changing temperature was observed for the crystalline Pt_2Si_3 with a positive temperature coefficient of 3×10^{-4} °C^{-1}. The crystalline Pt_2Si_3 is a metastable phase which decomposes into an equilibrium two-phase mixture of PtSi and Si at temperatures above 550°C. The average grain size of the crystalline Pt_2Si_3 is of

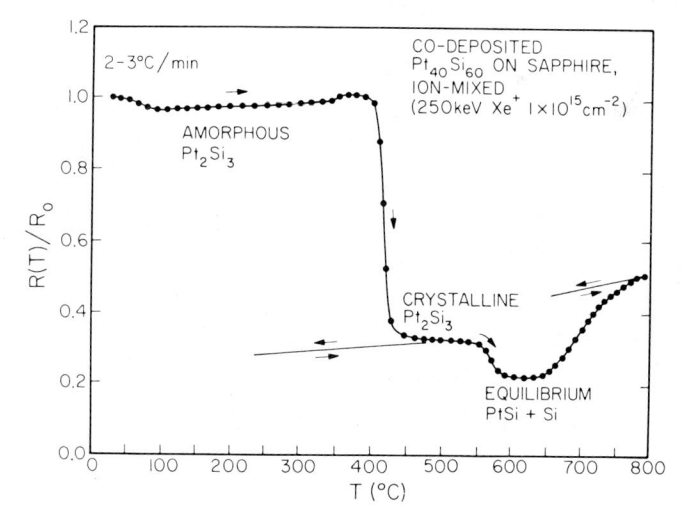

Fig. 3. Resistivity annealing curve for the codeposited, ion-mixed amorphous Pt_2Si_3 alloy upon heating at about 2-3°C/min. Two distinct stages of transformation: (1) from the amorphous to a metastable crystalline and (2) from the metastable to the equilibrium two-phase structure, are observed as revealed by rapid change of resistivity at temperatures around 400 and 550°C, respectively.

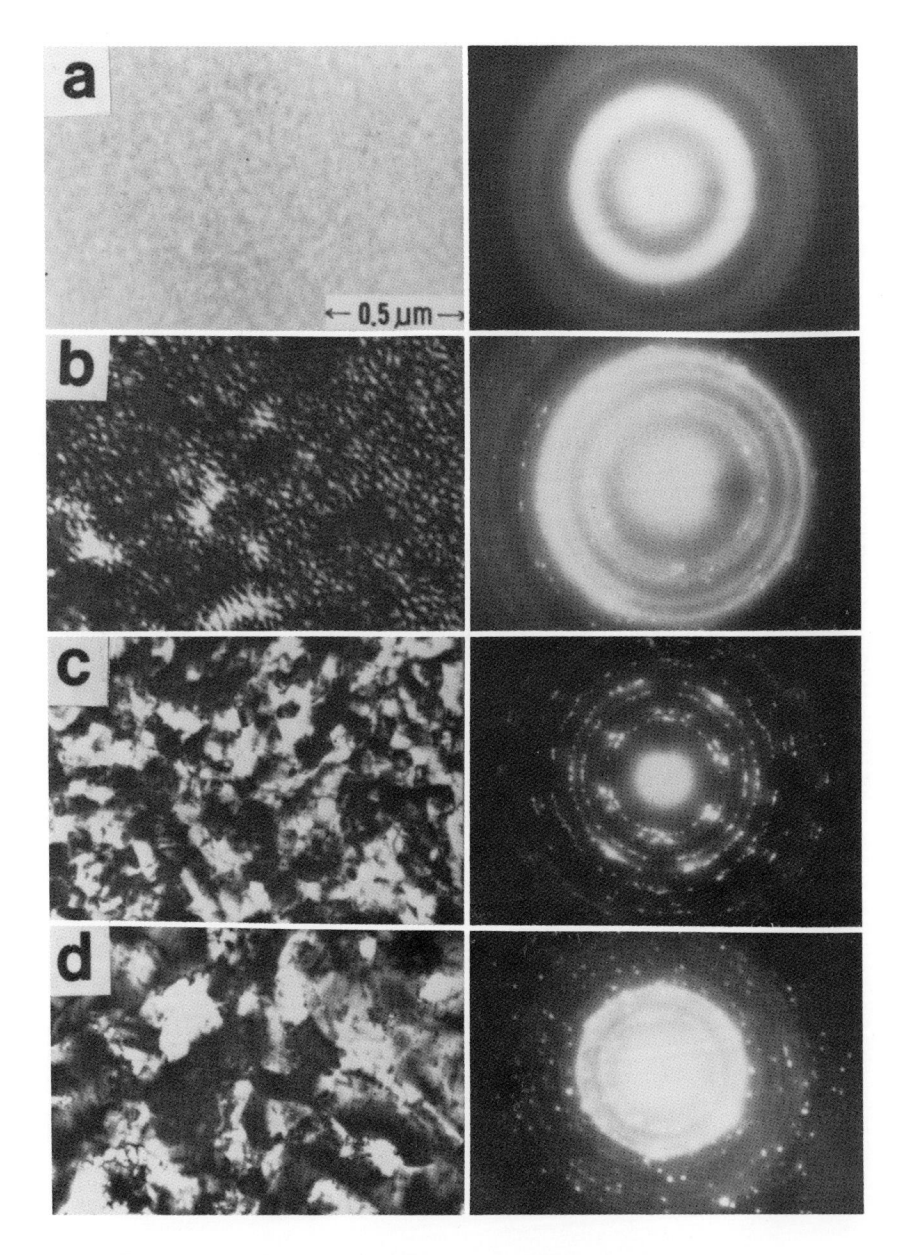

Fig. 4. Transmission electron micrographs and diffraction patterns for the codeposited, ion-mixed Pt_2Si_3 alloy: (a) as-implanted, amorphous, (b) annealed at 384°C for 30 min, dark-field, showing the early stage formation of Pt_2Si_3 crystallites. (c) annealed at 450°C for 30 min, metastable crystalline Pt_2Si_3 phase, (d) annealed up to 600°C, equilibrium two-phase (PtSi and Si mixture).

the order of 2000Å as measured by TEM and it increases slightly during the annealing between 450 and 550°C.

Figure 4 shows typical electron micrographs and diffraction patterns of a codeposited, ion-mixed Pt_2Si_3 alloy films. The diffraction pattern for the as-implanted film [Fig. 4(a)] reveals a series of diffused halos, indicating an amorphous-like structure. The first intense halo has a shoulder and the second halo splits into two broad peaks. Dark field techniques revealed a featureless image pattern within our detection limit (~10Å). However, some density or thickness variation was present in the amorphous layer as revealed by the bright field image. Figure 4(b) shows the electron micrographs and diffraction patterns for samples annealed for 60 min at 384°C. As can be seen in the diffraction pattern of Fig. 4(b), crystalline reflections were found superimposed on the background amorphous halos. Dark field techniques reveal coherently diffracting particles of about 100Å in diameter probably due to the Pt_2Si_3 crystallites which appear as white rounded spots in the micrograph. The film exhibited a resistivity of 145$\mu\Omega$cm,. Figure 4(c) shows the same film in the metastable crystalline state after passing through the annealing (450°C, 30 min). Crystalline diffraction rings and dots (due to the large crystals) can be seen in the diffraction pattern. The pattern consists of strong crystalline reflection belonging to the metastable Pt_2Si_3 phase. The micrograph shows an average grain size of about 1000Å. Finally, Fig. 4(d) shows the equilibrium structure after continuous annealing up to 600°C. The diffraction reveals dots and weak rings due to the large PtSi crystals and small randomly oriented Si precipitates, respectively. The Si precipitates, appearing as irregular white spots, are found mostly near the PtSi grain boundaries.

IV. AMORPHOUS Rh-Si ALLOY FORMED BY SOLID STATE INTERDIFFUSION[5]

Thin film bilayers of Rh and Si were deposited by sputter gun deposition onto 1 to 2μm thick AZ photoresist kept at room temperature during a single pumpdown. The sequence of deposition was first Rh and then Si. The thickness of Rh varied from 30 to 100Å and that of Si from 100 to 200Å. The bilayers were studied by TEM as self-supporting films on W grids after the AZ photoresist was dissolved in acetone.

In the as-deposited state, the Rh thin films were found to be crystalline and continuous down to 30Å in thickness and the Si films were amorphous for all thicknesses prepared. Figure 5(a) shows an electron diffraction pattern and the corresponding bright-field image of a bilayer of ~190Å Si and 60Å Rh in the as-deposited state. The diffraction rings belong to Rh, the weak halo within the rings belongs to amorphous Si, and no other amorphous phase can be detected within our resolution limit. The overall composition of the bilayer, if homogenized, is about $Rh_{30}Si_{70}$. Upon *in situ* heating in the electron microscope to 300°C, we observed the formation of an amorphous Rh-Si alloy. The diffraction pattern and bright-field image of this alloy film and the remaining amorphous Si are shown in Fig. 5(b). The strong halo belongs to the alloy, the weak halo to Si, and no reflections of any crystalline phase can be observed. Upon further heating to 400°C, the amorphous alloy was observed to crystallize into RhSi, whose diffraction pattern and bright-field image are shown in Figs. 5(c). The spotty

208

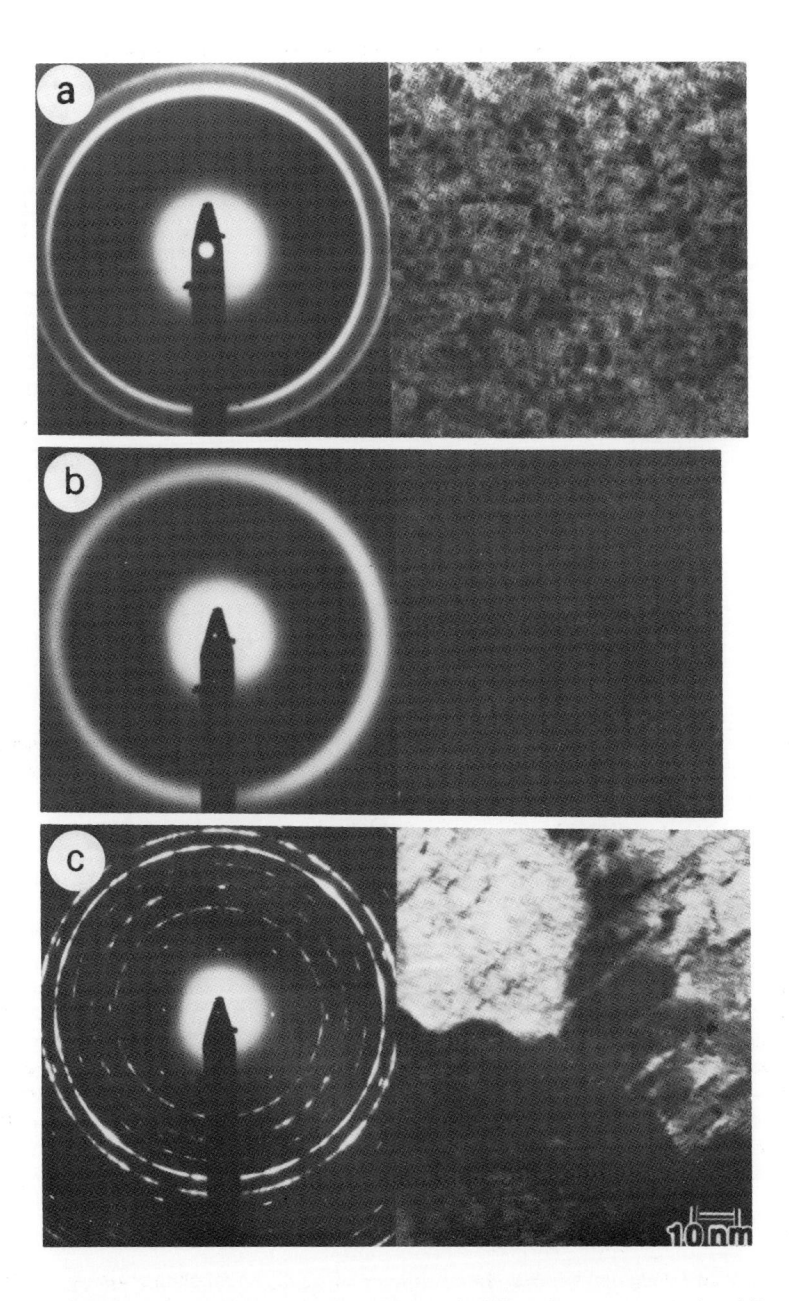

Fig. 5. Transmission electron micrographs and diffraction patterns of a bilayer film of 190Å of amorphous Si on top of 60Å of crystalline Rh. (a) as-deposited, (b) upon in-situ heating in the electron microscope (Phillips EM301) to 300°C and (c) upon in-situ heating in the microscope to 400°C.

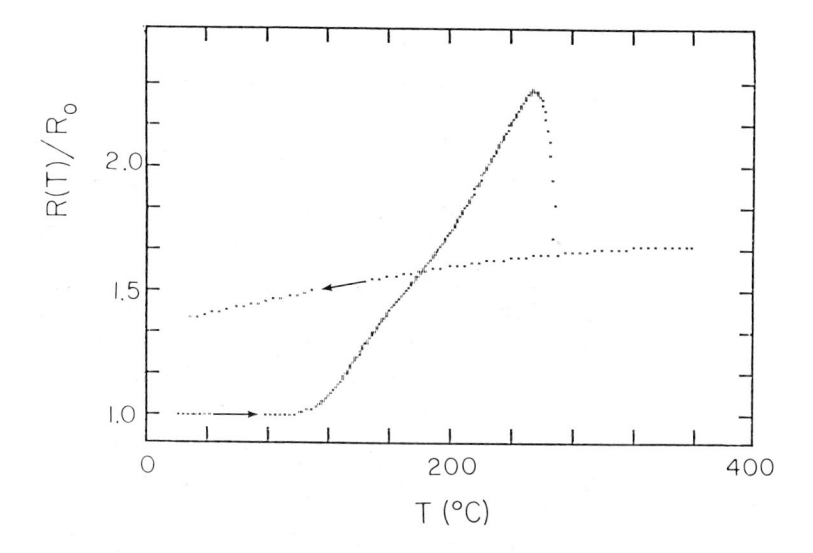

Fig. 6. In-situ resistivity curve as a function of annealing temperature at 2°C/min heating rate for the multi-layered amorphous Si and crystalline Rh film sample.

rings belong to cubic RhSi (low-temperature form) and the spots to the orthorhombic RhSi (high-temperature form). Both phases are present and separated by an interphase boundary in Fig. 5(c). The remaining amorphous Si was not observed to crystallize until heated above 730°C.

For resistivity measurements, a multi-layered structure consisting of ten alternating 44Å Rh and 230Å of Si were prepared on sapphire substrates.[11] The preparation condition was the same as for preparing the single bilayer structure. Fig. 6 shows the in-situ resistivity behavior as a function of anneal temperature. Upon heating, the resistivity remained almost constant at 19$\mu\Omega$cm to about 150°C, thereafter it increased rapidly up to 280°C where it peaked at a value of 2.5 times the original one and then dropped precipitously. After the drop, it changed very slowly with temperature showing a metallic behavior with a small positive slope. We note that the rapid increase in resistivity indicates the formation of an amorphous phase and the drop the crystallization of the amorphous phase. When we arrested the annealing near the peak, the resistivity curve of cooling showed a small negative slope.

V. SUMMARY

We have illustrated that the formation of amorphous silicide alloys can be achieved by rapid quenching as in the methods of co-evaporation and ion beam mixing, yet it can also be achieved by slow heating as in solid state interdiffusion. The criteria for forming amorphous alloy by the last method are unclear yet, the one which is certain is the prevention of nucleation of any crystalline phase.

ACKNOWLEDGEMENT

The authors are grateful to R.D. Thompson for obtaining the curve as shown in Fig. 6.

REFERENCES

1. P. Duwez, R. Williams and R. Crewdson, J. Appl. Phys. *36*, 2267 (1965).
2. S. Mader and A.S. Nowick, Acta Metall. *15*, 215 (1967).
3. B.Y. Tsaur, in "Thin Film Interfaces and Interactions" edited by J.E.E. Baglin and J.M. Poate, Electrochemical Society Proceedings Vol. 80-2, p. 205 (1980).
4. J.M. Poate, chapter 6 in "Preparation and properties of Thin Films", edited by K.N. Tu and R. Rosenberg, Academic Press, New York (1982).
5. S. Herd, K.N. Tu and K.Y. Ahn, Appl. Phys. Lett. *42*, 597 (1983).
6. R.B. Schwarz and W.L. Johnson, Phys. Rev. Lett., *51*, 415 (1983).
7. T. Tien, G. Ottaviani and K.N. Tu, J. Appl. Phys. *54*, 7074 (1983).
8. B.Y. Tsaur, J.W. Mayer and K.N. Tu, J. Appl. Phys. *51*, 5326 (1980).
9. B.Y. Tsaur, J.W. Mayer, J.F. Graczyk and K.N. Tu, J. Appl. Phys, *51*, 5334 (1980).
10. J.F. Graczyk, K.N. Tu, B.Y. Tsaur and J.W. Mayer, J. Appl. Phys. *53*, 6772 (1982).
11. K.N. Tu, R.D. Thompson and S.R. Herd, unpublished.

THE CHARACTERISATION OF INTERFACIAL DISLOCATION STRUCTURES

W. A. T. Clark
Department of Metallurgical Engineering
The Ohio State University
116 West 19th Avenue
Columbus, OH 43210

ABSTRACT

Experimental observation of many interfaces in metals indicate the presence of a dislocation-like structure. The geometrical basis for such a structure to exist is discussed, and the suitability of the transmission electron microscope (TEM) to characterise it considered. It is shown by reference to examples of grain and phase boundary structures in metals that such geometrical models may be used as a first step to predicting interfacial structures, and that the dislocations observed may play a role in thermo-mechanical behaviour of polycrystals. The general applicability of the applicability of the approach to non-metallic systems, such as semiconductors and ceramics is also indicated.

INTRODUCTION

The advances made in electron optical instruments during the past twenty years have led to considerable improvements in our understanding of the role of microstructure in determining the thermo-mechanical properties of metals, semiconductors, and ceramics. In particular, the ability of instruments such as the TEM and the field ion microscope (FIM) to resolve detail at a scale close to the atomic level has made it possible to examine finely spaced dislocation arrays, for example, as well as small, dispersed, precipitates. When these imaging techniques are combined with microchemical analytical methods, such as energy dispersive spectrometry (EDS) or electron energy loss spectrometry (EELS) in the case of the TEM, or time-of-flight mass spectrometry on the FIM, the correlation between the microstructure and local compositional variations can often be established. One area which has benefited from these advances is the understanding of the structure and properties of interfaces. This paper will outline some of the features of the geometry of interfaces between like or unlike phases, and show how TEM can be used systematically to investigate the validity of such geometrical models.

INTERFACIAL DISLOCATIONS

There is increasing evidence that a wide range of interfaces in pure metals (i.e. grain boundaries), alloys (phase boundaries), semiconductors, and ceramics contain well-defined networks of linear defects which can be identified as grain boundary dislocations (gbds) or interphase boundary dislocations (ibds) (for a review see e.g. ref. [1]). Some of the types of interface in which dislocations have been observed are shown in Table I-- the wide variety of interphase boundaries should be especially noted. The role of dislocations in these boundaries may be loosely categorised (Table II) according to their role in the interface, as structural dislocations, required by the geometry of the two crystals meeting at the boundary (see next section), transformational dislocations which are required to generate a change of crystal structure or orientation, and extrinsic dislocations

TABLE I. Typical Interfaces Containing Dislocations

Grain Boundary	Epitaxial Boundary	Interphase Boundary
(a) CSL (esp. cubic)	Coherent	Precipitate/Matrix(γ-γ', θ', γ-α)
(b) Random	Non-coherent	Martensitic
		Eutectic
		Massive
		FCC/BCC
		BCC/HCP, etc.

TABLE II. Interfacial Dislocations Classified by Role

Structural	Transformational	Extrinsic
Intrinsic GBDs	Martensitic	Recrystallisation
Intrinsic IBDs	Twinning	Creep
Epitaxial	Massive	Deformation
	Diffusional Growth	
	DIGM	

which enter the boundary from either crystal as a result of boundary motion or an applied stress. The list in these tables is not exhaustive, nor is it always possible to discriminate so clearly between the types of dislocations, as they may be necessary to maintain the optimal boundary structure, yet still be able to participate in transformations. Rather the Tables show the wide variety of interfacial phenomena in which structure has been seen to have a part.

THE GEOMETRY OF INTERFACIAL STRUCTURE

One starting point for predicting the likely structure of a given interface is the geometrical matching of the two crystals that meet to form a given interface. The geometry of interfacial structure has been extensively described [2-4] in many review articles and books and will not be given in detail here. However, some of the principal features of such models are indicated.

Firstly, it is proposed that a low-energy (and hence favoured) interfacial structure is one which maximises the degree of "good fit" between the two crystals which meet at the boundary. The relative orientations of any two crystals between which such structures are found are those in which some fraction of the lattice sites of the two crystal lattices, each considered infinite and allowed to interpenetrate, coincide. These are the so-called coincidence site lattice (CSL) orientations; an example is shown in Figure 1 in which the crystals are rotated by 38.94° about a common [011] direction normal to the page. One lattice site in nine is in coincidence and a parameter characterising the boundary, Σ (the reciprocal of the fraction of sites in coincidence) is thus set equal to nine for this case. A more general approach to the idea of coincidence is given by the O-lattice [2], in which all points within the lattices which have the same internal coordinates are examined for coincidence. This allows boundaries in low symmetry crystals, and between different crystal structures to be characterised. The resulting boundary structures are predicted to be arrays of dislocations, with crystal lattice Burgers vectors, sometimes referred to as "primary" dislocations. In the case of grain boundaries these dislocations are so closely spaced as to have only a formal significance for all but low angle (< 15°) boundaries. For phase boundaries they may have much larger separations, and can often be observed directly in the TEM. By analogy with low angle boundaries small angular deviations from the exact CSL orientation may be accommodated by the introduction of an array of interfacial dislocations, sometimes called "secondary" gbds, which exactly account for the deviation (see e.g. ref. [5] for a description of the geometry of such networks). It is these networks which are of particular interest in the case of grain boundaries. The Burgers vectors of these secondary gbds are related to the particular CSL which they preserve and hence change from one coincidence system to another. A useful tool for determining them is the DSC lattice [2] which may be constructed either graphically or analytically [6], and is shown in Figure 1 for the Σ9 orientation. The three shortest translation vectors which relate sites of crystal 1 to those of crystal 2 have the form:

$$\underline{b}_1 = a/18<411>, \quad \underline{b}_2 = a/9<221>, \quad \underline{b}_{31} = a/6<211> \tag{1}$$

and form a basis for the DSC lattice. Experimentally, it remains to establish whether the observed dislocations have the Burgers vectors, spacing, and line direction predicted by this model.

In the case of interphase boundaries, the interfacial energy may be minimised by orienting the two crystals, and selecting a boundary plane, so as to optimise the fit between them. The residual misfit is again accommodated by dislocations, generally with crystal lattice Burgers vectors, and it can be seen that this procedure is essentially one of obtaining the minimum interfacial dislocation density. Kurdjumov-Sachs, Nishyama-Wassermann, and the Bain orientation are some examples of orientations which bring low index directions into correspondence and have been considered to achieve this goal. Various parameters have been used to predict the most favourable orientations, and there is no overall agreement on which is the most physically realistic [7]. It has been shown [8], however, that from the geometrical point of view the traditional orientations referred to above do not, in general, represent the lowest interfacial energy, but rather starting points from which that may be calculated. The optimum orientation is found to depend sensitively on the ratio of the lattice parameters of the two phases, and takes a wide variety of values in practice. Furthermore, since orientation relationships in precipitation reactions are generally established during nucleation, and may even be followed by changes of crystal structure of the precipitate, it is difficult to predict the driving forces for creating such favoured orientation relationships.

ANALYSIS OF INTERFACIAL DISLOCATIONS IN THE TEM

Determination of Interfacial Parameters

The procedure for analysing interfacial structure in the TEM relies on obtaining complementary electron diffraction and imaging information, and has been outlined in ref. [9]. There are two principal tasks involved: (a) determining the relative orientation of the two crystals and the boundary plane between them, and (b) obtaining the Burgers vectors of the dislocations present in the interface. The former can be accomplished from the analysis of Kikuchi patterns by any one of several essentially equivalent methods (see e.g. [10]), and with care a precision of ~ ±0.1° obtained. It should be noted, however, that for calculating secondary gbd networks it is required to determine a small angular deviation (denoted in rotation matrix terms $\underline{R}_{02}$) from an exact CSL orientation, $\underline{R}$. The measured error is in the experimentally observed orientation R' which is composed of a rotation to exact coincidence plus the small deviation from it [11], i.e.

$$\underline{R}' = \underline{R}(O2)\ \underline{R} \qquad (2)$$

$\underline{R}$ is known exactly (see e.g. [6]) so that any measured error in $\underline{R}'$ resides solely in $\underline{R}(O2)$. If $\underline{R}(O2)$ is small, (and it is frequently ~ 1° or less) then an error of ± 0.1° becomes significant in the calculation of the most likely boundary dislocation network. This error may be estimated by overdetermining the rotation matrix and obtaining some measure of the confidence which may be placed in the result [12,13].

It should also be pointed out that this orientation determination allows the relevant CSL and DSC lattices to be identified (e.g. from [6]), and the likely Burgers derived. This information can also be obtained directly from a stereographic analysis conducted at the microscope, which has the advantage in terms of time required. The stereogram also provides a simple means of selecting and setting up diffracting vectors for imaging the boundary.

The dislocation network may also act as a diffraction grating, giving rise to extra spots in a diffraction pattern, some of which arise due to scattering by the gbds and some of which are due to wave matching effects at an interface inclined to the electron beam. Careful analysis of such patterns, whether they are obtained by electron or X-ray diffraction [14-16] contributes additional information about the boundary, and has been used, for example, to estimate the thickness of certain boundaries.

Burgers Vector Determination

Once the boundary parameters have been obtained, it remains to determine the Burgers vectors of the observed dislocations. The standard technique for doing so is the application of the $\underline{g}.\underline{b} = 0$ invisibility criterion [17]. While this technique is relatively straightforward to apply to crystal lattice dislocations, considerable care has to be exercised in the case of interfacial dislocations. This is because of the relatively complicated imaging conditions which must be satisfied for the analysis to be valid, and because the Burgers vectors of the dislocations involved are generally much smaller than those of crystal lattice dislocations and the images are superimposed on a strong pattern of thickness fringes (the problem has been more fully treated in refs. [18] and [19]. It has, however, been shown that, provided sufficient attention is paid to satisfying these conditions, the invisibility criterion may be successfully applied to the analysis of interfacial dislocations (e.g. [9]). The boundary is imaged by (a) setting up two-beam conditions in one crystal while setting the other as far from any strong diffracting position as possible. This procedure is repeated for reflections from both crystals. In some boundaries imaging conditions may be set up by (b) setting planes of the CSL into a two-beam condition. In diffraction terms the boundary is then "invisible" in that no thckness fringes are observed and the dislocation contrast resembles that from a low-angle boundary in a single crystal. Such conditions are generally restricted to a few reflections in low-Σ boundaries. Some success has been obtained in lattice imaging interfacial structures, especially in semiconductors [20]. From the point of view of directly resolving extra half planes of gbds, the Burgers vectors are generally below the resolution limit of the TEM. It has been demonstrated, however [21,22], that a series of lattice fringe images taken using low-index planes does allow Burgers vectors and boundary plane steps associated with interfacial dislocation cores to be measured. Further progress in the area of high resolution imaging will be watched with great interest.

OBSERVATIONS OF INTERFACIAL STRUCTURE

Three examples of interfacial structures analysed in the TEM are given; they are selected as representing a selection of the categories of boundaries listed in Tables 1 and 2, and illustrate the applicability of the experimental approach outlined above to a variety of interfacial structures.

Grain Boundary in Stainless Steel

A coincidence related grain boundary in solution treated type 304 stainless was analysed in the TEM [23]. The misorientation across the boundary was determined to be 37.36° about the $[1\ 70\ 71]_1$, $[1\ 70\ 71]_2$ axis, where the subscript refers to the particular crystal). The angular deviation from the exact $\Sigma = 9$ CSL disorientation of $38.94°/[011]_{1/2}$, which is illustrated in Fig. 1, was 1.61°(±0.2°) about the $[7\ \bar{40}\ \bar{32}]_1$, $[\bar{1}44]_2$ axis, and the boundary plane $(232)_1/(223)_2$. The three smallest Burgers vectors for the gbds are:

$$\underline{b}_1 = a/18\ [41\bar{1}]_1\ ,\ a/18\ [4\bar{1}1]_2$$

$$\underline{b}_2 = a/9\ [\bar{1}2\bar{2}]_1\ ,\ a/9\ [12\bar{2}]_2 \quad (3)$$

$$\underline{b}_{31} = a/18\ [172]_1\ ,\ a/6\ [121]_2$$

although in this CSL system three other equivalent $\underline{b}_{31}$ type vectors may be found:

$$\underline{b}_{32} = a/18\ [\bar{1}27]_1\ ,\ a/6\ [\bar{1}12]_2$$

$$\underline{b}_{33} = a/6\ [112]_1\ ,\ a/18\ [127]_2 \quad (4)$$

$$\underline{b}_{34} = a/6\ [\bar{1}21]_1\ ,\ a/18\ [\bar{1}72]_2$$

Any linear combination of these vectors may be used to define larger Burgers vectors--the $a/2\ <110>$ fcc lattice translation vectors are an example. This combination process is more important when considering how the dissociation of a crystal lattice dislocation which enters the boundary might proceed [24,25], and how the accompanying reduction in elastic energy is to be estimated.

The boundary geometry is sketched in Fig. 2, in which the sense of inclination of the boundary is shown, together with a number of crystal slip planes in either crystal. Three sets of gbds, designated A, B, and C in Figure 2 could be observed in the interface. Some images of the boundary, taken using two-beam diffracting conditions in crystal 2 only, are shown in Fig. 3; these form part of a more extensive set of images from both crystals used in the $\underline{g}.\underline{b} = 0$ analysis. Figure 4 shows the same region of the interface imaged in a diffracting plane which is also a plane of the CSL; the reduction in the strength of thickness fringes is especially noticeable. The $\underline{g}.\underline{b} = 0$ analysis led to the conclusion that gbd set A could be identified as having Burgers vector $\underline{b}_1$, set C as having Burgers vector $\underline{b}_{33}$, and set B being consistent with $\underline{b}_2$. Using these Burgers vectors, and the angular deviation from exact coincidence, R(O2), already given, the gbd network may be calculated using either the O-lattice or Frank's formula [5]. In either case the result shows excellent agreement with the observed network although some discrepancy in gbd spacing is found. This is most likely due to the residual error in the determination of R(O2) referred to earlier. It is concluded, however, that in this case the predictions of the geometrical model can be confirmed by careful experiment-it is to be anticipated that similar results hold for near-CSL boundaries (up to some limiting value of Σ) in other crystalline materials.

FCC-BCC interface in Co(Al)-CoAl. This example examines the interface between the CsCl-type CoAl phase and the fcc Co-Al solid solution (referred to as "Co") formed by directional solification of a Co-rich Co-Al eutectic alloy [26]. The typical orientation relationship between the phases was about 0.2° from Kurdjumov-Sachs (KS) orientation, in which the {111}"Co" and {110}Co-Al planes meet at the boundary. A small rotation about this common direction accounted for most of the deviation from KS. The observed interfacial structure, shown in Fig. 5 for various reflections in the two phases, was two sets of orthogonal line defects approximately 4.5 nm apart. Once again a $\underline{g}.\underline{b} = 0$ analysis was carried out for images formed in reflections in the two phases; the most likely Burgers vectors were found to be:

$$\underline{b}_1 = a[1\bar{1}0]\text{"Co"} , a[100]\text{CoAl}$$
$$\underline{b}_2 = a[0\bar{1}1]\text{"Co"} , a[\bar{1}1\bar{1}]\text{CoAl}$$
(5)

It should be noted that the magnitudes of three of these four Burgers vectors are twice that expected. This is because the CoAl has CsCl type lattice, so that a Burgers vector twice the length of that in bcc is required to avoid the creation of antiphase boundary in the interface.

Since the dislocation Burgers vectors all lie (within the limit of experimental error) in the conjugate planes forming the boundary, it is concluded that they are misfit dislocations required to match the two crystal structures. Calculation of the appropriate dislocation configuration and optimum orientation relationship (0.28° from KS) agree closely with the observations. It therefore appears that in well-equilibrated interfaces such as this, the geometrical approach is able to model the interfacial structure satisfactorily.

Massive transformation boundaries in Cu-Zn. At the other extreme from the highly stable interface in the CoAl alloy is the third example of observed interfacial structure. In some alloys two different crystal structures are stable at two different temperatures for the same composition. Then only short range diffusion is necessary to grow the low temperature phase on quenching the high temperature one. This is true in the Cu-Zn system for a small range of compositions around 38 W/o Zn, where an fcc α_m phase will grow out of the metastable β (bcc) one, at rates observed to be up to 1 cm/second. Such rapid interfacial migration has previously been interpreted as indicating a disordered interfacial structure, so that no preferred crystal orientations exist to impede growth. Optical microscopy has, however, shown the interfaces to contain facets, in apparent contradiction of this proposal. Fig. 6 shows the interface between metastable β and the α_m phases, using an α_m reflection, and reveals the interface to be clearly facetted, one facet orientation containing a closely spaced (~2.5nm) array of line defects. The two facet planes lie close to 111 in the β_m phase, indicating that the boundary may prefer to lie along close-packed β planes in the absence of any special orientation relationship with the ^{m}β crystal into which the α_m is growing. It was not possible in this case to determine the α_m:β orientation at the prior β:β interface from which the α_m crystal grew (a consequence of the relative size of the α_m crystal and the electron transparent region of the specimen). A combined TEM/electron channeling study in the Ag-Al (bcc-hcp) system [27] showed that almost all the massive crystals exhibited a special orientation relationship with one or other contiguous parent crystal--again in contrast to the established view of the massive transformation.

Burgers vector analysis showed the line defects to be a/2 <110> type (in α_m) dislocations, almost sessile in the long interface facet, and sessile in the short one. Their motion depends on short-range trans-interphase boundary diffusion, while the step at their core will cause interfacial migration, and hence α_m growth, to be associated with their motion. The dislocations may then be considered to act in a manner similar to disordered ledges often invoked in diffusional phase transformations. This example illustrates that the observation of such interfacial dislocations in a wide variety of phase transformations has important implications for many well-established views of the transformation mechanisms.

CONCLUSIONS

Some methods of using the TEM to characterise interfacial structures have been outlined. It has been shown that systematic experiments allow accurate orientation determinations and Burgers vector analyses, with which the predictions of the geometrical models of boundary structure may be compared. The examples presented show some of the wide range of intefacial structures which can be investigated in this way, and illustrate the generality of the method. The implications of interfacial dislocation based models of many thermo-mechanical processes in all crystalline materials are far-reaching, and as yet still in an early stage of development.

ACKNOWLEDGMENTS

Financial support for some of this work from the ARO (Grant No. DAAG29-77-90019) and NSF (Grant No. DMR77-11374) is gratefully acknowledged. Thanks are also due to Drs. D. A. Smith, H. I. Aaronson, R. C. Pond, and A. G. Guha for many helpful discussions during the course of this work.

REFERENCES

1. Grain Boundary Structure and Kinetics, R.W. Balluffi,ed.(ASM,Ohio,1980).
2. W. Bollmann, Crystal Defects and Crystalline Interfaces (Springer-Verlag, Berlin 1970).
3. D.A. Smith and R.C. Pond, Int. Metal. Reviews, 21, 61 (1976).
4. R.C. Pond and W. Bollmann, Phil. Trans. Roy. Soc. Lond., 292, 449(1979).
5. S. Amelinckx and W. Dekeyser, Solid St. Phys., 8, 325 (1959).
6. H. Grimmer, W. Bollmann and D.H. Warrington, Acta. Cryst, A30,197(1974).
7. K.M. Knowles, D.A. Smith and W.A.T. Clark, Scripta Met., 16, 413 (1982).
8. D.A. Smith, K.M. Knowles, H.I. Aaronson and W.A.T. Clark in: Solid-Solid Phase Transformations (AIME, Warrendale, PA, 1983) pp. 587-590.
9. W.A.T. Clark, D.Phil. Thesis, University of Oxford (1976).
10. P. Heilmann, W.A.T. Clark and D.A. Rigney, Ultramicroscopy, 9,365(1982).
11. W. Bollmann, B. Michaut and G. Sainfort, Phys.Stat.Sol.(a),13,637(1972).
12. J.K. Mackenzie, Acta. Cryst., 10, 61 (1957).
13. D.A. Smith and M.J. Goringe, Phil. Mag., 25, 1505 (1972).
14. S.L. Sass and P. D. Bristowe, in ref [1], pp. 71-114.
15. C.B. Carter, A.M. Donald and S.L. Sass, Phil. Mag., A41, 467 (1980).
16. E. Hall, J.E. Walter and C.L. Briant, Phil. Mag., A45, 753 (1982).
17. P.B. Hirsch, A. Howie, R.B. Nicholson, D.W. Pashley and M.J. Whelan, Electron Microscopy of Thin Crystals (Butterworths, London 1965).
18. W.A.T. Clark, R.C. Pond and D.A. Smith in: Proceedings of EMAG 1975 (Academic Press, London, 1976) pp. 433-36.
19. W.A.T. Clark and D.A. Smith in: Grain Boundaries (Inst. of Metallurgists, London, 1976) pp. A35-A39.
20. O.L. Krivanek, S. Isoda, and K. Kobayashi, Phil. Mag., 36, 931 (1977).
21. D.A Smith and P.J. Goodhew, Phil. Mag., A46, 161 (1982).
22. D.A. Smith, this volume.
23. W.A.T. Clark and D.A. Smith, Phil. Mag., A38, 367 (1978).
24. W.A.T. Clark and D.A. Smith, J. Matls. Sci., 14, 776 (1979).
25. R.C. Pond, D.A. Smith, and P.W.J. Southerdon, Phil.Mag., A37, 27(1978).
26. A.G. Guha, W.A.T. Clark, and H.I. Aaronson, Met. Trans., in press.
27. M.R. Plichta, Ph.D. Thesis, Michigan Technological University, 1979.

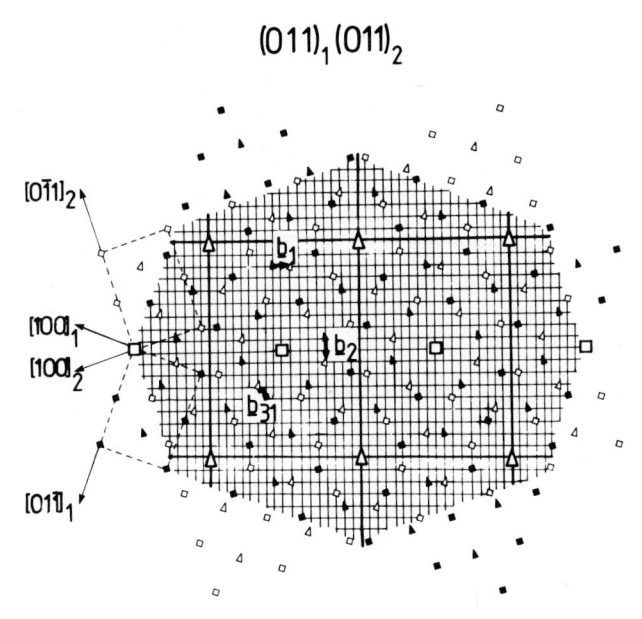

FIG. 1. Lattice plot of two fcc crystals in the Σ9 orientation; the fine mesh is the DSC lattice.

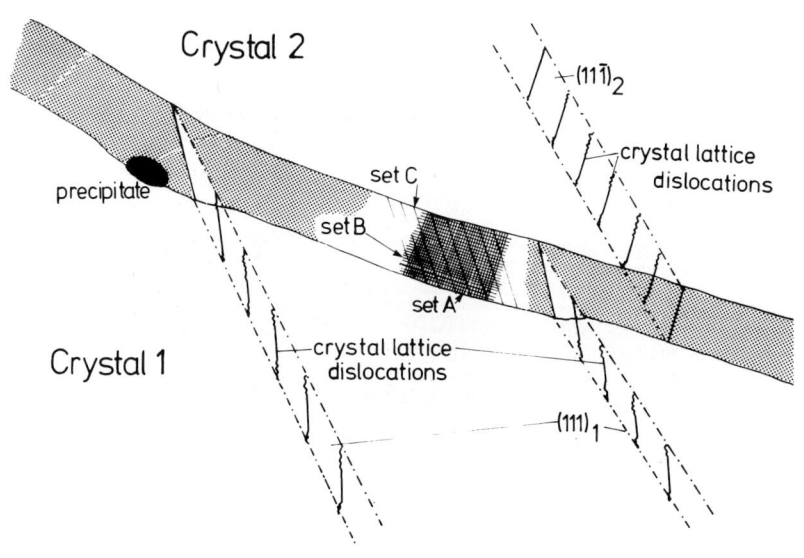

FIG. 2. Sketch of boundary geometry for Σ9 boundary in stainless steel.

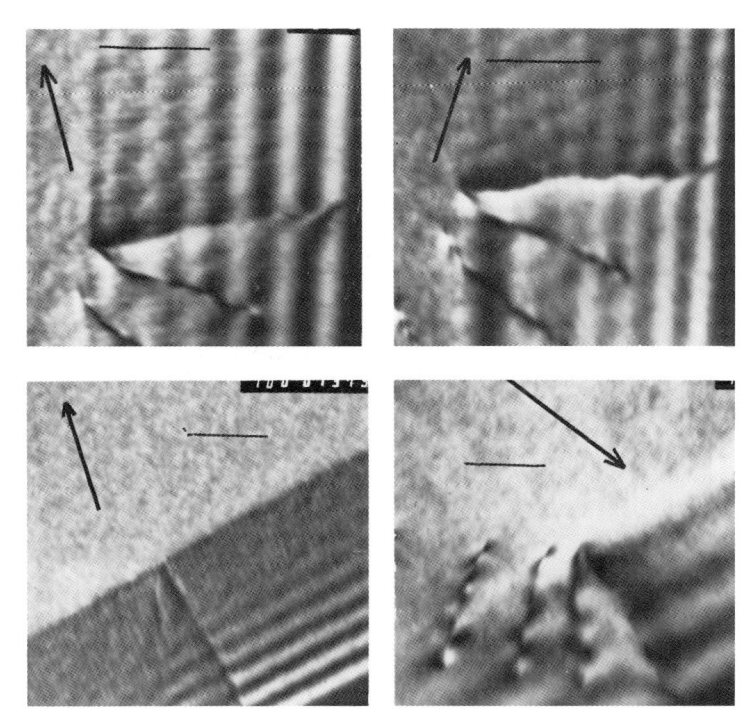

FIG. 3. Two beam micrographs of Σ9 boundary in stainless steel. All dark field, (a) $\underline{g} = 00\bar{2}_2$, (b) $\underline{g} = 1\bar{1}\bar{1}_2$, (c) $\underline{g} = \bar{1}1\bar{1}_2$, (d) $\underline{g} = 2\bar{2}0_2$. Scale markers 250 nm.

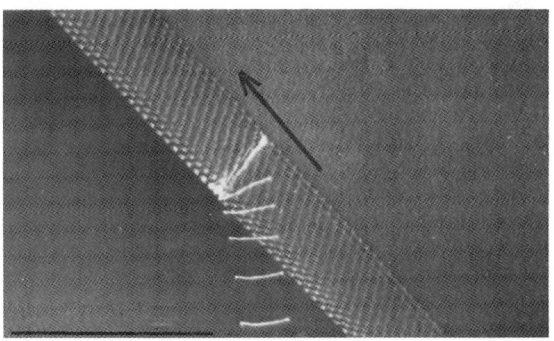

FIG. 4. Same area in simultaneous two beam condition. DF, $\underline{g} = 1\bar{3}1_1/1\bar{1}\bar{3}_2$, scale markers 500 nm.

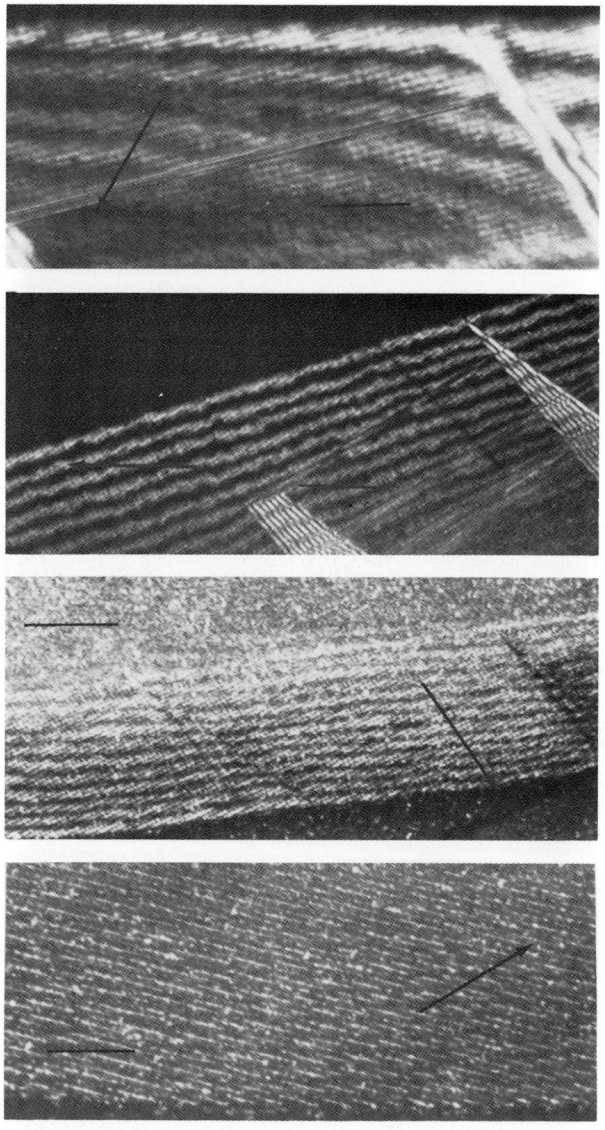

FIG. 5. Micrographs of "Co" (bottom) and CoAl interface in Co-Al eutectic alloy.
(a) DF, $\underline{g} = 200_{"Co"}$ (b) DF, $\underline{g} = \bar{1}1\bar{1}_{"Co"}$
(c) DF, $\underline{g} = 01\bar{1}_{CoAl}$ (d) DF, $\underline{g} = 1\bar{1}0_{CoAl}$
Scale markers 50 nm.

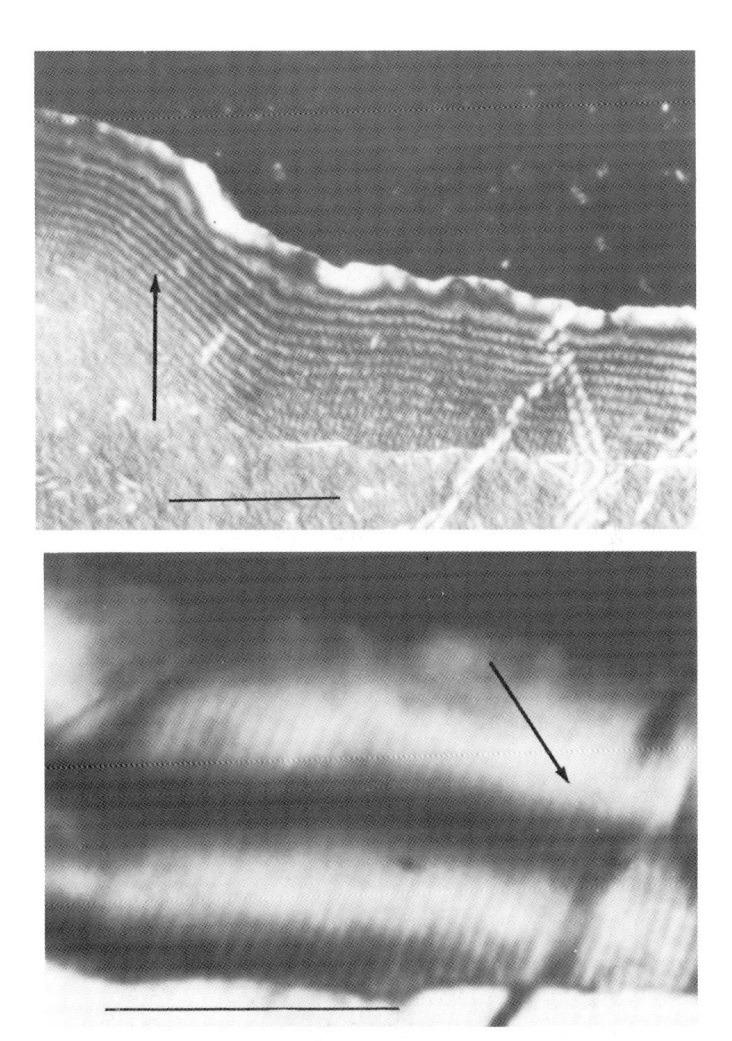

FIG. 6. DF images of boundary between α_m (bottom) and β phases in Cu-Zn. (a) $\underline{g} = 2\bar{2}0_\alpha$, (b) $\underline{g} = 110_\beta$. Scale markers 50 nm.

DISLOCATIONS, STEPS AND INTERPHASE BOUNDARY MIGRATION

D.A. SMITH
IBM T.J. Watson Research Center, P.O. Box 218, Yorktown Heights, N.Y. 10598

ABSTRACT

Interface migration is the fundamental process in numerous phase transformations. For those which are thermally activated the basic event is addition of atoms to a defect in an interface. Interfacial defects can have a range of topographic and elastic characteristics which complicates investigations by transmission electron microscopy, it is difficult to make quantitative in-situ studies of interface migration because of problems of resolution and control of driving forces in specimens where surface effects dominate.

INTRODUCTION

The dislocation theory has been extremely successful in accounting for the mechanical properties of crystalline materials. Now that it is widely recognized that interfaces can have ordered structures one may look for a correlation between interface properties and their defect structure. This raises the questions of interfacial defect classification and characterization together with investigation of their mobility. This paper is devoted to selected aspects of each of these issues.

NATURE OF INTERFACE DEFECTS

The classification of interfacial defects is rather unsatisfactory; in particular the terms steps, ledge and dislocation are used without consistent regard for the physical properties of what is being described. Linear features are often observed during transmission electron microscopy of interfaces and it will be argued that interfacial steps, ledges and dislocations are more closely related defects than appears at first sight. This discussion elaborates on the work of Hirth and Balluffi, [1] and Pond and Smith, [2].

The nature and interrelationship of steps and dislocations is illustrated in figure 1. The transformation of a strain free surface *step* into an edge *dislocation* in the α-phase, figures 1a and b subsequent propagation of the dislocation across the α-β phase boundary by a reaction: $b_\alpha \rightarrow b_\beta + \Delta b$, figures 1c and d, and the formation of a slip step on the surface of the β- phase, figure 1e, are shown. The α-β interface becomes stepped as a result of these processes, figures 1d and e, and the step has an associated elastic field which gives it dislocation character, figure 1e. An interface structure identical with that in figure 1e results when the slabs of α and β phases shown in figure 1f are joined and relaxed. The terms dislocation and step may thus be used to convey information about the *elastic* and *topographic* properties of a defect, respectively. Linear features, in interphase boundaries particularly, may exhibit dual character. In practice interfacial line defects show a spectrum of characteristics which is illustrated in Table 1. The nature of a line defect in an interface can thus be specified in principle by defining as appropriate, its Burgers vector and step height but the experimental and interpretive problems are considerable in practice.

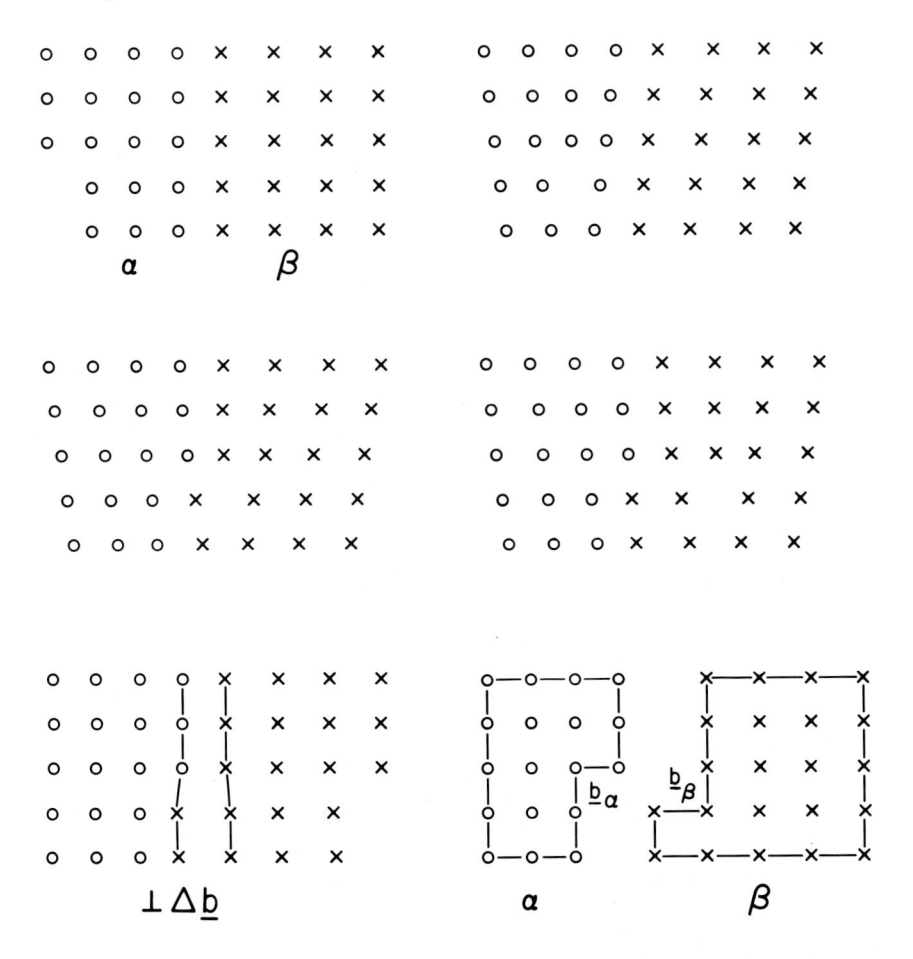

Figure 1 illustrates the interrelationship of steps and dislocations. A surface step on the phase in fig 1a transforms into an edge dislocation in fig. 1b and propagates across the α-β phase boundary by a dislocation reaction figs. 1c and 1d and ultimately produces a surface step on the β phase.

TABLE I: The Spectrum of Defects From Dislocation to Step

Lattice dislocation[a]

twist boundary, epitaxial 001 γ' - NiCr

Shockley dislocation in a twin or fcc-hcp Co or Ag_2Al-Al

Defects in θ'-Al interface

3 d_{111} steps on annealing twin

Cleavage steps[a]

---

[a] The lattice dislocation is a pure topological defect and the cleavage step is a pure topographical defect.

IDENTIFICATION OF INTERFACIAL DEFECTS

Interphase boundaries may be imaged under a variety of contrast conditions in the electron microscope. These include two beam using either or both phases, weak beam, common **g** and lattice imaging. In the first two situations thickness fringes are produced with a period that depends on the appropriate extinction distance ξ_g and the deviation parameter **s**. The effect of an abrupt change in the thickness of the diffracting crystal is to cause a discrete offset of the thickness fringes. Although the geometry of this effect is straightforward it must be remembered that interfacial dislocations also give rise to perturbations to thickness fringes by changing **s** locally. (This effect is the basis of a method for determining the *magnitude* of the Burgers vectors of interfacial dislocations from weak beam images Ishida, Ishida, Kohra and Ichinose [3]). Whilst there is little difficulty in distinguishing a large isolated strain-free interfacial step from a single interfacial dislocation in a planar interface, the correct identification of features with possible dual character and present in dense arrays is much more problematical. A further complication arises because of the range of Burgers vectors possible for interfacial dislocations. The methods for characterizing crystal lattice dislocations are well established; the problem is to select the correct Burgers vector from a small number of previously known possibilities usually by a process of elimination. Whilst the analysis of image extinction can be misleading computer simulation techniques have reached a high degree of sophistication. However, for interfacial dislocations the principles governing the permissible Burgers vectors are not known in general, extinctions are not possible with low index diffraction vectors and computer image matching for interphase boundaries is in its infancy. Even the relatively simple case of dislocations in high angle grain boundaries in cubic metals poses problems. This is because the misorientations at which low energy structures exist (reference states) are not known generally, the energetics of grain boundary dislocations are not known so that the probability of non-primitive DSC dislocations occurring is difficult to assess and various partial dislocations having Burgers vectors which cannot be predicted crystallographically may occur. In addition, even when the Burgers vectors are known, their magnitudes may be small and the corresponding dislocations closely spaced; as a consequence the contrast for at least some members of an array is low. Indirect methods such as the investigation of dislocation reactions or correlation of dislocation spacing with an appropriate model for the interface structure sometimes make it possible to circumvent the difficulties of direct determination of Burgers vectors. The following examples illustrate some of the problems which complicate the analysis of interphase boundary structure by transmission electron microscopy.

Fig. 2 shows two views of crystallographically equivalent orthorhombic Cr_3C_2 fibers in a cubic Ni-Cr-Al matrix which were produced by directional solidification, Spiller and Smith, [4]. The precipitate matrix interface consists, as far as can be resolved, of planar facets, fig. 2a, which contain similar densities of dislocations, fig. 2b. This is a case where the $(117)_m$ $(110)_f$ and $(335)_m$ $(110)_f$ fiber, f, matrix, m, interfaces are semicoherent but the nature of the structure which is being conserved is unclear. In addition further dislocations are geometrically necessary but not resolved.

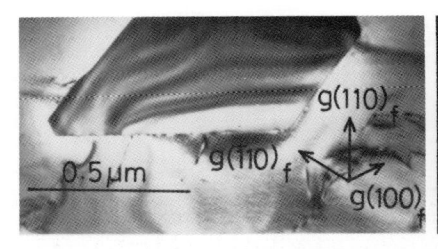

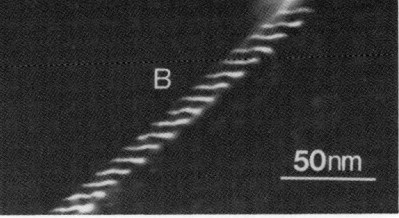

Figure 2 shows two views of crystallographically equivalent orthorhombic Cr_3C_2 fibers in a cubic Ni-Cr-Al matrix. Note the pronounced facetting and the interfacial dislocations. Courtesy G.P.T. Spiller and J. Cryst. Growth.

Fig. 3 shows some results of an *in situ* heating experiment on a twin defect The incoherent facets (no long range strain field) rearrange; this appears to occur by the emission and absorption of 1/6 <112> dislocations (core steps and long range strain field) which propagate along the coherent segments of the boundary plane. This is an example of an interface which can be plausibly described in two different ways. The incoherent facets may be described as closely spaced 1/6 <112> dislocations on successive {111} planes, alternatively they may be regarded as free of dislocations and the only evidence somewhat supporting the physical validity of the dislocation description is the observed mechanism by which the facet configuration changes. In the electron microscope image the step character of the facet is dominant whilst conversely the image of a single 1/6 <112> dislocation derives mainly from the existence of an elastic field and the core step must be inferred. The interface between fcc and hexagonal cobalt is crystallographically very similar to a twin boundary and may propagate in the same way.

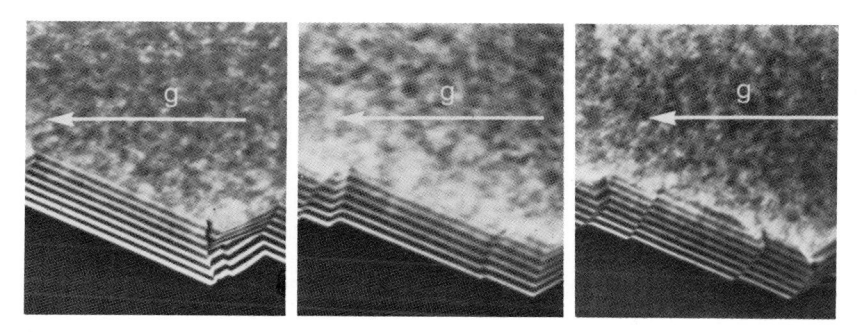

Figure 3 shows three stages in the motion of a facetted twin in aluminum. The steps appear to rearrange by the emission and absorption of twinning dislocations. Courtesy C.M.F. Rae and Philosophical Magazine.

Fig. 4 shows a facetted interface between bcc parent and hexagonal Ti-Mn martensite. The crystallographic theory of martensite predicts that the parent-martensite interface is the invariant plane of the transformation (as it is macroscopically) but the facetting is not expected. Crystallographic analysis predicts a misfit in the parent-martensite interface but there is no indication of any interfacial dislocations Knowles and Smith, [5].

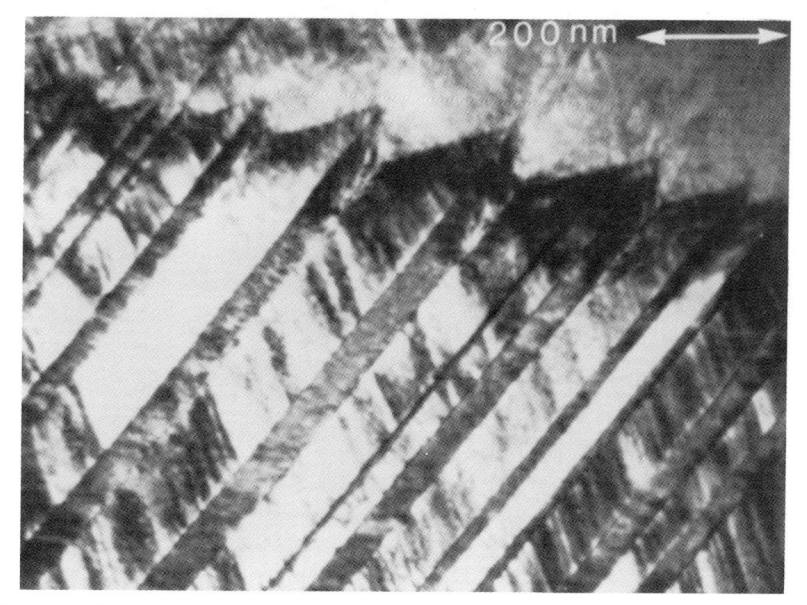

Figure 4 shows a facetted interface between bcc parent and hexagonal Ti-Mn martensite. Courtesy K.M. Knowles

Fig. 5 shows a lattice image of a Si-Pd$_2$Si interface. The hexagonal Pd$_2$Si grew epitaxially on the Si {111} planes with <0001> Pd$_2$Si parallel to <111>$_{Si}$. Despite the almost atomic resolution, many questions with respect to the interface structure and possible interface dislocations remain unanswered. A pseudo-Burgers circuit including a large portion of the interface shows the presence of more Pd$_2$Si {2240} lattice fringes than corresponding Si {111} fringes. It is not immediately clear with what kind of defect these ending fringes correlate. A recent analysis of the fringe configuration in high resolution images of interfaces shows how the number of fringes meeting a reference length of interface depends on the fringe spacing, the orientation of the fringes relative to the interface and the nature of any interface dislocations, Smith and Goodhew, [6].

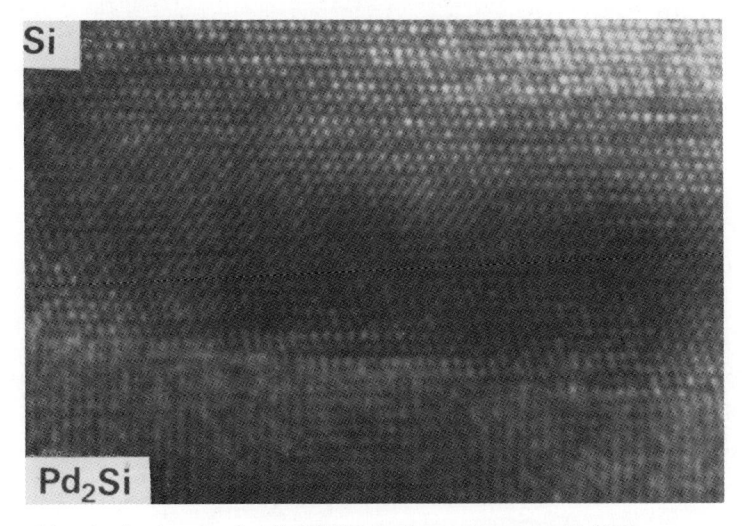

Figure 5 is a lattice image of a Si-Pd$_2$Si interface. Courtesy H. Föll.

ANALYSIS OF LATTICE FRINGE IMAGES OF INTERFACES

It is known that the lattice image of an end-on edge dislocation, with Burgers vector **b** formed from atom planes with normal **g** has **g · b** terminating fringes, Hirsch, Howie, Nicholson, Pashley and Whelan, [7]. There is a one-to-one correspondence for planar specimens between the *numbers* of terminating atom planes in the structure and of terminating fringes in the image providing the microscope has adequate resolution. The question of how faithfully the information concerning the position and configuration of the atom planes is transferred to the image does not affect the argument that follows, Cockayne, Parsons and Hoelke [8]. The gross geometrical features of the lattice image of a defect may be deduced by marking the traces of appropriate atom planes on projections of atomic models of the selected defects. The novel feature for the case of an interfacial dislocation is that the atom planes which form the image of each grain or phase are not necessarily parallel.

It is assumed that three beams are selected to contribute to the image, the transmitted beam and one diffracted beam from each crystal. The "simulation" method is illustrated by the example of a $\frac{1}{6}$<112> twinning dislocation in a $\Sigma = 3$ related 70.53°/<110> coherent boundary in an f.c.c. metal. The dislocation is sketched in Fig. 6 where the boundary is stepped down from left to right by one (111) plane. The occurrence and character of such steps are important in boundary migration Rae and Smith [9] and certain phase transformations, Christian [10], Laird and Aaronson [11]. The traces of the 002 planes are indicated. There is a characteristic chevron pattern of lattice fringes which is interrupted at the dislocation core. Inspection of the fringe patterns shows that there are two extra fringes in the upper grain. The number of fringes cutting the trace of a grain (or interphase) boundary depends on the relative orientation of the fringes, the trace of the boundary, which need not be linear, and the Burgers vectors of any dislocations present. For a symmetrical case with the extra half planes in the upper crystal as illustrated in Fig. 6 the number of extra fringes is $g_1 \cdot h_1 + g_1 \cdot b_1 - g_2 \cdot h_2 = 2$, where subscripts 1 and 2 refer to the upper and lower grain respectively and the sign conventions are given by Smith and Goodhew [6].

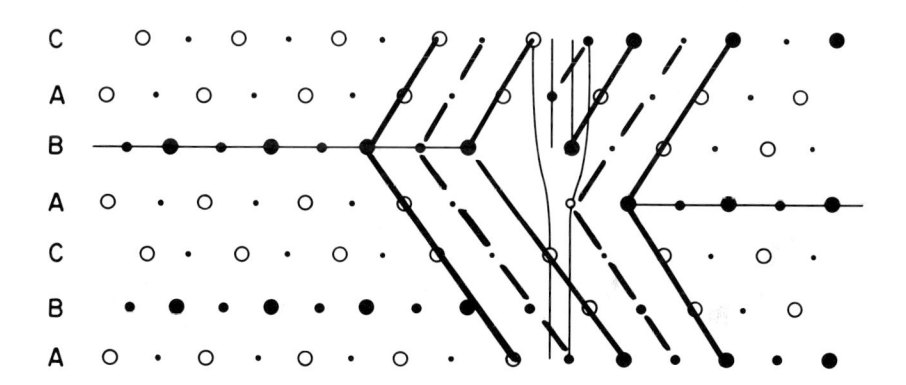

Figure 6 is a sketch of a lattice fringe image from a twinning dislocation.

To emphasize the point that terminating fringes need not imply the presence of dislocations Fig. 7 shows a strain free step on a coherent twin in an f.c.c. metal. Note the 4 terminating 002, fringes.

When a boundary is imaged asymmetrically it is difficult to distinguish unambiguously between a step and a dislocation by fringe counting alone.

Supplementary observations of strain fields, Cosandey and Bauer, [12] or steps can sometimes be helpful. It is invalid to count fringes with a view to determining Burgers vectors of interfacial dislocations unless proper allowance is made for the effects of steps and the orientation of the interface relative to $\underline{g}_1$ and $\underline{g}_2$. Additional discussion of the electron microscopy of interfaces is given in the preceeding paper, Clark [13].

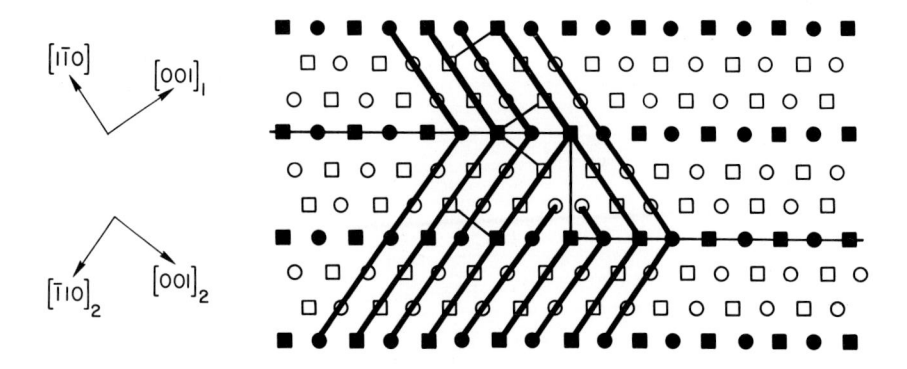

Figure 7 is a sketch of a lattice fringe image of a step in a coherent twin.

INTERFACE MOTION IN THIN FOILS

Interface motion in thin foils is dominated by the influence of surfaces. Table 2 gives order of magnitude estimates of some driving and resistive forces for grain boundary migration in these specimens. Surface energy and grooving are major factors, in contrast to bulk behavior, and additional effects arise because of the capacity of a surface to act as a sink for point defects, dislocations and elastic stresses. Consequently the results of experiment or thin films give results which may differ sharply from the behavior of bulk specimens. For example recrystallization textures are dominated by surface energy, the crystallography of martensitic transformation changes, grain boundary migration is almost invariably intermittent and the grain growth kinetics change according to whether experiments are done with films on or off substrates.

ROLE OF DEFECTS IN THE MOTION OF INTERPHASE BOUNDARIES

Although the present knowledge of the defect structure of interphase boundaries is rudimentary it may well be that the nature and behavior of defects is the key to understanding the mechanisms by which phase transformations occur. For the usual magnitude of free energy change during a transformation a defect mechanism for interface propagation is necessary [14]. The nature of the defect predicates that of the transformation in turn. For example, if any mechanism involving atom attachment to strain free ledges is proposed then the influence of a stress will be quite different than if a mechanism involving dislocation motion operates.

It is a striking observation that the growth of hexagonal Ag_2Al, the martensitic transformation of f.c.c. to h.c.p. Co, deformation twinning and the movement of the epitaxial Pd_2Si-Si interface can all be accounted for by the motion of 1/6 <112> dislocations. In each case motion of dislocations causes the lattice trans-

TABLE II: Forces on a Grain Boundary in a Thin Film

Origin	Formula	Order of Magnitude Nm^{-2}
Stored energy	$1/2\ \varepsilon^2 E$ [a]	5.10^5
Surface energy	$2\Delta\gamma_s f/h$	5.10^5
Grooves	$\gamma_b^2/h\gamma_s$	10^7
Curvature	$2\gamma_b (1/r_1 + 1/r_2)$ [b]	10^7

[a] $\varepsilon = \Delta\alpha\Delta T$ on substrate [b] $= 0$ for catenoid

ε = strain, E = Youngs modulus, $\Delta\gamma_s$ = averaged difference in surface energy, f = area fraction of low surface energy grains, h = film thickness, γ_b = grain boundary energy, r_1 and r_2 = principal radii of curvature.

formation. Any change of composition is accomplished by diffusion. It is not necessary that a dislocation mechanism of interfacial motion results in macroscopic deformation. Often there are crystallographically equivalent dislocations which can operate together to give no macroscopic deformation. This is similar to martensitic transformations where if the lattice invariant transformation is twinning the alternative twinning modes occur in volume fractions such that macroscopically there exists an invariant plane.

A goal of studies of interface migration is to correlate the motion of interface defects with the overall migration of the interface. For example if there are parallel steps of height h moving with a velocity v_s the velocity of the interface, v_i, is $v_i = nhv_s$. This form of behavior has been verified for the thickening of Ag_2Al precipitates and is the basis of the "ledge" theory of precipitate coarsening. It is reemphasized here that the interface defects concerned have dual step and dislocation character. However, the growth kinetics are strictly diffusion limited in contrast for example to the f.c.c.-h.c.p. transformation in cobalt which is athermal.

Whilst migration of grain boundaries has been investigated in-situ in a number of instances it has never, so far been possible to resolve a sufficient density of moving grain boundary dislocations to account for the observed migration rate.

CONCLUSION

It is possible to determine the step and dislocation character of isolated interfacial defects by conventional transmission electron microscopy. However when the defects are closely spaced lattice fringe imaging is more useful. The number and location of terminating fringes permits the characterization of interfacial defects as steps, dislocations or of mixed nature. It remains an elusive goal of investigations of interface migration to correlate fully the motion of an interface with that of individiual defects.

REFERENCES

1. J.P. Hirth and R.W. Balluffi, Acta Met. *21*, 929 (1973).

2. R.C. Pond and D.A. Smith, *6th European Congress on Electron Microscopy*, Jerusalem, 1976 p.500.

3. Y. Ishida, H. Ishida, K.Kohra and H. Ichinose, Phil. Mag.A *42*, 453 (1980).

4. G.D.T. Spiller and D.A. Smith, J. Cryst. Growth *50*, 445 (1980).

5. K.M. Knowles and D.A. Smith, Acta. Met. *29*, 101 (1981).

6. D.A. Smith and P.J. Goodhew, Phil. Mag.A. *46*, 161.

7. P.B. Hirsch, A. Howie, R.B. Nicholson, D.W. Pashley and M.J. Whelan, Electron Microscopy of Thin Crystals, Butterworth, London, 1965, p.371.

8. D.J.H. Cockayne, J.R. Parsons and C.W. Hoelke, Phil. Mag. *24*, 139 (1971).

9. C.M.F. Rae and D.A. Smith, Phil. Mag.A, *41*, 477 (1980).

10. J.W. Christian, *The Theory of Transformations in Metals and Alloys*, C. Laird and H.I. Aaronson, Acta Met., *15*, 73 (1967).

11. C. Laird and H.I. Aaronson, Acta Met. *15*, 73 (1967).

12. F. Cosandey and C.L. Bauer, Phil. Mag. A, *44*, 391 (1981).

13. W.A.T. Clark, this volume.

14. J.W. Cahn, Acta Met. *8*, 556 (1960).

MICRODIFFRACTION STUDIES OF SMALL GOLD METALLIC PARTICLES

MIGUEL JOSE-YACAMAN, ALFREDO GOMEZ, AND KRYSTYNA TRUSZKOWSKA
Instituto de Física, Universidad Nacional Autónoma de México,
Apartado Postal 20-364, Delegación Alvaro Obregón,
México 01000, D. F.

ABSTRACT

The crystalline structure of small gold particles in the size range between 50-200 Å is studied using STEM in microdiffraction.
It is found that some single-crystalline particles showed an anomalous diffraction pattern. In some cases these pattern be indexed as an hexagonal lattice with a ratio c/a= 2.46. A number of models to explain the hexagonal diffraction are discussed. In some other cases the patterns correspond to fcc structures but with a large splitting effect. The results reveal the strong dynamical character of the diffraction by small gold particles.

1. INTRODUCTION

The crystal structure of small metallic particles is still not well known. Although has been direct evidence of departure from the bulk structure in multiple twinned particles[1-2], it is normally assumed that small particles have the same crystal symmetry than the bulk. In modern electron microscopes microdiffraction can be used to study the crystalline structure of particles between 10-300 Å diameter in individual basis. STEM electron optics can be used to focuss a beam in region of few manometers in diameter on a region of the sample. In practice it is possible to grow small particles which are isolated enough so as to obtain diffraction from an individual particle.
In the present study we reporte microdiffraction studies of gold small particles obtained by vacuum evaporation onto a KCl substrate at 250°C. The patterns were obtained using a Jeol 100-CX Electron microscope. In this instrument patterns from an areas between 50-400 Å can be obtained with a beam divergency (α), between 3.45×10^{-4}, 1.3×10^{-3} rad. This, produces sharp spot patterns that can be readely analyzed by the standard methods.

2. ANOMALOUS DIFFRACTION EFFECTS IN SMALL PARTICLES

In the past TEM studies of the early stages of growth of thin films produced selected area diffraction patterns from areas of about 5000 Å in size. The patterns contained many extra spots as shown in figure 1. The most popular interpretation of those spots was given by Pashley and Stowell[3] in terms of twin reflections and double diffraction at twin boundaries. In the case of thin films explanation required either a buckled or very thin sample. The forbidden reflections have been also observed in small metallic particles[4]. With use of microdiffraction it is possible to obtain a more detailed information about the origin of the forbidden reflections. In the present paper we will discuss only some anomalous patterns that can not be interpreted in terms of normal fcc structure. Figures 2-4 show examples of anomalous patterns which were obtained from small particles.
The particles do not have any twin contrast and can be considered as "normal" single-crystalline. From the corresponding diffraction patterns shown in the figures, it is immediately apparent that the pattern can not be simply indexed as fcc. There are many reflections that correspond to fractional fcc indexes. On the other hand the angle between spots are also different from the expected fcc values.
The patterns can be indexed assuming an hexagonal diffraction symmetry.

The observed reflections can be reproduced with an hexagonal net with the following parameters:

$$a = a_s = 2.86 \text{Å}, \quad b = a_s = 2.86 \text{Å}, \quad c = ba_s = 7.05 \text{Å}$$

were a_s is the nearest neighborg distance in the gold lattice. The resulting reflections are shown in Table 1 in hexagonal miller indexes. In terms of the hexagonal lattice the patterns in figure 2, 3 and 4 correspond to <00.1>, <$\bar{1}$1.0> and <011> zone axis respectively. The described net is just the hexagonal contained in the fcc lattice. In a normal situation most of the reflections will cancel out by scattering in successive layers of the fcc stacking sequence. It is therefore necessary to explain the breaking of the fcc diffraction geometry in small metallic particles.

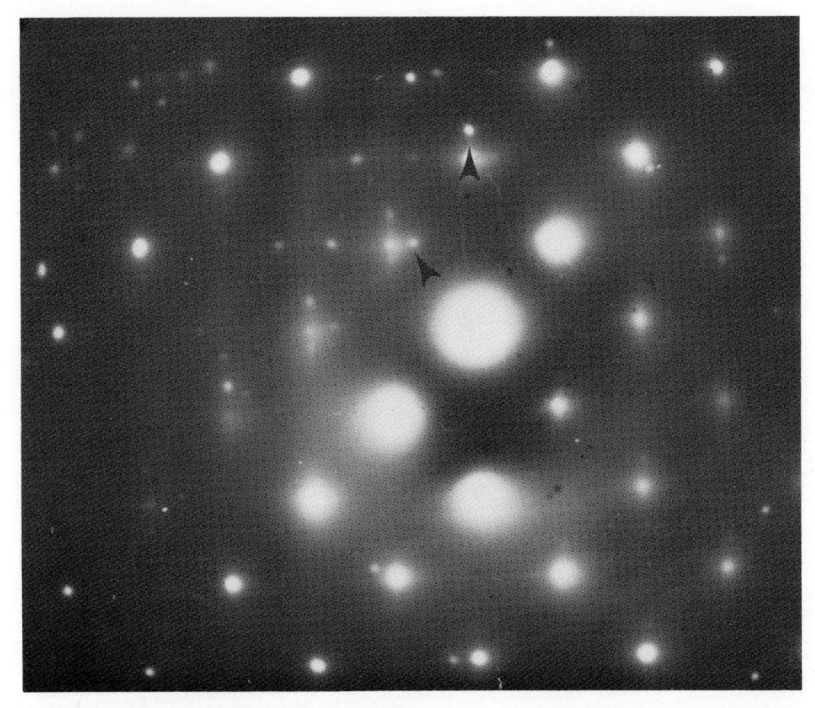

Fig. 1. Typical selected area diffraction from gold particles. Many extra reflections are observed (indicated by arrows).

In the case of thin gold films extra reflections has been observed by several authors. Cherns[5] and Krakow[6] have explained them as the result of monoatomic steps which break the stacking sequence in surface regions. This model implies a purely surface diffraction and the extra spots are expected to be weak and insensitive to tilting. However the extension of those ideas for small metallic particles is not obvious. The amount of scattering centers at the surface of the particle do not seem to be enough to produce the strong forbidden intensity observed. The pattern seems to suggest a bulk rather than a surface effect.

The same argument can be applied to the model of Heyraud and Metois[7] and Tanishiro et al[8] that have attributed extra reflections in larger gold particles ($\sim$ 1000 Å) to surface reconstruction effects[9]. Again the

235

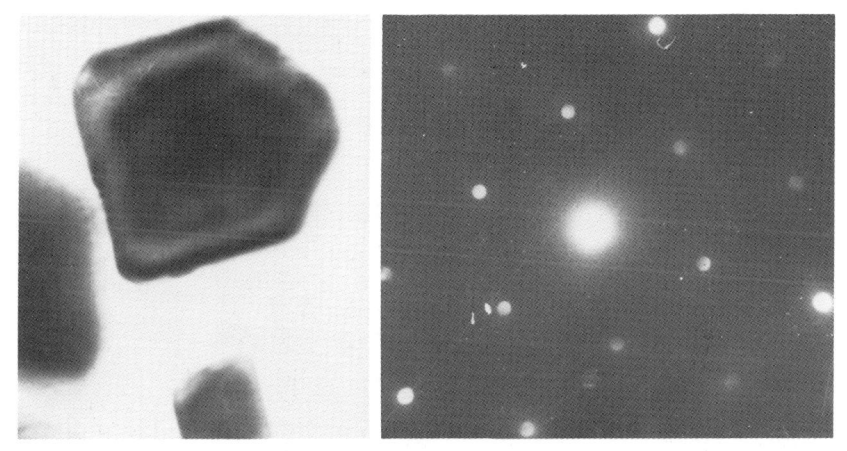

Fig. 2. Bright field image of a gold particle grown by evaporation onto CKl substrate, and corresponding microdiffraction patterns showing six forbidden reflections corresponding to an interplanar distance of ∿ 2.46 Å.

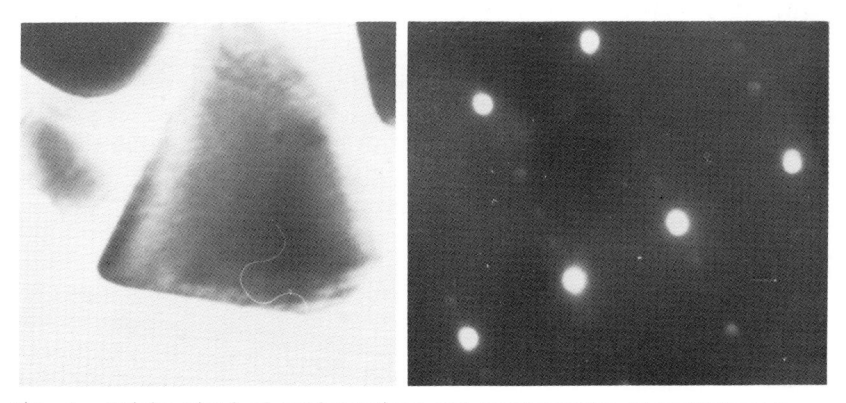

Fig. 3. Bright field of gold particle and corresponding microdiffraction pattern. The pattern corresponds to a <11.0> zone axis.

"bulk" behavior of the hexagonal reflections rules out this model. In any case the reconstruction will have to be considered extended through the whole particle.

Models based on twinning or repeated faulting[4] are also inadequate. No twin contrast is observed in weak-beam images of the particles and some of the reflections observed do not correspond with the ones expected for repeated faulting[11].

On the other hand twin reflections due to <112> boundaries parallel to the electron beam ie; double possitioning boundaries[12] will not be seen in a diffraction setting close to the Laue condition. A tilting close to 10° is necessary to show up those reflections. Therefore this mechanism will not explain the symmetric forbidden spots seen in fig. 2-4.

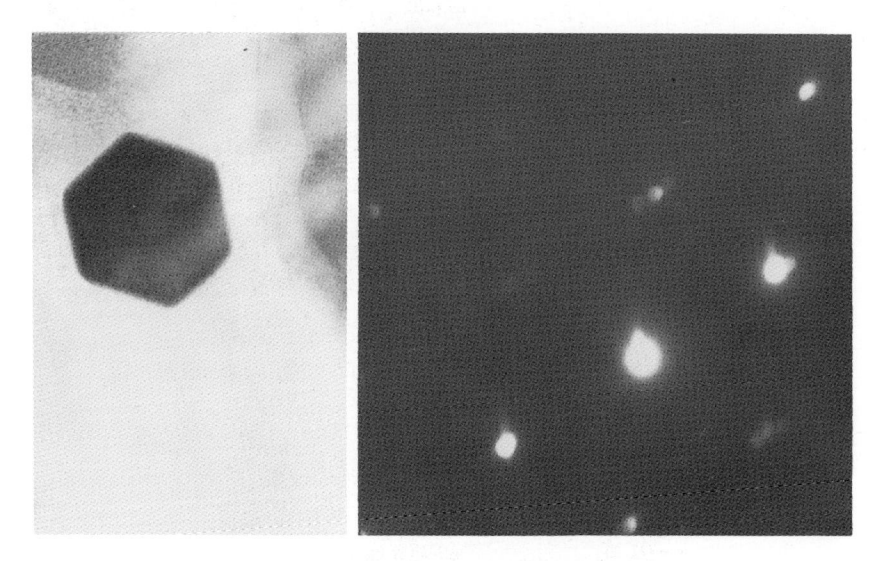

Fig. 4. Bright field and corresponding microdiffraction pattern of a gold particle. The symmetry corresponds to a <011> hexagonal zone axis.

3. SPLITTING EFFECTS

An additional effect observed in diffraction of small particles is shown in figure 5. When the particle is diffracted with the beam located at the center of the particle, the pattern shows four reflections corresponding to interplanar distances of 1.44 and 1.20 Å in a perpendicular array. This type of pattern can not be explained in an fcc basis. If now the beam is located over one of the edges of the particle new spots are observed in the pattern. The new array of spots can be now indexed as a <100> fcc zone axis in which some of the spots are splitted by 0.138 A^{-1} along a <110> direction. In real space this splitting corresponds to periodicity of 7.2Å.

This type of spot splitting can not be explained in terms of the theory of Gómez et al[13], since the effect is equal in $\vec{g}$ and $-\vec{g}$ reflections and its value is too large for the one expected for a wedge.

A very relevant point is that the observed periodicity corresponds to $\frac{5a}{2}$s. It is well known that {100} gold crystal surfaces are reconstructed with a (5X20) super net[14]. However the basic unit cell can be consider as (5X1). It is very likely that in the case of diffraction along the

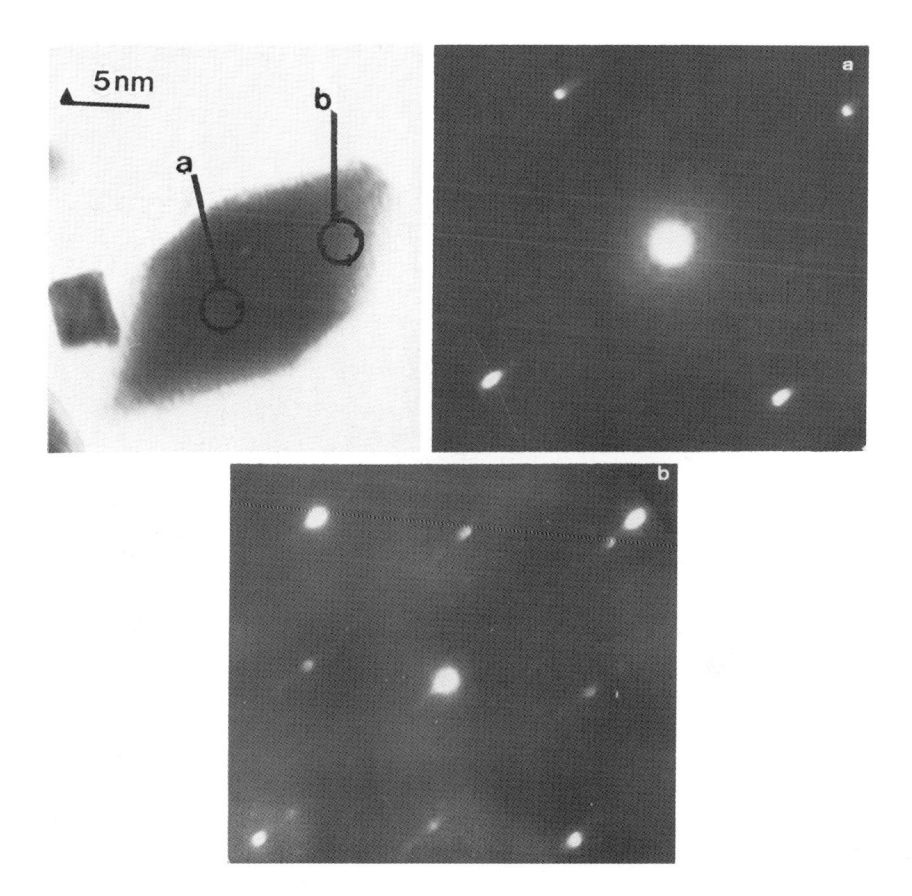

Fig. 5. Bright field and corresponding microdiffraction patterns of a gold
particle from the areas indicated with a and b.

thinnest part of the particle surface effects become more conspicuous. It
is not clear however why the extra spots appear so strong if they correspond
to a second order diffraction of the super net.

It is clear that strong dynamical diffraction effects are taking place
in small gold particles. A full dynamical calculation is necessary in order
to obtain a model of the particle structure. A very important question to
address is if the reconstruction is restricted to the surface or it propagates through the bulk of the particle.

4. DISCUSSION

From the preceeding sections it has been shown that small particles present two types of anomalies; either hexagonal diffraction or splitted fcc patterns. In the latter case the origin of the extra spots seems to be correlated with super periodicities on the particle symilar to the ones observed in LEED patterns of large single crystals.

The origin of the hexagonal reflections is still not clear. An effect to be consider is that such reflections are connected with excitation of higher order Laue zones (holz). For a particle of 50 Å thickness the reciprocal lattice spots will spread by about 2×10^{-2} A^{-1}. Thus if in the <111> zone axis the spreading of the <$\bar{1}$11> spot contained in the first zone will be larger than its excitation error in the Laue condition (3.4×10^{-3} A^1). Therefore the reflection can be excited and will be proyected on the plane of the zero zone as $\frac{2}{3}$ {422} or 2.46 Å interplanar distance. However it is difficult to explain that the intensity of some of the spots of the first zone can be even larger that the ones in the zero zone. Again a full dynamical calculation taking into account Holz appears necessary.

Another possibility to explain the hexagonal spots migth be in terms o particle surface roughness. Calculations of Pérez et al[15] have indicated that the particle contains many kinks and micro-facets. Microthermodynamic models[15] of particles seem to indicate that this migth correspond to an equilibrium shape of the particle. Surface roughness will produce many incomplete unit cells in the particle producing an effect symilar to the one described by Cherns[5] and Krakow[6] but with a larger intensity.

Finally distorsions in their unit cell that break the fcc stacking sequence. Lateral displacements on the rows of atoms will produce hexagonal spots.

5. CONCLUSSION

STEM microdiffraction appears to be an invaluable tool to study crystal structure of small metallic particles. However the strong dynamical character of scattering in noble metals makes the interpretation of the patterns not straigth-forward. The implications of the observed diffraction structures can only be understood by multi-beam dynamical diffraction theory. Calculations of this type are being carried out in our laboratory.

A full understanding of the patterns migth provide important information on the particle structure which could relevant to their catalytic properties.

TABLE 1. Hexagonal reflections present in Au films a= 2.88 Å   b= 2.88 Å c= 7.05 Å

(hkl)	d (Å)	(hkl)	d (Å)
(00.1)	7.05	(01.4)	
(00.2)	3.52	(21.1)	1.43
(10.0)		(12.0)	
(01.0)	2.47	(11.2)	
(11.0)			1.33
(10.1)		(12.2)	
(01.1)	2.35	(21.4)	1.15
(11.1)		(20.2)	1.17
(00.3)			
(01.2)	2.03		
(11.2)			
(01.3)	1.71		
(11.3)			

REFERENCES

1. C.Y. Yang - J. of Cryst. Growth 47, 274 (1979).
2. A. Gómez, P.S. Schabes and M.J. Yacamán - Thin Solid Films, 98, L95 (1982).
3. D.W. Pashely and M.J. Stowell, Phil. Mag. 8, 1605, (1963).
4. T. Hayashi, T. Ono, S. Yatsuya and R. Uyeda - Jap. Journ. of Appl. Phys. 16, 705, (1977).
5. D. Cherns - Phil. Mag. 30, 549, (1974).
6. W. Krakow, Surf. Sci. 111, 503 (1981).
7. J.C. Heyraud and J.J. Metois - Surf. Sci. 100, 519, (1980).
8. Y. Tanishiro, H. Kanamori, K. Takayanagi, K. Yagi and G. Honjo. Surf. Sci. 395 (1981).
9. M.A. Van Hove, R.J. Hoestiner, P.C. Stair, J.P. Biberian, L.L. Hesmodel, I. Barios and C.A. Somorjai, Surf. Sci. 103, 189, (1981).
10. J.E. Davey and R.F. Deiter, J. of Appl. Phys. 36, 284; (1965).
11. M.J. Stowell in epitaxial growth, edited by J. Matthews Academic Press, London, 1975.
12. A. Gómez, P. Schabes, M.J. Yacamán and T. Ocaña, Phil. Mag. 47 A. 169 (1983).
13. D.G. Fedak and N.A. Gjostein - Surf. Sci. 8, 77, (1967).
14. O.L. Pérez, D. Romeu and M.J. Yacamán - Appl.of Surf. Sci. 10, 135, (1982).
15. A. Searcy. Jour. of Sol. Stat. Chem. 48, 93, (1983).

HIGH RESOLUTION STUDY OF THE RELATIONSHIP BETWEEN MISFIT ACCOMMODATION AND GROWTH OF $Cu_{2-x}S$ IN CdS

T. SANDS, J. WASHBURN AND R. GRONSKY
Materials and Molecular Research Division,
Lawrence Berkeley Laboratory, Berkeley, CA 94720

ABSTRACT

The growth of $Cu_{2-x}S$ in the (0001) basal face of single crystal cadmium sulfide has been studied with high resolution transmission electron microscopy (HRTEM) and diffraction. Cross-sectional $Cu_{2-x}S$/CdS specimens and plan-view $Cu_{2-x}S$ separated films were prepared from heterojunctions formed by the aqueous conversion of CdS to topotaxial $Cu_{2-x}S$. Low chalcocite films were found to form in two principal stages; 1) coalescence of low chalcocite islands and 2) an accelerated reaction localized at pores in the $Cu_{2-x}S$ film. Several possible misfit accommodation mechanisms involving twinning, cracking, and a second copper sulfide phase were identified.

INTRODUCTION

Nearly thirty years of research and development of the photovoltaic copper sulfide-cadmium sulfide heterojunction have culminated in the present-day thin-film frontwall devices with efficiencies approaching ten percent [1]. However, viability as a terrestrial solar cell also requires reproducibility and stability, two areas in which the performance of the $Cu_{2-x}S$/CdS solar cell is still questionable. Further breakthroughs in these two areas will require a detailed understanding of both the growth mechanisms of Cu_2S in CdS and the evolution of the interfacial region during heat treatment and long-term exposure at operating temperatures. In this paper, the results of a TEM study of the as-prepared $Cu_{2-x}S$/CdS heterojunction are presented. Within the framework of these results, an understanding of the relationship between misfit accommodation and growth of $Cu_{2-x}S$ in the cadmium face of basal CdS is developed.

Structural Aspects of the $Cu_{2-x}S$/CdS Thin Film Heterojunction

The polycrystalline n-CdS layer of the $Cu_{2-x}S$/CdS solar cell is composed of columnar grains of the wurtzite structure with the c-axes preferentially oriented normal to the substrate (see the review of CdS layer fabrication procedures by R. Hill [2]). The wurtzite structure, being non-centrosymmetric, exhibits two basal faces, each with its own chemistry. Consequently, the HCl texture etch which precedes $Cu_{2-x}S$ layer formation has radically different effects on the cadmium (0001) and sulfur (000$\bar{1}$) faces. In this paper, only the flat basal portion of the etched cadmium face is considered. The growth of $Cu_{2-x}S$ in the terraced and faceted faces of (0001) etch pits and (000$\bar{1}$) hillocks is considered in a second paper [3].

After etching, the CdS film is dipped into an aqueous solution of CuCl at 95-98°C for several seconds. A large negative change in free energy [2] drives an exchange reaction between cations yielding a p-type copper sulfide absorber layer approximately 150 nm thick. The optimum phase and composition has been determined to be low chalcocite (a structure based upon the interstitial ordering of copper within a distorted h.c.p. sulfur network [4,5]) with [Cu]/[S] $\cong$ 1.997 [2]. Unfortunately, several stable and metastable phases complicate the structure of the $Cu_{2-x}S$ layer [6]. However, only one of these phases, the f.c.c.-based tetragonal phase, is important to the following discussion. A summary of the relevant structural data is given in Table 1.

TABLE 1. Comparison of Lattice Parameters

phase	a_{eff}[nm](%misfit)	c_{eff}[nm](%misfit)
CdS	$a_o = 0.4137$	$c_o/2 = 0.3358$
low chalcocite [4]	$d_{030} = 0.3961$ (-4.38)	$d_{004} = 0.3374$ (+0.50)
tetragonal phase [7]	$d_{100} = 0.4008$ (-3.20)	$d_{102} = 0.3268$ (-2.77)

TEM SPECIMEN PREPARATION

In order to avoid the complicating effects of grain boundaries and high dopant levels, experiments were performed with undoped single crystal CdS (Eagle-Picher). Boules of CdS were cut into basal slices and mechanically polished. The final polish was accomplished with 1 μm diamond paste. Work-damaged surface material was removed during a texture etch in fresh 37 % HCl at room temperature for 30-60 seconds. At this stage, CdS cross-section specimens were prepared by ion milling (as described below) and examined by electron diffraction to verify the presence of the wurtzite structure in the near-surface region. No sphalerite CdS was detected. The copper sulfide layer was formed in an aqueous solution of CuCl (6g/l) and NaCl (2g/l) (substrate preheated to conversion temperature) at 98°C for 8-12 seconds [8].

Plan-view separated $Cu_{2-x}S$ films were formed by selectively etching away the CdS in HCl. Copper sulfide films were then floated onto grids and examined in the TEM within 24 hours of preparation. $Cu_{2-x}S$/CdS cross-sections were prepared by argon ion milling (4kV, 30 μA specimen current, 14° tilt). Ion milling heating effects were kept to a minimum by utilizing an LN_2 cold stage. "Conventional" microscopy and diffraction and high resolution lattice imaging were performed with a Siemens 102 TEM ($C_s \cong 1.9$ mm) at 100 kV.

RESULTS

Micrographs of separated films at the earliest observable stage indicated that the initial reaction produced small islands of low chalcocite. As the reaction proceeded the islands grew laterally and coalesced. The pores remaining between the islands were, in many cases, bounded by inclusions of the tetragonal phase (Fig. 1). In addition, six variants of microtwins were found in both the interior and at the edges of the low chalcocite islands. Dark-field micrographs (not shown) using the $1\bar{1}02$-type twin reflections showed that the twins were thin plates inclined to [0001] and elongated along $<2\bar{1}\bar{1}0>$. The best fit with the diffraction and microscopy data is provided by the $\{0\bar{1}13\}$ reflection twin which effectively expands the copper sulfide lattice by 10% in a $<01\bar{1}0>$ direction [3].

The next stage of the reaction began at the pores. These pores acted as short-circuit diffusion paths for the reacting species resulting in a locally accelerated reaction. As the penetrations deepened, the pores developed into cracks on $\{\bar{2}1\bar{1}0\}$, thereby fascilitating further reaction with the solution (Fig. 2(a)). A film in the final heavily textured form is imaged in Fig. 2(b). Cross-section micrographs revealed that the deeper penetrations consisted of a mixture of low chalcocite and the tetragonal phase (Fig.3).

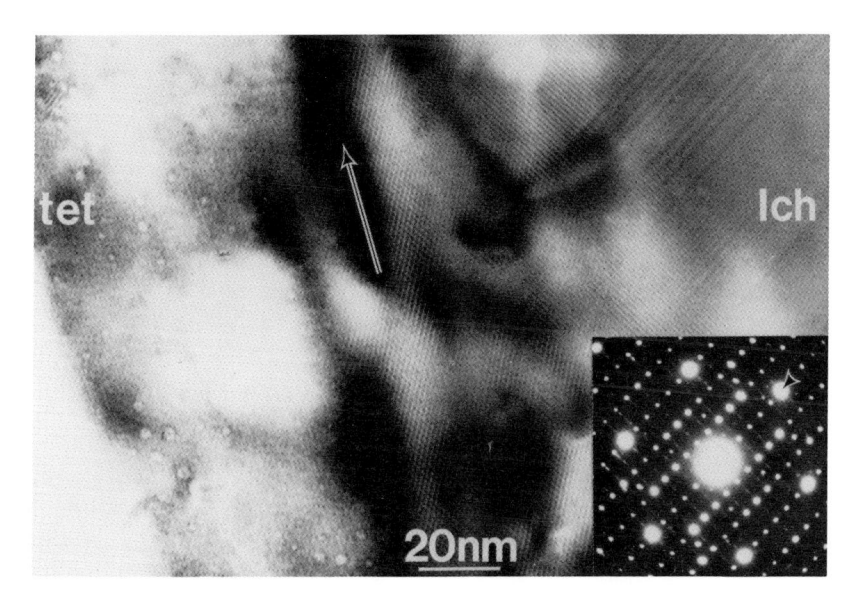

FIG.1. Tetragonal phase inclusion at island edge in low chalcocite separated film. Arrow indicates a <$\bar{2}110$> sublattice direction. Note antiphase boundaries on (200) low chalcocite (upper right). These boundaries are commonly found in low chalcocite films grown in basal CdS substrates [5]. Low chalcocite and the tetragonal phase in [001] and [40$\bar{1}$] zone-axis orientations, respectively. The 060 lch. and 020 tet. spots are located at arrow in inset diffraction pattern.

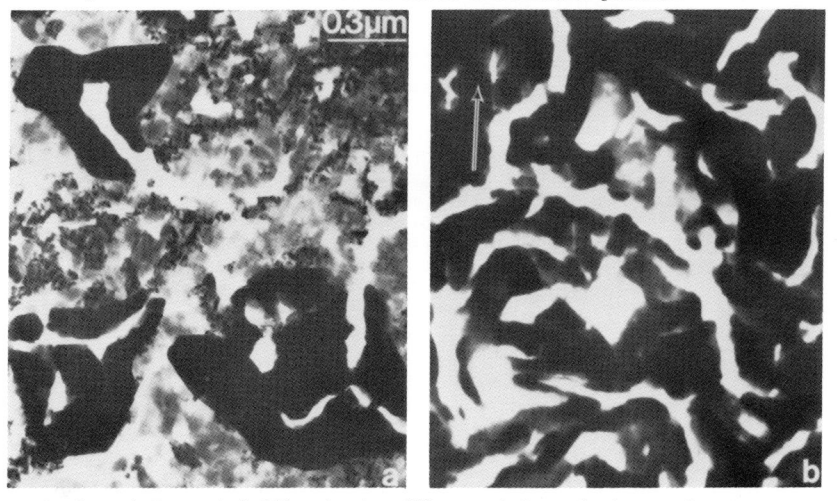

FIG. 2. a) Separated film showing film morphology during early stages of reaction. b) Film in final highly textured state. Arrow indicates a <0$\bar{1}$10> sublattice direction.

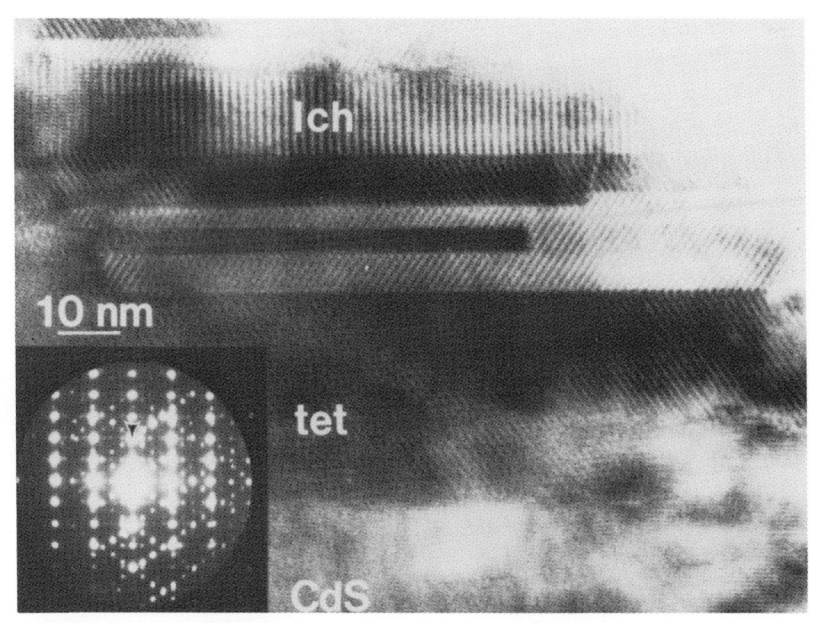

FIG.3. Cross-section image of deep penetration near crack. Slight misorientation of specimen accounts for black/white contrast of twinned tetragonal phase. CdS is in [$\bar{2}$110] zone-axis orientation. L.ch. and tet. are in [010] zone axis orientation. Arrowed spot in diffraction pattern is 0002 CdS, 204 l.ch. and 110 tet.

DISCUSSION

The observed reaction sequence can be rationalized as follows: The initial exchange reaction forms small coherent islands of low chalcocite. At this stage, the nucleation of the tetragonal phase (in the absence of large stresses) is inhibited by the fact that the surface is parallel to the glide plane of the required transformation dislocations (1/3<10$\bar{1}$0> on every other close-packed sulfur plane). As the islands grow laterally and vertically, the 4.4 % misfit between low chalcocite and CdS can no longer be accommodated by elastic deformation alone. The stress is partially relieved by the introduction of microtwins and martensite-like tetragonal phase inclusions at the island edges were the stress build-up is high.

The pores remaining in the film now act as conduits for diffusion of the reacting species. Consequently, the CdS adjacent to the pores is rapidly converted to copper sulfide. These deeper penetrations cause further stress build-up due to the mismatch in the basal plane. To accommodate this stress the pores become elongated and form microcracks on {$\bar{2}$110}. The thicker penetrations then merge to form the heavily textured film described in the results. It should also be noted that as the microcracks develop they expose {$\bar{2}$110} planes which can serve as nucleation sites for transformation dislocations and misfit dislocations with Burgers vectors in the basal plane. A combination of 15 % tetragonal phase and the remainder, low chalcocite, results in a perfect match with unstrained CdS in the $\vec{c}$ direction. This fact, combined with the favorable value of a_{eff} for the tetragonal phase (Table 1), may explain the presence of the tetragonal phase in the deeper penetrations of the copper sulfide adjacent to the cracks [9].

REFERENCES

1. A. M. Barnett, J. A. Bragagnolo, R. B. Hall, J. E. Phillips and J. D. Meakin, Proc. 13th Photovoltaic Spec. Conf. (Washington, D.C., IEEE New York 1978) p. 419.
2. R. Hill, Solid-St. and El. Dev. 2, S49 (1978).
3. T. Sands, R. Gronsky and J. Washburn, submitted to Thin Solid Films.
4. H. T. Evans Jr., Zeitschrift fur Krist. 150, 299 (1979).
5. T. D. Sands, J. Washburn and R. Gronsky, phys. stat. sol. (a) 72, 551 (1982).
6. R. W. Potter, Econ. Geology 72, 1524 (1977).
7. S. Djurle, Acta Chem. Scand. 12, 1415 (1958).
8. B. Baron, A. W. Catalano and E. A. Fagen, Proc. 13th Photovoltaic Spec. Conf. (Washington, D.C., IEEE New York 1978). p. 406.
9. This work was supported by the Director, Office of Energy Research, Office of Basic Energy Sciences, Materials Sciences Division of the U.S. Department of Energy under Contract No. DE-AC03-76SF00098.

EPITAXY OF BCC AND HEXAGONAL METALS ON FCC(001) SUBSTRATES.

C.R.M. GROVENOR[*] AND D.A. SMITH[**]
[*] Dept of Metallurgy and Science of Materials, Parks Road, Oxford.
[**] IBM T.J. Watson Research Centre, Yorktown Heights, NY.

ABSTRACT

A variety of bcc and hexagonal metals have been deposited onto fcc(001) surfaces held at between 150 and 350°C. The epitaxial relationships developed between deposit and fcc substrate have been investigated, and the application of current theories of the energy of interphase boundaries to explaining these results is critically discussed.

INTRODUCTION

Epitaxial relationships between a substrate and deposit film have been studied extensively by transmission electron microscopy and X-ray techniques, and in particular fcc_s/fcc_d and bcc_s/fcc_d epitaxies are well characterized [1] (s and d as subscripts identify substrate and deposit respectively). It has been shown that fcc deposits often grow homoepitaxially on fcc substrates with an interfacial net of dislocations being nucleated to accommodate the lattice misfit as the deposit thickens. Fcc deposits on bcc (110) substrates [2] have been used to investigate the character of the Nishiyama-Wasserman and Kurdjumov-Sachs epitaxial relationships that are observed during numerous bulk phase transformations. The development of energetic criteria to explain the stability of observed epitaxial relationships has occurred in two ways; one where a purely geometric model of epitaxial interfaces (and more general interphase boundaries) is used to define a net dislocation content, and a second where the total potential energy of the epitaxial interface is calculated from geometric and elastic parameters. Both approaches can successfully explain some features of experimentally observed epitaxies.

This paper will briefly discuss the problems involved in using either model to predict preferred expitaxial relationships, and present some new results on the epitaxies between fcc (001) substrates and bcc and hcp deposits. The use of the energetic criteria in predicting the experimentally observed epitaxies in the fcc/bcc systems will be discussed in the greatest detail.

ESTIMATING THE ENERGY OF INTERFACES

The geometric model of interfaces is based on the Frank-Bilby equation [3] and the formally equivalent O-lattice theory [4]. The net Burgers vector content of any interface can be ascertained from the equation

$$\underline{B} = (\underline{I} - \underline{S}^{-1}) \underline{X} \qquad (1)$$

where $\underline{X}$ is a vector in the interface phase, $\underline{B}$ the total Burgers vector that intersects unit length of $\underline{X}$, and $\underline{S}$ the deformation matrix that transforms the crystal lattice on one side of the interface into that on the other side. However there are numerous problems in the application of this theory to calculating the energy of any particular interface. Firstly there are an infinite number of ways of defining $\underline{S}$ for any two lattices

impinging to form an interface. Different deformation matrices will result
in the calculation of different net Burgers vector contents when equation 1
is applied to formally identical interfaces. This has necessitated the
assumption that the minimum calculated Burgers vector content gives the
most physically significant description of the interface. However for low
angle symmetric tilt grain boundaries it is always possible to define $\underline{S}$
such that $\underline{B}$ =0 [5], even though there is extensive experimental evidence
that there is a net dislocation content accommodating the misorientation in
these interfaces. The minimum Burgers vector criterion is thus not a
satisfactory method of deciding on the correct value of $\underline{S}$ to apply to any
particular interface. Even should a more consistent criterion be developed
it is still necessary to then estimate the energy of an interface. It is
normally assumed that the sum of the energy of the elastic fields of the
dislocation content of the interface gives a reasonable approximation to
the interfacial energy. This however ignores the 'core' energy of the
interface, the energy associated with the potential energy of atomic misfit
in regions of the interface between the dislocations. The geometric model
for interfacial energy allows for no contribution from this source. The
last ambiguity in the estimation of energy from $\underline{B}$ lies in the form of the
relationship between net Burgers vector content and energy. A number of
parameters have been suggested to define this relationship, and there is
some evidence [6] that the Read-Shockley relation developed for low angle
grain boundaries [7]

$$E_I \propto \frac{b^2}{d} \qquad (2),$$

where b is the Burgers vector of discrete dislocations and d their spacing,
is the most satisfactory estimate of the energy of a dislocation array for
reasonably small values of angular or dilatational misfit. $\underline{B}$ is formally
equivalent to $\frac{b}{d}$, and so the energy of an interface can be estimated by

$$E \propto |\underline{B}|^2 \cdot |\underline{d}| \qquad (3)$$

where $|\underline{d}|$ is the length of the misfitting crystal lattice vectors in
substrate and deposit.

The Van der Merwe model [8] of epitaxial interfaces is based on the
calculation of the net potential energy of an interface treated as two
interacting atomic meshes. The potential variation across an atomic
surface has been related to the activation energy for surface diffusion and
the adsorption energy, both measureable quantities. In this way the
relationship between interface parameters, such as misfit factor f and
angle of misorientation θ, and the potential energy of the interface can be
calculated. Such calculations show variations in the mean energy per atom
with θ and f predicting low energy interfaces at critical values of these
two parameters. Two particularly interesting features are shown by such
calculations; around the 'ideal' epitaxies there are limited areas in (θ,
f) space of decreased interfacial energy, and in between these epitaxial
relationships an energy plateau is predicted. In particular a decreased
energy is expected for interfaces with values of f that deviate less than
9% from an ideal epitaxial relationship. In between ideal relationships
the interfacial energy varies smoothly in distinction to the energy
predicted by the geometric model where the net energy surface is formed by
the intersection of parabolic surfaces. At the ideal epitaxial
relationships the calculated energy is not necessarily zero, and this
energy component can be considered an estimate of the core energy of the
interface. Thus this model may be used to calculate both the core and the
elastic (dislocation) contributions to the interfacial energy. It must be
remembered however that the calculations of the variation of energy in
(f, θ) space have been carried out for deposit thicknesses of only 1 atomic

layer. Thicker deposits, such as those normally studied experimentally, will show a larger number of ideal configurations [8] and so caution must be taken in comparing the theory and experimental results. Values of the core energy will be altered both by the variation in elastic constants for different materials, and by possible translational states of the interface. The model does not allow for such transitions, and thus the calculated values of core energy may not be accurate. The use of the Van der Merwe model to compare the energies of different epitaxial interfaces is difficult to use quantitatively at present, but the prediction that ideal epitaxial relationships are the result of atomic row matching (plane matching) is a criterion that has been shown to satisfactorily explain some observed fcc/bcc epitaxies.

In this paper the two simple criteria for minimizing interfacial energy, minimizing $|\underline{B}|^2 \cdot |\underline{d}|$, and plane matching, will be considered in relation to experimental evidence on prefered epitaxies in the $fcc_s(001)/bcc_d$ and $fcc_s(001)/hcp_d$ systems. It is shown that neither criterion can as yet explain the observed epitaxial relationships, although the $|\underline{B}|^2 \cdot |\underline{d}|$ energy factor is reasonably accurate given the limitations in formulating satisfactory parameters for the interface.

EXPERIMENTAL DETAILS AND RESULTS

Fcc single crystal substrates were prepared by the evaporation of Au, Pd and Ni onto rocksalt substrates held at 350°C in a vacuum of 10^{-6}Torr. Bcc and hcp metals were then evaporated onto these substrates in the same pumpdown. The fcc (001) substrate temperature was chosen to try and ensure a satisfactory epitaxial growth of the deposit films. The bilayer films were grown such that the fcc substrates were about 50nm thick and the deposits about 20nm thick. The films were removed from the rocksalt in distilled water and observed directly in a conventional transmission electron microcope.

fcc/bcc samples

Although substantial evidence has been collected on the epitaxial relationships between fcc deposits and bcc substrates [2] little work has been carried out using the fcc layer as the substrate. Most of the experiments on bcc deposits have been to study their pseudomorphic growth on fcc surfaces. In this work four bcc metals have been deposited onto the Au, Pd and Ni substrates, Cr, V, Mo and Nb so that the effect of varying the lattice parameter ratio, a_{bcc}/a_{fcc}, on the resulting preferred epitaxial relationship could be investigated. Table I shows the experimentally observed epitaxial relationships between the substrates and deposits. Three epitaxies have been observed; the well known Baker Nutting relationship [9], a 45° rotated varient of the Baker Nutting relationship, and an epitaxy, labelled A, that has not been observed before for fcc/bcc interfaces to the authors' knowledge. This epitaxy is rotated 11° from the Bain relation fcc(001)//bcc(011), fcc[00$\bar{2}$]//bcc[01$\bar{1}$]. Figures 1a-c show diffraction patterns typical of each of these 3 epitaxial relationships, and indexed diagrammatic representations of each pattern. There are 4 geometrically equivalent varients of the A epitaxy identifed in Figure 1b, and a dark field micrograph made using a diffraction spot from one variant is shown in Figure 1d. This micrograph shows domains of one of the four equivalent epitaxial variants.

fcc/hcp samples

Co and Zr have been deposited only onto Au substrates. Figure 1e shows an indexed diffraction pattern from a Co/Au film where 2 variants of

the fcc(001)//hcp(1$\bar{2}$13), fcc[200]//hcp[10$\bar{1}$0] epitaxial relationship are
identified. Co films deposited on Ag (111) and Cu (111) surfaces have been
investigated but in these cases the simple fcc(111)//hcp(0001) epitaxies
were found[10,11]. The observed Co epitaxy is a particularly interesting
case to which to apply the Van der Merwe plane matching criterion since Co
has plane spacings of 2.035Å (0002) and 2.17Å (10$\bar{1}$0), both of which could
potentially be matched with the Au 2.039Å (002) planes.

Zr deposits on Au (001) showed no epitaxial growth even when the
substrate was at 350°C.

DISCUSSION

fcc/bcc samples

The variation of the energy factor $|\underline{B}|^2 \cdot |\underline{d}|$ with the lattice misfit
r ($r = a_{bcc}/a_{fcc}$) can be simply calculated for these interfaces by
considering the near coincidence of the atomic positions on the fcc (100)
and bcc (001), and (011), planes for Baker-Nutting and A epitaxial
relationships respectively. Figure 2 shows constructions of these near
coincidence relations for BN and A epitaxial relationships for two
arbitrarily chosen values of r. $|\underline{B}|^2 \cdot |\underline{d}|$ is calculated from the lattice
misfit along two orthogonal directions, $\underline{d}_1$ and $\underline{d}_2$, and $\underline{B}$ is formally
equivalent to the misfit. These energy factors when plotted against r will
fall on the two intersecting parabolas, one for each epitaxial
relationship, since $\underline{B}$ varies linearly with r. In this way comparison can
be made of the variation of energy factor with r for two or more epitaxies
in the range of misfit. As explained in Section 2 comparison of the
absolute values of energy factors calculated for two different epitaxies
will not be meaningful unless the deformation matrix, $\underline{S}$, for each epitaxial
relationship is chosen in a consistant manner. It is proposed here that
the two simple near coincidence nets defined in Figure 2 are mutually
consistant descriptions of the interface structure

The energy of dislocation arrays at the interfaces between two metals
will depend on the elastic constants of the fcc and bcc layers. Barnett
and Lothe [12] have developed an equation for the energy of an edge
dislocation in the interface between two dissimilar materials.

$$E \propto \underline{b}^2 \cdot G \qquad G = \mu_1 \mu_2 \left[\frac{\mu_2 (k_1 + 1) + \mu_1 (k_2 + 1)}{(\mu_2 k_1 + \mu_1) \cdot (\mu_1 k_2 + \mu_2)} \right] \quad (4)$$

where μ_1 and μ_2 are the shear moduli and $k_i = 3 - 4v_i$ where v_i is
poissons ratio of the ith material. Figure 3 shows a plot of $|\underline{B}|^2 \cdot |\underline{d}| \cdot G$
against r for tthe fcc/bcc epitaxial systems investigated here for BN and A
type epitaxies. These calculated values of the energy factors are given in
Table 1. The energy factor variation with r can be considered to lie
within a region the extent of which is determined by the effective
interfacial shear modulus, G, and the length of the misfit vectors $\underline{d}_1$ and
$\underline{d}_2$. The elastic energy stored in the Cr/Pd system is very much larger than
that in the V/Au system for A epitaxial interfaces, even though the lattice
mismatch is extremely similar. This is a reflection of the greater
rigidity of the Cr and Pd lattices compared to those of V and Au. From the
criterion of minimization of $|\underline{B}|^2 \cdot |\underline{d}| \cdot G$, A epitaxial interfaces would be
expected to be preferred in systems with r > 0.8, and Baker Nutting
epitaxies when r < 0.785, as is shown in Figure 3. Table 1 shows that the
experimentally observed epitaxial relationships agree with this prediction
except for V/Pd which shows BN epitaxy and not A type as expected, and
Nb/Au which shows no epitaxy at all. The value of r at which BN epitaxy is
predicted to be energetically unstable with respect to A epitaxy is
r ~ 0.775 in Figure 3, but the experimentally determined value lies between
0.785 and 0.8. As has been discussed above the $|\underline{B}|^2 \cdot |\underline{d}|$ criterion contains
no core contribution to the interfacial energy. The discrepancy between

the theoretical and experimental results can be rationalized if core energy of the A epitaxial interfaces is higher than that of a Baker Nutting interface. This is not a unrealistic assumption since the A epitaxial interface contains a lower density of near coincident atoms than an interface with Baker Nutting epitaxy. A further contribution to the net energy of an A type epitaxial interface will come from the interfaces between the domains of differently orientated geometric variants. As can be seen in Figure 1d the density of these interfaces is very high, the average domain size is only 40-60nm, and so might provide a substantial increase in the net energy of such an epitaxial interface. Table 1 also shows that the $|B|^2 \cdot |d|$ calculations for the A type and BN 45 epitaxies for Nb/Ni, where r = 0.938, which indicate a substantially lower interfacial elastic energy for the BN 45 epitaxy, as is observed experimentally.

Application of the Van der Merwe approach to predicting the preferred orientation relationships in fcc (001)/bcc systems should ideally be done from calculations for each epitaxial relationship and pair of metals. However there may be problems with this method due to the fact that some interfaces show Baker Nutting epitaxial relationships with dilatational misfits 9.2% (Mo/Au) and 10.8% (V/Pd). The calculation of the energy per atom in an interface of intersecting square atomic nets shows a lowering in energy only within 9% misfit of the perfect matching situation, f = 0. Thus there would be expected to be no net energy gain for the formation of Baker Nutting interfaces with such high values of dilatational misfit in the Van der Merwe model. Some of the observed epitaxial relationships are roughly consistent with the criteria of low energy interfaces being associated with plane matching. In the 5 fcc/bcc systems that show A type epitaxial interfaces matching to within ~ 3% of plane spacing occurs on one set of planes in each case. No such plane matching occurs for Baker Nutting epitaxial interfaces for the same systems.

fcc (001)/hcp

The observed epitaxial relationship between Co and Au(001) appears to be a simple plane matching interface consistent with the Van der Merwe criteria. Au(200) and Co(10$\bar{1}$0) planes match with a misfit of 6.4%, within the range in which energetically preferred interfaces might be expected. However matching of the Au(200) and Co(0002) planes would result in far lower planar misfit (0.02%). In this case therefore it is not possible to predict the preferred epitaxial relationship purely on the basis of the best matching of a single set of planes. Core energy and the contribution to elastic energy from lattice misfit in the direction orthogonal to the well fitting planes must also contribute to determining the most energetically favourable epitaxy. Calculations of the energy variations with f and θ for this system might help explain the occurrence of the observed epitaxy. Further evidence on the preferred orientations of hcp deposits on fcc (001) substrates could allow application of the energy factor criteria to explaining how the observed epitaxies change with the hcp lattice parameter and c/a ratio. For the Zr deposits onto Au the lack of preferred epitaxy may be understood by consideration of the plane spacings of deposit and substrate. None of the densely populated Zr planes have spacings close to those of Au(200) or (220), and so neither plane matching, nor low net dislocation density, interfaces are predicted. Under these circumstances a fine grained polycrystalline deposit is the preferred structure. (It is not necessarily true however that high misfit systems always show no epitaxy, epitaxial Au films can be grown on both NaCl and GaAs[13].)

CONCLUSIONS

It has been shown that bcc deposits grow with 3 epitaxial

TABLE I

System	fcc substrate T°C	Epitaxial relationship observed	$\|\underline{B}\|^2\cdot\|\underline{d}\|\cdot G$ A	$\|\underline{B}\|^2\cdot\|\underline{d}\|\cdot G$ BN	$\|\underline{B}^2\|\cdot\|\underline{d}\|$ BN45	r
Cr/Au	150	BN	1282	0*	–	0.707
Cr/Pd	150	BN	934	71*	–	0.744
V/Au	150	BN	602	52*	–	0.745
Mo/Au	200	BN	445	191*	–	0.77
V/Pd	150	BN	217	242*	–	0.783
Nb/Au	200-300	None	101	344	–	0.809
Mo/Pd	200	A	175*	635	–	0.811
Cr/Ni	150	A	141*	813	–	0.819
Nb/Pd	200	A	27*	795	–	0.85
V/Ni	150	A	37*	1100	–	0.862
Mo/Ni	200	A	223*	2479	328	0.893
Nb/Ni	300	BN45	390	–	114*	0.938

Baker Nutting (BN) fcc (001)//bcc (001), fcc [0$\underline{0}$1]//bcc [1$\underline{0}$0]
A epitaxy (A) fcc (001)//bcc (110), fcc [1$\underline{1}$0]//bcc [1$\underline{1}$2]
(BN45) fcc (001)//bcc (100), fcc [1$\underline{0}$0]//bcc [1$\underline{0}$0]
* Experimentally observed epitaxial relationship.

relationships with fcc (001) substrates, each epitaxy being preferred in a particular range of lattice parameter ratio, r. Application of the criterion of minimizing net Burgers vector content in the interfaces has been shown to give reasonable agreement with experimental results when the inaccuracies in estimating relative interfacial energy by this method are taken into account. Two possibly major contributions to interface energy are ignored, core energy and that of domain boundaries between regions of epitaxially equivalent geometric variants. The Van der Merwe model has the potential to provide more accurate values of relative interfacial energy since a contribution from core energy is included in the calculation. However those calculations must be carried out for each epitaxial system to allow direct comparison of relative energies of different epitaxial interfaces. The criterion of plane matching in one dimension providing a low energy interface has been shown to be inaccurate in predicting the preferred epitaxial interface in the Co/Au system.

REFERENCES

1. J.W.Matthews, Epitaxial Growth (Academic Press, New York 1975)
2. L.A.Bruce and H.Jaeger, Phil.Mag.A 38, 223 (1978)
3. B.A.Bilby, R.Bullough and E.Smith, Proc.Roy.Soc.(London) A231, 263 (1955)
4. W.Bollmann, Crystal Defects and Crystalline Interfaces (Springer-Verlag, Berlin 1970)
5. J.W.Christian and A.G.Crocker, Dislocations in Solids (ed. F.R.N. Nabarro, North Holland, Amsterdam 1980) p.615
6. K.M.Knowles, D.A.Smith and W.A.T.Clark, Scripta Met. 16, 413 (1982)
7. W.T.Read and W.Shockley, Imperfections in Nearly Perfect Crystals (ed. W.Shockley, Wiley, New York 1952) Chap. 13
8. J.H.van der Merwe, Phil.Mag.A 45, 127,145 and 159 (1982)
9. R.G.Baker and J.Nutting, Precipitation Proceses in Steels Iron and Steel Institute Spec. Rept.No 64 (1959)
10. J.E.Fisher, Thin Solid Films 17, 531 (1973)
11. E.Grumbaum and G.Kremer, J.Appl.Phys. 39, 347 (1968)
12. D.M.Barnett and J.Lothe, J.Phys.F. 4, 1618 (1974)
13. Y.Yoshiie and C.L.Bauer, to be published.

253

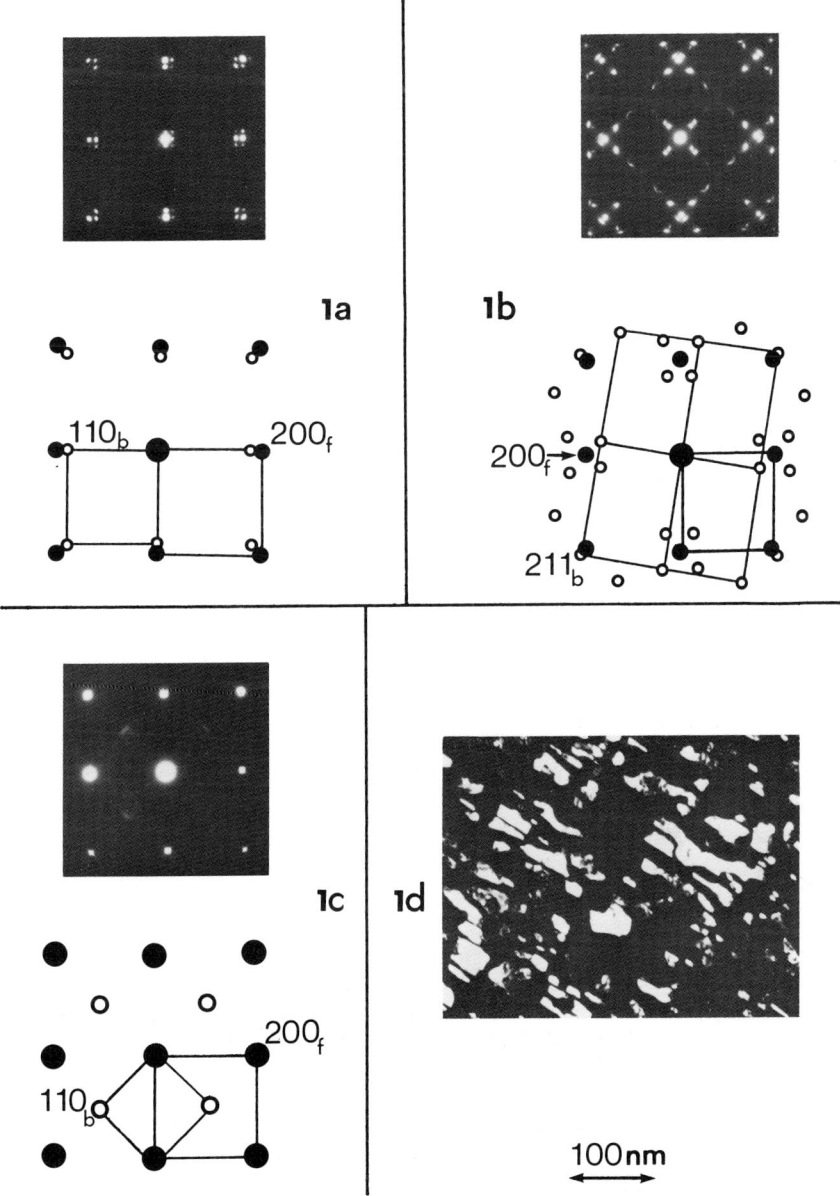

Figure 1. 1a-c are transmission electron diffraction patterns and labelled diagrammatic representations for typical Baker Nutting, A and BN45 epitaxial relationships. 1d is a dark field electron micrograph showing domains of one of the 4 variants of the V/Pd films. 1e is a diffraction pattern and representation of the epitaxy of Co on Au(001).(Double diffraction spots are not included in the diagrams , and fcc diffraction spots are filled symbols , bcc or hcp open symbols.)

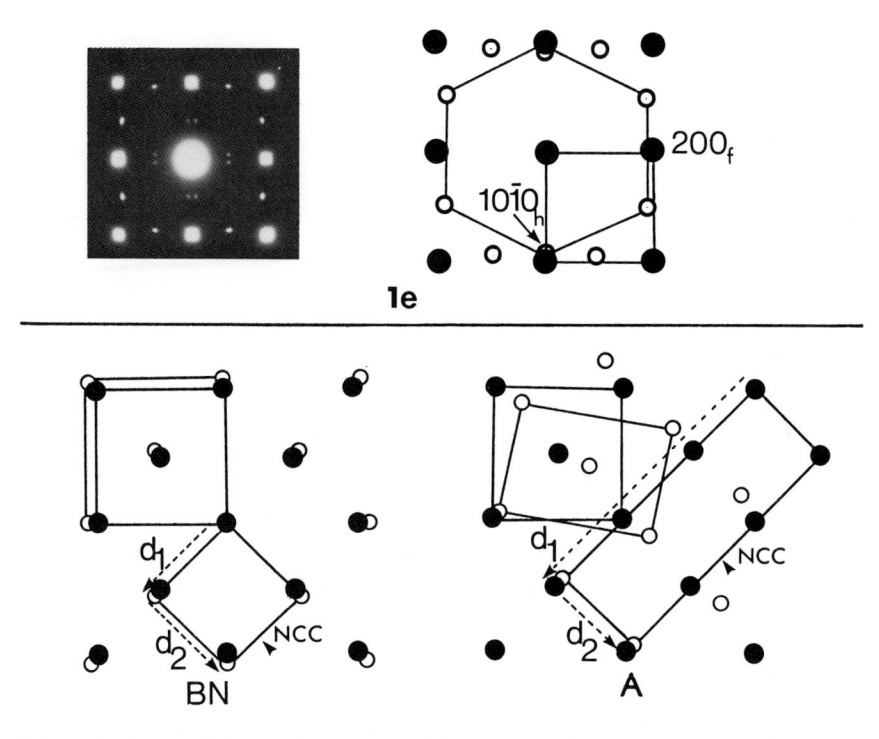

Figure 2. Constructions of near coincidence cells for Baker Nutting and A type epitaxial relationships. Fcc and bcc atom positions are represented by filled and open symbols respectively. This choice of ncc's for these epitaxies is equivilent to a choice of $\underline{S}$, the deformation matrix between the fcc and bcc lattices.

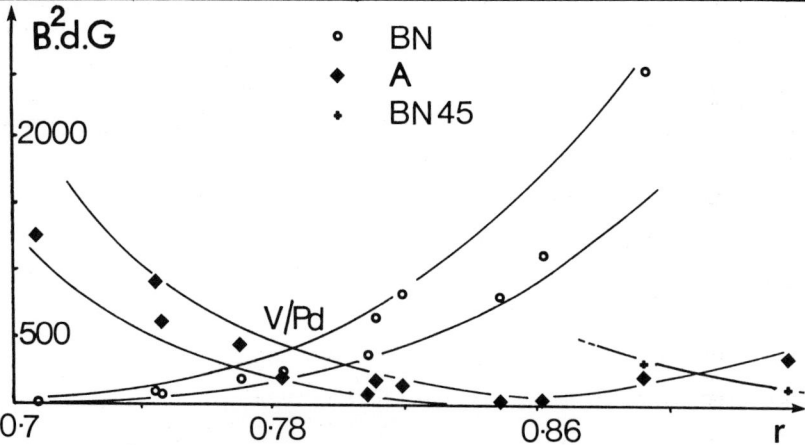

Figure 3. Plots of the energy factor against r for the 3 observed epitaxies in the fcc/bcc films. The V/Pd sample is the only one for which this simple energy criterion does not predict the correct epitaxy.

STEM Microanalysis of Hazed Polycrystalline
Silicon Layers

G.J.C. Carpenter,* and D.C. Houghton**

*Physical Metallurgy Research Laboratory, Canada Centre
for Mineral and Energy Technology, 568 Booth St., Ottawa,
Canada, K1A 0G1, **Bell-Northern Research Laboratories,
P.O. Box 3511, Station C, Ottawa, Canada, K1Y 4H7

ABSTRACT

The LPCVD of polysilicon on oxidized silicon
wafers sometimes results in defective material
having a "hazed" appearance. Examination of this
material by SEM has revealed surface nodules which
correlate on a one-to-one basis with anomalously
large silicon grains, observed by transmission
microscopy. Darkfield microscopy has shown that
this abnormal grain growth initiated during deposition. X-ray EDS analysis of isolated poly-Si
nodules revealed the presence of metallic impurities.

INTRODUCTION

Low pressure chemical vapour deposition (LPCVD) of poly
silicon on thermally oxidized silicon wafers sometimes results
in defective material which has a "hazed" appearance in oblique
illumination due to the scattering from surface irregularities.
This is caused by surface roughening which can cause serious
problems with lithography during device fabrication. In an
effort to determine the cause of the haze phenomenon, samples
of both defective and acceptable material have been examined
using a scanning electron microscope (SEM), a
transmission/scanning transmission electron microscope
(TEM/STEM) and a dedicated STEM. Here, we report the preliminary results of this study. A more detailed account of this
work will be published at a later date.

EXPERIMENTAL TECHNIQUES

The material used for this study was <100> Si/SiO2/poly-Si
where the oxide thickness was ~100 nM and ~300 nM poly Si was
deposited at a temperature of between 600 - 675°C by conventional low pressure chemical vapour deposition.

Bulk specimens were used for examination in the SEM.
Transmission microscopy was carried out on conventional 3 mm
disc specimens prepared by back-thinning to the SiO2/poly-Si
interface by jet chemical polishing with HNO3: CH3COOH: HF,
5:1:1 in volume ratio.

RESULTS

Examination in the SEM showed that the hazed appearance was caused by growth defects in the form of hemispherical nodules, typically ~500 nm in diameter and ~200 nm high, Figure 1

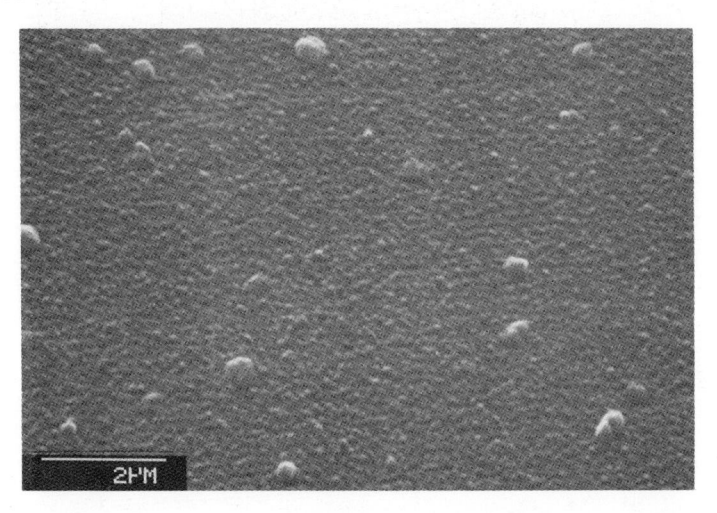

Figure 1 Secondary Electron SEM Image of the Surface of a Defective ("hazed") Wafer, Showing Anomalous Surface Nodules

SEM and TEM images of identical regions revealed a one-to-one correlation between the haze nodules and anomalously large grains of silicon (Figure 2). These large grains were not observed in TEM specimens made from normal, non-hazed material. The grains of normal polysilicon had diameters that were typically in the range 20 nm to 100 nm and were heavily twinned (1), (2). Certain rings in the electron diffraction patterns were seen to be split; this is tentatively attributed to double diffraction, caused by the microtwins. Extensive streaking of the diffraction spots making up the rings was also ascribed to the high density of microtwins.

The silicon nodules were also highly faulted and frequently had an obvious rosette appearance (Figure 3). Preliminary results suggest that the nodules may have formed with a preferred orientation, but further work is needed before a firm conclusion can be reached.

Tilting experiments in both bright field and dark field have shown that the nodules overlapped the smaller polysilicon grains. The large grains had therefore nucleated at some point during the deposition process and had grown in a radial fashion, giving the characteristic rosette appearance.

In one specimen, the chemical polishing solution had dissolved some of the polysilicon preferentially, forming a "taper section". This showed that the grain size of the poly

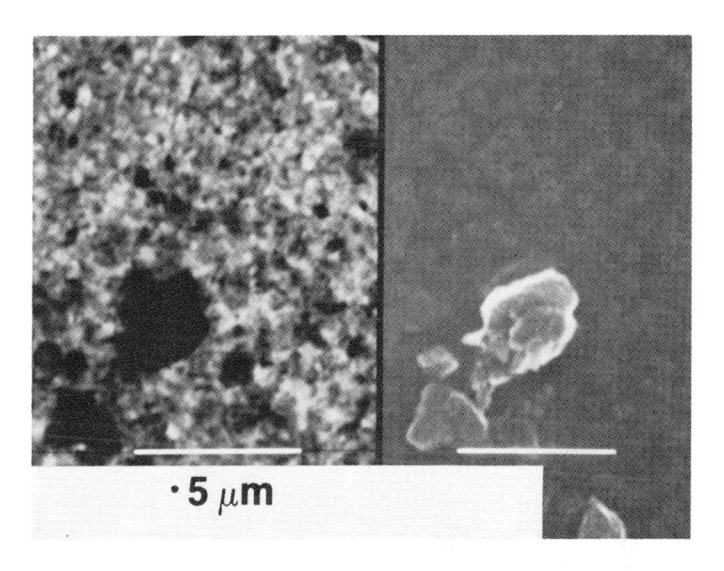

Figure 2. Comparison of Simultaneous SEM and STEM Images of the Same Nodule.

silicon varied with depth; the diameter of the crystals near the oxide interlayer was ~25 nm, which is significantly smaller than that of the crystals that made up the rest of the polysilicon layer (~70 nm). Exposed regions of the silica interlayer could also be observed. This film was amorphous and contained no resolvable structure or features that might have caused the hazed appearance or acted as nuclei for nodule formation.

EDS X-ray spectra, obtained from the nodules in the STEM, revealed the presence of significant concentrations of metallic elements, such as iron, titanium, and tantalum. An example is shown in Figure 4. These elements were not detectable in the adjacent polysilicon that had a normal grain size and structure.

DISCUSSION

The anomalous grain growth behavior that is sometimes encountered during the deposition of poly silicon on oxidized silicon wafers clearly arises during deposition and does not appear to be directly associated with incipient nuclei at the silicon substrate. This viewpoint is consistant with the absence of observable defects in the oxide interlayer.

It appears from the X-ray analyses that the presence of metallic impurities in the vapour deposition chamber during deposition are responsible for the abnormal grain growth.

Whether this is generally true for all types of haze phenomena requires further investigation.

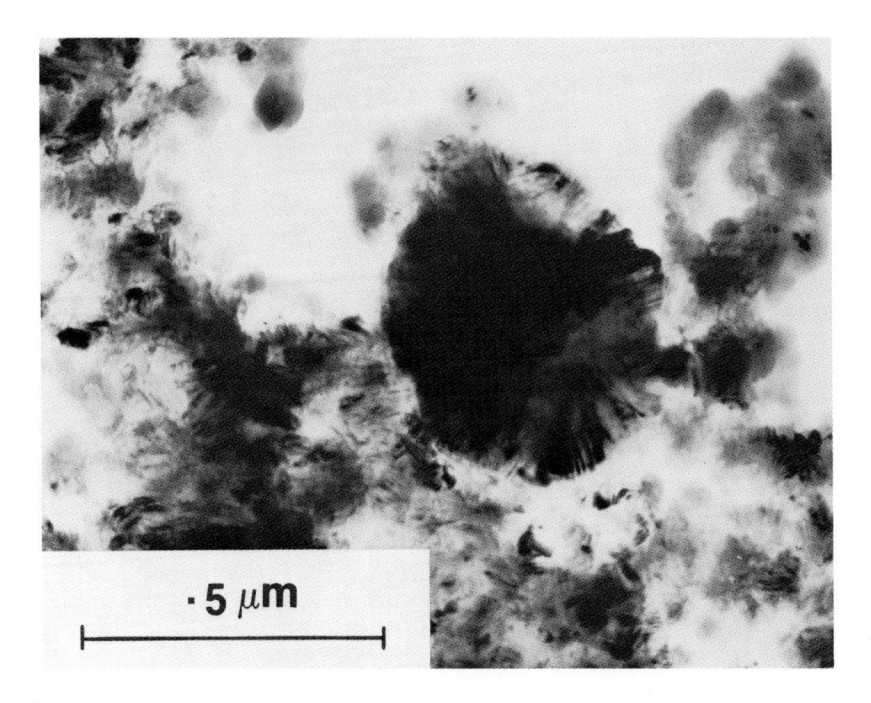

Figure 3 TEM Micrograph of a Single Nodule at the Edge of the Thin Film, Showing "Rosette" Structure

CONCLUSIONS

This example of hazed poly-Si was caused by an acceleration in the growth rate of isolated grains during the deposition process. Initiation of this abnormally high rate of grain growth occurs during the deposition cycle and not directly from sites on the SiO2 substrate. The hazed poly-Si nodules have a rosette structure containing radially symmetric heavily faulted segments which appear to have a common preferred growth direction. In addition STEM-EDX microanalysis of individual nodules has revealed the presence of contaminants such as Fe, Ti and Ta which are not detected in normal poly-Si grains.

ACKNOWLEDGEMENTS

The authors are indebted to Northern Telecom Electronics for furnishing material and technical support. We are grateful to our coworkers and for provision of laboratory facilities at the Advanced Technology Laboratory (BNR) and Physical Metallurgy Research Laboratory (CANMET).

REFERENCES

1. T.I. Kamins, M.M. Mandurah and K.C. Saraswat, J. Electrochem Soc. (1978), <u>125</u>, 927.

2. J.C. Bravman and R. Sinclair, Mat Res Soc Symposium Proc (1982), <u>18</u>, 153.

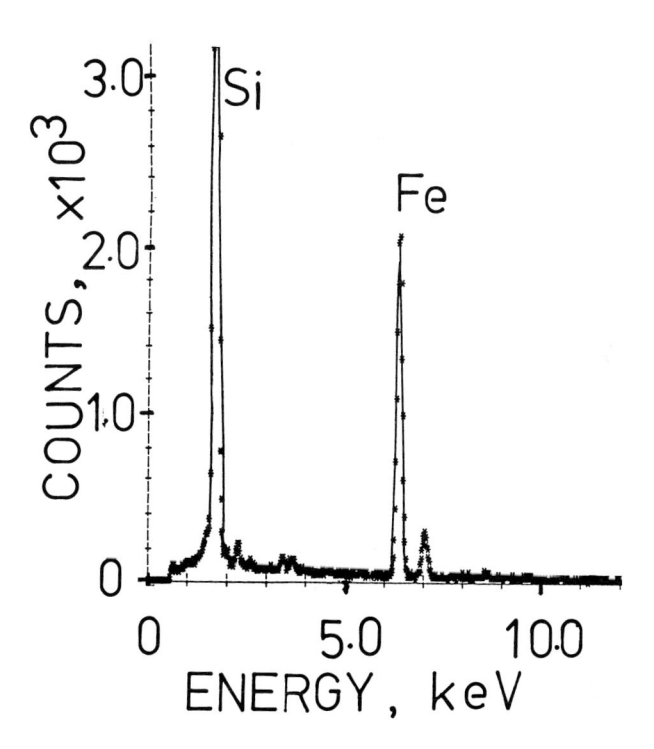

Figure 4 An EDX Spectrum Obtained Using a Dedicated STEM from a Single Nodule.

MISFIT DISLOCATIONS IN THE INTERFACE BETWEEN THE METALLIC AND INSULATING
PHASES IN Cr DOPED V_2O_3

NOBUO OTSUKA AND HIROSHI SATO

School of Materials Engineering, Purdue University,
West Lafayette, IN 47907, U.S.A.

ABSTRACT

The formation of misfit dislocations in the interface between the metallic and insulating phases which occur in 1.2% Cr doped V_2O_3 near room temperature was confirmed by TEM. Orientations of Burgers vectors of misfit dislocations were determined by the weak beam technique. The Burger vectors were parallel to principal axes in the basal plane (<100>) and inclined somewhat to interfaces. The Burgers vectors of misfit dislocations are, however, close enough to the direction of the maximum misfit in the interface. Only one or two misfit dislocations appear in an interface regardless of the thickness of specimens, and these were always located near the surface.

INTRODUCTION

Chromium doped V_2O_3 undergoes a first order metal-insulator transition near room temperature in addition to that at low temperatures [1]. The transition is known to occur in a step-wise fashion and to be accompanied by a large temperature hysteresis [1,2]. Despite extensive studies of this transition, the origin of the discontinuities was not well understood until the present authors showed crystallographic features of the transition by TEM [3]. Here, the origin was ascribed to a lattice misfit at the interfaces between the two phases; both phases crystallize in the corundum structure, but their lattice parameters are slightly different from each other [4]. (Lattice parameters of 1% Cr doped V_2O_3 are 4.9540Å and 13.9906Å for the metallic phase and 4.9974Å and 13.9260Å for the insulating phase, respectively [4].) Since the symmetry does not change at the transition, the transition proceeds simply by the migration of the interface.

As is well known, the lattice misfit gives rise to misfit dislocations in interfaces when it becomes too large to be accommodated by elastic strain [5]. Indeed, we observed images of misfit dislocations in interfaces at this transition [3]. However, interfaces at this transition are basically different from those where misfit dislocations are usually observed. Unlike ordinary interfaces, interfaces in the present case form through small atomic displacements and move rapidly. No observation of misfit dislocations in this type of interfaces had been reported. Therefore, we report a detailed analysis of dislocation images including the determination of Burgers vectors by the weak beam technique [6].

EXPERIMENTAL PROCEDURE

Single crystals with the composition $(V_{0.988} Cr_{0.012})_2O_3$ were grown by the triarc melting method at the Central Material Preparation Facility of Purdue University. At this composition, the metal-insulator transition occurs at 40°C by heating and at -30°C by cooling. The crystals were cut into thin slices by a diamond saw and etched with a 75% HNO_3 and 25% HF solution. After etching, thin specimens were cooled to -50°C in cold alcohol. A JEM 200CX electron microscope with a side-entry goniometer was used for the observation. All observations were made at room temperature.

RESULTS

Since specimens were cooled prior to the observation, all observed areas initially were in the low temperature phase. Diffraction patterns did not show any sign of the coexistence of the two phases. During the observation, however, the high temperature phase was found to nucleate at thin edges of specimens due to the heating by the electron beam as stated in the earlier study [3]. Areas of the high temperature phase then expanded rapidly towards thicker parts by increasing beam intensity. Figure 1(a) is a bright field image showing the coexistence of the two phases. The image was taken nearly under the Bragg condition (110 reflection). (Indices are based on the hexagonal lattice throughout this paper.) The thinner part of the specimen (the upper side of the image in Fig. 1(a)) has already transformed to the high temperature phase, while the thicker part still remains in the low temperature phase. The two parts are divided by an interface which is inclined to the beam direction and the interface shows nearly periodic contours similar to those for ordinary grain boundaries. The number of contours at the interface coincides with that of thickness contours seen between the edge of the specimen and the interface. This indicates that the appearance of contours in the interface is due to the difference of the spacing or orientation of the lattice planes between these two parts, but not to the existence of a stacking fault [7].

Figure 1(b) is a bright field image taken after tilting the crystal from the orientation of Fig. 1(a) by about 0.5 degrees. The image shows two dark lines at the position of the interface. Contrasts of these dark lines were very sensitive to the deviation from the Bragg condition, and their appearance depended on reflections utilized for the imaging as described later. Therefore, these lines can be regarded as dislocation images. Similar dark lines were also observed at other interfaces. Figure 2(a) and (b) are bright field images taken from another interface under imaging conditions similar to those of Fig. 1(a) and (b). Figure 2(b) shows two dark-lines along the interfaces which shows contours in Fig. 2(a). These observations show the existence of misfit dislocations in interfaces between the insulating and metallic phases. In the following the analysis of the image of dislocations in Fig. 1(b) is given.

First, the crystallographic orientation of the interface is determined in order to find a possible relation between the lattice misfit and the geometry of dislocations. The two parts of the crystal have the common orientation. Since the observed area is thin and flat, two unique axes, the long axis which is parallel to the surface, and the short axis which is normal to the long axis can be defined in the interface. Here, the long axis is nearly parallel to [201], and the short axis to [120]. Based on this orientation and the lattice parameters of the two phases, the size of the lattice misfit along each direction in the interface can be estimated; the largest misfit, 0.85%, is along the short axis, while the smallest one, 0.08%, is along the long axis as was confirmed earlier [3]. Since the lattice misfit along the long axis is almost zero, the interface has a misfit only along the short axis. This result is consistent with the geometry of dislocation lines because such an uniaxial lattice misfit can be most efficiently accommodated by edge dislocations which lie along the long axis and have Burgers vectors parallel to the short axis [5]. The crystallographic orientation of the interface is not unique but is determined mainly by the condition of minimum strain energy in the interface [3].

The orientations of Burgers vectors of dislocations were determined by the weak beam technique [6]. Seven reflections were examined. No more than two dislocations were observed in any case, and this indicates that

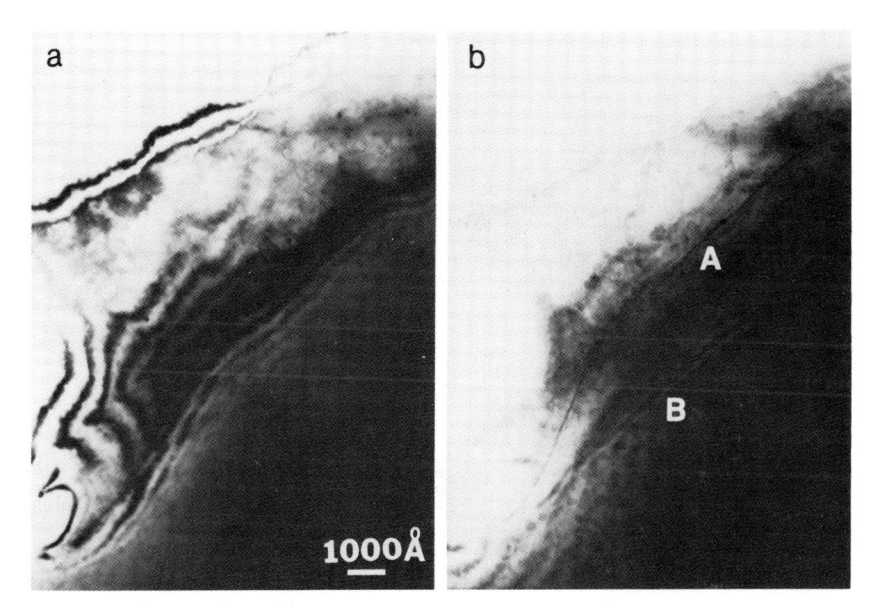

FIG. 1(a) and (b). Bright field images of an interface in a 1.2% Cr doped V$_2$O$_3$ thin crystal.

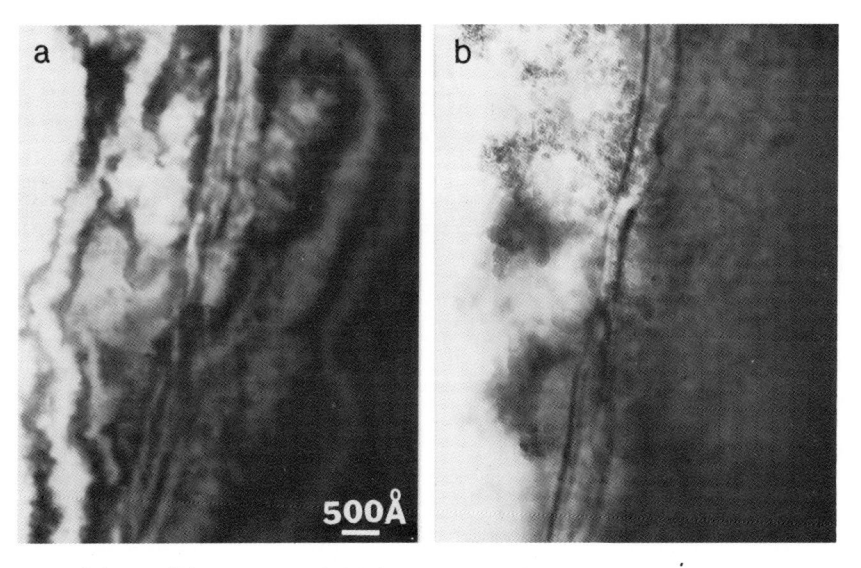

FIG. 2(a) and (b). Bright field images of an interface in a 1.2% Cr doped V$_2$O$_3$ thin crystal.

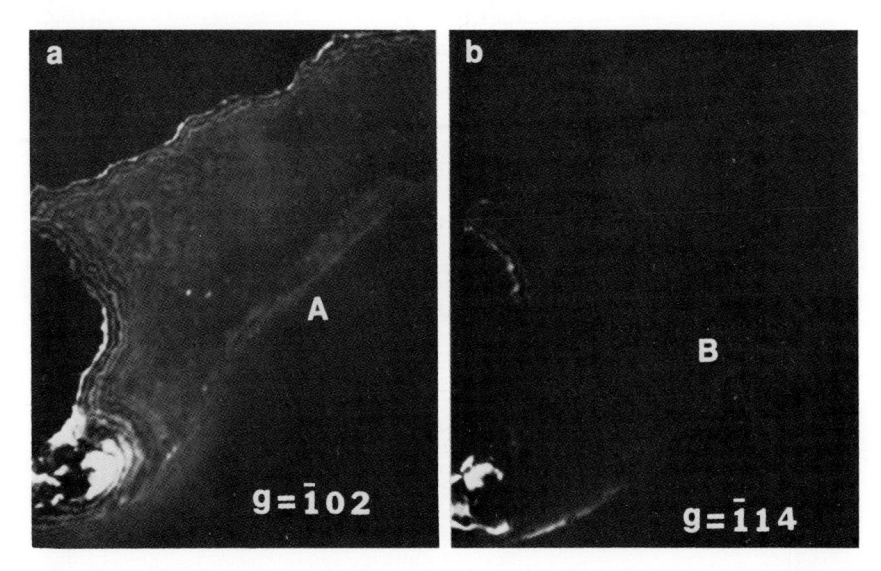

FIG. 3(a) and (b). Weak beam dark field images of an interface with $\bar{1}02$ (a) and $\bar{1}14$ (b) reflection.

only two dislocations existed in the interface. From the sharpness of the dislocation images in Fig. 1(b), it seems that one dislocation (A) is located near the bottom surface, while the other (B) exists near the top surface [7]. From the number of thickness contours, the thickness of the crystal at this position is estimated to be about 5400Å. Figure 3(a) and (b) are examples of weak beam dark field images of the same location as Fig. 1 taken with $\bar{1}02$ and $\bar{1}14$ reflection, respectively. These images show only one dislocation each as shown as A and B respectively in Fig. 3. This indicates that the two dislocations have different Burgers vectors. The results of these observations with respect to the visibility are summarized for each dislocation in Table I. Since these dislocations are not purely of the screw type, images taken under the condition, $\vec{g}\cdot\vec{b} = 0$, may show weak dislocation lines [7]. Here, $\vec{g}$ and $\vec{b}$ represent a reflection vector and Burgers vector, respectively. Based on the results shown in the table, it can be concluded that the orientations of Burgers vectors for the dislocations A and B are [110] and [010], respectively. This means that neither Burgers vectors are parallel to the short axis. Therefore, they are not pure edge dislocations expected for ordinary misfit dislocations [8]. These Burgers vectors, however, are both parallel to two of the <100> type principal axes which are nearest to the short axis. Since the angles between the Burgers vectors and the short axis are 30° in both cases, primary components of the Burgers vectors are parallel to the short axis. Similar results are also obtained for dislocations in Fig. 2(b). In this case, the Burgers vectors for both dislocations are parallel to [110], while the short axis of the interface is slightly inclined to the [110] direction. Finally, the effect of the beam intensity on positions of dislocations was investigated. Bright

TABLE I. Appearance of dislocations in weak beam dark field images.

Reflection	Dislocation A	Dislocation B
110	0*	0
0$\bar{1}$2	0	0
$\bar{1}$02	0	x
$\bar{1}$14	x**	0
113	0	0
006	x	x
2$\bar{1}$3	0	0

*A circle indicates the appearance of a clear dislocation image.
**A cross is either the appearance of a very weak image or the disappearance of a image.

field images in Fig. 4(a) and (b) were taken under the same imaging condition, but the beam intensity was reduced from the image (a) to (b). Therefore, the interface for (b) moved from the position for (a) towards the thinner part of the specimen, reducing the area of the high temperature phase. It can be seen that dislocations appear at different positions in the two images. These positions coincide with those of the interface in both cases. Because we could not directly observe moving dislocations, the motion of the interface being too fast to be observed on the screen, we cannot say that these dislocations move with the interface. At the same time, the above observation rules out the possibility that the interfaces are simply trapped by dislocations which have originally existed in the specimen, independent of the motion of the interface.

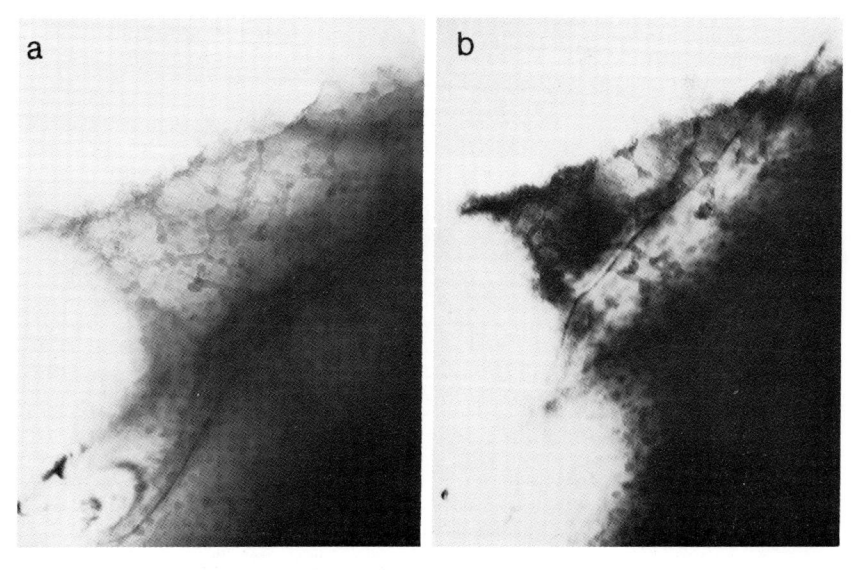

FIG. 4(a) and (b). Bright field images of a dislocation with the high (a) and low (b) beam intensity.

DISCUSSION

The existence of misfit dislocations in interfaces between the insulating and metallic phases in Cr doped V_2O_3 are observed and their relations with lattice misfit in the interfaces are analyzed. The lattice misfit exists only in the direction perpendicular to the surface of thin specimens, while misfit dislocations lie parallel to the surface and their Burgers vectors are parallel to one of the <100> type principal axes which are nearest to the interface plane. In other words, the Burgers vectors are somewhat inclined to the interface. The possibility of partial dislocations can be ruled out because no indications of stacking faults were found in the observation. Therefore, Burgers vectors of the misfit dislocations are considered to coincide with the unit cell vectors along these principal axes. Many examples of misfit dislocations with inclined Burgers vectors were also found in studies of thin films and precipitates [9,10]. The fact that interfaces always appear practically perpendicular to the surface of thin specimens and the misfit is confined only in the direction perpendicular to the surface indicates the importance of the condition of minimum strain energy in choosing the orientation of the interface [3]. This is due to the fact the crystal symmetry of the both phases is the same and the lattice parameters of these phases are only slightly different. Under this condition, the inclination of the Burgers vector to the interface is inevitable.

The role of misfit dislocations in interfaces which move rapidly through crystals is yet a mystery. Figure 4(a) and (b) show that misfit dislocations appear at positions before and after the motion of the interface. Since the primary component of the Burgers vector of the misfit dislocation is parallel to the interface, it is unlikely that the misfit dislocations move with the interface by glide. They may be able to move by climb, but the motion by this mechanism at low temperature is expected to be too slow to explain the observation in Fig. 4. On the other hand, misfit dislocations are only formed near surfaces of the crystals as seen in Fig. 1(b) and Fig. 2(b). It was also found that interfaces were accompanied by only one or two misfit dislocations regardless of the thickness of crystals. It seems that misfit dislocations are introduced from the surface when interfaces stop for some reason [3], and that they are left behind and eventually escape when interfaces move. In other words, misfit dislocations seem to serve the role of self trapping, since they can release strain due to lattice misfit.

ACKNOWLEDGMENTS

The authors are indebted to Professor G. L. Liedl for his helpful discussions and comments. This work was supported by NSF Grant DMR-83-04314. Equipment provided through NSF Grant DMR-78-09025 was crucial for the sucdess of this work.

REFERENCES

1. D. B. McWhan and J. P. Remeika, Phys. Rev. B., 2, 3734 (1970).
2. H. Kuwamoto, J. M. Honig and J. Appel, Phys. Rev. B., 22, 2626 (1980).
3. N. Otsuka, H. Sato, and G. L. Liedl, J. Solid State Chem., 45, 241 (1982).
4. W. Robinson, Acta Cryst.B, 31, 1153 (1975).
5. J. W. Matthews in: Dislocations in Solids, F.R.N. Nabarro, Ed., (North-Holland 1979), Vol. 2, pp. 461-545.
6. J. H. Cockayne, I.L.F. Ray, and M. J. Whelan, Phil. Mag., 20, 1265 (1969).
7. P. B. Hirsh, A. Howie, R. B. Nicholson, D. W. Pashley, and M. J. Whelan, Electron Microscopy of Thin Crystals (Butterworth & Co., London 1965).
8. F. C. Frank and J. H. Van der Merwe, Proc. Roy. Soc. A, 198, 216 (1949).
9. J. W. Matthews, Phil. Mag., 13, 1207 (1966).
10. G. C. Weatherly and R. B. Nicholson, Phil. Mag., 17, 801 (1968).

SECTION IV
Ceramic Materials

THE STRUCTURE OF GRAIN BOUNDARIES AND PHASE BOUNDARIES IN CERAMICS

C. B. CARTER
Department of Materials Science and Engineering
Bard Hall, Cornell University, Ithaca, NY 14853

ABSTRACT

The structure of different grain boundaries and phase boundaries are discussed by reference to a series of specific examples. The chosen examples of grain boundaries include the <110>{1$\bar{1}$0} tilt boundary, the (001) twist boundary and the first-order twin boundary in spinel and the basal twin and the (11$\bar{2}$3) twin in Al_2O_3. The phase boundaries discussed are the β'''-alumina/spinel interface, the wustite/spinel interface and the spinel/alumina interface. It is shown that these interfaces do have properties which result specifically from the ionic nature of the material and the large unit cells involved in each case.

INTRODUCTION

Interfaces in ceramic materials are important in such processes as sintering, deformation and solid-state reactions. The behavior of both grain boundaries and phase boundaries will be dependent on the structure of the interface not only because of intrinsic geometrical reasons but also because the chemical composition of the material will differ at the interface. This compositional variation can be understood by remembering that in many ionically bonded materials the accommodation of cations is particularly sensitive to the availability of suitable sites [1]. For example in α-Al_2O_3 the oxygen ions form a pseudo-hexagonal close-packed lattice and the Aluminum ions occupy two thirds of the octahedral interstices: in normal Mg-Al spinel the oxygen ions form a near face-centered cubic lattice and the Aluminum and Magnesium ions occupy octahedral and tetrahedral interstices respectively.

If a large cation such as K+ were present in the material it would significantly distort the oxygen sublattice since its ionic radius is considerably larger than that of either Mg^{2+} or Al^{3+}. K+ ions are actually often present in commercial Al_2O_3 and are expected to segregate to the grain boundary where larger interstices will usually be present. A relation between interface structure and chemical composition (segregation) at the interface is thus predicted. This concept has actually been used in the discussion of chemical twinning, a process whereby very large quantities of "impurity" ions can be accommodated at a twin plane because new types of site are created for larger anions; if the twin plane is repeated periodically a new compound can be formed [2, 3].

The present paper will review a number of specific examples of grain boundary structures in ceramic materials and compare these with similar boundaries in metals and elemental semiconductors. Experimental observations will be presented to demonstrate that the structures can be different and that these differences can result in suitable sites for the accommodation of impurity ions. All the transmission electron microscopy (TEM) was performed using a Siemens 102 microscope operating at 125 kV.

GRAIN BOUNDARY STRUCTURE

Example 1: The <110>{1$\bar{1}$0} Tilt Boundary in Spinel [4,5]

The low-angle <110>{1$\bar{1}$0} tilt boundary in Mg-Al spinel is particularly interesting because it can be compared directly with the same type of boundary in Si [6] and Ge [7] where the dislocation structure has been examined by high-resolution TEM (HRTEM). The boundary is composed of an array of edge dislocations with Burgers vector a/2 <1$\bar{1}$0> as shown by the weak-beam image in Fig. 1, but as can be seen in this micrograph each of the dislocations is dissociated into two partial dislocations with the same a/4 <1$\bar{1}$0> Burgers vector. The high resolution image presented in Fig. 2 was recorded with the electron beam parallel to the <110> tilt axis and shows that the stacking-fault connecting the two partial dislocations is on the {1$\bar{1}$0} plane. This dissociation was originally predicted by Hornstra [8] and has been observed by diffraction contrast imaging by several groups working on spinel materials [eg 9-12].

Dislocations in spinel can dissociate by climb because, to a good approximation, the stacking fault so produced does not disturb the oxygen sublattice and the strain energy is reduced according to the Σb^2 criterion. This type of structural effect is different from that found in Si, Ge and Au for several reasons each of which stems from the fact that, in common with many other ceramics, the unit cell of spinel is large because of the need to accommodate two different cations. The differences include: i) dislocations in spinel can dissociate by pure climb: climb dissociation can occur in Si on the 111 plane [13], for example, but in spinel not only are there are several other available stacking-fault planes this dissociation also occurs for lattice dislocations. ii) As a result of this climb dissociation the stoichiometry may be changed in the vicinity of the stacking-fault. iii) The grain boundary thickness will be affected by this dissociation [5].

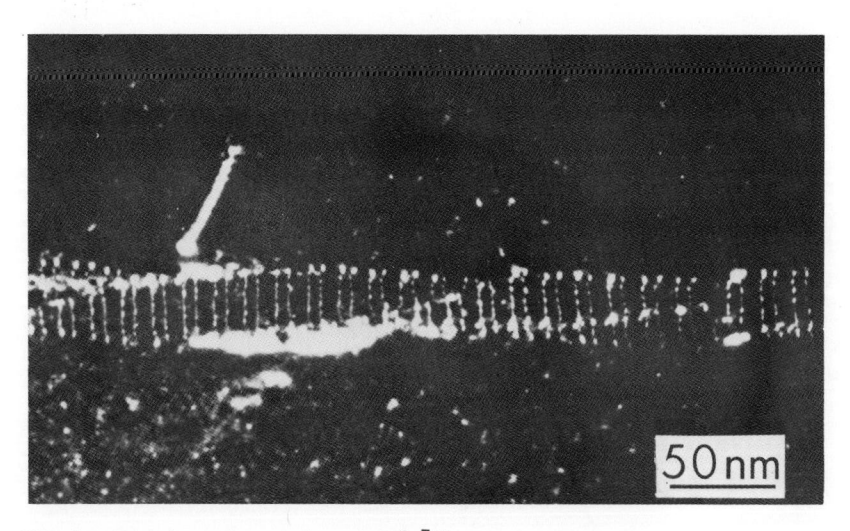

FIG. 1. Weak-beam image of a <110>{1$\bar{1}$0} tilt boundary in stoichiometric spinel.

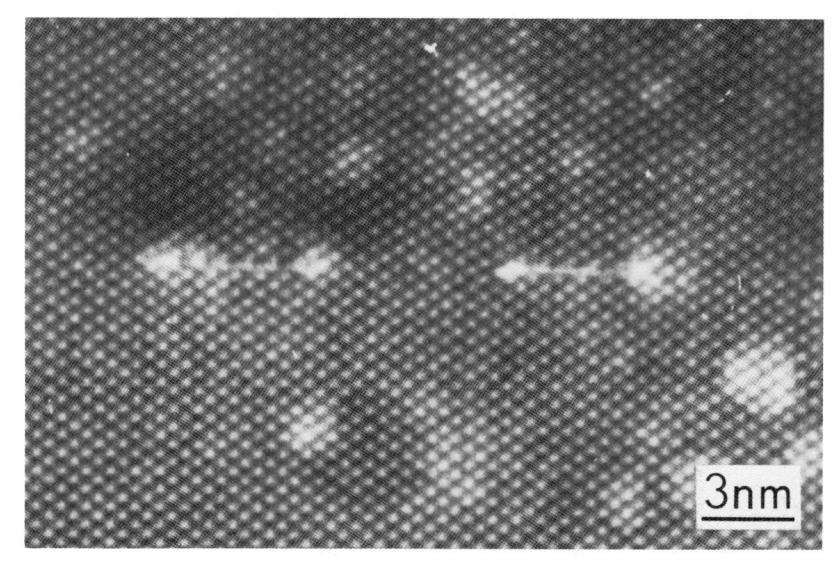

FIG. 2. High-resolution image of the boundary shown in Fig. 1 viewed along the tilt axis.

Example 2: The [001] Twist Boundary in Spinel [4,19]

The [001] twist boundary in face-centered cubic materials has been extensively studied for Au [eg. 14], Si [15,16], MgO [17] and NiO [18]. The predicted structure is a network of two orthogonal sets of screw dislocations with a/2 <110> Burgers vectors. A low angle [001] twist boundary in stoichiometric (i.e. equimolar) Mg-Al spinel has recently been examined in detail [4,19]. The boundary was formed during deformation at ∿1600°C. A weak-beam image of this boundary is shown in Fig. 3; standard diffraction contrast techniques showed that the orthogonal dislocations were pure screw in character and a comparison of the angle of twist and dislocation spacing confirmed the Burgers vectors to be a/2 <110>. There are many new features present in this twist boundary (see [4] for details) but only one, namely the dislocation dissociation, will be discussed here.

Each of the perfect dislocations is dissociated into two partial dislocations both having the same a/4 <110> Burgers vector and being pure screw in character; the stacking-fault plane is {001}. The structure of the low-angle [001] twist boundary in spinel is therefore similar to that in Au, Si etc. with the important difference that the dislocations can dissociate by glide on the (001) plane itself; the reason is again that the stacking-fault does not, to a first approximation, affect the oxygen sub-lattice. Actually when the stacking-fault energy is low or the dislocations are close together, the structure can consist of two orthogonal arrays of equi-spaced screw partial dislocations with appropriate stacking-faults. The presence of the stacking-fault will affect the structure since it may introduce local changes in stoichiometry in a manner analogous to that discussed for the tilt boundaries.

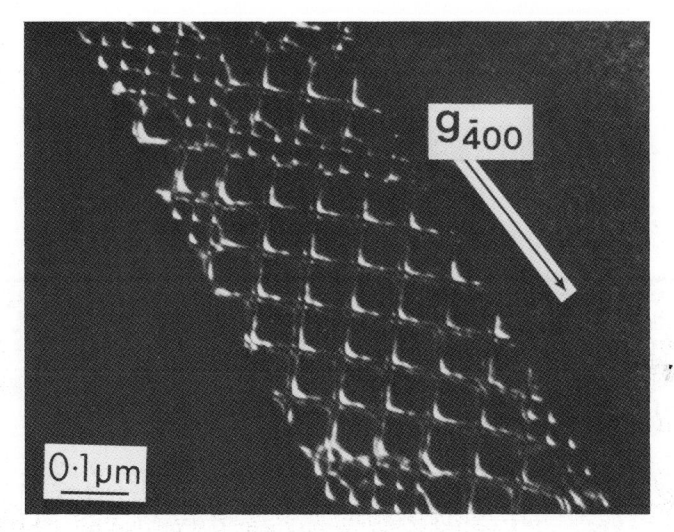

FIG. 3. Weak-beam image of a low-angle [001] twist boundary in stoichiometric spinel.

Example 3: The First-Order Twin in Spinel [20]

The first-order twin is one of the most familiar grain boundaries having been studied in metals eg. Al [21], semiconductors, Si [22] Ge [23], and more recently in ceramic oxides including MgO [25], spinel [21] and α-Al$_2$O$_3$ [25]. The structure of this twin in spinel has, like the tilt and twist boundaries, been found to be very different from the structure of macroscopically equivalent twins in Al and Si. The term 'macroscopically equivalent' indicates that both the diffraction pattern and low magnification TEM images of these interfaces would appear similar for spinel, fcc and diamond-cubic materials. However, when examined by HRTEM, the structure of the spinel twin differs both for the coherent twin (parallel to the common {111} plane) and for the incoherent twin (containing the common <1$\bar{1}$0> rotation axis). In both cases the reason for the difference is again partly due to the large unit cell accommodating two cations. The structure of the incoherent twin boundary is also affected by the ionic nature of the bonding.

The images presented in Fig. 4 show two different structures for, what is macroscopically, the same grain boundary: one image shows white spots along the interface while the other shows a translation at the boundary plane. The material is MgO.2Al$_2$O$_3$. These and other images have been analyzed in detail [26] and the difference is a direct result of a difference in the exact location of the twin plane in the oxygen sublattice with respect to the different cation layers. The two images correspond to twin interfaces which differ both in structure and in composition i.e. in stoichiometry. These two images are actually from the same micrograph and it is therefore known that the contrast effects are not due to imaging conditions, thickness effects or specimen preparation (see [26] for further details).

The short segment of incoherent twin interface shown in Fig. 5 illustrates the principal feature of all such interfaces in this material, namely the 'white-spot' defects; these defects correspond to regions of lower density and are one unit cell high: i.e. 6 {111} oxygen planes. They can be considered as dislocation-free defects or as two pairs of vacancy dipoles [20]. All such incoherent twins interfaces consist of different arrangements of these 'white-spot' defects. When such structures are compared with postulated structures of macroscopically similar interfaces in Al [21] the importance of the different bonding becomes evident: large localized difference in density found in the spinel boundary would not occur in aluminum because of the properties of an electron gas (metallic bonding) compared to the properties of ionic bonding. This interface structure is therefore unique to the ionically bonded material.

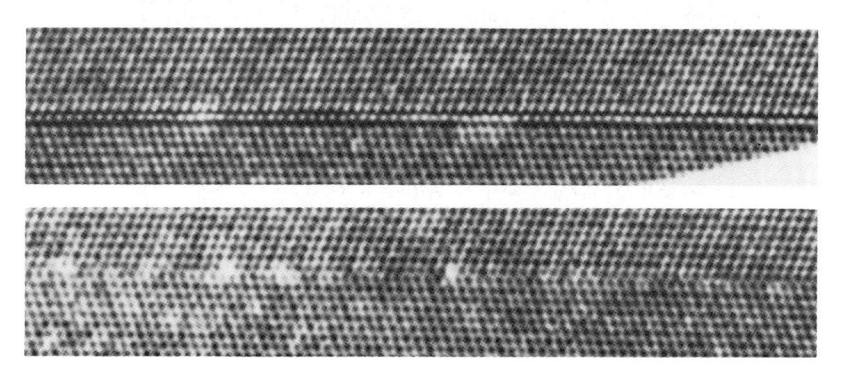

FIG. 4. High-resolution images of two coherent twin interfaces in $MgO \cdot 2Al_2O_3$.

Example 4: The Basal Twin in Al_2O_3 (25)

The basal twin in Al_2O_3 has been discussed in several recent papers and is particularly interesting because, to a first approximation the pseudo-hexagonal close-packed oxygen sublattice is not affected by the presence of the grain boundary. The twin boundary does however facet and behave in a similar manner to first-order twin boundaries in Al, Si etc. An illustration of such a boundary in Al_2O_3 is shown in Fig. 6. The boundary has been shown to fit the model of the coincident site lattice (CSL) theory both with respect to the type of observed grain boundary facet plane and with respect to the type of dislocations present in the grain boundary.

The feature of this boundary which is unique to this ceramic material is that one sublattice is unaffected by the presence of the grain boundary. However some effects are felt by the oxygen sublattice: in particular lattice dislocations can dissociate in the grain boundary into perfect grain boundary dislocations when the same dissociation in the lattice would produce partial dislocations and an associated stacking-fault (26).

Example 5: Special Grain Boundaries in Al_2O_3 [27,28]

Not only have special grain boundaries in Al_2O_3 recently been found to exist, but they have also been found to be quite numerous. The analysis of several experimentally observed interfaces has revealed that they can often be regarded as special twin boundaries. Such twin boundaries were also briefly discussed by Hyde et al. [3] in relation to the concept

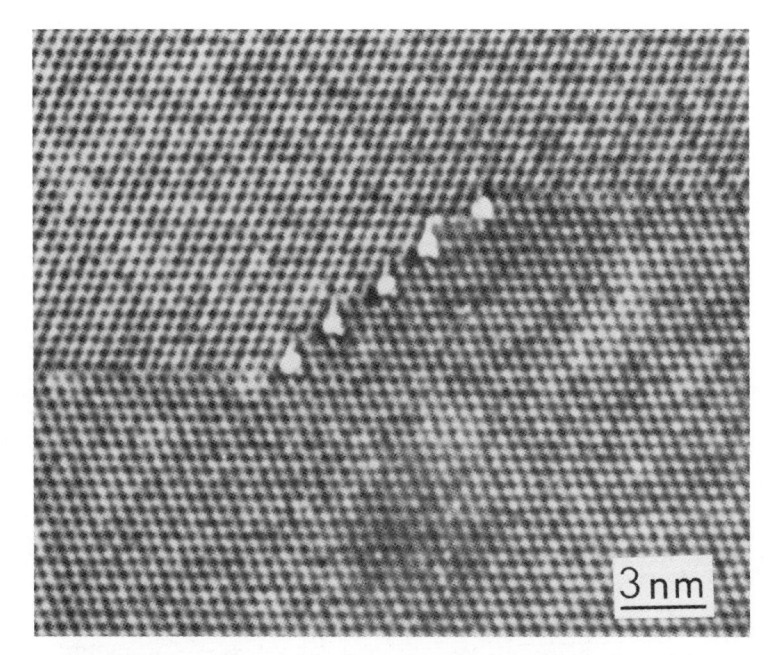

FIG. 5. A short incoherent twin interface in $MgO \cdot 2Al_2O_3$ imaged by HRTEM.

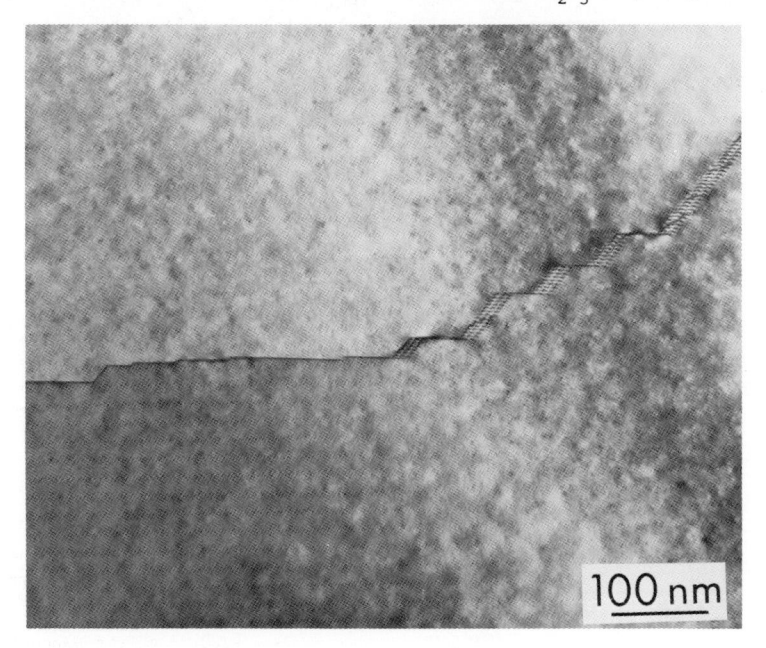

FIG. 6. Basal twin boundary in Al_2O_3 imaged using bright-field diffraction conditions at a common pole.

of chemical twinning. A schematic diagram of such a grain boundary is presented in Fig. 7a. It can be formed either by a rotation about the [$\bar{1}100$] axis or by mirroring across the ($11\bar{2}3$) plane. This type of grain boundary has been observed experimentally as illustrated by the image shown in Fig. 7b. Following the earlier discussion [3] the new oxygen polyhedra can accommodate impurity or dopant ions and segregation to such an interface is therefore anticipated. Similarly, it is possible that such a boundary could provide a fast diffusion path for neutral species [29].

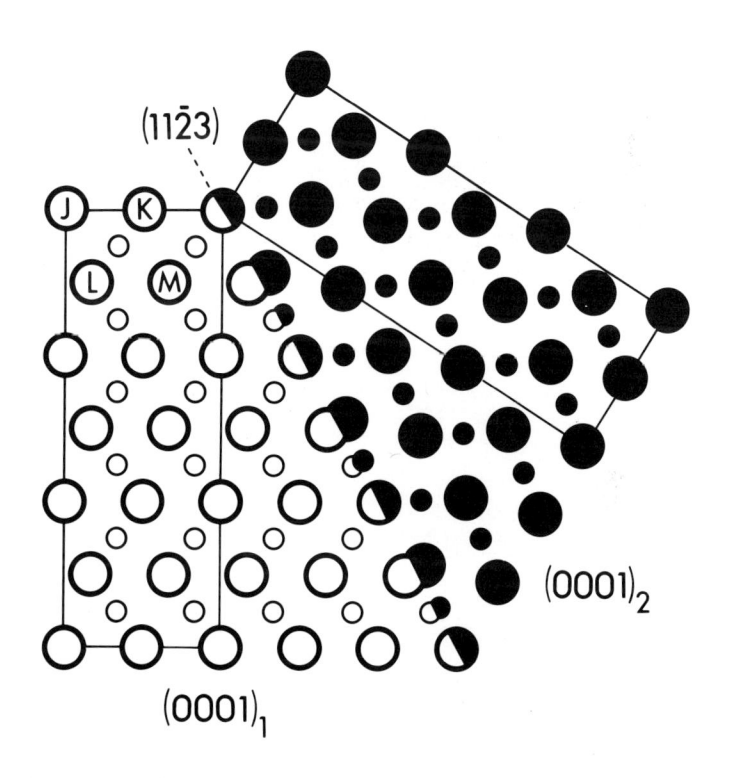

FIG. 7. a) Schematic diagram of a $11\bar{2}3$ twin in Al_2O_3.

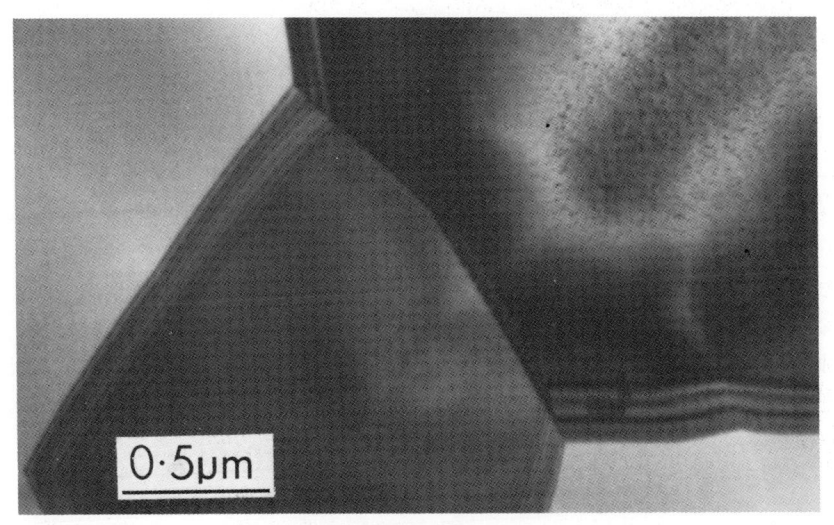

FIG. 7. b) experimental image of the $11\bar{2}3$ twin in Al_2O_3.

PHASE BOUNDARY STRUCTURE

The choice of the following examples was made to illustrate the different types of interface which may develop between structurally related ceramic materials as a result of solid-state reactions.

Example 1: β'''-Al_2O_3/Spinel [30]

This type of interface occurs during the growth of β'''-Al_2O_3 into a spinel matrix, and has been observed in a wide range of commercial aluminas and hot-pressed spinel containing Na and K as dopant or impurity. The growth mechanism is illustrated by Fig. 10. The β'''-Al_2O_3 can be identified by the rows of white spots which correspond to twin planes and contain the Na or K ion. The uniform region is Mg-Al spinel viewed in the <110> direction. A more complete treatment of the appearance of high-resolution images of β'''-Al_2O_3 is given elsewhere in these proceedings [31]. The distance between twin planes in this material is 15.9 Å and the c lattice parameter is 31.8 Å. Thus when a microtwin such as shown in Fig. 10, moves along the surface of the β'''-Al_2O_3 phase and at a distance of 15.9 Å from it, the β'''-Al_2O_3 phase boundary advances into the spinel along its c-axis by one unit cell distance. The driving force for the movement of the microtwin is the need to accommodate excess Na or K in the material. This process is therefore an example of chemical twinning and has been termed chemical microtwinning [30].

An alternative view of this same process is that the image shows a ledge (two ledges in this case) on the phase boundary and that the β'''-Al_2O_3 phase grows by the movement of this ledge. It should be stressed that this interpretation is equivalent to the model of chemical microtwinning and may be more amenable to conventionally phase transformation

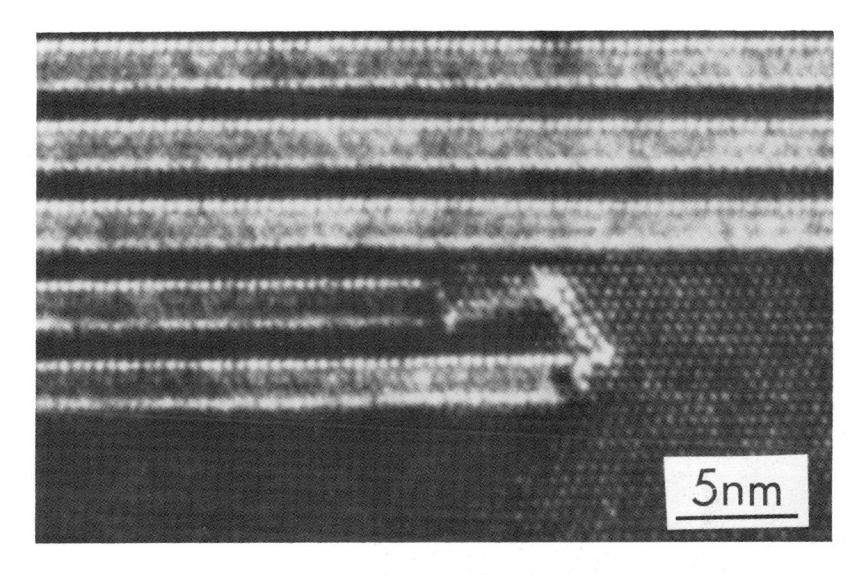

FIG. 8. The growth of β'''-Al_2O_3 in spinel as observed in a HRTEM image.

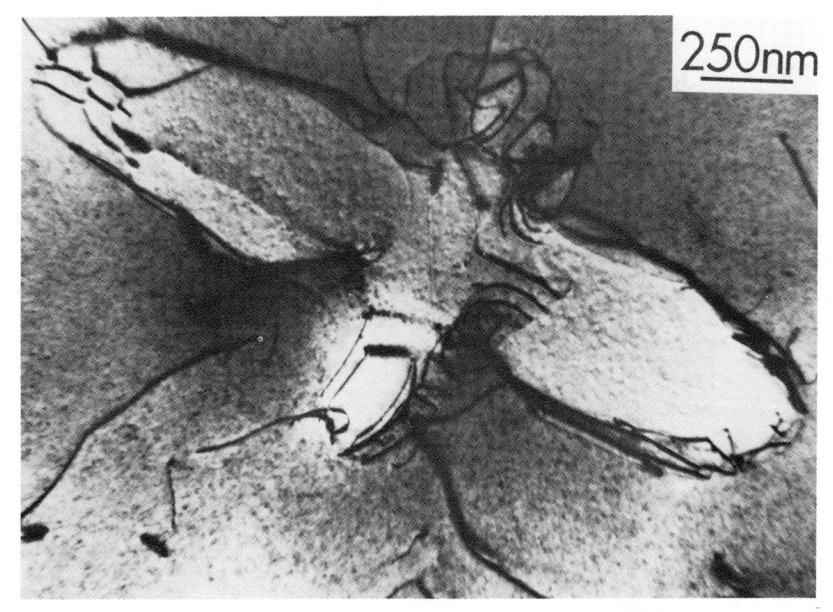

FIG. 9. A particle of $NiFe_2O_4$ in a NiO matrix.

theory. However it has been found that the chemical microtwinning model is actually more useful because it can be extended to situations where the microtwin is not 15.9 A wide [32].

Example 2: NiO/NiFe$_2$O$_4$ [33]

The phase boundary in this system separates two similar oxides both of which have the oxygen ions forming a face-centered cubic (fcc) sub-lattice. The NiFe$_2$O$_4$ has a spinel structure and its lattice parameter is very close to being twice that of the NiO (a is 4.18 A and 8.34 A for NiO and NiFe$_2$O$_4$ respectively). There is thus a very good match between the oxygen sublattices and the ferrite can grow in the NiO with a topotactic relationship. However the phase boundary appears to be structurally quite abrupt as illustrated by the example shown in Fig. 11.

This system provides a particularly well-defined situation for studying solid-state reactions since to a first approximation the growth of the ferrite only involves cation diffusion. However as can be seen in the example, misfit dislocations do form and growth of the particle must involve some oxygen diffusion to allow these edge dislocations to climb as the interface migrates. Again the special feature shown by this system is that one sublattice is almost unaffected by the movement of the phase boundary. The oxygen sublattice does however play a role in deciding the direction and rate of growth through its influence on the elasticity parameters and on the diffusion paths of the migrating anions.

Example 3: Spinel/Al$_2$O$_3$ [34]

This phase boundary is interesting because it separates two oxides where the oxygen ions are in both cases nearly close-packed but differ in that spinel would correspond to fcc while Al$_2$O$_3$ would correspond to hcp. (See discussion of each material above.) Thus in order for the interface to advance in either direction the stacking sequence of the "closed-packed" oxygen ions must change from ABCABC to ABABAB or vice versa. It has been suggested that this could occur by the movement of transformation dislocations similar to Shockley partial dislocations [35]. An image of the CoAl$_2$O$_4$/Al$_2$O$_3$ phase boundary is shown in Fig. 12. The images only rarely show this type of Moire pattern and more frequently show the type of image seen in Fig. 13. A new model for this phase boundary has been developed which take account of the experimentally observed rotation of the spinel {111} plane relative to the (0001) plane of the alumina [34]. The reason for this rotation is that it reduces the need for transformation dislocation and misfit dislocations.

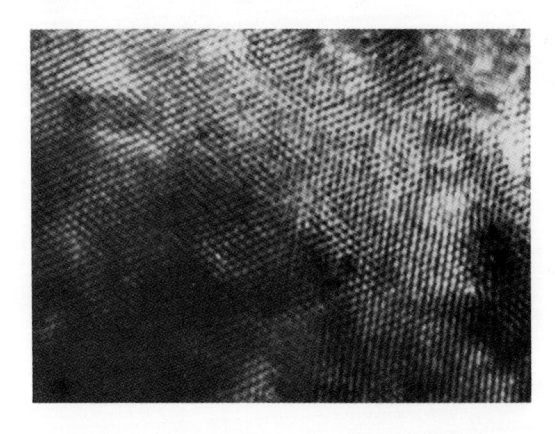

FIG. 10. Moire pattern across a CoAl$_2$O$_4$/Al$_2$O$_3$ interface caused by lattice mismatch.

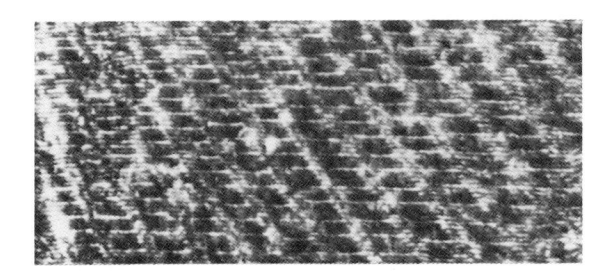

FIG. 11. Dislocations and steps on a $CoAl_2O_4/Al_2O_3$ interface imaged using weak-beam conditions.

SUMMARY

Grain boundaries and phase boundaries in ceramic materials can show features which do not occur in metals with similar crystal symmetry. The principal reasons for these newly reported observations are

i) In many cases the unit cell of the ceramic crystal lattice may be large because of the need to accommodate different cations. The illustration presented for spinel shows that many features of the boundary can then be associated with a smaller anion-sublattice "unit cell."

ii) Grain boundaries can form in one sublattice and leave the other almost unaffected as illustrated by the basal twin in Al_2O_3.

iii) The actual structure of the grain boundary can be very different from that of analogous boundaries in metals due to the ionic rather than metallic bonding.

iv) Phase boundaries can form in a simple manner corresponding to an interface between two polytypes accommodating small changes in chemical composition: the new feature particular to ceramic materials is the creation of new anion polyhedra which can accommodate cations of different size.

v) Phase boundaries can occur as interfaces which almost solely affect one sublattice as illustrated by the $NiO/NiFe_2O_4$ system (cf i and ii).

vi) Phase boundaries may form by a rotation of the crystal lattice to accommodate both the change in crystal structure and part of the lattice misfit.

ACKNOWLEDGMENTS

The author would like to acknowledge the students and postdoctoral associates at Cornell who have been involved in this research. Thanks are particularly due to K. J. Morrissey and K. M. Ostyn. Collaboration with Prof. H. Schmalzried and Dr. T. M. Shaw is also gratefully acknowledged. A useful discussion with Prof. V. Vitek is also acknowledged. This research on grain boundaries is supported by DOE under contract No. DE-AC02-82ER12076 and that on phase boundaries is supported by NSF under grant No. DMR81-02294. Earlier research on these topics was supported by

NSF through the Materials Science Center at Cornell University; the MSC also supports the Electron Microscopy Facility which is maintained by Mr. Ray Coles.

REFERENCES

1. W. D. Kingery, H. K. Bowen, and D. R. Uhlmann, Introduction to Ceramics (Wiley, New York) (1976).
2. S. Andersson and B. G. Hyde, J. Solid State Chem. 9, 92 (1974).
3. B. G. Hyde, S. Anderson, M. Bakker, C. M. Plug and M. O'Keeffe, Proc. Solid St. 10. 12, 273 (1974).
4. K. M. Ostyn, H. Wendt and C. B. Carter, Beitr. elektronenmikroskop. Direktabb. Oberfl. 16, 225 (1983).
5. K. M. Ostyn and C. B. Carter, Advances in Ceramics 5 in press (1982).
6. A. Bourret and J. Desseaux, Phil. Mag. A39, 405 (1979).
7. W. Skrotski, H. Wendt, C. B. Carter and D. L. Kohlstedt, Proc. MRS Symp. on Thin Films and Interfaces (1983).
8. J. Hornstra, J. Phys. Chem. Solids, 15, 311 (1960).
9. M. H. Lewis, Phil. Mag. 13, 777 (1966).
10. T. E. Mitchell, L. Hwang and A. H. Heuer, J. Mats. Sci. 11, 264-272 (1976).
11. P. Veyssière, J. Rabier, H. Garem and J. Grilhé, Phil. Mag. 33, 143 (1976).
12. R. Duclos, N. Doukhan, and B. Escaig, J. Mats. Sci. 13, 1740 (1978).
13. C. B. Carter, Mat. Res. Soc. Symp. Proc. 5, 33 (1982).
14. T. Schober and R. W. Balluffi, Phil. Mag. 21, 109-123 (1970).
15. H. Foll and D. G. Ast, Phil. Mag. A39, 589 (1979).
16. C. B. Carter, H. Foll, D. G. Ast and S. L. Sass, Phil. Mag. A43, 441 (1981).
17. C. P. Sun and R. W. Balluffi, Phil. Mag. A46, 49, 63 (1982).
18. K. M. Ostyn and C. B. Carter, in preparation.
19. K. M. Ostyn and C. B. Carter, Proc. 41st Ann. Meeting EMSA, p. (1983).
20. T. M. Shaw and C. B. Carter, Scripta. Met. 16, 1431 (1982).
21. R. C. Pond, D. A. Smith, and W. A. T. Clark, J. Microsc. 102, 309 (1974).
22. B. Cunningham, Proc. 40th Ann. Meeting CMSA, p. 434 (1982).
23. Z. Elgat and C. B. Carter, Proc. MRS Symp. on Thin Films and Interfaces (1983).
24. S. B. Newcomb, D. J. Smith and W. M. Stobbs, J. Microsc. 130, 137-1465 (1983).
25. K. J. Morrissey and C. B. Carter, Advances in Ceramics 5, in press (1982).
26. T. M. Shaw and C. B. Carter, in preparation.
27. K. J. Morrissey and C. B. Carter, Proc. 10th Int. Conf. on Electron Microscopy (Hamburg), p. 343 (1982).
28. K. J. Morrissey and C. B. Carter, Adv. in Ceramics, 12, in press.
29. F. A. Kroger, Proc. MIT Symposium on MgO/Al_2O_3 Structure Property Relations. Boston (1983).
30. K. J. Morrissey and C. B. Carter, Advances in Materials Characterization (Plenum), p. 297 (1983).
31. K. J. Morrissey, Z. Elgat, Y. Kouh and C. B. Carter, these proceedings.
32. K. J. Morrissey and C. B. Carter, in preparation.
33. K. M. Ostyn, M. Köhne, H. Falke, C. B. Carter and H. Schmalzried, submitted for publication.
34. C. B. Carter, K. M. Ostyn and H. Schmalzried, in preparation.
35. H. G. Sockel and H. Schmalzried, Mat. Sci. Res. 3, 61 (1966).

APPLICATIONS OF HIGH RESOLUTION ELECTRON MICROSCOPY IN CERAMICS RESEARCH

G. VAN TENDELOO
Universiteit Antwerpen, RUCA, Groenenborgerlaan 171,
B-2020 Antwerp, Belgium

ABSTRACT

High resolution microscopy of ceramics has its specific problems due to the difficulties in preparing thin foils. Most of these problems however can be overcome and recent years a lot of effort has been concentrated on electron microscopy of ceramics because of their high technological importance. Some results on important ceramics such as SiC, Mg-Si-O, ZrO_2-ZrN and Si-Al-O-N obtained by high resolution microscopy will be briefly discussed here.

INTRODUCTION

Micro-structural research on ceramics or minerals using electron microscopy techniques has been tempered for a long time due to problems in the sample preparation. Electropolishing techniques, so popular for metals and alloys can no longer be used for the poorly conducting ceramics and until ion beam thinning became popular ten to fifteen years ago the only way to prepare samples was by crushing. This technique however is quite useless when crystal defects or precipitation have to be studied since the electron transparent areas are usually very limited and moreover crystals have the tendency to break at grain boundaries or other crystal imperfections. By the time that adequate preparation techniques became available the metallurgical electron microscopy research was already much further advanced and few materials scientists wanted to do the step back. However the high technological interest of some ceramics as well as the knowledge that the important mechanical properties strongly depend upon the microstructure created a tremendous regain of interest in the electron microscopy of ceramics during recent years.

A second and more general reason for the increased interest in the structural features of materials is the improvement of commercial electron microscopes to a resolution level of 0.2 or 0.3 nm allowing to obtain atomic scale information from about all crystal structures under different low order zone orientations.

In the present overview we will first deal with some specific problems associated with ceramics (or non-metallic) electron microscopy research and subsequently treat some examples of microstructural research of technologically important ceramics.

PROBLEMS ASSOCIATED WITH ELECTRON MICROSCOPY OF CERAMICS

Apart from the problems in preparing suitable thin samples, ceramics may cause other problems when introduced in the electron microscope. While metallic specimens are hardly sensitive to the electron beam (apart from radiation damage when the accelerating voltage exceeds the threshold voltage for atom displacement) some ceramics may undergo transformations due to the influence of the electron beam; the effect being known as radiolysis. More details about these problems can be found in the recent review of L.W. Hobbs [1]. We will briefly treat three different examples which are the most common failures for ceramics.

A crystalline to amorphous transformation

It is well known that a number of silicates undergo an amorphisation process or a loss of long range order under the electron beam, even at voltages of 100 kV or below. An example is shown in fig. 1 for ortho-enstatite $MgSiO_3$. The amorphisation sets in from the surface and gradually eats away the thin areas. An explanation suggested for SiO_2 quartz by Hobbs and Pascucci [2] is that peroxy formation breaks the Si-O-Si bonds in the network silicates and at the end provides enough freedom to the SiO_4 tetrahedra to rotate and rebond. A similar mechanism is probably acting in most silicates. Sometimes an equilibrium can be reached between amorphisation and recrystallisation; this happens in quartz at higher temperatures [3] or in CdTe where the process has been observed under high resolution conditions [4].

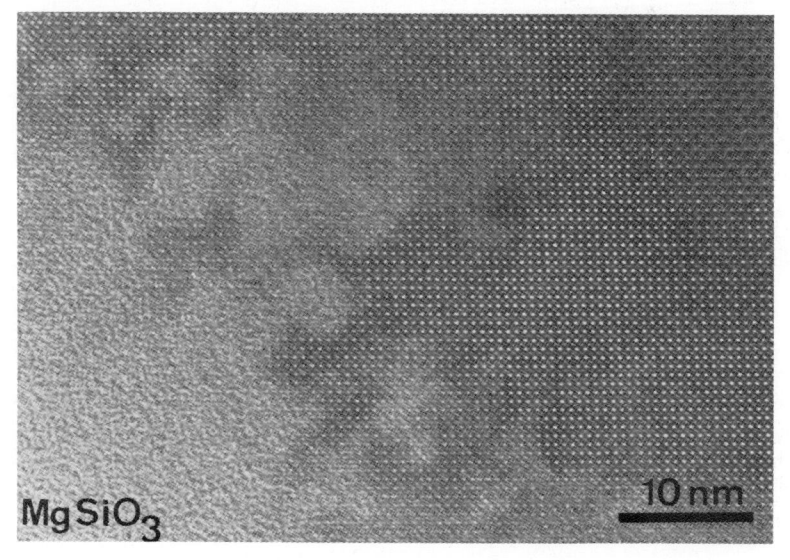

FIG. 1. High resolution micrograph of ortho-enstatite in a [001] orientation. The crystalline phase gradually vanishes away from the specimen edge.

Loss of a superstructure

ZrO_2-ZrN alloys containing at least 2.5% ZrN form a complex rhombohedral oxynitride superstructure of the $Zr_5Sc_2O_{13}$ type [5]. The diffraction pattern along the trigonal axis is reproduced in fig. 2a. When focussing the electron beam to the usual extent for high resolution microscopy the superstructure reflections gradually weaken and within less than half a minute they have totally gone (fig. 2b). This makes it extremely difficult to produce decent high resolution images of this material. A bright field image along the trigonal axis is reproduced in fig. 2c. The image is far from ideal and in some areas the superstructure has already gone. The incident electrons probably produce enough energy to cause a disordering of the vacancies on the oxynitride sublattice.

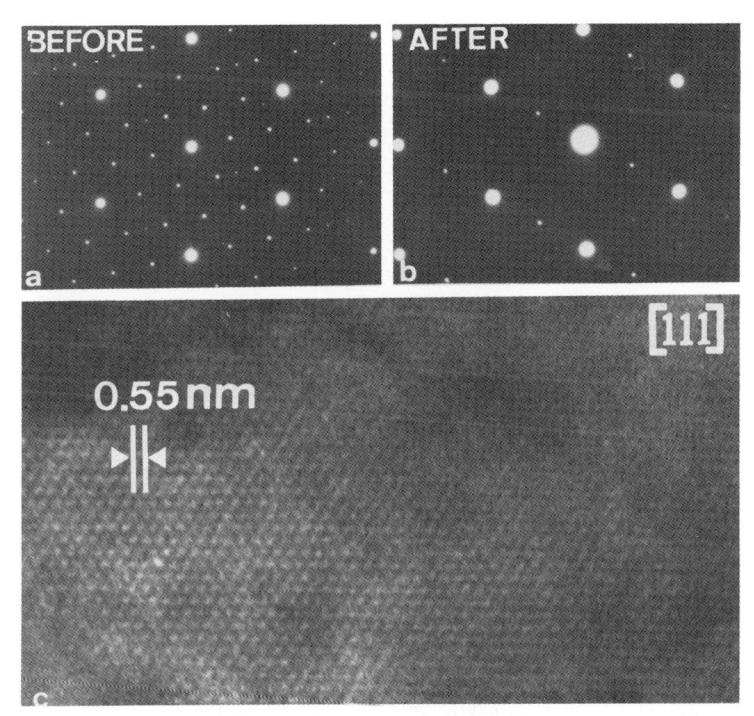

FIG. 2. a) Electron diffraction pattern of ZrO_2-15% ZrN along the trigonal axis of the rhombohedral Zr-O-N superstructure which corresponds to the [111] direction in the ZrO_2 cubic structure.
b) Diffraction pattern from the same region as in fig. a after focussing the electron beam for high resolution microscopy conditions for 15 sec.
c) Resulting high resolution image under the conditions of fig. a. Note that in some areas the superstructure has already gone.

Ion bombardment effects

The possible artifacts arising from ion beam thinning are well known (see e.g. [6]). The penetration depth of the ions is only a few nanometers and for conventional microscopy this is just a small portion of the volume but specifically for high resolution microscopy where the useful thickness is limited to 20-30 nm it can become important. Apart from the well known surface roughness and ion-radiation damage, the ion beam can introduce artifacts which especially for ceramics can be highly confusing.

a) Amorphisation of very thin areas. Very often the thin borders of the specimen will be amorphous. Whenever such effects are not expected e.g. in SiC (fig. 3) they are annoying but not confusing; however when amorphous phases are among the expected stable products e.g. in Mg-Si-O compounds one has to be extremely careful.

b) Thin foil phase transformations. The information from very thin areas, only a few atomic layers thick, is always doubtful in electron microscopy, especially after ion beam preparation. In fig. 4 an example is shown of monoclinic ZrO_2 which close to the edge transforms back to the cubic high temperature phase. Note the twinned area (marked b) which gradually disappears near the edge of the wedge shaped crystal.

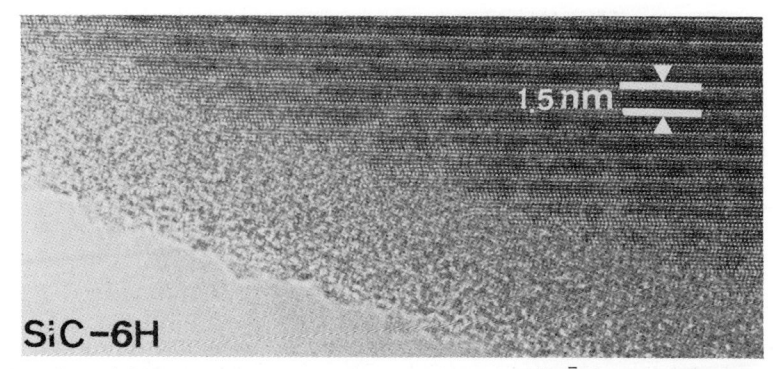

FIG. 3. High resolution image of 6H-SiC along a [11$\bar{2}$0] direction. All samples prepared by ion bombardment show an amorphous edge of at least 10 nm larger (courtesy H. Bender).

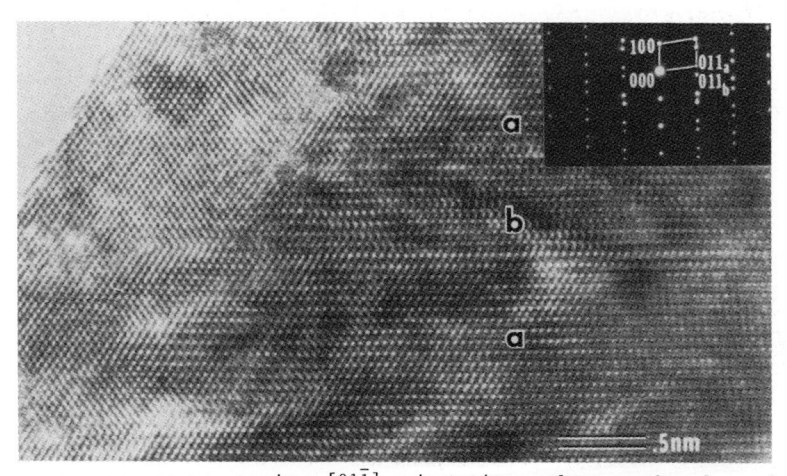

FIG. 4. Monoclinic ZrO_2 in a [01$\bar{1}$] orientation. Close to the edge and as a result of ion beam bombardment the material has undergone a phase transformation back to the cubic phase. Note the twin lamella (marked b) vanishing close to the edge.

HIGH RESOLUTION IMAGING MODE

The information to be expected from the lattice imaging technique largely depends upon the resolving power of the microscope. Apart from instrumental parameters such as stability, aberrations and objective lens excitation the final results are also influenced by more physical parameters such as sample thickness and exact crystal orientation e.g. [7]. Another important parameter is the size of the objective aperture; i.e. which reflections are included to form the final image. The choice is related to the performance of the microscope as well as to the physical problem. We will illustrate this for the 15R polytype of SiC. The diffraction pattern oriented along [1$\bar{1}$20] is reproduced in fig. 5a. Intense reflections in the central row marked by arrows correspond to the 0.25 nm spacing between subsequent SiC layers.

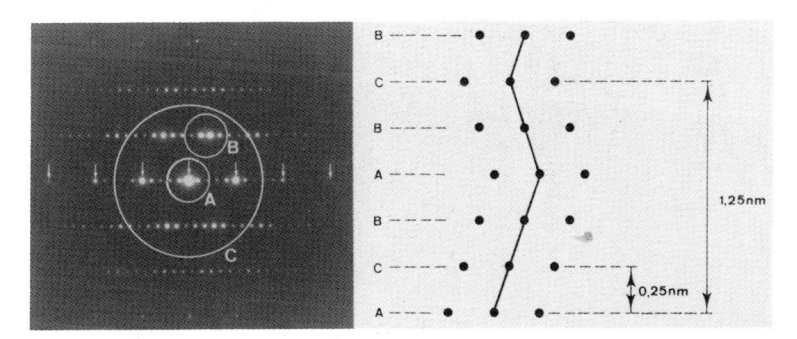

FIG. 5. a) Electron diffraction pattern of 15R-SiC taken along [11$\bar{2}$0]. The central row reflections resulting from the close packed stacking are indicated by arrows while the objective apertures used to obtain fig. 6a and 6c are indicated by A and C respectively.
b) Schematic representation of part of the 15R-SiC polytype (see also [8]). Note the 5 layer repeat and the 32 (Zhdanov notation) stacking.

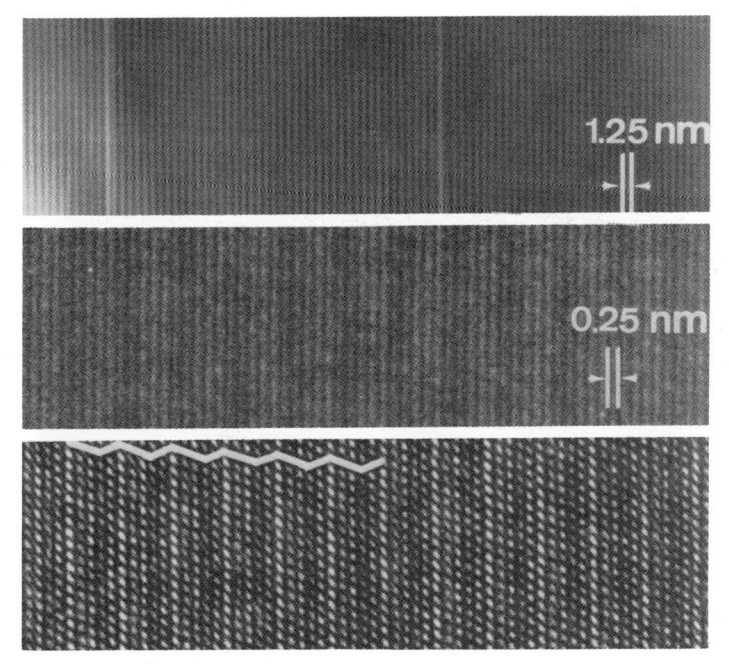

FIG. 6. High resolution of the 15R-SiC structure under different experimental conditions.
a) Using a small objective diaphragm (A in fig. 5a) which is the only useful size in a microscope with low resolution possibilities.
b) Using a larger objective aperture including at least one of the basic (arrowed) reflections of the central row. The orientation however is different from [11$\bar{2}$0] and noncentral rows are too far to be resolved.
c) Using a large objective aperture (C in fig. 5a). A structure image directly related to the projected crystal structure (fig. 5b) can be obtained.

When restricting the objective aperture to a fundamental reflection and two or three satellites on both sides (aperture A in fig. 5a for the bright field or aperture B for a non-central dark field), information can be expected about the unit cell. The 1.25 nm fringe period is revealed (fig.6a) and defects in the stacking sequence are revealed as local deviations in the fringe spacing. For a long time this method has been succesfully applied e.g. to identify different polytypes in SiC [9-13].

With recent microscopes it makes sense to use larger apertures including the fundamental (arrowed) reflections. If non-central rows are excluded the 0.25 nm spacing of the individual SiC layers can be resolved (fig. 6b) but detailed subunit cell information is lacking.

Under suitable diffraction conditions (e.g. along [11$\bar{2}$0]) and including also non-central reflections (aperture C in fig. 5a) the HREM image will provide detailed information about the stacking sequence. The relation between bright dots and the structural representation of fig. 5b is obvious (see also [8]).

The most appropriate resolution will depend upon the physical problem and in spite of the continuous struggle for a decreased resolution one is sometimes better off with a "reduced" high resolution image. A striking example is illustrated in [14] for an Au-Mn alloy where basic fcc reflections are deliberately excluded by the objective aperture in order to produce images allowing a straightforward interpretation.

POLYTYPISM IN SIALON CERAMICS

Because of its possible high temperature application in ceramics technology the Si_3N_4-AlN-Al_2O_3-SiO_2 quasi-equilibrium phase diagram has been widely studied (see e.g. [15]). The aluminium nitride polytype group is situated in the phase diagram between β' sialon and AlN [15, 16] and has been identified as $Al_{x+y}Si_{6-x}O_xN_{8-x+y}$

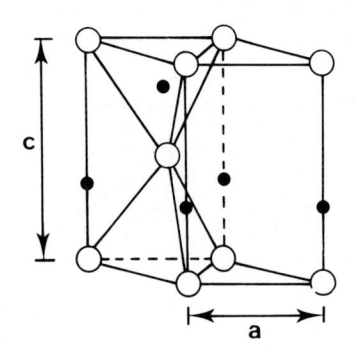

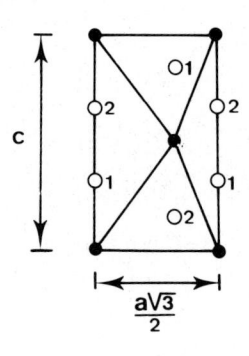

FIG. 7. a) Spacial view of the basic 2H wurtzite structure. The metal atoms are represented as open circles; anions are black dots.
b) Projection along [10.0]. The centers of two different tetrahedra are indicated by 1 (the ones pointing down) and 2 (the ones pointing up).

Each phase extends along lines of constant metal to non-metal ratio and all have a very narrow range of homogeneity [15][17]. The structure of the AlN-polytypes is based on the 2H structure (wurtzite type) which is reproduced in fig. 7, together with its [10.0] projection. The metal atoms fill one half of the available tetrahedral sites (e.g. all tetrahedra pointing downwards; noted 1 in fig. 7)

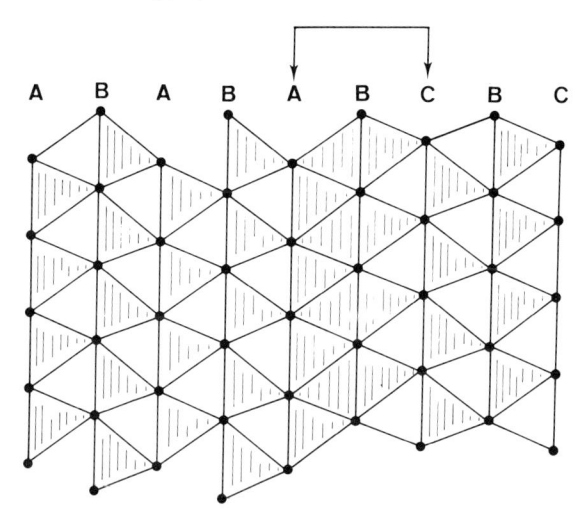

FIG. 8. [10.0] projection of a stacking fault in the basic 2H structure locally creating an ABC cubic stacking. Hatched tetrahedra are filled by anions.

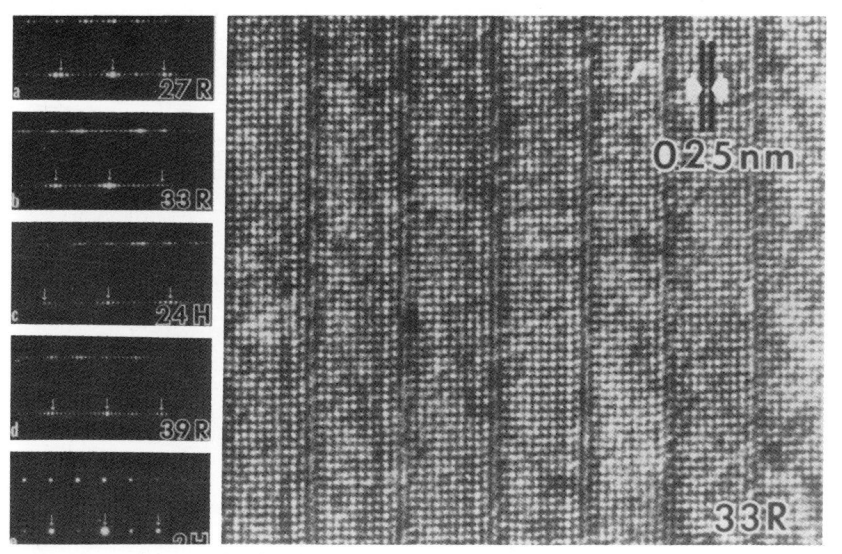

FIG. 9. High resolution image of a 33R polytype in AlN-SiO$_2$. The diffraction pattern of different polytypes along [10.0]$_{2H}$ is shown at the left.

giving rise to a composition Al_1N_1. When the metal to non-metal ratio (M/X) decreases, the non-metal ions have to fill adjacent tetrahedra which share common faces (1 and 2 in fig. 7b). This would give rise to impossibly short interatomic distances between non-metal atoms which can only be avoided if the metal atom configuration is locally changed from hexagonal (ABAB ...) to cubic (ABC ...) i.e. by introducing a stacking fault in the metal configuration. Such a stacking fault configuration projected along [10.0] is depicted in fig. 8; filled tetrahedra are hatched. An excess of non-metal atoms can be taken up in the faulted region (indicated by arrows) where the filled tetrahedra only share an edge, increasing significantly the interatomic non-metal distance. It is clear that the further M/X deviates from 1, the closer the stacking fault spacing. Identified polytypes are 8H, 15R, 12H, 21R, 27R, 33R, 24H and 39R [15][17] corresponding with M/X values varying between 4/5 and 13/14 respectively. All these polytypes are built on the same stacking principle just considered. The shorter polytypes 8H up till 27R have been discovered and their structure determined by X-rays [15] while the longer polytypes 33R, 24H and 39R have been analysed using electron diffraction and high resolution microscopy. An example of a 33R polytype is reproduced in fig. 9.

ANALYSIS OF DOMAIN STRUCTURES IN ZrO_2 - ZrN ALLOYS

There is a significant technological interest in ZrO_2-based ceramics mainly since it was discovered that the mechanical properties can be improved by alloying ZrO_2 with other ceramic matrices, mostly oxides [18][19]. In a cubic matrix metastable tetragonal precipitates are formed which transform martensitically to the monoclinic ZrO_2 under applied stress, retarding the propagation of cracks. ZrO_2-ZrN alloys were also proposed as possible candidates but were quickly rejected. ZrO_2-ZrN alloys containing between 2.5 and 75% ZrN consist of a rhombohedral matrix (no cubic matrix) in which monoclinic as well as tetragonal precipitates are dispersed. The monoclinic precipitates (already formed without applied stress) are martensitic products and may occur in different orientation variants (see fig.10 at low magnification). The fragmentation into variants is on such a small scale that

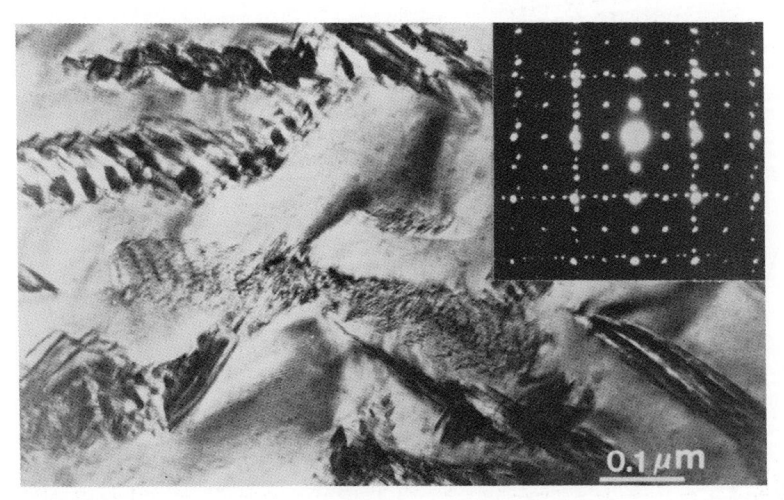

FIG. 10. *Low magnification of ZrO_2-ZrN. Heavily twinned monoclinic precipitates are present in a rhombohedral matrix. The complex diffraction pattern is shown as an inset.*

conventional microscopy or diffraction are unable to determine the orientation of the different variants; the diffraction pattern (inset of fig. 10) looks terribly complex due to the presence of several variants. One way to overcome this problem (apart from microdiffraction using convergent beam) is by high resolution microscopy. This technique also allows to gain information about the fine structure of the twin boundaries. We will first consider a simple situation where all monoclinic variants have a common $\bar{c}$-axis parallel to the electron beam and then we will treat a more general case.

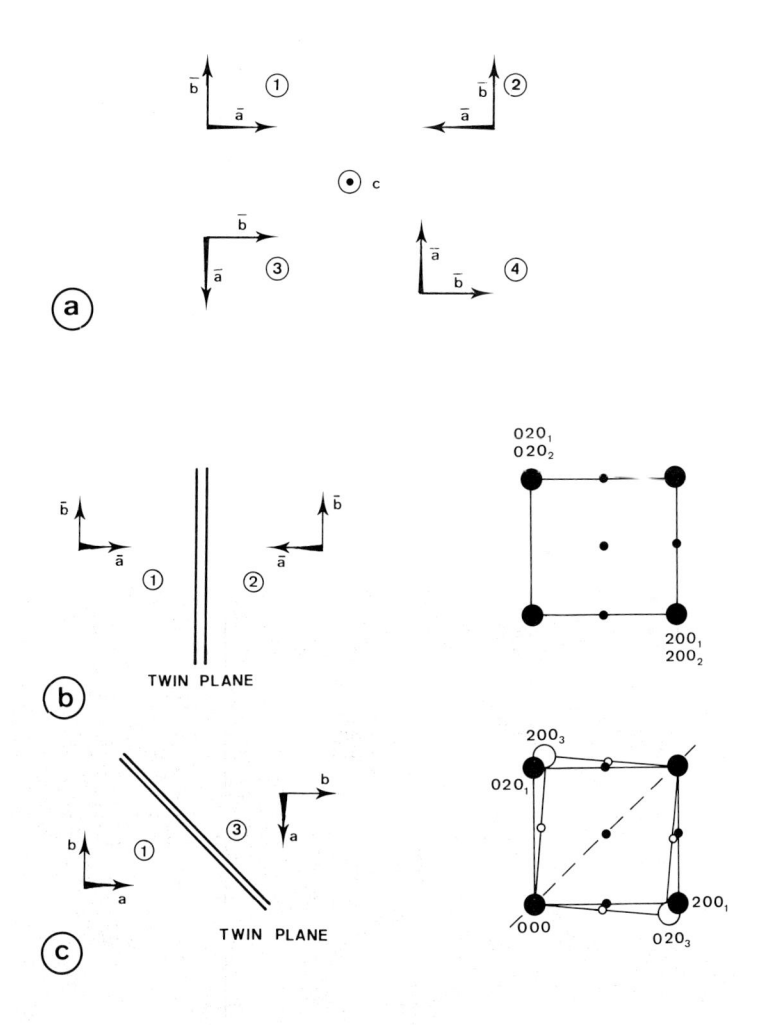

FIG. 11. a) Schematic representation of four different orientation variants having a common $\bar{c}$-axis.
b) c) representation of two possible twin boundaries between these four variants. The schematic diffraction pattern is shown at the right side.

Four different orientation variants are possible having a common $\bar{c}$-axis; [20][21] they are reproduced in fig. 11a. Variants 1 and 2 (and 3 and 4) have a parallel $\bar{b}$-axis but the $\bar{a}$-axis pointing in the opposite direction. Between these four variants two different interfaces can be formed; they are represented schematically in fig. 11b,c. The first one, having a (100) twin plane, forms the boundary between variants 1 and 2 (or 3 and 4) and will not produce

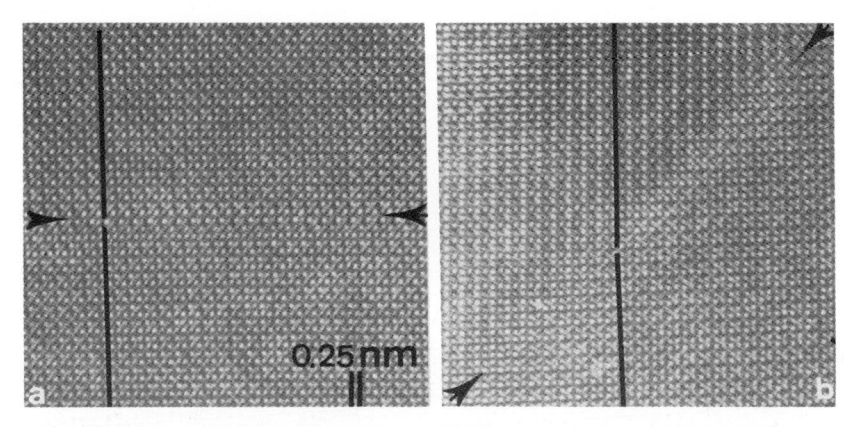

FIG. 12. High resolution image of two different twin boundaries having a common $\bar{c}$-axis in monoclinic ZrO_2.
a) a (100) twin boundary (see fig. 11b).
b) a (110) twin boundary (see fig. 11c).

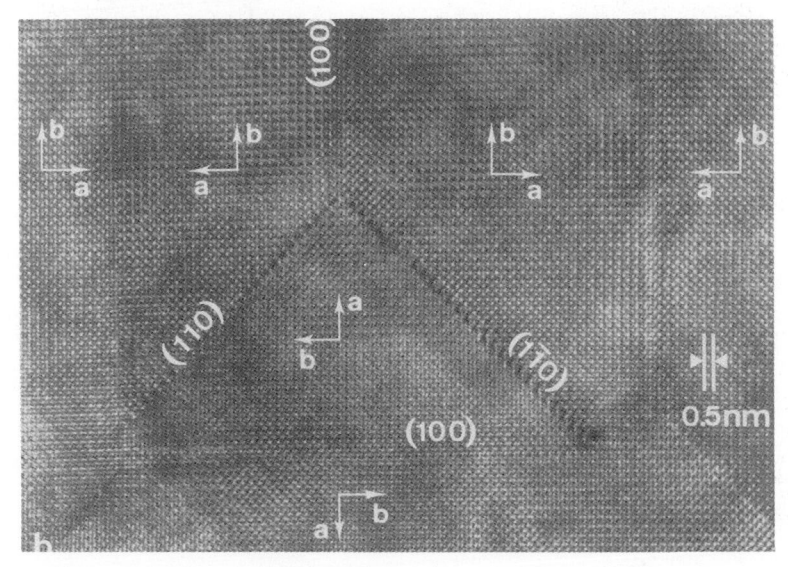

FIG. 13. A heavily twinned monoclinic ZrO_2 precipitate. All variants have a common $\bar{c}$-axis parallel to the electron beam. The orientation of the axes as determined from HREM and optical diffraction is indicated.

any diffraction effect (apart from a streaking due to the presence of the
boundary). The second one having a (110) or (1$\bar{1}$0) twin plane forms the
boundary between variants 1 and 3 or 1 and 4. In the diffraction pattern
such twin boundaries produce a small spot splitting parallel to the [220]*
or [2$\bar{2}$0]* direction (fig. 11c right side). On a high resolution image both
boundaries can be easily discerned due to the slight orientation change associated with the latter twins (fig. 12). This knowledge allows to analyse
in detail the complex microdomain structures only 20 to 30 nm wide (fig. 13)
and to propose atomic scale models for the twin boundaries. However since
details on such high resolution images are very much function of focus and
exact crystal orientation it is not possible to draw straightforward conclusions without computer simulations of the boundaries.

Another technique, very useful when combined with high resolution microscopy is the optical diffraction. It allows to obtain micro-diffraction information from very small areas down to a few square nanometers (e.g. [22]
[23][24]). We will illustrate here the power of the optical diffraction
technique to analyse the complex configuration of monoclinic precipitates
in fig. 10. Conventional diffraction patterns (inset fig. 10) mostly indicate
superposition of several variants. From the high resolution and the optical

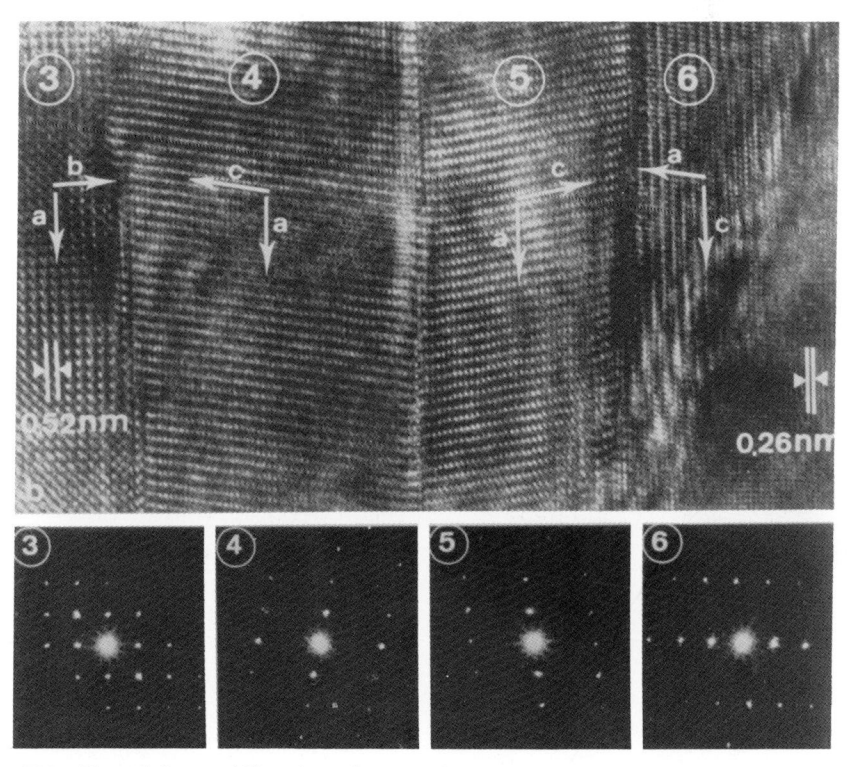

FIG. 14. High magnification of part of a monoclinic precipitate in ZrO_2-
ZrN (see fig. 10). Laser optical diffraction of the areas indicated 3 to
6 allows to determine the orientation relationship and to unscramble the
complex diffraction pattern (fig. 10 inset).

diffraction patterns (fig. 14) it is possible to deduce unambiguously the orientation relationship between the different orientation variants.

This example clearly illustrates the importance of the well known optical diffraction as a microdiffraction technique when combined with high resolution microscopy.

REFERENCES

1. L.W. Hobbs in : "Proceedings of the 41st Annual Meeting of the EMSA" ed. G.W. Bailey (San Fransisco Press 1983) pp. 346.
2. L.W. Hobbs and M.R. Pascucci, J. Physique 41, C6, 237 (1980).
3. G. Van Tendeloo, J. Van Landuyt and S. Amelinckx, Phys. stat. sol. (a) 33, 723 (1976).
4. R. Sinclair, D.J. Smith, S.J. Erasmus and F.A. Ponce in "Electron Microscopy 1982" volume 2, ed. The Congress Organizing Committee p. 47.
5. G. Van Tendeloo and G. Thomas, Acta Met. 31, 1611 (1983).
6. P.J. Goodhew in : Practical Methods in Electron Microscopy, volume 1, ed. A.M. Glauert (North Holland, Amsterdam, London 1972) pp 83-91.
7. J.H.Spence "Experimental High Resolution Electron Microscopy" (Clarendon Press, Oxford 1981).
8. J. Van Landuyt, G. Van Tendeloo and S. Amelinckx in "Progress in Crystal Growth and Characterisation" ed. P. Krishna (1983).

9. S. Shinozaki and K.R. Kinsman, Acta Met. 26, 769 (1978).
10. J. Van Landuyt and S. Amelinckx, Coll. Int. CNRS n°205, p.87 (1972).
11. D.R. Clarke, J. American Ceramic Soc. 62, 236 (1979).
12. A.H. Heuer, V. Lou, L. Ogbugji and T.E. Mitchell, J. Micr. Spectr. Electr. 2, 475 (1977).
13. M. Dubey, G. Singh and G. Van Tendeloo, Acta Cryst. A33, 276 (1977).
14. G. Van Tendeloo and S. Amelinckx, Phys. stat. sol. (a) 65, 73 (1981); 65, 431 (1981).
15. K.H. Jack, J. Mater. Sci. 11, 1135 (1976).
16. L.J. Glauckler, H.L. Lucas and G. Petzow, J. Amer. Ceram. Soc. 58, 346 (1975).
17. G.Van Tendeloo, K.T. Faber and G. Thomas, J. Mater. Sci. 18, 525 (1983).
18. E.C. Subbarao in : "Science and Technology of Zirconia" eds. A.H. Heuer and L.W. Hobbs (The American Ceramic Society, Inc. Columbus, Ohio 1981) pp. 1.
19. M. Rühle and W.M. Kriven, Ber. Bunsenges Phys. Chem. 87, 222 (1983).
20. G. Van Tendeloo, L. Anders and G. Thomas, Acta Met. 31, 1619 (1983).
21. E. Bischoff and M. Rühle, J. Amer. Ceram. Soc. 66, 123 (1983).
22. T. Mulvey, J. of Microscopy 98, 232 (1973).
23. T. Tanji and H. Hashimoto, Acta Cryst. A34, 453 (1978).
24. G. Van Tendeloo and J. Van Landuyt, J. Spectr. et Microsc. Electr. (1983), in the press.

THE DEFECT STRUCTURES OF Fe_9S_{10}

THAO A. NGUYEN and LINN W. HOBBS
Department of Materials Science and Engineering
Massachusetts Institute of Technology
Cambridge, MA 02139, USA

ABSTRACT

The defect structures of Fe_9S_{10} have been studied by high-resolution transmission electron microscopy. Lattice images of the 3C and 4C superstructures and at least one other phase, which has not been previously reported, were observed. It has been found that the 4C superstructure transforms into the 3C superstructure rather than the MC phase as previously suggested. Intrinsic stacking faults in the sulfur sublattice and two different types of vacancy-ordering antiphase domains were also observed. Evidence from optical diffratograms of areas containing these defects suggests that complex features in the electron diffraction pattern may be artifactual.

INTRODUCTION

Pyrrhotite, $Fe_{1-x}S$, is a series of iron sulfide compounds with composition ranging from FeS to Fe_7S_8. Large deviations from stoichiometry in pyrrhotite are accommodated by iron vacancies which order and form superstructures at low temperatures. Although the phase relations and defect structures of the vacancy-ordered pyrrhotite have been investigated extensively, many results from these studies are conflicting and inconclusive. The disagreements can be attributed to the uncertain thermal history of samples, complex intergrowth of superstructures, and the microscopic inhomogeneities in individual defect structures. The latter is compounded by the spatial limitation of the investigative tools usually employed, which at best average over significant inhomogeneities in the defect structure. High-resolution transmission electron microscopic studies of pyrrhotite have been carried out by Nakazawa and Morimoto [1-3] and by Pierce and Buseck [4,5]. These investigators not only confirmed the ordering of vacancies in the iron sulfide compounds but also observed a number of other defects such as pervasive twins, antiphase domains with an apparent translation vector of $\frac{1}{2}a$ [4,5], and out-of-step domains with a displacement vector of $\frac{c}{4}$ [1]. They disagreed with each other, however, on the structural models of pyrrhotite which give rise to the commensurate and incommensurate superstructures.

In the first part of this paper we present observations of significant vacancy rearrangement and planar defect generation as a function of increasing temperature in crushed Fe_9S_{10} crystals during high-resolution transmission electron microscopy. Next, observations of antiphase domains with translation vectors $\frac{1}{4}(a + b)$ and $\frac{c}{8}$, and stacking faults in the sulfur sublattice are discussed. We show evidence from optical diffractograms that these planar defects give rise to complicated diffraction patterns.

MATERIALS AND PROCEDURES

The samples used in this study were $Fe_{0.9}S$ synthetic crystals provided by Dr. R. A. McKee of the Oak Ridge National Laboratory. The crystals were grown by a modified Bridgman Technique [6]. The composition of the crystal is nominally Fe_9S_{10}; however, powder x-ray measurements of the d_{102} spacing (Arnold and Reichen, 1962) [7] indicated a small compositional variation. Therefore the local composition of the studied samples is not known precisely.

The $Fe_{1-x}S$ crystals were crushed and deposited on copper grids coated with a very thin layer of lacy carbon film. Attempts to prepare disks by ion thinning were unsuccessful because of unacceptable heating during ion bombardment. In this study, a side-entry JEM200CX transmission electron microscope with spherical aberration coefficient C_s = 2.9mm was used. At 200kV, the Scherzer defocus is -105.4nm. A 200μm condenser aperture was used to ensure that the beam divergence was less than 0.45m rad. The 40μm objective aperture used admits the smallest lattice spacing of 0.285nm. The electron optical parameters were calibrated using methods described by Spence [8]. Lattice images of $Fe_{1-x}S$ were obtained along the [100] or [110] directions. The iron-iron separation distance is too small to be resolved under the microscope operating conditions. Hence, the dark contrast corresponds only to a high density of iron atoms; the light contrast corresponds, however, to individual projected iron-site columns containing a significant fraction of iron vacancies, since the separation of such vacancy-rich columns is well within the experimental resolving power.

SUPERSTRUCTURES OF $Fe_{1-x}S$

We review only the 3C and 4C superstructures which are relevant to this study. Other proposed superstructures are discussed in the accounts of Power and Fine [9] and Morimoto [10]. The 3C and 4C superstructures have been associated with many different compositions and temperature ranges [1-3, 10-13]. In this paper, we refer to these structures only as models for the ordered vacancies, i.e., no specific composition is associated with these superstructures.

The $Fe_{1-x}S$ compounds have a parent NiAs structure in which the anions form a close-packed hexagonal sublattice and the cations occupy all the octahedral interstices in the close-packed array (Fig. 1). A special feature of the NiAs structure is that $[FeS_6]$ octahedra share faces along the c-axis and edges perpendicular to the a-axis. As indicated above, the nonstoichiometry in iron sulfide compounds is accommodated by iron vacancies which order at low temperature.

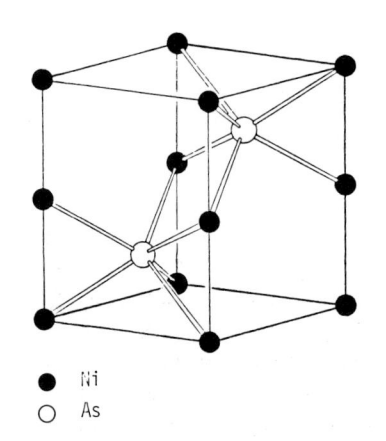

Fig. 1. The NiAs structure showing linkage of ions in the unit cell.

● Ni
○ As

In the 3C and 4C superstructures, the vacancies order in alternate iron layers normal to the c-axis (Fig. 2). The 4C superstructure has the unit cell dimension of a = 2A, b = 2√3 A, c = 4C, where A and C are the dimensions of the parent NiAs unit cell. The projections of the 4C superstructure along the [010] and [110] directions are shown in Fig. 3. In the [010] projection, the vacancy rows form a zig-zag pattern along the c-axis, and in the [110] projection a zig-zig-

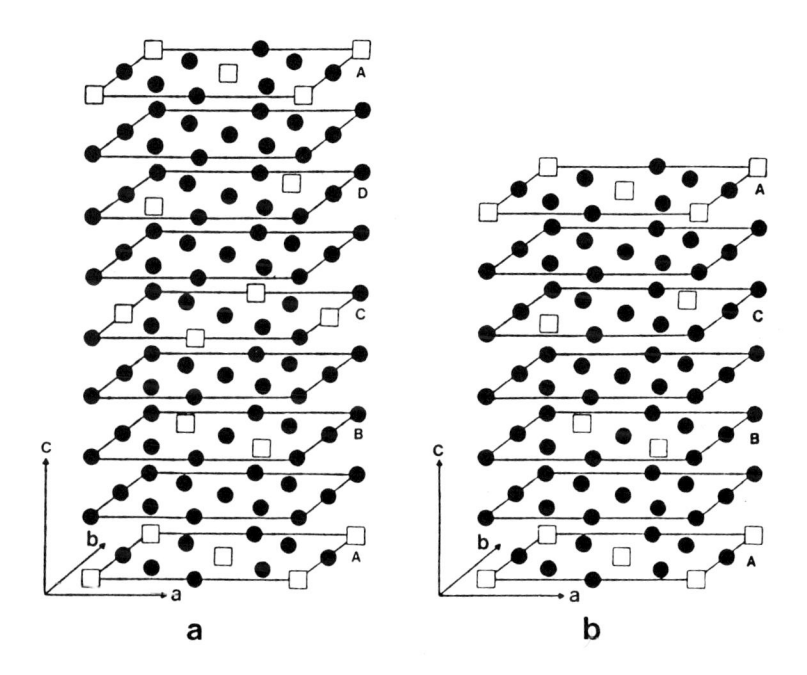

Fig. 2. The iron vacancies in (a) 4C-type superstructure and (b) 3C-type superstructure. Only iron atoms are shown here. Squares are the iron vacancy sites.

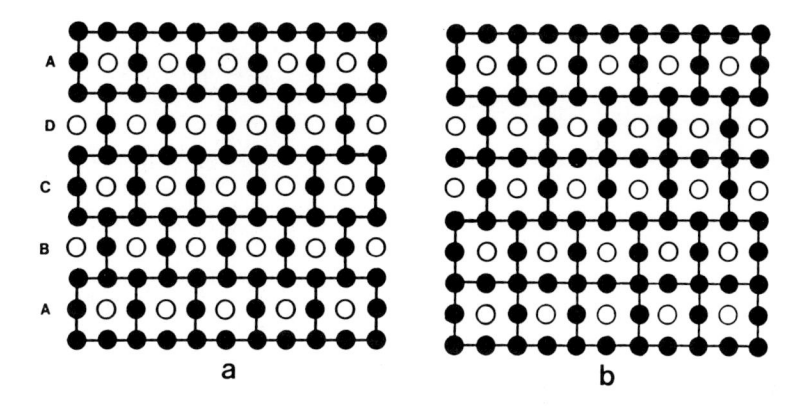

Fig. 3. Projected structure of the 4C type along the (a) [010] and (b) [110] directions. The open circles correspond to highly-vacant iron columns.

zag-zag pattern. The unit cell dimensions of the 3C superstructure are a = b = 2A, c = 3C. In both the [100] and [110] projections, the vacancy rows arrange in a zig-zig-zag pattern (Fig.4). If all structural vacancy sites were indeed unoccupied, the composition of the 4C and 3C superstructures would be Fe_7S_8.

EXPERIMENTAL RESULTS

1. Phase Transformation of Fe_9S_{10}

During high resolution transmission electron microscopy, electron beam heating increases the temperature of crushed samples significantly. This temperature rise in Fe_9S_{10} crystals induces a phase transformation. Lattice images of these phase transitions were obtained in both through-focus series and in series of micrographs with a given fixed focus taken approximately two minutes apart. Figures 5a,b show dynamic phase transformations where considerable vacancy rearrangement, planar defect generation and transformation between phases have taken place. Different arrangements of the highly-vacant iron columns (henceforth referred to as 'vacancy rows') can be seen across areas A-B (Fig. 6), C-D (Fig. 7) and E (Fig. 8). In area A (Fig. 6), the vacancy rows arrange themselves into a zig-zag pattern along the c-axis direction, corresponding to the [010] projection of the 4C superstructure. Vacancy rows in area B form a zig-zig-zag pattern similar to vacancy arrangement in the 3C superstructure. Growth of this 3C-type of vacancy arrangement at the expense of the 4C superstructure can be clearly seen in Fig. 9. Hence, we have direct evidence that the 4C superstructure transforms into the 3C superstructure rather than the MC phase [3]. It is significant to note that the vacancies rearrange themselves row-by-row in the basal plane, and this occurs in at least two or more successive planes. This is clearly shown in Figures 6 and 8. The lattice spacing along the c-axis of areas A, B, C and E is somewhat shorter than 0.57nm, the oft-reported value in the literature.

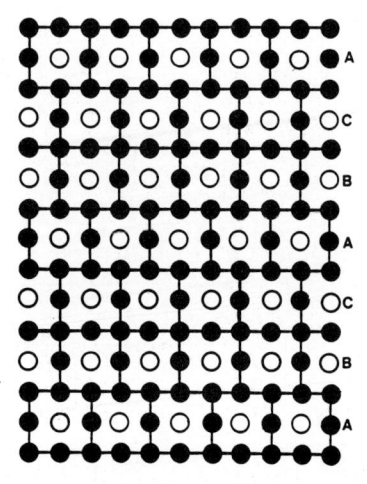

Fig. 4. Projected structure of the 3C superlattice along the [101] direction.

Figures 5a,b also show that the boundary between areas C and D advances toward area C. The vacancy arrangements in these areas are very complex (Fig. 7) and still being investigated. The lattice parameters in area D are, however, significantly different than those in areas A, B, C and E. Lattice images similar to that in area D were again observed in a sample which was annealed at 573 K (well beyond the vacancy disordering temperature) and quenched. This suggests that the structure in area D is at least metastable. Finally, the electron beam heating also generated another type of defect (Fig. 10) whose nature has not been identified.

2. Antiphase Boundaries

A prominent feature in Figure 5a is the antiphase boundaries running parallel to the c-axis. The adjacent antiphase domains are formed by a $\frac{1}{4}(a + b)$ translation of the iron atoms and vacancies of the upper domain with respect to the lower (Fig. 11).

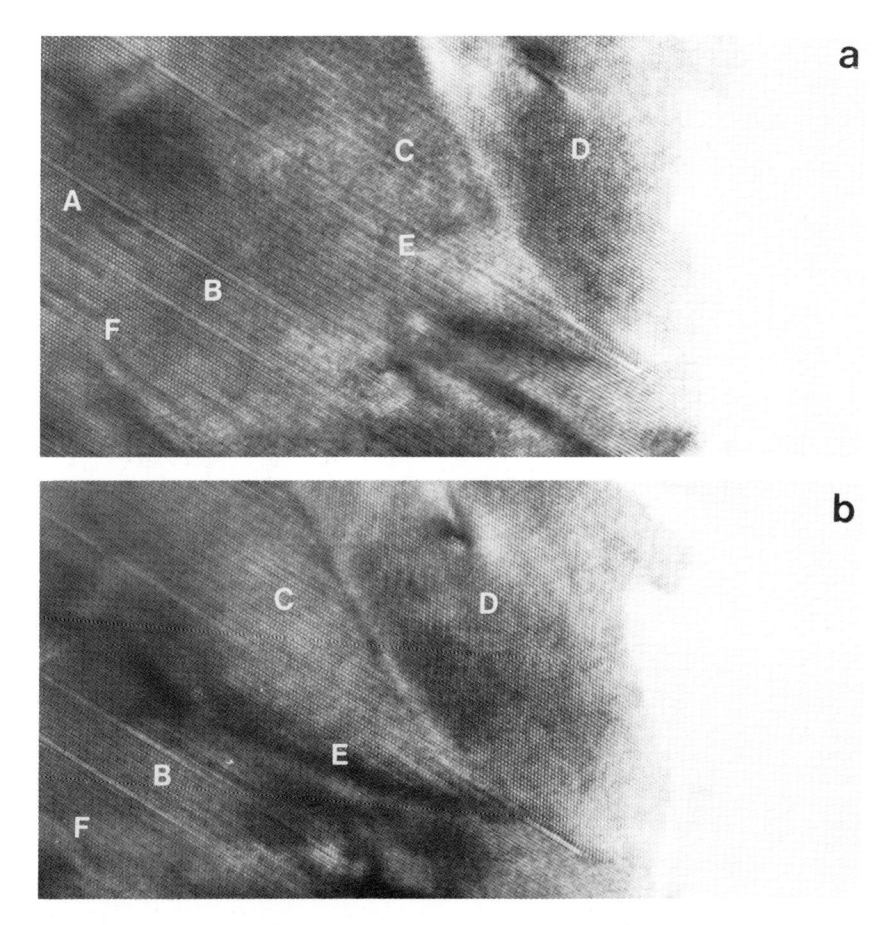

Fig. 5. Lattice images of vacancy rearrangement (areas A and B), and transformation between phases in areas C and D. The phase transition is induced by electron beam heating.

At the antiphase boundary, the vacancy rows stack on top of each other rather than in the normal zig-zag sequence. These boundaries migrate along the c-axis, forming steps indicated by arrows in Figure 11. The presence of these antiphase domains also gives rise to splitting of the superstructure reflections. The genuine nature of this splitting, shown in the inserted optical diffractogram of area F (Fig. 11), is confirmed by optical diffraction from antiphase domain models (Fig. 12).

A second type of antiphase domain was also observed in pyrrhotite. Figure 13 shows domains which are displaced from each other by a vector $\frac{c}{8}$. Here, filled iron planes (normal to the c-axis) in one domain are joined by highly-vacated iron planes of the adjacent domain. The zig-zig-zag-zag vacancy pattern in this lattice image corresponds to the [110] projection of the 4C superstructure.

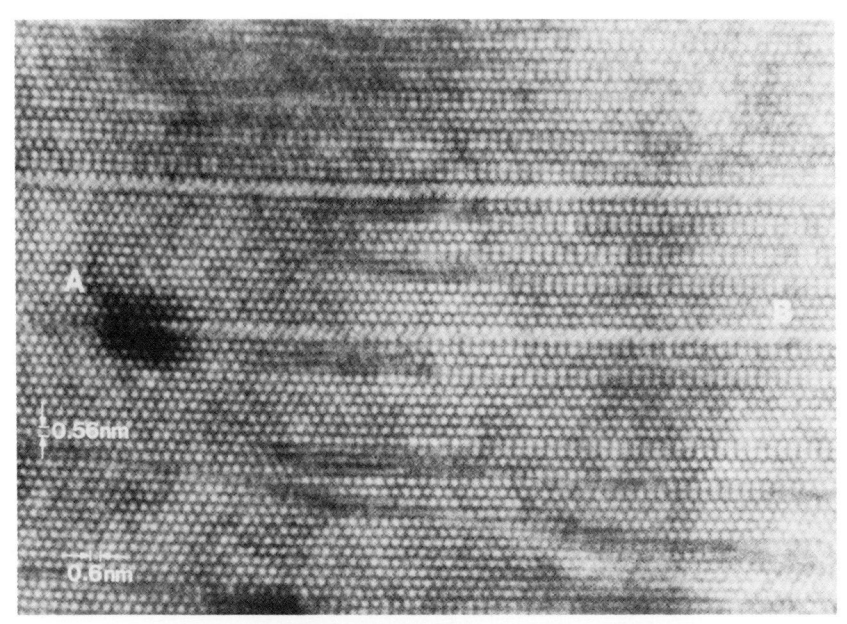

Fig. 6 Lattice image of a dynamic phase transition between the 4C-superstructure (area A) and 3C-superstructure (area B).

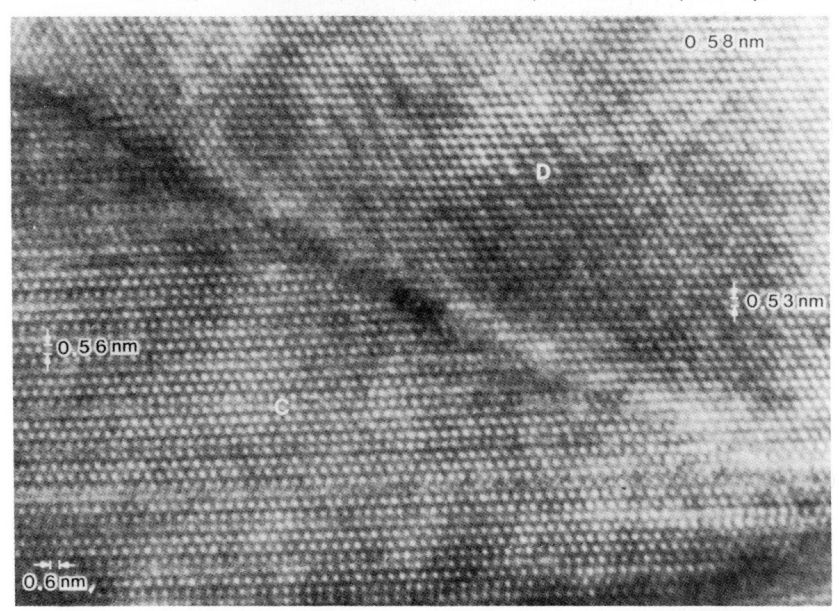

Fig. 7 The structures of areas C and D have not been reported previously. Lattice spacing of area D is significantly different from that in area C.

297

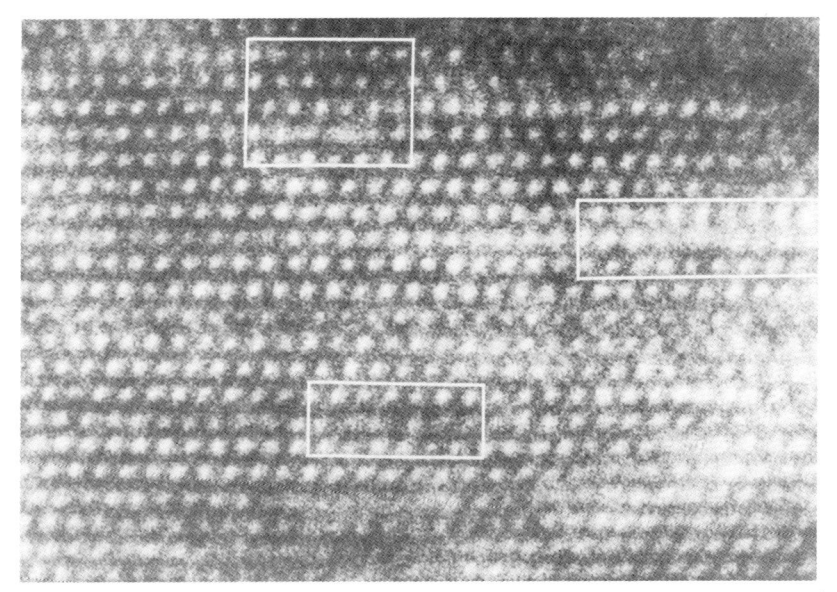

Fig. 8 Vacancies rearrange themselves row-by-row in the basal plane. This rearrangement occurs on at least two or more successive planes.

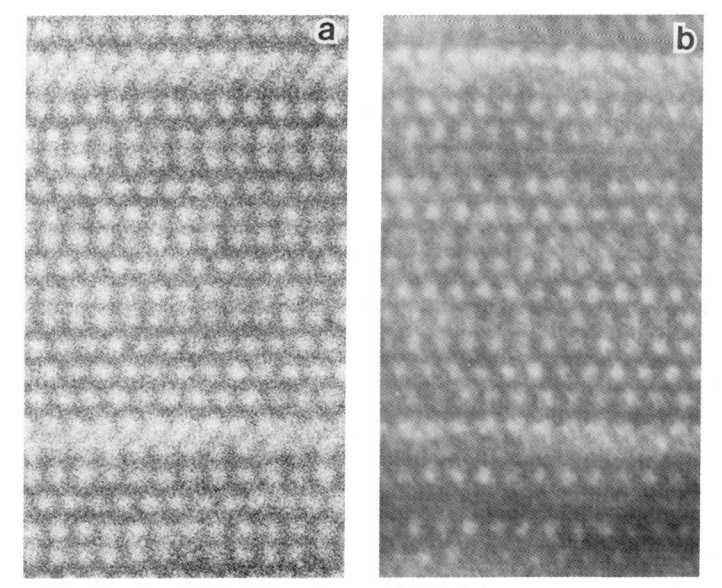

Fig. 9 Growth of the 3C-phase at the expense of the 4C-superstructure. The two micrographs were taken approximately two minutes apart.

3. Stacking Faults

Also evident in Fig. 5a,b are somewhat irregular linear features running along the basal plane. These lines are intrinsic stacking fault interfaces in the sulfur sublattice. The presence of these stacking faults is not at all obvious, since neither the cation nor the anion positions can be resolved from the lattice images. Nevertheless, from careful consideration of the detailed geometrical features of lattice images above and below these interfaces, one can unambiguously determine these defects to be stacking fault interfaces in the sulfur sublattice. The limited resolution of the lattice images does not allow a clear determination of the intrinsic stacking fault as type I (ABABACAC) or type II (ABABCAC), however. A detailed description of geometrical features of type I intrinsic stacking fault follows below. The geometry of type II stacking fault is identical except at the interface, where the atomic arrangements are different.

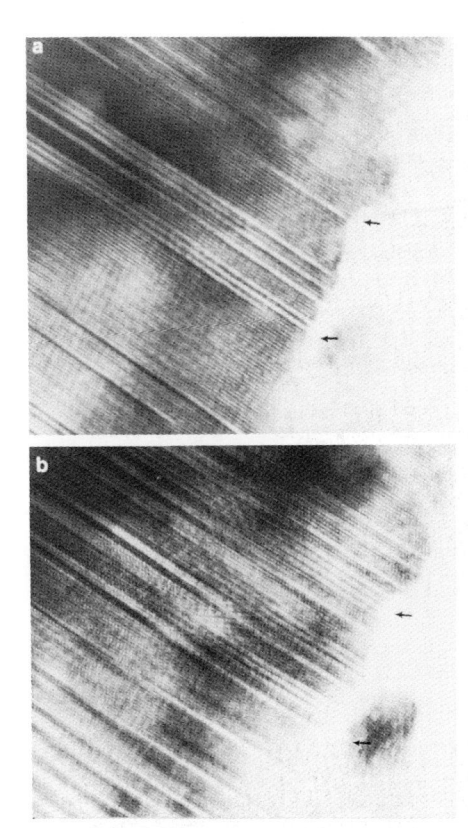

Fig. 10 Defects are being generated by electron beam heating. The nature of these defects has not been identified.

Consider the packing geometry of type I intrinsic stacking fault (ABABACAC) in the [001] direction (see Fig. 14). Across the fault interface, the packing of the sulfur layers changes from an AB to an AC stacking sequence. This also means that the iron and iron vacancy sites in the AC stacking sequence are displaced relative to sites in AB layers. (The positions of octahedral interstices associated with an ABAB stacking sequence are different from those of an ACAC stacking sequence). The displacement can be discerned in the lattice image of a stacking fault. In order to quantify the observation, it is necessary to delineate the geometrical features of three crystallographic sites in an ACAC stacking sequence which can be occupied by a vacancy.

1) Assuming the vacancy occupies site 1 (Fig. 15), the positions of the vacancies between AC layers are related to those between AB layers by a translation vector of $(\sqrt{3} - 1)/2\sqrt{3}\ a$. Along the [110], [1$\bar{1}$0] and [001] directions, the iron and vacancy rows associated with AB layers shift down, up and to the right respectively as they cross the stacking fault interface.

2) Assuming sites 2 are occupied by the vacancies, the translation vector relating the vacancy sites for AB to those for AC sequences is $(1/4\sqrt{3})a + (1/4)\ b$. Across the interface, the iron or vacancy rows of the AB layer shift up in the [110] and [1$\bar{1}$0] directions and to the left in the [001] direction.

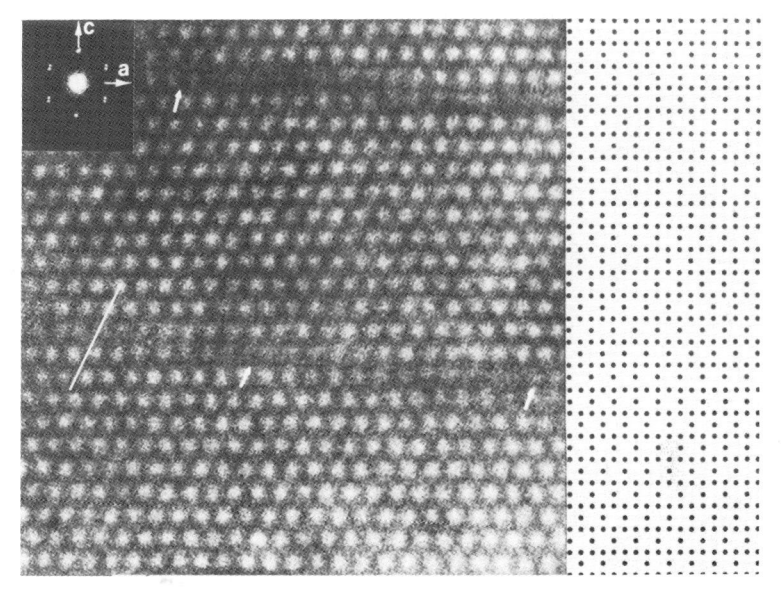

Fig. 11 Lattice image of antiphase domains with an out-of-step vector of $\frac{1}{4}(a + b)$. These boundaries migrate along the c-axis, forming steps indicated by arrows.

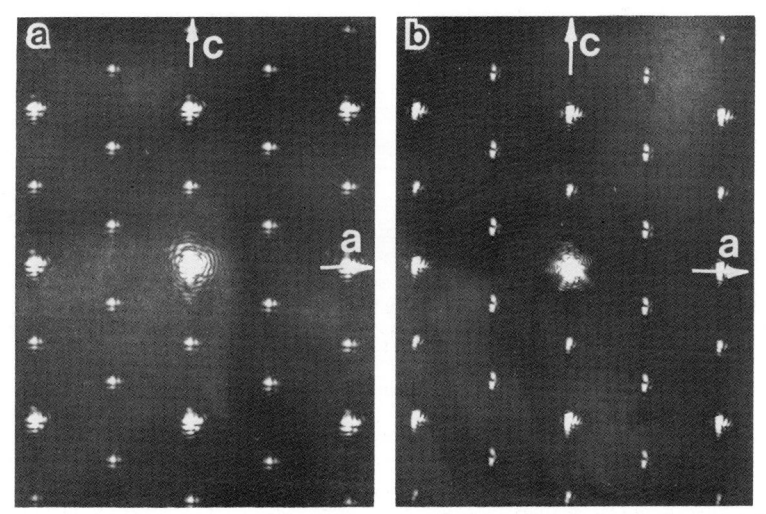

Fig. 12 Optical diffraction patterns of (a) the [100] projection model of the 4C superstructure; (b) the [100] projection model with an antiphase translation of $\frac{1}{4}(a + b)$. The presence of the antiphase boundary gives rise to splitting in the superstructure reflections. The satellite modulations in the main and superstructure reflections are from the shape of the 60 x 60 iron-atom model.

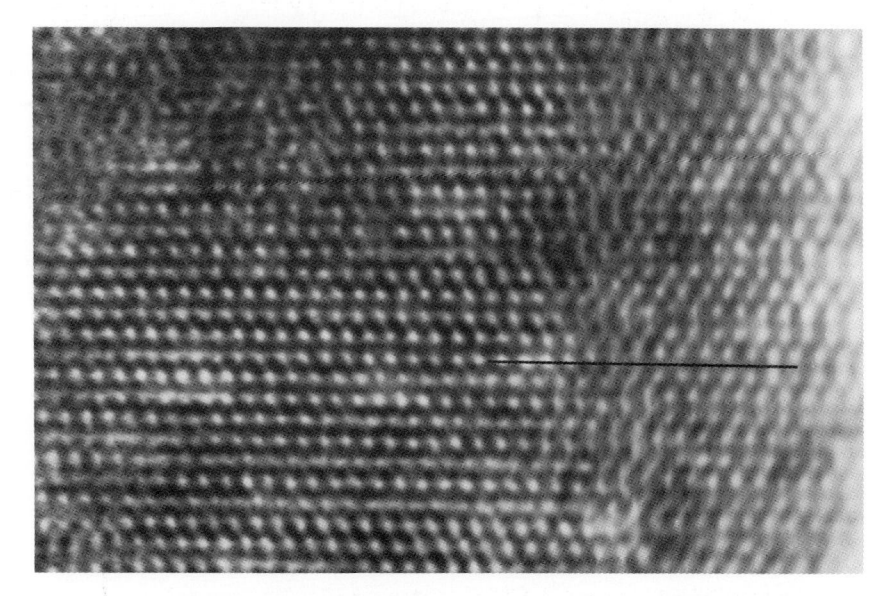

Fig. 13 Lattice image of antiphase domains with a translation vector of $\frac{c}{8}$. The zig-zig-zag-zag stacking sequence corresponds to the [110] projection of the 4C-superstructure.

3) Occupation of sites 3 by the vacancies results in an upward shift of iron and vacancy rows in the [110] and [1$\bar{1}$0] directions as they cross the interface. The cation sites associated with AB layers are displaced from those associated with AC layers by a translation vector $(1/4\sqrt{3})\ a-(1/4)\ b$.

The geometrical features described above enable not only identification of the stacking fault but also determination of the site occupied by the cations and cation vacancies. We now examine the lattice image in Fig. 16. Across the interface the cation and vacancy rows shift down and up respectively in the [110] and [1$\bar{1}$0] directions. These shifts are those expected of site 1, indicating the presence of an intrinsic stacking fault and occupation of site 1 by iron vacancies. A Type II intrinsic stacking fault also gives the same displacement with a translation vector of $-(1/4\sqrt{3})a$. This position is related to site 1 by a two-fold rotation. Therefore these faults cannot be distinguished unless the iron packing geometry at the interface can be resolved.

DISCUSSION

The problem of electron-beam heating of the crushed pyrrhotite samples was found to be much more severe than expected. The extent to which the temperature changes is a strong function of the thermal contact between sample and carbon film. The temperature rise is very fast, making its

○ Vacancy column
● Iron column

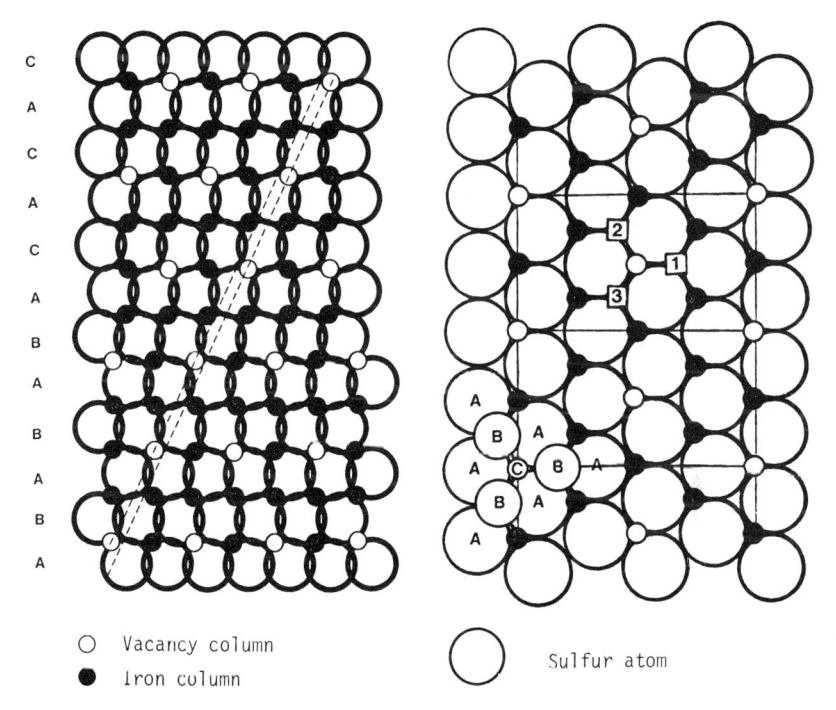

◯ Sulfur atom

Fig. 14 Type I intrinsic stacking fault of $Fe_{1-x}S$ projected in [010] direction. Dotted lines show displacement of iron and vacancy rows across fault interface.

Fig. 15 The [001] projection of the sulfur and highly-vacated iron layers. Across the stacking fault interface the vacancy can occupy one of the three sites indicated in this figure.

Fig. 16 Lattice image of stacking fault $Fe_{1-x}S$. The geometrical features in this image indicate the presence of an intrinsic stacking fault and occupation of site 1 by iron vacancies.

effects extremely difficult to observe dynamically, since for the most part only subtle changes in vacancy rearrangement are involved. Only after careful examination of several through-focus series were these subtle changes detected.

Two significant findings in this study are the presence of the 3C superstructure and the transformation of the 4C superstructure into the 3C phase. Further study using *in situ* high-temperature transmission electron microscopy is being carried out and will be reported separately. A new phase (area D), which has not been previously reported, was also observed. The vacancy arrangement of this phase is under investigation.

Finally, two different vacancy-ordering antiphase boundaries and an intrinsic stacking fault in the sulfur sublattice were observed. The presence of these defects can greatly complicate observed electron diffraction patterns and render phase assignment dangerous. Selected area optical diffractograms suggest a splitting of the superstructure reflections. A detailed report of these effects will appear elsewhere.

ACKNOWLEDGEMENTS

This work has been supported by the National Science Foundation through MIT's Center for Materials Science and Engineering and research grant DMR-8309461. The authors are grateful to Dr. R. A. McKee, Oak Ridge National Laboratory, for supplying the $Fe_{0.9}S$ single crystals.

REFERENCES

1. H. Nakazawa, N. Morimoto and E. Watanabe, Amer. Miner. 60, 359-366 (1975).

2. *idem.*, in Electron Microscopy in Minerology, ed. H. R. Wenk, et al., Springer-Verlag, Berlin (1976).

3. N. Morimoto, Recent Prog. Nat. Sci. Japan 3, 183-206 (1978).

4. L. Pierce and P. Buseck, Science 186, 1209-1217 (1974).

5. *idem.*, in Electron Microscopy in Minerology, ed. H. R. Wenk, et al., Springer-Verlag, Berlin (1976).

6. J. V. Cathcart, R. E. Druschel, L. C. Manley and G. F. Peterson, J. Cryst. Growth 62, 299-308 (1982).

7. R. G. Arnold and L. E. Reichen, Amer. Miner. 47, 105-111 (1962).

8. L. F. Power and H. A. Fine, Minerals Sci. Eng. 8, 106-128 (1976).

9. B. J. Wuensch, Miner. Soc. Am. Special Paper 1, 156-163 (1963).

10. S. A. Kissin and S. D. Scott, Econ. Geol. 77, 1739-1754 (1982).

11. A. Nakano, M. Tokonami and M. Morimoto, Acta Cryst. B35, 722-724 (1979).

12. M. E. Fleet, Acta Cryst. B27, 1864-1867 (1971).

13. J. C. H. Spence, Experimental High Resolution Electron Microscopy, Clarendon Press, Oxford, 1981.

TEM of Dislocations in Sapphire (α-Al$_2$O$_3$)

K.P.D. Lagerlöf, T.E. Mitchell, and A.H. Heuer
Department of Metallurgy and Material Science
Case Western Reserve University
Cleveland, Ohio 44106

Abstract

Dissociation of both basal and prism plane dislocations in sapphire, α-Al$_2$O$_3$, is common and the partial dislocations can be imaged using conventional transmission electron microscopy and weak beam dark field imaging techniques. At elevated temperatures the dissociation takes place by conservative self-climb, a process involving short range diffusion, whereas at low temperatures the dissociation can occur by glide. Dissociation of a dislocation can in some situations give rise to very strong contrast when using g vectors for which $\vec{g}\cdot\vec{b}=0$ for the undissociated dislocation. Those contrast conditions can be used to obtain information about the dislocation morphology and the stacking fault energy of the fault plane through determination of the separation distance.

1 Introduction

The use of conventional transmission electron microscopy (CTEM) has made it possible to relate the macroscopic mechcanical behavior of sapphire, α-Al$_2$O$_3$, to its dislocation substructure. Thus, the details of plastic deformation at elevated temperatures are reasonably well understood, primarily from observations using weak beam dark field (WBDF) techniques.

At elevated temperatures (T>1200°C), two major slip systems operate: (0001)1/3⟨11$\bar{2}$0⟩ basal slip (1,2) and {11$\bar{2}$0}⟨1$\bar{1}$00⟩ prism plane slip (3), basal slip being the easy slip system. {$\bar{1}$012}⟨10$\bar{1}$1⟩ pyramidal slip has also been observed in carefully oriented crystals (4) and in polycrystalline α-Al$_2$O$_3$ (5). Studies of work hardening and dynamic recovery at elevated temperatures (6-9) show the importance of diffusive processes on the mechanical behavior and the observed substructure, which can be explained in terms of various dislocations reactions. CTEM studies have also shown that dislocation dissociation by conservative climb is prominent at elevated temperatures (10).

Room temperature microhardness indentation studies (11) indicate that both basal and rhombohedral twinning (12,13) are major components of the low temperature plastic behavior. The twinning mechanisms are not well understood, although Hockey (14) suggested that basal twins could form by uniform shear and Scott (15) and Scott and Orr (16) suggested a shuffling mechanism for rhombohedral twinning. Plastic deformation below the conventional ductile-to-brittle transition temperature (carried out under a hydrostatic confining pressure (17)) show that prism plane slip is the easy slip system below 700°C while basal slip is the easy slip system above that temperature (18) (Fig. 1). Those experiments also confirm the major importance of twinning at low temperatures.

304

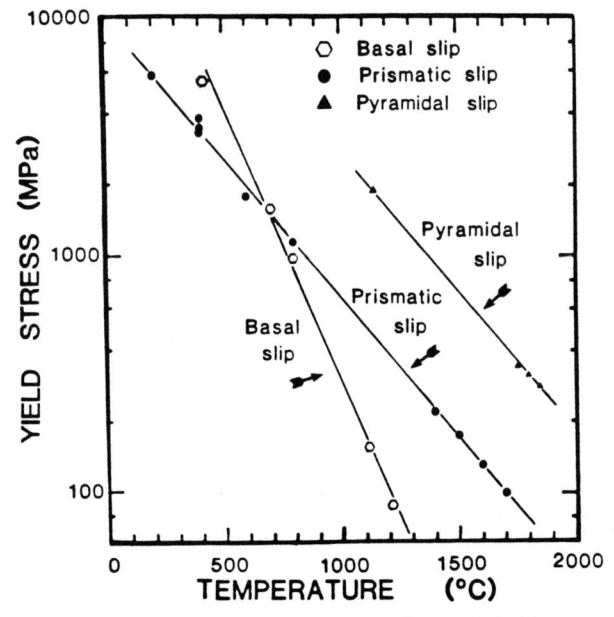

Figure 1: The yield stress for basal slip (1,2,17), prism plane slip (3,17) and pyramidal slip (4,5) versus temperature.

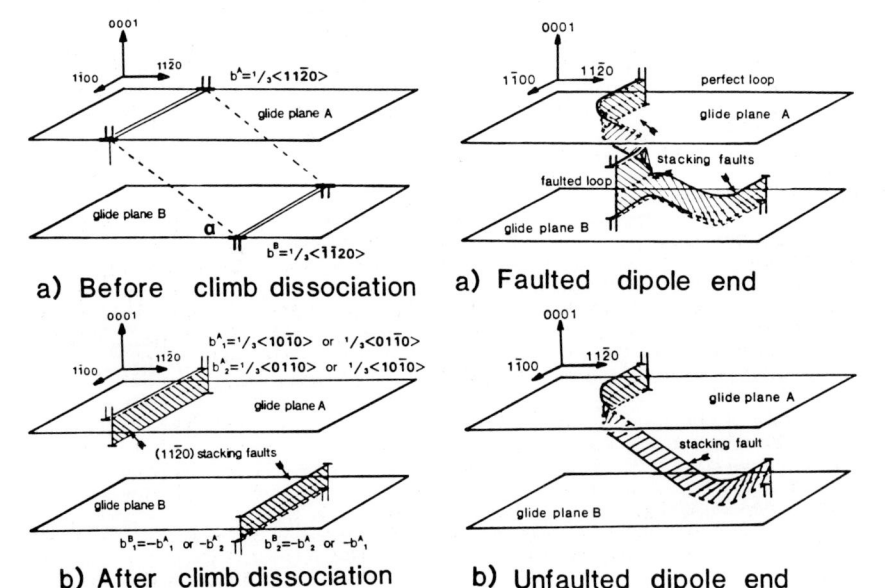

Figure 2: The geometry of an edge trapped dislocation dipole a) before and b) after dissociation by self-climb.

Figure 3: The geometry of a) formation of a faulted dipole from an unfaulted dipole and b) the end of an unfaulted dipole.

This paper will discuss some of the dislocation reactions in sapphire and a few of the microscopy techniques that can be used to image the dislocation substructure. Of particular interest is the WBDF techniques, which gives the resolution necessary to resolve the core of dissociated dislocations and the accompanying stacking faults (19).

2. Dislocation Dissociation and Dislocation Reactions

2.1 Basal Slip

Kronberg (12) suggested that a $1/3\langle 1120\rangle$ basal dislocation could dissociate into half partials according to

$$1/3\langle 11\bar{2}0\rangle \longrightarrow 1/3\langle 10\bar{1}0\rangle + 1/3\langle 01\bar{1}0\rangle \qquad [1]$$

with further dissociation into quarter partials according to

$$1/3\langle 10\bar{1}0\rangle \longrightarrow 1/9\langle 11\bar{2}0\rangle + 1/9\langle 21\bar{1}0\rangle \qquad [2]$$

Dissociation reaction [1] leaves a stacking fault which only involves the cation sublattice, since $1/3\langle 10\bar{1}0\rangle$ corresponds to a perfect translation in the anion sublattice. On the other hand, dissociation reaction [2] leaves a stacking fault involving both sublattices, and would give a fault with a relatively large stacking fault energy. Since the separation distance between two partials is determined by the stacking fault energy, dissociation according to equation [1] is more likely to be observed than the possible dissociation into the quarter partials described by equation [2].

Dissociation of a perfect dislocation can take place by glide, which would give a stacking fault on the glide plane, or by conservative climb, for which the plane of the stacking fault depends on the line direction of the perfect dislocation. Since conservative climb requires diffusion, albeit short range, this process is likely to take place only at elevated temperatures (10), while dissociation by glide can take place without the aid of diffusive processes.

During basal slip of sapphire, the dislocation debris consists predominantly of dislocations with edge character. Further, edge-trapped dislocation dipoles, formed when two edge dislocations of opposite sign gliding on parallel glide planes are trapped in their mutual stress field, are commonly observed after basal slip (6). The individual dislocations in an edge-trapped dipole may then dissociate into half partials, as illustrated in Fig. 2. For the case where the inner partials of a dipole have co-linear Burgers vectors, the formation of a faulted dipole is possible, as shown in Fig. 3a. These faulted dipoles also tend to rotate into perfect edge orientation after formation. In fact, the rather complex substructure observed after basal slip arises from decomposition of edge-trapped dislocation dipoles by cross-slip or by pipe and bulk diffusion (20). Further, edge-trapped dislocation dipoles and faulted dipoles have been observed to emit small prismatic loops by a dipole fluctuation process (20,21); the tip of such a dipole after break-up is illustrated in Fig. 3b. These loops play a significant role in the work-hardening behavior of sapphire undergoing basal slip (7).

2.2 Prism plane slip

A prism plane dislocation can dissociate into three colinear partial dislocations which, as for the case of basal dislocations, are unfaulted with respect to the anion sublattice, and the dislocation reaction can be written as

$$\langle 1\bar{1}00 \rangle \longrightarrow 1/3\langle 1\bar{1}00 \rangle + 1/3\langle 1\bar{1}00 \rangle + 1/3\langle 1\bar{1}00 \rangle \qquad [3]$$

This dissociation reaction can also occur either by glide or by climb. Two types of stacking faults must form, since three partials are involved, and a discussion of the possible geometries for climb dissociation of a perfect edge dislocation can be found in a recent paper by Heuer and Castaing (22).

Prism plane dislocations can also decompose into two basal dislocations, according to the reaction

$$\langle 1\bar{1}00 \rangle \longrightarrow 1/3\langle 2\bar{1}\bar{1}0 \rangle + 1/3\langle 1\bar{2}10 \rangle \qquad [4]$$

which takes place without the formation of any stacking faults. The basal dislocations, however, can dissociate into $1/3\langle 1\bar{1}00 \rangle$ half partial dislocations according to reaction [1].

$1/3\langle 1\bar{1}00 \rangle$ dislocations are also believed to be involved in the formation of basal twins (14). Basal twins can be described crystallographically as a rotation of the lattice by 180° around [0001], which can be obtained by macroscopic shear along $\langle 10\bar{1}0 \rangle$.

2.3 Dislocation Reactions

Dislocations may interact and produce new dislocations and new dislocation arrangements. Two crystallographically equivalent basal dislocations may react and form the third type of basal dislocation (9) according to the reaction

$$1/3\langle 11\bar{2}0 \rangle + 1/3\langle 1\bar{2}10 \rangle \longrightarrow 1/3\langle 2\bar{1}\bar{1}0 \rangle \qquad [5]$$

Similarily, two different prism plane dislocations may react to form the third prism plane dislocation, according to the reaction

$$\langle 1\bar{1}00 \rangle + \langle 01\bar{1}0 \rangle \longrightarrow \langle 10\bar{1}0 \rangle \qquad [6]$$

All other dislocation reactions involve dislocations with different Burgers vectors, for example, the reaction

$$1/3\langle \bar{1}101 \rangle + 1/3\langle 1\bar{2}10 \rangle \longrightarrow 1/3\langle 0\bar{1}11 \rangle \qquad [7]$$

which has been observed after plastic deformation (23) and after precipitation of TiO_2 in Ti-doped sapphire (24).

Prism plane dislocations may also react with basal dislocations according to the reaction

$$\langle 1\bar{1}00 \rangle + 1/3\langle \bar{2}110 \rangle \longrightarrow 1/3\langle 1\bar{2}10 \rangle \qquad [8]$$

which, of course, leads to a very large reduction in energy. The possibilities for prism plane dislocations to decompose into basal dislocations (equation [4]) and react to form basal

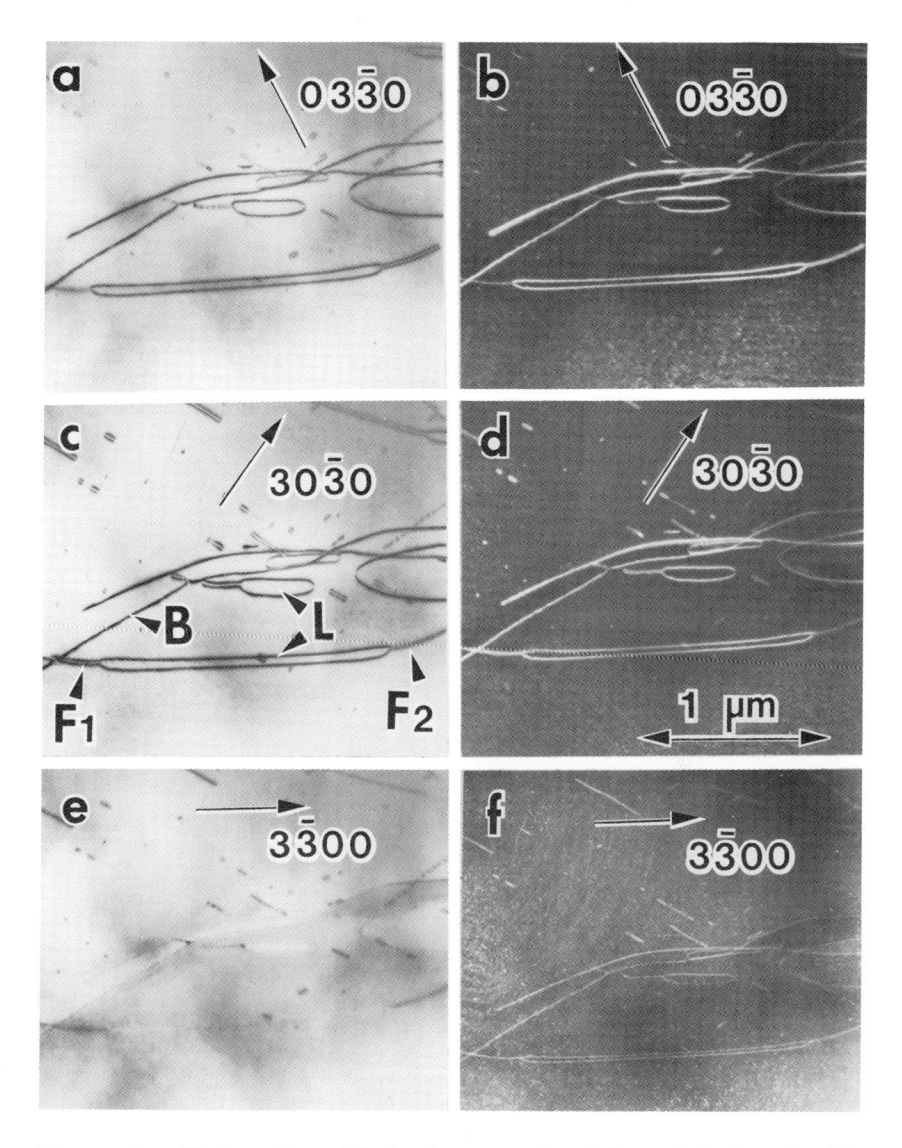

Figure 4: Dislocation debris in a sample deformed by basal slip at 1400°C to 5% strain; a) is a BF and b) is a WBDF image using $\vec{g}=03\bar{3}0$; c) is a BF and d) is a WBDF image using $\vec{g}=30\bar{3}0$; and e) is a BF and f) is a WBDF image using $\vec{g}=3\bar{3}00$. The dislocation marked B is a $1/3[11\bar{2}0]$ basal dislocation, the loops marked L are $1/3[11\bar{2}0]$ dislocation loops and the features marked F1 and F2 are faulted dipoles with $1/3[10\bar{1}0]$ and $1/3[01\bar{1}0]$ fault vectors respectively.

dislocations (equation [8]), account for the large density of basal dislocations present after prism plane slip (9,25).

3 TEM of Dislocations and Stacking Faults

Weak beam dark field (WBDF) provides a useful TEM technique with adequate resolution in order to resolve partial dislocations, and by using g vectors with a short extinction distance both good contrast and good resolution of dislocations can be obtained (19).

The retained dislocation substructure in a [0001] basal foil of sapphire deformed by basal slip at 1400°C is shown in BF and WBDF pairs using {30$\bar{3}$0} g vectors in the basal zone (Fig. 4). The dislocation marked B is a 1/3[11$\bar{2}$0] dislocation, while the dislocations marked L are 1/3[11$\bar{2}$0] elongated loops formed from break-up of dislocation dipoles (20,21). The features marked F 1 and F 2 are 1/3[10$\bar{1}$0] and 1/3[01$\bar{1}$0] faulted dipole segments respectively, formed at the ends of an elongated 1/3[11$\bar{2}$0] loop.

It can be seen in Fig. 4f, where $\vec{g}\cdot\vec{b}=0$ for B and L, that the residual contrast in WBDF is rather strong, although no apparent dissociation of the 1/3[11$\bar{2}$0] basal dislocations is detectable. This is due to the fact that the 1/3[11$\bar{2}$0] dislocations and loops are dissociated by climb rather than by glide. Thus, since the dislocations are dissociated and give strong residual contrast in WBDF, a $\vec{g}\cdot\vec{b}$ analysis is best carried out in BF, where the residual contrast is less apparent (Fig. 4e). The strong residual contrast in WBDF, however, can be used for tilting experiments around elongated loops and edge-trapped dipoles, since the g vector parallel to the line direction, for which $\vec{g}\cdot\vec{b}=0$, can be used as a rotation axis. Fig. 5 shows such a high angle tilting experiment of an elongated loop in residual contrast. The geometry of the loop can easily be observed, and it can be seen that the ends of the loop are more bulbous and have a different inclination angle than the rest of the loop (10,20).

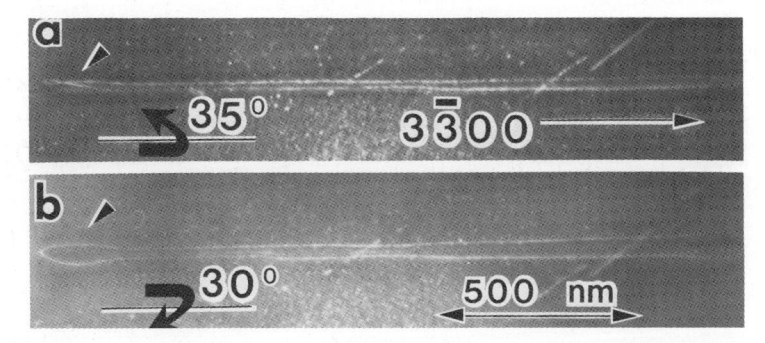

Figure 5: An elongated dislocation loop viewed in residual contrast, illustrating a high angle rotation experiment. The micrographs show the loop rotated away from the basal plane toward the [11$\bar{2}$0] zone by a) 35° and b) -30° respectively, where the rotation axis is parallel with the line direction. Note that the ends of the loop are bulbous and have a different inclination angle compared to the mid-section.

The climb-dissociated partials of a 1/3[11$\bar{2}$0] dislocation can be observed in a [11$\bar{2}$0] prismatic foil, in which the electron beam is normal to the fault plane. Fig. 6 shows a pair of WBDF images using ±g=3$\bar{3}$00. If $\vec{g}\cdot\vec{b}$=0 for the perfect dislocation, the residual contrast would be very weak if the dislocation was undissociated. However, since $\vec{g}\cdot\vec{b}$=±1 for the respective partials, a dissociated dislocation gives rise to inside/outside contrast. Overlap of the intensities for inside contrast results in one strong line, while two distinct lines are observed for outside contrast. Fig. 7 shows an enlargement of dislocations 1 and 2 of Fig. 6, using BF and WBDF, along with the computer simulated BF images of the two dislocations using the Head-Humble simulation method (26). Comparison between the micrographs and the simulations gives an accurate determination of the separation distance between the partials, and thus the stacking fault energy of the fault plane (27). It can also be seen in Fig. 6 that the small prismatic loops, L, are dissociated.

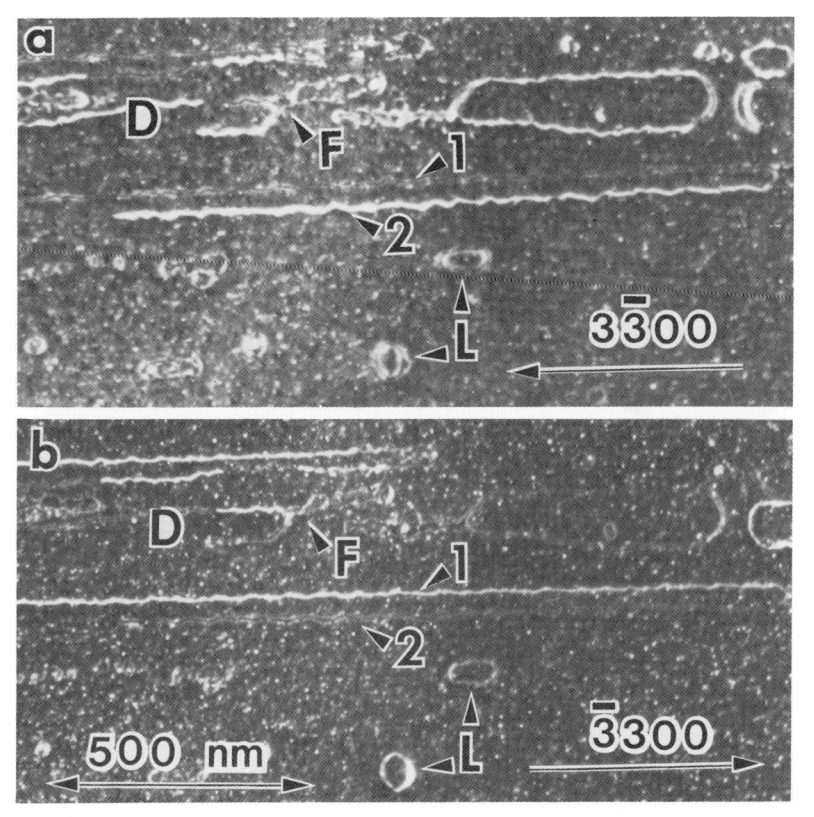

Figure 6: Two WBDF images of the same area in a [11$\bar{2}$0] foil using a) $\vec{g}$=$\bar{3}$300 (g/2.8g) and b) $\vec{g}$=3$\bar{3}$00 (g/2.7g). The dislocations in a dipole marked 1 and 2 and the loops marked L show inside/outside contrast due to dissociation by climb. Note the geometry of the faulted dipole segment F being formed from the dipole D.

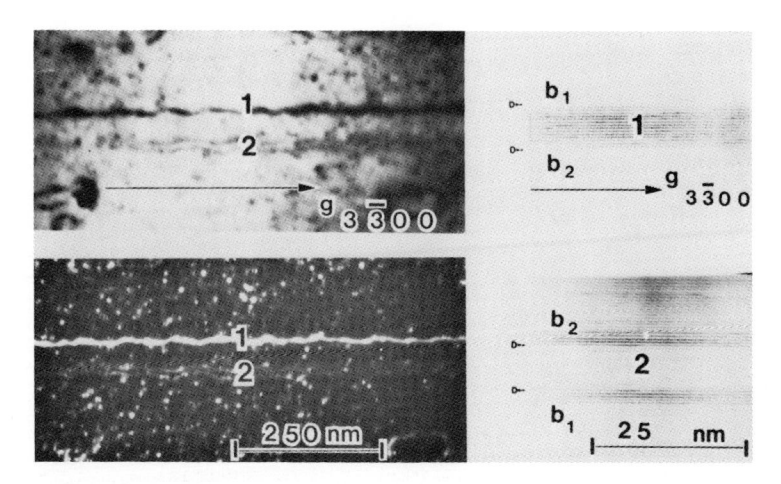

Figure 7: Two micrographs of the same area where a) is a BF image and b) is a WBDF image of a dislocation dipole (dislocations 1 and 2) using g=3$\bar{3}$00. Dislocation 1 shows inside contrast while dislocation 2 shows outside contrast due to dissociation by climb; c) and d) show the computer simulated image of dislocation 1 and 2 respectively.

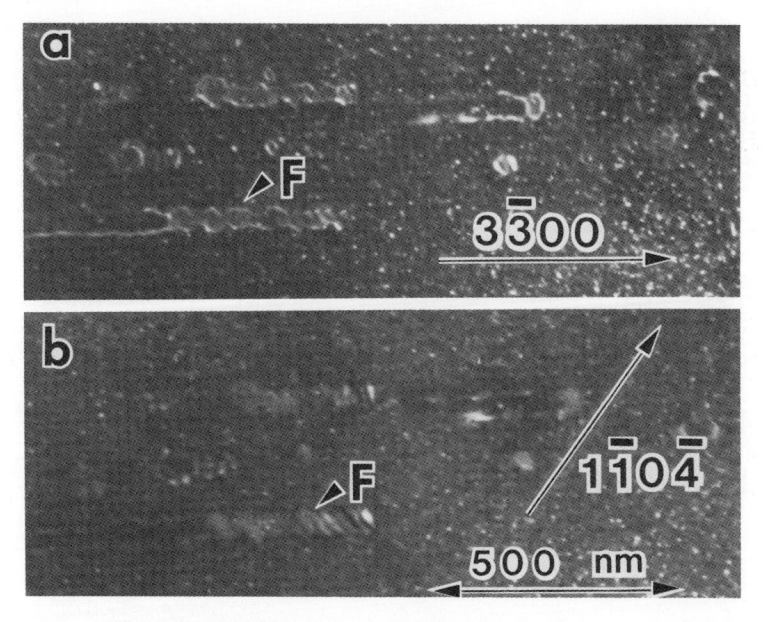

Figure 8: Two WBDF images of the same area in a [11$\bar{2}$0] foil using a) $\vec{g}$=3$\bar{3}$00 (g/2.7g) and b) $\vec{g}$=1$\bar{1}$0$\bar{4}$ (g/3.2g). The dislocation segment marked F is a 1/3[10$\bar{1}$0] faulted dipole. Stacking fault fringes can be observed in b) as $\vec{g}\cdot\vec{R}$=1/3 while no fringes can be observed in a) as $\vec{g}\cdot\vec{R}$=1.

The formation of faulted dipoles from an elongated 1/3[11$\bar{2}$0] loop is also shown in Fig. 6, marked F (compare with the illustration in Fig. 3a). By using a $\vec{g}$ vector for which $|\vec{g}\cdot\vec{b}|<1$, the stacking fault will show stacking fault fringes. Fig. 8 shows two WBDF images of the same area using $\vec{g}=3\bar{3}00$ and $\vec{g}=1\bar{1}0\bar{4}$, for which $\vec{g}\cdot\vec{R}=1$ and 1/3, respectively. The stacking fault fringes are apparent in the 1/3[10$\bar{1}$0] faulted dipoles, but are difficult to resolve in the climb-dissociated 1/3[11$\bar{2}$0] dislocations, since the separation distance between the partials is smaller than for the faulted dipole. The faulted dipoles are also found to break up into strings of small faulted loops by a dipole fluctuation mechanism (20,21), which explains the geometry of the dipoles in Fig. 8.

Measurements indicate that the stacking fault energy is surprisingly isotropic in sapphire (27), and assuming isotropic stacking fault energy dissociation by climb is energetically favorable compared to dissociation by glide (22,28). Dissociation by conservative climb, however, requires diffusion, and at low temperatures, where diffusion is very slow (29), one would expect that dissociation by glide would be dominant. Fig. 9 shows a WBDF micrograph of a sample deformed by basal slip at 800°C under hydrostatic confining pressure. The dislocation marked B, a 1/3[11$\bar{2}$0] basal dislocation lying in the basal plane, shows distinct double contrast which may be due to glide dissociation. The faulted dipoles and loops, F, are believed to form by a jog-dragging process, where a superjog has been formed along the trailing partial (30).

Similar observations have been made in samples deformed at 400°C, as seen in Fig. 10. The basal dislocations in the 400°C sample are not as crystallographic as in the samples deformed at 800°C, which could be a result of dislocation interactions between basal and prism plane dislocations. It can be seen in Fig. 1 that the critical resolved shear stress for prism plane slip is a factor of two lower than that of basal slip at 400°C, so prism plane slip is expected to be activated, together with basal slip, in a sample oriented for basal slip. Also, it can be seen that a faulted dipole segment, F, is formed from one of the partials of the 1/3[11$\bar{2}$0] basal dislocation, again believed to be due to a jog-dragging mechanism.

Another area of a sample deformed at 400°C is shown in Fig. 11. Two types of rhombohedral twins are present: one of the twins has ⟨1$\bar{1}$00⟩ prism plane dislocations at the twin-matrix interface. It can be seen that the ⟨1$\bar{1}$00⟩ prism plane dislocations are dissociated into three colinear partials, according to dissociation reaction [3]. It is not clear whether the prism plane dislocations are formed at the twin boundary due to twin growth, or if they are piled up against the boundary as a resut of slip. In addition, tendencies to form a dislocation network can be seen at the twin-matrix interface.

The microstructure of sapphire deformed at low temperatures is very complex, since it contains a large density of both basal and rhombohedral twins (30). Further, since the mechanics of twin formation and growth are not well understood, the interpretation of the observed dislocation substructure is difficult.

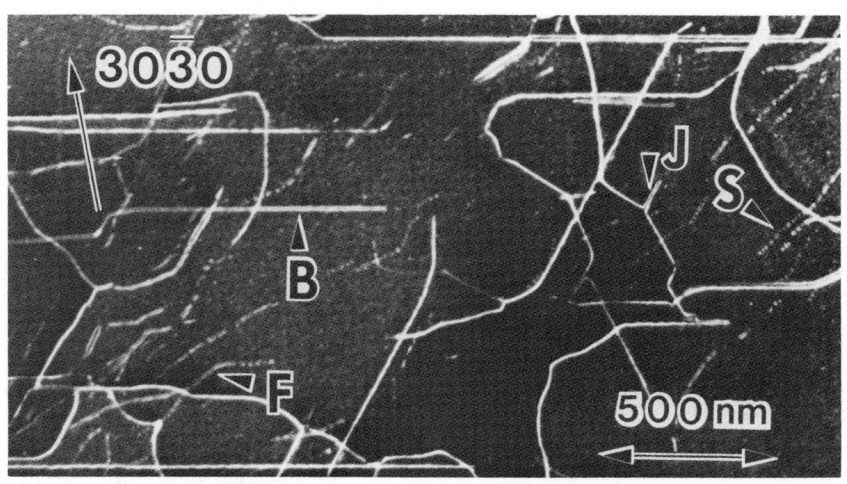

Figure 9: A WBDF micrograph using $\bar{g}=30\bar{3}0$ (g/2.4g) of the dislocation substructure in a basal foil of a sample oriented for basal slip and deformed at 800°C under hydrostatic confining pressure. The dislocations are very crystallographic and the dislocation marked B, a 60° 1/3[11$\bar{2}$0] dislocation, show a distinct double contrast, which is believed to be due to dissociation by glide. Note the pinned dipole segments J, the faulted dipoles F and the strings of faulted loops S.

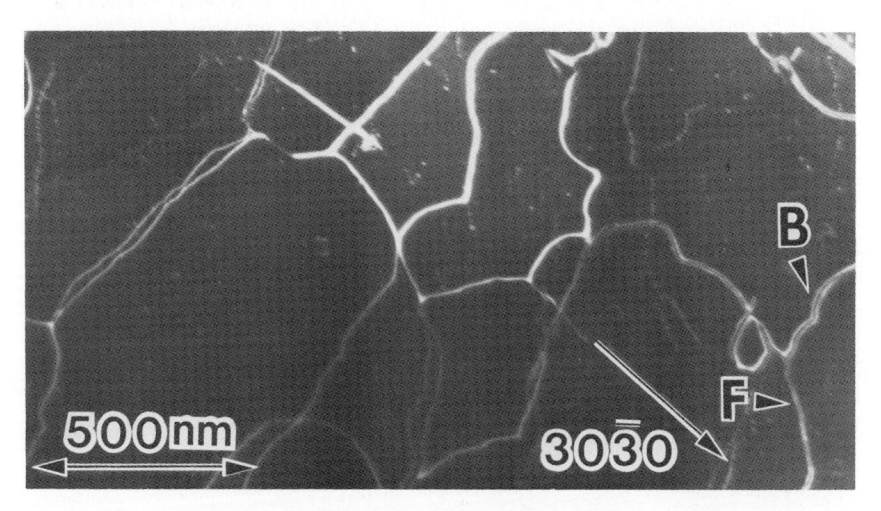

Figure 10: A WBDF micrograph of the dislocation substructure in a basal foil of a sample deformed at 400°C, using $g=30\bar{3}0$ (g/2.4g). The dislocation B is a 1/3[11$\bar{2}$0] dislocation showing distinct double contrast, believed to be a result of dissociation in the glide plane. Note the formation of the faulted dipole segment F from one of the partial dislocations in B.

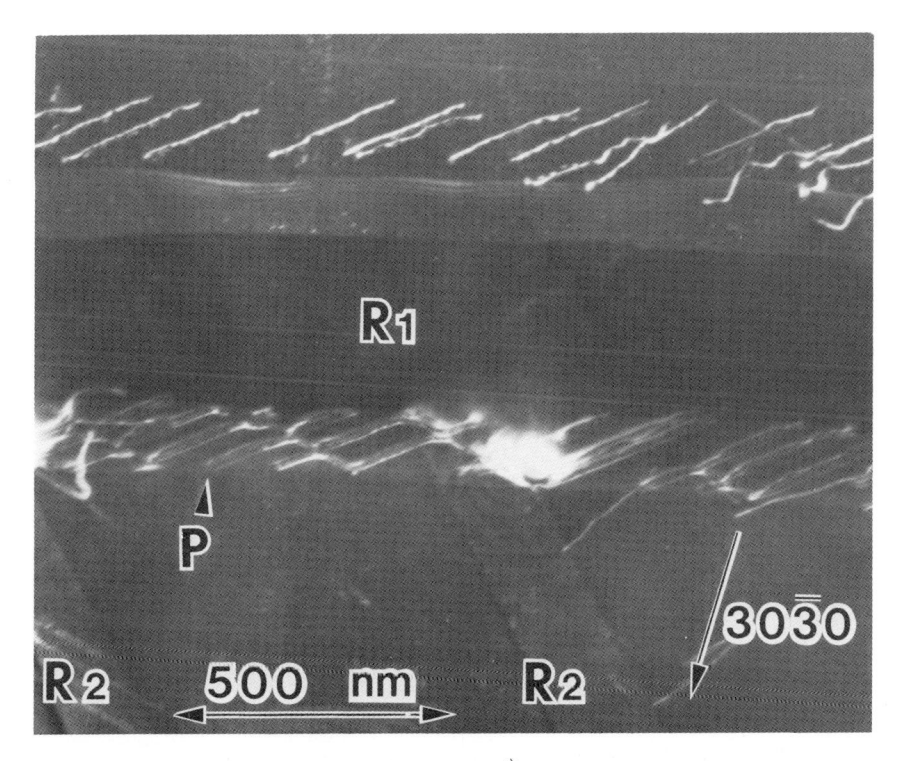

Figure 11: A WBDF micrograph using $\vec{g}=30\bar{3}0$ (g/2.3g) of the deformation substructure of a sample deformed at 400°C. R1 and R2 indicates two types of rhombohedral twins formed during the deformation and P show several prism plane dislocation close to the twin- matrix interface. The prism plane dislocations show distinct triple contrast due to dissociation, and it can be seen that some of the dislocations have started to form a dislocation network at the twin-matrix interface.

4 Discussion

The use of CTEM and particularly WBDF imaging techniques give, in general, adequate resolution to study dislocations in sapphire, in which most dislocations are dissociated. The stacking faults arising from the various dissociation reactions can be imaged using $|\vec{g}\cdot\vec{b}|<1$, a condition producing stacking fault fringes. The stacking fault fringes, however, tend to wash out the contrast from the partial dislocations, and make it difficult to measure the separation distance. This problem can easily be avoided by the use of $\vec{g}\cdot\vec{b}=n$ conditions, for which the stacking fault fringes disappear. Further, in order to get a good estimate of the separation distance between two partial dislocations, it is desirable to view the stacking fault along its normal. For the case of climb-dissociated $1/3[11\bar{2}0]$ basal dislocations viewed in a $[11\bar{2}0]$ foil, inside/outside contrast for the partials using $g=\pm 3\bar{3}00$, together with computer simulation of the images, has proven useful in determining the separation distance. If the stacking fault is inclined in the foil, the inclination angle has to be determined from stereo microscopy in order to obtain the true separation distance. It should be recognized, however, that this procedure can be quite difficult since the separation distances between the partials is small.

Since all the partials of a dissociated $\langle 1\bar{1}00 \rangle$ prism plane dislocation have the same sign, the image shifts are identical for a given g vector, which simplifies the image interpretation. Other dislocation reactions, however, may complicate the interpretation.

Although the presence of $1/9\langle 11\bar{2}0 \rangle$ Kronberg quarter partial dislocations has not yet been verified, the possibility of their existence has to be considered when interpreting the dislocation substructure. This type of partial dislocation may also be present during the formation of basal twins, according to the synchroshear model suggested by Kronberg (12).

The exact nature of the dislocation core of $1/3\langle 11\bar{2}0 \rangle$ basal dislocations, $\langle 1\bar{1}00 \rangle$ prism plane dislocations, and $1/3\langle 10\bar{1}0 \rangle$ half-partial dislocations has yet to be determined. For this purpose, high resolution electron microscopy (HREM) techniques is required (31).

5 Summary

WBDF microscopy of dislocations in α-Al_2O_3 has proven to be a very useful technique since the resolution is adequate to resolve dissociated dislocations. The use of CTEM has provided insight into the mechanical behavior at elevated temperatures, both in terms of dislocation slip, work hardening, and dynamic recovery. The evolution of the dislocation substructure is well understood for both basal slip and prism plane slip, and can be explained by various dislocation reactions.

The mechanical behavior at low temperatures (T<900°C) is more complex, due to the formation of both basal and rhombohedral twins. The twinning mechanics are not well understood at present, which complicates the interpretation of the dislocation substructure. The dislocations may be a result of slip, as at elevated temperatures, but could also be a result of the twinning mechanisms operating, where dislocations may be pushed out into the matrix as the twins grow.

Kronberg $1/3\langle 10\bar{1}0 \rangle$ half partials have been observed in samples deformed by both basal slip and prism plane slip. The existence of $1/9\langle 11\bar{2}0 \rangle$ quarter partials, however, is yet to be verified. The resolution using CTEM is probably not good enough to resolve quarter partials, and other techniques, such as high resolution electron microscopy, may be required (31).

Acknowledgements

The authors would like to acknowledge Dr. J. Castaing, CNRS, Meudon, France, for providing samples deformed by basal slip below 950°C, and for profitable collaboration over a number of years. This research was supported by the National Science Foundation under Grant Number INT76-10829. KPDL has also been the recipient of an Alcoa Foundation fellowship.

References

1. J.B. Wachtman and L.H. Maxwell, J. Am. Cer. Soc. 40, 377 (1957).
2. M.L. Kronberg, J. Am. Cer. Soc. 45, 274 (1962).
3. R. Scheuplein and P. Gibbs, J. Am. Cer. Soc. 43, 458 (1960).
4. R.E. Tressler and D.J. Barber, J. Am. Cer. Soc. 57, 13 (1974).
5. J.D. Snow and A.H. Heuer, J. Am. Cer. Soc. 56, 156 (1973).
6. B.J. Pletka, T.E. Mitchell, and A.H. Heuer, J. Am. Cer. Soc. 57, 388 (1974).
7. B.J. Pletka, A.H. Heuer, and T.E. Mitchell, Acta Met. 25 25 (1977).
8. B.J. Pletka, T.E. Mitchell, and A.H. Heuer, Acta Met. 30, 147 (1982).
9. J. Cadoz, J. Castaing, D.S. Phillips, A.H. Heuer, and T.E. Mitchell, Acta Met. 30, 2205 (1982).
10. D.S. Phillips, T.E. Mitchell, and A.H. Heuer, Phil. Mag. A45, 371 (1982); T.E. Mitchell, B.J. Pletka, D.S. Phillips and A.H. Heuer, Phil. Mag. 34, 441 (1971).
11. B.J. Hockey, J. Am. Cer. Soc. 54, 223 (1971).
12. M.L. Kronberg, Acta Met. 5, 507 (1957).
13. A.H. Heuer, Phil. Mag. 13, 379 (1966).
14. B.J. Hockey, Deformation of Ceramic Materials, R.C. Bradt and R.E. Tressler, eds, p 167, Plennum Publishing Corp., 1975.
15. W.D. Scott, Deformation of Ceramic Materials, R.C. Bradt and R.E. Tressler, eds, p 151, Plennum Publiching Corp., 1975.
16. W.D. Scott and K.K. Orr, J. Am. Cer. Soc. 66, 27 (1983).
17. J. Cadoz, J. Castaing, and J. Philibert, Revue Phys. Appl. 16, 135, (1981).
18. J. Castaing, J. Cadoz, and S.H. Kirby, J. Am. Cer. Soc. 64, 504 (1981).
19. D.J.H. Cockayne, M.L. Jenkins and I.L.F. Ray, Phil. Mag. 24, 1382 (1971).
20. K.P.D. Lagerlöf, T.E. Mitchell, and A.H. Heuer, to be published.
21. D.S. Phillips, B.J. Pletka, A.H. Heuer, and T.E. Mitchell, Acta Met. 30, 491 (1982).
22. A.H. Heuer and J. Castaing, Adv. in Ceramics (in press).
23. J. Cadoz and P. Pellissia, Scripta Met. 10, 597 (1976).
24. D.S. Phillips, A.H. Heuer, and T.E. Mitchell, Phil. Mag. A42, 285 (1980).
25. J. Cadoz, D. Hokim, M. Meyer, and J.P. Rivière, Rev. Phys. Appl. 12, 473 (1977).
26. A.K. Head, P. Humble, L.M. Clarebrough, A.J. Morton, and C.T. Forword, Computed Electron Micrographs and Defect Identification, North Holland, Amsterdam, 1973.
27. K.P.D. Lagerlöf, T.E. Mitchell, A.H. Heuer, J.P. Rivière, J. Cadoz, J. Castaing, and D.S. Phillips, Acta Met. 32, 97 (1984).
28. T.E. Mitchell, W.T. Donlon, K.P.D. Lagerlöf, and A.H. Heuer, Plastic Deformation in Ceramic Materials, (in press).
29. K.P.D. Lagerlöf, B.J. Pletka, T.E. Mitchell and A.H. Heuer Rad. Eff. 74, 87 (1983).
30. K.P.D. Lagerlöf, T.E. Mitchell, A.H. Heuer, and J. Castaing, to be published.
31. J.M. Cowley and S. Iijima, Physics Today 30, 32 (1977).

TEM OBSERVATIONS OF GRAIN BOUNDARIES IN CERAMICS

M. RÜHLE
Max-Planck-Institut für Metallforschung, Institut für Werkstoffwissenschaften, Seestraße 92, 7000 Stuttgart 1, West-Germany.

ABSTRACT

Grain boundaries in ZrO_2-containing alumina, ZrO_2, and in $Ni_{1-y}O$ (containing different impurities) were investigated by TEM including analytical studies. Quite frequently, interphases exist at the boundaries. The interphases are composed mainly by impurities present in the materials. The "width" of boundaries in $Ni_{1-y}O$ depends on intrinsic defects as well as on impurities; the dependency will be discussed. The observations are related to properties of the different ceramics.

INTRODUCTION

Grain boundaries and regions between grains (interphase interfaces) have an important, sometimes controlling influence on many properties of the materials. The properties are often controlled by the structure of the grain boundaries. However, it is expected that the structure of grain boundaries in ceramics differs essentially from that of metals (Kingery [1]). Recently, theoretical and experimental studies were performed for the determination of the structures of "clean" grain boundaries. Duffy and Tasker [2] showed by molecular static computer simulations that symmetrical tilt boundaries in MgO and NiO (with stoichiometric composition) possess a rather "open" structure, large gaps in the boundary plane are calculated. The configurations are much less dense than those of the corresponding grain boundary in metals. Structures and energies of <001> twist boundaries are calculated by Wolf [3] and Duffy and Tasker [4]. Wolf contained a rather high boundary energy which continuously increased with increasing twist angle. The energies of the twist boundary result in values which are close to the doubled free surface energy, which implies that the twist boundaries should be extremely unstable. Tasker and Duffy [4] found that for a Σ5 twist boundary a very stable structure can be reached by the introduction of Schottky defects into the plane of the boundary.

All the calculations are performed for "clean" grain boundaries where neither intrinsic defects nor impurities are present. However, further computations [5] showed that intrinsic defects as well as impurities can segregate at the grain boundary (negative binding energy). Those defects may also influence the structure of the boundary. However, the modifications of the structures are not yet calculated.

If the density of segregated impurities is (essentially) larger than one atomic layer, then an interphase interface or an amorphous (vitreous) film is formed on the grain boundary which should invalidate all geometrical descriptions of the "clean" boundary, e.g. the CSL and the DSC lattice [6].

The experimental observations on the structure of grain boundaries in ceramics were reviewed recently [7]. In this paper, observations on zirconia-containing ceramics, zirconia, and on $Ni_{1-y}O$ are described.

EXPERIMENTAL DETAILS

Fine-grained tetragonal zirconia polycrystals were prepared [8] from very fine powders (grain sizes < 0.1 μm and mean diameter of powder particles ~ 1 μm) which contained 2.2 mol% Y_2O_3. The solute prevents the tetragonal (t) to monoclinic (m) transformation of ZrO_2. The zirconia ceramics were sintered at 1400°C and slowly cooled to room temperature. The thermal expansion of $t-ZrO_2$ is anisotropic which results in local strains and stresses at RT.

The Al_2O_3/15 vol.% ZrO_2 ceramic was prepared by mixing fine-grained powders (of pure Al_2O_3 and pure ZrO_2) and by a subsequent sintering of the dispersion ceramic at 1450°C [9]. The fine-grained Al_2O_3 ceramic contained ZrO_2 crystallites which were intercrystalline particles with respect to the alumina matrix (Heuer et al. [10]).

$Ni_{1-y}O$ single crystals as well as polycrystalline specimens of $Ni_{1-y}O$ were grown in a commercial mirror furnace which had the shape of two intersecting ellipsoidal shells. Xenon high pressure lamps were mounted in one focal point of the shells. The specimen was located in the other focus (common to both intersecting ellipsoidal shells. The specimen could be melted by the energy emitted from the xenon high pressure lamps. Crystals with diameters up to 10 mm could be grown by the floating zone technique. Large growth velocities result in polycrystalline $Ni_{1-y}O$. Pure $Ni_{1-y}O$ crystals as well as Cr_2O_3 doped materials were produced.

All specimens were prepared for the TEM studies in the conventional way. It was essential to polish the specimens to very low foil thicknesses prior to ion thinning. The TEM studies were performed in a Siemens 102 electron microscope or in a JEOL 200 CX electron microscope. An EDS system was linked to the TEM. The chemical composition for elements with Z>12 could be determined.

EXPERIMENTAL RESULTS AND INTERPRETATION

Zirconia-containing alumina.- All grain boundaries were covered with an amorphous layer in which Si and Mg could be detected. The amorphous layer became clearly visible by taking a dark field image from an area of the diffraction pattern where (i) no reflexions of the crystals were positioned and (ii) scattering from amorphous regions was expected [11]. Fig. 1 shows a bright field image and such a dark field image. The amorphous layer can easily be detected even for inclined boundaries. The vitreous film possesses a thickness of > 3 nm.

Tetragonal zirconia polycrystals.- Similar observations were obtained for the tetragonal zirconia polycrystals (TZP) (Fig. 2) where the amorphous grain boundary phase contains yttria, silica and alumina in the composition of a low melting eutectic. The edges of the grains were rounded by milling. It must be assumed that the ceramics sintered by a liquid phase sintering process.

The amorphous layer of TZP materials which was sintered of ultrafine, not milled powders was much thinner. In the edges of the grains large local strains developed and the grain boundary broke quite frequently. Frequently, microcracks could be observed at the grain boundaries (Fig. 3).

Results on $Ni_{1-y}O$.- Small angle grain boundaries in $Ni_{1-y}O$ were studied systematically and an experimental method was developed which allowed the determination of the change in the mean inner potential at the dislocations of the boundary. For the analysis the specimen must be tilt in an edge on orientation and strong diffraction contrast has to be avoided. It is then observed that the residual contrast increases strongly by imaging in a defocusing mode, c.f., the image plane does not coincide with the lower foil surface. Examples of the defocusing series are shown in Fig. 4. A careful experimental testing proved that no artifacts caused the defocusing contrast (Rühle and Sass [12]). Fig. 4 demonstrates that on going from the overfocussed condition (image plane between lower foil surface and objective lens) to underfocussed condition, the image appearance changes from black spots surrounded by white rings to white spots surrounded by black rings. For a certain defocusing distance the contrast of the dislocation disappears nearly for a limited region of the grain boundary corresponding to a certain thickness of the foil. For the interpretation of the observation a model is assumed where the mean inner potential V along a tube (radius r_o) parallel to the dislocation line is changed compared to the mean inner potential V_o of the perfect crystal, with $V = V_o + \Delta V_o$. A theory was developed [12] which includes the influence of defocusing on the contrast formation and also the dependence of the contrast on changes in mean inner potential, widths of the disturbed regions and other imaging conditions [12]. The change in mean inner potential, ΔV_o, and the width of the circled disturbed region, $2r_o$, can be determined and it was found that the values

$$\Delta V_o = - (0.09...0.15) V_o$$
$$\text{and} \quad r_o = (0.32...0.25) \text{ nm} \quad (1)$$

result in calculated contrast profiles which agree quite well with the experimental observations.

Calculations by Puls [13] for stoichiometric NiO predict that the dilatated region of the dislocation core results in a change of $\Delta V_o \sim -0.05 V_o$ with $r_o = 0.25$ nm. The residual change in ΔV_o (eq.(1)) can be caused by (i) a space charge at the boundary [1], (ii) change in composition of $Ni_{1-y}O$ at the grain boundary or (iii) by impurities which segregate at the boundary.

The influence of the segregation of Cr_2O_3 (at the grain boundary) on the change in mean inner potential was studied. Specimens which contained nominally 0.4 and 0.8 wt.% Cr_2O_3, respectively, were investigated and it was observed that the defocusing contrast became wider and stronger with an increasing content of impurities. The content of Cr at the boundary was determined by EDS methods. With the analytical attachment the ratio $X = N^{Cr}/N^{Ni}$ could be determined, where N^{Cr} (N^{Ni}) means the number of counts under the Cr (Ni) K_α peak of the EDS spectrum. The results (Fig. 5) of the analytical TEM studies show clearly that Cr segregated at the boundary. A simple geometrical configuration (Fig. 6) allows the determination of the thickness δ of the segregated layer:

$$\delta = \frac{\pi}{4} d \cdot \frac{X}{1 + X}.$$

For the simple evaluation it is assumed that no Cr exists outside the segregated layer. Then δ is determined to $\delta \approx 2.5$ Å for 0.8% Cr_2O_3 and probe diameters of 30 nm and 15 nm, respectively.

At large angle grain boundaries in $Ni_{1-y}O$ the contrast changes in a similar way as for small angle grain boundaries (Fig. 7). The contrast of the edge-on boundary was bright (dark) for an underfocussed (overfocussed) imaging. No impurities could be detected at the boundary. The widths and the intensities of the defocusing contrasts increased with increasing Cr content.

DISCUSSION

The TEM studies in the zirconia-containing Al_2O_3 ceramics and in ZrO_2 show that an amorphous film is present on all grain boundaries. This amorphous film should invalidate both the CSL and the DSC lattice [6] construction. However, it is interesting to know that all strong and tough ceramics contain an amorphous grain boundary film. It may well be that the amorphous film (with the proper thickness) lowers the energies of the general grain boundary and increases thereby the strength and toughness of the material.

The observations in nickel oxide show that the structure of the grain boundaries and of the grain boundary dislocations are strongly dependent on the presence of intrinsic defects, space charges and on impurities. It could be verified that Cr impurities segregate at grain boundaries and produce an area with a strongly reduced mean inner potential at the region of the boundary. The reduction in mean inner potential is equivalent to a reduction in density, that means, that the volume of the grain boundary increases if Cr segregates. The formation of a layer of about 0.25 nm could be observed which is equivalent to a monoatomic layer. Under this condition a formation of an interphase starts already. Further work, including diffraction studies and high resolution studies, will hopefully give more information on the atomic structure of grain boundaries and also the dependence of the structure on the concentration of segregated impurities.

REFERENCES

1. W.D. Kingery, J. Amer. Ceram. Soc. 57 (1974) 1; 74.
2. D.M. Duffy and P.W. Tasker, Phil. Mag. A47 (1983) 817; A48 (1983) 155.
3. D. Wolf, J. de Physique 43 (1982) C6-45.
4. P.W. Tasker and D.M. Duffy, Phil. Mag. A47 (1983) L45.
5. D.M. Duffy and P.W. Tasker, Adv. in Ceramics, Vol. 12 (1984) in the press.
6. D.A. Smith and R.C. Pond, Int. Met. Rev. 205 (1976) 61.
7. M. Rühle, J. de Physique 43 (1982) C6-115.
8. H. Schubert, N. Claussen and M. Rühle, Adv. in Ceramics, Vol. 11 (1984) in the press.
9. N. Claussen, J. Amer. Ceram. Soc. 61 (1978) 85.
10. A.H. Heuer, N. Claussen, W.M. Kriven and M. Rühle, J. Amer. Ceram. Soc. 65 (1982) 642.
11. D. Clarke, Ultramicroscopy 4 (1979) 33.
12. M. Rühle and S.L. Sass, Proc. 10th Int. Congress Electron Microscopy 2 (1982) 99; Phil. Mag., to be published.
13. M.P. Puls, unpublished research. See also: M.P. Puls and M.J. Norgett, J. Appl. Phys. 47 (1976) 466.

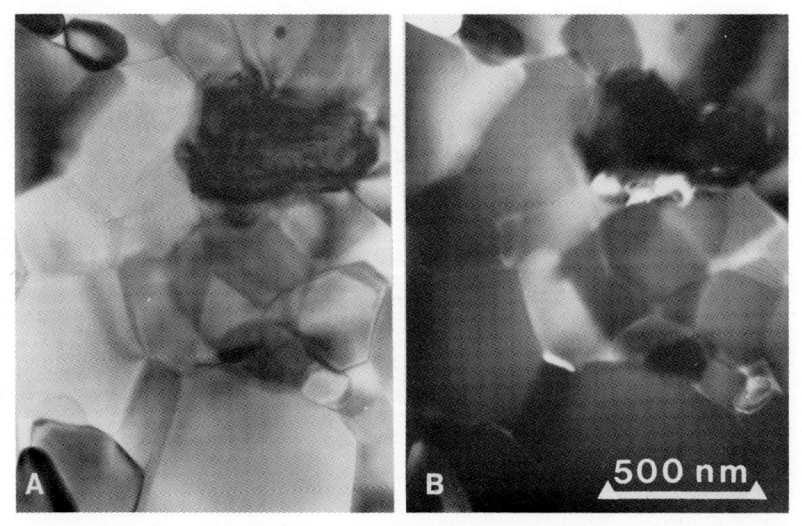

FIG. 1.: Al_2O_3/15 vol.% ZrO_2 dispersion ceramic. ZrO_2 grains are distributed intercrystalline in the fine-grained Al_2O_3 matrix. (A) Bright field image; (B) dark field image taken with a fraction of the diffuse scattering [11]. The bright lines in (B) are caused by the amorphous grain boundary film.

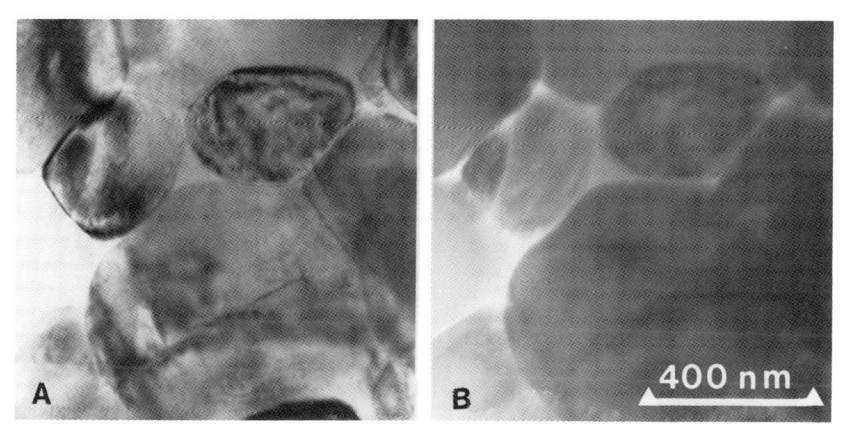

FIG. 2.: Fine-grained tetragonal zirconia polycrystal. The ZrO_2 powder was milled before sintering. The edges of the grains are rounded. (A) Bright field image; (B) dark field image (see Fig. 1 and [11]). The bright areas correspond to the amorphous grain boundary film.

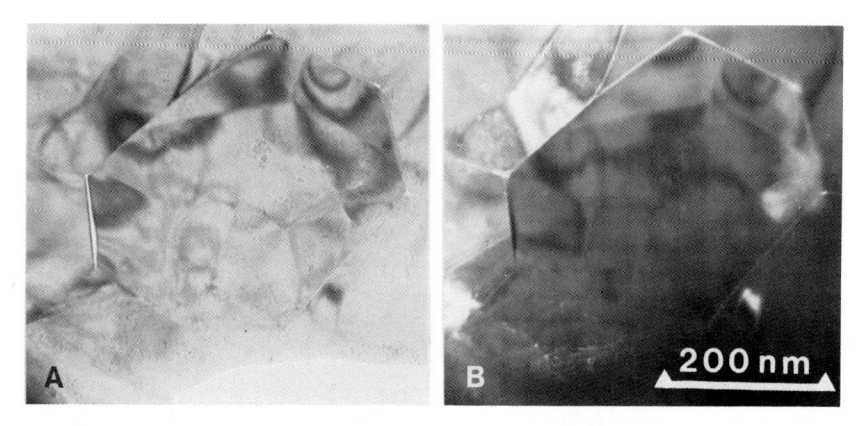

FIG. 3.: Fine-grained tetragonal zirconia polycrystals. (A) Bright field image; (B) dark field image. The bright lines correspond to a thin amorphous grain boundary film. The dark line represents a microcrack.

322

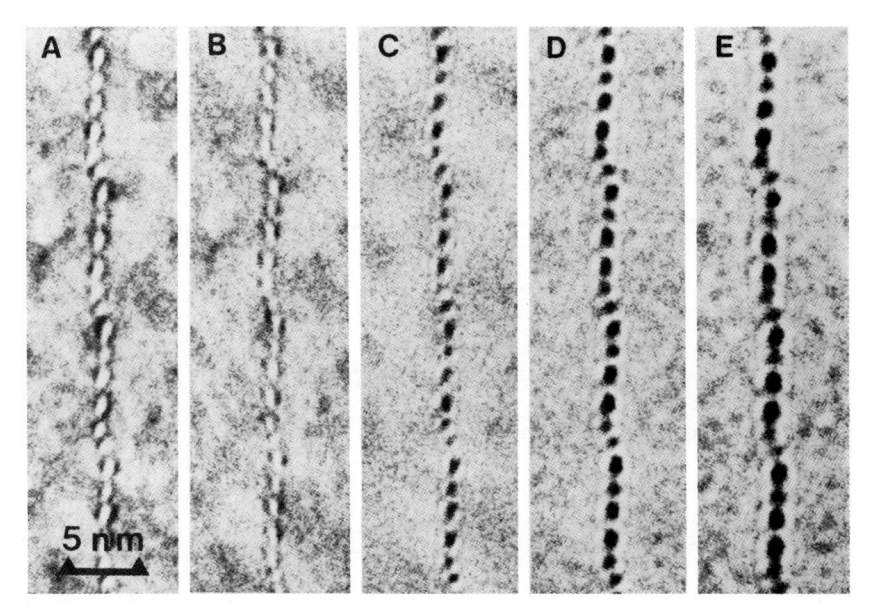

FIG. 4.: Grain boundary in nickel oxide. Edge on configuration. Dependence of contrasts on defocusing distance ζ. (Only a small section of the boundary is shown). (A) ζ=-198nm;; (B) ζ=-110nm; (C) ζ=±0nm; (D) ζ=+110nm; ζ=+198nm.

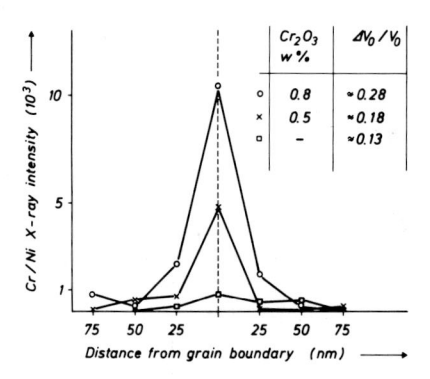

FIG. 5.: EDS studies of Cr segregation at grain boundaries in $Ni_{1-y}O$. The ratio of N^{Cr}/N^{Ni} is plotted as a function of distance from the boundary (for 3 different grain boundaries). N^{Cr} (N^{Ni}): counts under K_α peak of chromium (nickel).

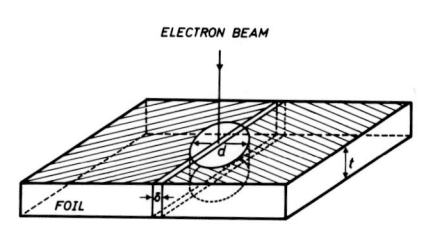

FIG. 6.: Geometry from the determination of thickness δ. d: probe diameter; t: foil thickness.

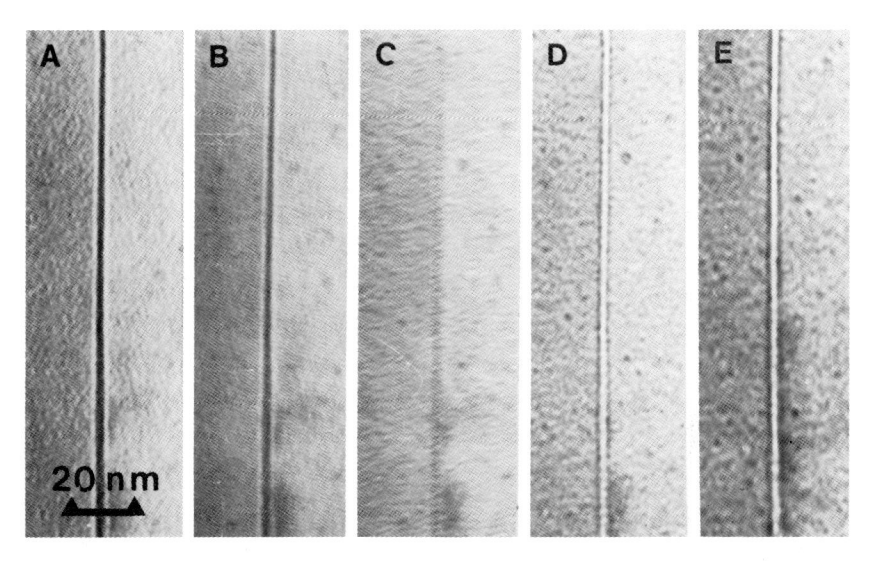

FIG. 7.: Large angle grain boundary in $Ni_{1-y}O$. Edge on configuration. Dependence of contrast on defocusing distance ζ. (A) $\zeta=-240nm$; (B) $\zeta=-120nm$; (C) $\zeta=\pm 0nm$; (D) $\zeta=+120nm$; (E) $\zeta=+240nm$.

FAULT STRUCTURES IN CVD SILICON NITRIDE

T.M. SHAW,[*] J.W. STEEDS[†] AND D.R. CLARKE[§]
[*]Structural Ceramics Group, Rockwell International Science Center, Thousand Oaks, California 91360
[†]Physics Dept., University of Bristol, England
[§]Materials Science Dept., MIT, Cambridge, Mass.

ABSTRACT

Fault structures occurring in crystalline CVD silicon nitride deposits have been examined using transmission electron microscopy. A variety of mixed rotation and displacement faults are observed. The faults have extensive basal plane facets but also extend onto other planes within the grains. Possible structures for the faults are discussed with reference fo the structures of the α and β silicon nitride polymorphs.

INTRODUCTION

Chemical vapor deposition (CVD) provides a method for preparing layers of high purity silicon nitride up to several millimeters thick. The deposits have attracted attention both as a passivating layer for electronic devices[1] and as potential structural materials.[2-4] Several methods, based on the reactions of silicon halides with ammonia on a hot substrate, have been described for depositing silicon nitride.[1-4] The type of silicon nitride obtained depends on the gas flow rates, substrate temperatures and reaction times used for deposition. With a low temperature substrate an amorphous deposit is obtained. Higher temperatures lead to a crystalline deposit that is usually highly textured. Crystalline deposits in most cases consist entirely of the α form of silicon nitride, although, recent reports[5,6] indicate that deposits of mixed α and β crystals can be produced by "seeding" with a third gas stream.

The morphology of α Si_3N_4 deposits have been described in detail and related to growth conditions employed for deposition.[4] Microscopic characterization of the internal structure of deposits, however, has been limited.[7] The objective of the present work was to determine the nature of crystal defects present in as-grown crystalline deposits. Where possible the defects are related to the crystallography of the α Si_3N_4 structure and to growth mechanisms for the crystalline deposits.

MATERIALS

Materials prepared by chemical vapor deposition techniques at two different laboratories were investigated.[*] The precise conditions under which the materials were prepared are proprietary, but the samples made at UTRC were deposited onto a graphite substrate using the gaseous reaction.

$$3SiF_4 + 4NH_3 \rightarrow Si_3N_4 + 12HF$$

in the temperature range 1100 to 1550°C.

---

[*]See Acknowledgements.

FIG. 1. General view of the fault structures that occur within single α-Si₃N₄ grains in CVD silicon nitride materials.

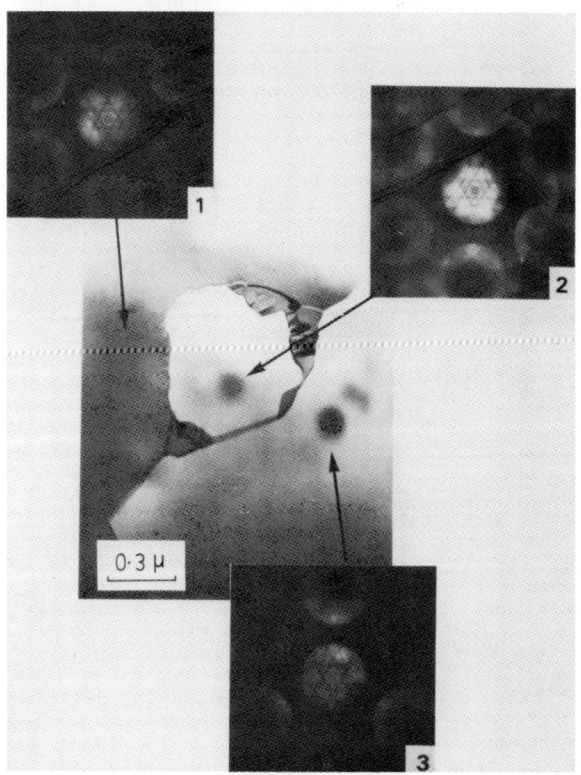

FIG. 2. The effect of tilting on the contrast in a single grain α-Si₃N₄ grain. The three-fold symmetry of convergent electron beam diffraction patterns recorded from different regions of the grain indicate a rotation of 60° between regions 1 and 2, but not between 2 and 3.

Both materials had the "cone"-like microstructure previously reported by investigators[4] which consists of large parallel columnar grains. The grain size was 50 - 100 μm. X-ray diffraction indicated that both materials were entirely alpha silicon nitride and no evidence of additional second crystalline phases was found. Standard techniques were used to prepare thin foils for T.E.M. observations.

OBSERVATIONS

The microstructure of the two materials studied were similar, both containing numerous faults. A transmission electron micrograph of a typical area is shown in Fig. 1. The single grain in the micrograph is divided into separate "domains" by a number of curved faults or boundaries. Diffraction and tilting experiments indicate that several different types of faults occur within the grains.

When examined in an [0001] orientation, the grains showed uniform contrast across all the faults (see Fig. 1 for example), where as strong contrast differences arose across certain faults when reflections of the type UVW (W ≠ 0) were excited (Fig. 2). Convergent beam diffraction patterns from the three largest domains in Fig. 1 are shown in Fig. 2 with the grain in an [0001] orientation. The patterns indicate that there is a 60° rotation of the trigonal α Si_3N_4 structure associated with faults that exhibit contrast differences. At least two types of faults therefore occur within the grains, i.e., rotation or twin boundaries and displacement faults. The bright field contrast produced by tilting provides a simple method for distinguishing the two.

Several faults visable in Fig. 2 are out of contrast when the foil is in an exact [0001] orientation (Fig. 1). Disappearance of the faults indicates that there is no displacement parallel to the basal plane associated with them. Contrast differences seen in the tilted image in Fig. 2, however, indicate the presence of a 60° rotation. Faults of this type are therefore simple rotation boundaries. (It is possible that an [0001] displacement component is associated with such boundaries, but this could not be determined by contrast experiments.) Rotation (R) boundaries of this sort were frequently faceted on the {1̄100} planes as can be seen from the lattice image in Fig. 3. The three sets of {1̄100} fringes visible in the image are perfectly aligned across the rotation boundary confirming that any displacement associated with the boundary is perpendicular to the basal plane of the α Si_3N_4 structure.

Figure 4 shows a lattice fringe image of another fault identified as a rotation boundary by tilting. The crystals on either side of the boundary are again in an [0001] orientation, but unlike the boundary in Fig. 3, the 10̄10 fringes, although parallel, are not aligned across the boundary. The misalignment indicates that there is a displacement associated with the boundary that has a component in the basal plane of the α-Si_3N_4 structure. Rotation displacement (R/D) boundaries of this type generally were highly curved and showed no tendency to facet.

Two further types of fault were found in the CVD materials; both had a displacement, but no rotation associated with them. For the first type of fault (D1), misalignment of the 10̄10 fringes in a lattice image of the boundary indicated that its displacement vector had a basal plane component. For the second type of displacement fault (D2), there was no displacement of the 10̄10 fringes on crossing the fault, indicating that the diplacement vector of the fault was normal to the basal plane of the α-Si_3N_4 structure. Generally, D1 type faults followed curved surfaces in the material and were similar in appearance to R/D boundaries, whereas D2 type faults were sometimes faceted on the (10̄10) prism planes and resembled R boundaries.

orientation it could be seen that the R, R/D, D1 and D2 boundaries all form extensive facets on the basal plane of the α-Si₃N₄ structure.

DISCUSSION

Both forms of silicon nitride are made up of layers of SiN₄ tetrahedra linked at the corners by nitrogens.[8] Each nitrogen links three tetrahedra such that one out of four nitrogens link tetrahedra in the same layer and the remaining three bond the layers together. If each layer is positioned directly above each other the β Si₃N₄ structure is produced. The α Si₃N₄ structure is formed by inverting alternate layers and positioning them so that the nitrogens link the layers and alternate layers are in exact coincidence. It is also possible, however, to position the alternate layers so that adjacent layers are linked as in the α Si₃N₄ structure but the alternate layers are not in coincidence. Three layers of such an arrangement constitute a stacking fault in the (0001) plane and across which the α Si₃N₄ is rotated by 60° and displaced by 1/3 [1$\bar{1}$00]. A displacement vector of 1/3 [1$\bar{1}$00] offsets all three sets of (1$\bar{1}$00) planes by 1/3 of their spacing. Fringe offset measurements made on the rotation boundary in Fig. 3 are in good agreement with this prediction suggesting that the R/D faults observed are of the type described above. The R and D1 faults can be accounted for by similar structures consisting of combinations of two and three R/D faults, respectively. D2 type faults may arise if a layer of SiN₄ tetrahedra were to deposit as in the β Si₃N₄ structure. In this way all the faults observed can be accounted for by structures in which bonding between adjacent (0001) layers is the same as in one of the polymorphs of Si₃N₄ and only second nearest layers of SiN₄ tetrahedra are misplaced. Such faults would be expected to have particularly low energies accounting for their frequent occurrance during deposition and the extensive (0001) plane facets observed.

Texturing of the CVD samples examined indicates that the deposition occurs by the growth of α Si₃N₄ grains in an [0001] direction. If islands α Si₃N₄ structure containing different (0001) plane faults form on the (0001) growth front then where the islands grow together faults extending off the basal planes will be formed. In this way faults can be extended out of the basal plane even when such a fault configuration is not particularly energetically favorable and therefore unlikely to form by direct deposition. A mechanism of this sort may account for the extensive non-basal plane fault areas found in the deposits.

CONCLUSIONS

Crystalline α Si₃N₄ deposits prepared by CVD contain a large number of structural faults. Using transmission electron microscopy, four different types of faults with combinations of rotational and displacement components have been identified. The faults form extensive facets on the basal plane of the α Si₃N₄ structure but also extended onto non basal planes and curved surfaces within grains. The basal facets of each type fault can be accounted for by structures in which bonding is the same as that in one of the polymorphs of silicon nitride and only second nearest neighbors are misplaced. The non basal plane segments of the faults are believed to arise where faulted regions on the deposit growth front grow together.

ACKNOWLEDGEMENTS

The authors are indebted to the U.S. Department of Energy, Division of Materials Sciences, for financial support under Contract DE-AC03-78ER01885

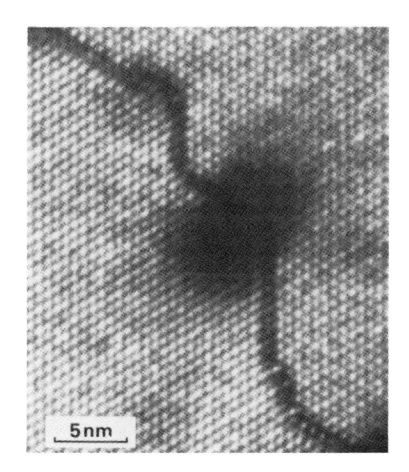

FIG. 3 Lattice image of a rotation boundary occurring in an α-Si$_3$N$_4$ grain. The fringes are continuous across the boundary, indicating that there is no displacement component to the boundary.

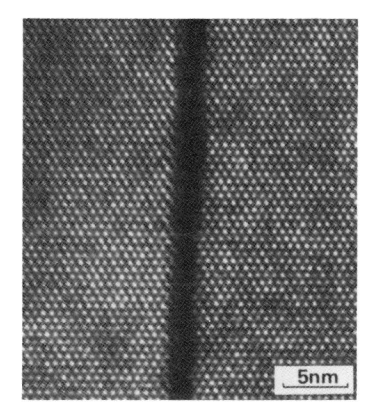

FIG. 4 Lattice image of a rotation displacement boundary in α-Si$_3$N$_4$. Displacement of the fringes occurs on crossing the boundary.

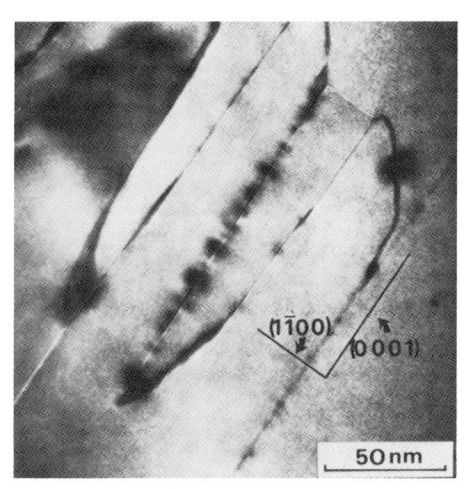

FIG. 5 General view of faults in a single α-Si$_3$N$_4$ grain with the grain in an ⟨11$\bar{2}$0⟩ orientation. Extensive facets on the (0001) planes occur on several of the faults.

Pronounced texturing of the CVD material made it difficult to tilt grains to major zone axis other than the (0001). A few grains were found that could be oriented so that the basal plane was edge on and the crystal was in an ⟨11$\bar{2}$0⟩ orientation; an example is shown in Fig. 5. In this to Dr. F. Galasso, United Technologies Research Center, and to Chemetal Inc., for kindly providing the CVD silicon nitride materials.

REFERENCES

1. C.E. Morosanu, Thin Solid Films 65 171-208 (1980).
2. F. Galasso, U. Kuntz and W.J. Croft, J. Am. Ceram. Soc. 55, 431-432 (1972)
3. A.C. Airey, S. Clarke and P. Popper, Proc. Brit. Ceram. Soc. 22, 305 (1973).
4. K. Kiihara and T. Hirai, J. Mater. Sci. 11, 593-603 (1976).
5. T. Hirai and S. Hayashi, J. Amer. Ceram. Soc. 64, C88 (1981).
6. T. Hirai and S. Hayashi, J. Mat. Sci. 17, 1320 (1982).
7. K.H. Jack in "Progress in Nitrogen Ceramics," proceedings of NATO Advanced Study Institute, Ed., F.L. Riley published Martinus Nijhoff (1983).
8. S. Wild, P. Grieveson and K.H. Jack in "Special Ceramics," 5 385-395, ed. P. Popper, published British Ceramic Research Association (1970).

HIGH RESOLUTION TEM STUDIES OF β-ALUMINA TYPE STRUCTURES

K. J. MORRISSEY, Z. ELGAT, Y. KOUH AND C. B. CARTER
Department of Materials Science and Engineering, Bard Hall, Cornell University, Ithaca, NY 14853

ABSTRACT

High resolution transmission electron microscopy (HRTEM) has been used to study structures found in second-phase particles in commercial alumina compacts. Analytical electron microscopy has been used to identify elements present in the particles. Computer image simulation has been used for both the structural interpretation of high resolution images and predicting the effect which the presence of other elements would have on the observed structures.

INTRODUCTION

The preparation of polycrystalline alumina compacts often involves the addition of a densification aid such as MgO. When the additive exists in amounts exceeding the solid solubility limit the formation of second-phase particles takes place. Impurities existing in the system can combine with the additive in the particles. An extensive study of the formation of second-phase particles has been presented elsewhere [1,2].

Blanc et al. reported the presence of particles containing K-β'''-alumina and Mg-Al spinel, but the spinel and β'''-alumina were not always found together [3]. In the present investigation β'''-alumina and spinel were always found in the same particles implying either that large concentrations of impurities were present or that less MgO had been added. Besides these two phases a number of other different structures have been found in the second-phase particles and these have been attributed to local variations in chemical composition.

Two other related phases have been studied in this program for purposes of comparison; they are β- and β''-alumina. These materials, along with β'''-alumina, are important because of their superionic conducting properties. β-alumina is a hexagonal material whose structure is composed of two distinct features: spinel-like blocks and open conduction planes. The unit cell consists of two aluminate spinel blocks related by a twin plane. β''-alumina has a rhombohedral structure containing three such magnesium-aluminate spinel blocks but related in this case by a 3-fold symmetry axis, the spinel blocks are again separated by less-dense conduction planes. The spinel blocks in β- and β''- are actually composed of 4 close-packed layers of O with the aluminum and aluminum and magnesium ions in the interstices respectively. Like β-alumina, β'''-alumina is also a hexagonal material with the twin planes separating the spinel blocks, but in this case the spinel blocks contain 6 close-packed layers of oxygen ions with Mg and Al in the interstices. The conduction planes and twin planes contain the alkali metal ion (K and Na) and bridging O ions.

Detailed studies have previously been made on crushed samples of these materials primarily to determine their defect structure. De Jonghe and Bovin have both used HREM to study the crystal structure of β''-alumina and the structure of defects present in this material [4, 5]. Matsui also studied β''-alumina and discussed both defect structures found by De Jonghe and Bovin [6]. These authors have shown that a full understanding of the complex nature of these materials and their defect structures requires the use of computer image simulation as a guide to determining the positions of the ions.

The use of computer image simulation is especially important in the study of the phases present in the second-phase particles, not only because there are a variety of impurities present in the compacts and it is therefore necessary to be able to understand the effect different elements may have on HRTEM images but also because the structure or chemical composition may be altered either during sample preparation or observation in the TEM. This report deals with both experimental images and computer simulated images and combines these with elemental information gained from using energy dispersive x-ray spectroscopy (EDS).

Experimental

The observations discussed in this paper were made on commercial alumina compacts which had been prepared using MgO as a densification aid. Both hot pressed and conventionally sintered materials have been studied. The materials were prepared for examination in the transmission electron microscope (TEM) by mechanically polishing samples to a thickness of 20-50 µm and then cutting out discs using an ultrasonic drill. The discs were then thinned to perforation using a ion thinner with argon gas and an operating voltage of 5 kV.

The high resolution observations were made using a Siemens 102 TEM operating at 125 kV. Compositional analysis was carried out in a JEM200CX scanning transmission electron microscope (STEM) fitted with a Tracor Northern energy dispersive x-ray spectroscopy (EDS) system incorporating a high take-off angle (70°) detector and operating at 200 kV.

Experimental Observations and Discussion

The typical β'''-alumina particles in the hot-pressed material are much longer than they are wide as illustrated in Fig. 1 this particle is approximately 7 µm long, the large aspect ratio is due to the fast growth of β''' parallel to its basal plane. Triangular shaped regions are always present on the particles and are composed of Mg-Al spinel; they are always associated with grain boundaries. Fig. 2 is a high resolution image obtained from a similar second-phase particle: a diagram of the structure is given in Fig. 2b for comparison. The black spots in the image correspond to the projection of one of spinel blocks, and the rows of white spots correspond to the twin planes in the structure.

In order to assess the effect of foil thickness, objective lens defocus and local chemical composition, a large number of images have been calculated. The examples presented in this paper have been chosen to illustrate particular features of these calculated images. Fig. 3 is a computer simulation image of K-β'''-alumina using the atomic coordinates given by Bettman and Peters [7] a Cs value of 2.1 mm and an objective aperture of .3 recip A diameter. The simulation is in very good agreement with the experimental image. The same program has been used to produce simulated images of β'''-alumina substituting different elements for the Potassium so as to asses the sensitivity of the image to the presence of elements other than K. The experimental study has shown for example that Na can rapidly leave the area of interest during electron-beam irradiation and that Ar ions may replace the alkali metal ion during ion-beam thinning.

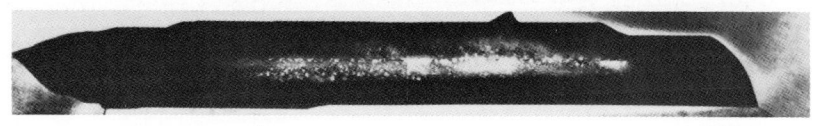

FIG. 1. Bright field image of a typical second-phase particle, the length is approximately 7µm.

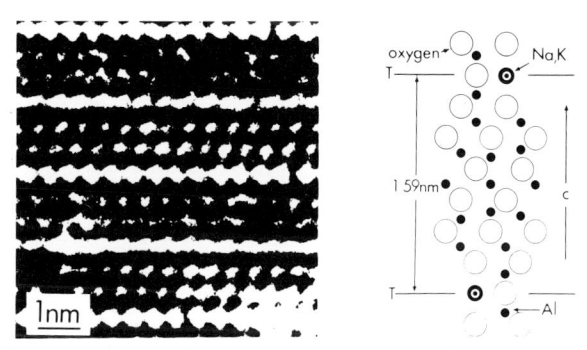

FIG. 2. a) HREM image of β'''-alumina, b) diagram of the β'''-alumina structure.

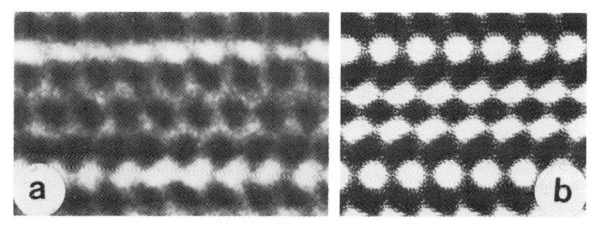

FIG. 3. A comparison of the β'''-alumina HREM image (a) and a simulated image (b).

Fig. 4a-c are simulated images of β'''-alumina with a) K, b) Na, and c) without any element other than oxygen in the conduction planes. The simulated images are for a defocus value of -250 Å and a thickness of 90 Å. The results suggest that there is relatively little difference between the Na three structures although the white dots of the twin planes when K is present are somewhat smaller than those present in the other simulated images. Simulated images using different defocus values but keeping other

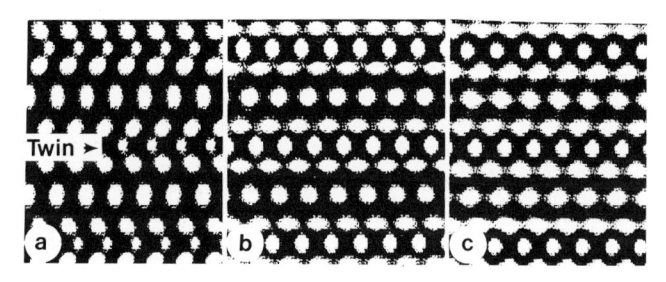

FIG. 4. Simulated image of β'''-alumina with a) K, b) Na, and c) without any element other than oxygen in the conduction planes.

variables constant are shown in 5(a-f). As expected very small changes in defocus can cause dramatic changes in the image. The simulated image in Fig. 5b corresponds closely to the experimental image in Fig. 2. The twin planes in all the simulated images continue to consist of white spots, but the white spots themselves change shape. In Fig. 5d the twin plane would appear to be represented by a row of black dots although the white dots can still be detected. This result is important for low magnification images such as have been used in the study of phase transformations and conduction plane collapse.

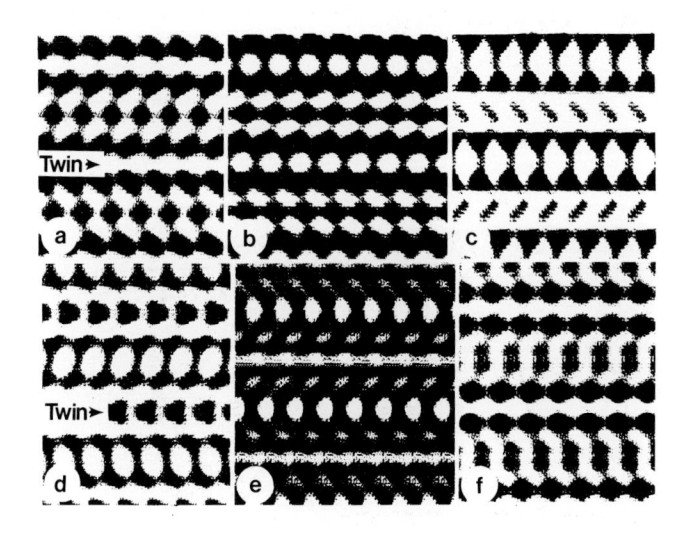

FIG. 5. Simulated images of the 20th slice with varied defocus: a) -750° Å, b) -850 Å, c) -1150 Å, d) 150 Å, e) -400 Å, f) -600 Å.

Fig. 6a shows the projected potential for β''' used in the actual calculations, while Fig. 6 (b-c) shows simulated images for a thickness of 45 Å and a defocus of -300 Å for β'''-alumina containing a) no ion in the conduction plane (other than oxygen) and b) sodium. The reason for investigating such images is to assess the possibility of identifying particular thickness/defocus values where Na, K and an empty twin plane can be distinguished. This type of information would be helpful in understanding the structure of second phase particles.

FIG. 6. a) The projected potential for β'''. b) Simulated images of the 20th slice, defocus of -300 Å for β''' containing b) no ion in the conduction plane, c) sodium in the conduction plane.

During EDS studies of crushed samples of β-alumina it was found that due to its high mobility in the conduction planes, the Na signal decreases during irradiation by the electron beam. Fig. 7a is an EDS spectrum from a grain of β-alumina prepared by crushing and does show the presence of Na although this signal decreases with time. K does not move as rapidly as Na and therefore is more easily detected. If the structure of particles is studied by HRTEM prior to EDS studies the Na may already have been driven off.

The nature of the study of β'''-alumina involves a structural analysis of the Al_2O_3/particle interface and for this reason the samples cannot be crushed. The samples are thinned using Ar ions with an energy of 5 keV. The spectrum in 7b shows a peak of Ar. The argon peak is not found on either the Al_2O_3 matrix grains or on spinel. As noted in the discussion of Fig. 7a, Ar is not present in the spectrum from crushed material. It is therefore suggested that Ar ions enter the twin planes of the β'''-alumina during specimen preparation and that the ions then lose their charge and remain trapped in the twin planes. The Ar ions can actually replace the Na and K in the twin planes.

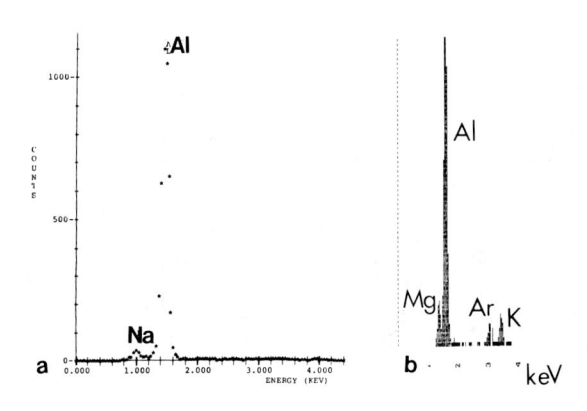

FIG. 7. EDS spectra from a) β-alumina, b) a second-phase particle.

Fig. 8 shows an image of a where the conduction planes have collapsed as seen for example at A. The collapse of the planes starts in the interior of the material and continues out toward the edge as the arrow indicates. This structural information is an important addition to the data taken using EDS. The collapse of β''' alumina has been discussed recently by Hull et al. [8]. Fig. 9a shows another image of a second-phase particle: it shows the type of damage which occurs in the spinel blocks of β'''-alumina due to the electron beam. The damaged regions are mainly restricted to the spinel blocks. This is very similar to the type of irradiation damage noted in Mg-Al spinel samples as shown in 9b. This related behavior of the two materials is further evidence of the similarities between the spinel block and spinel structure.

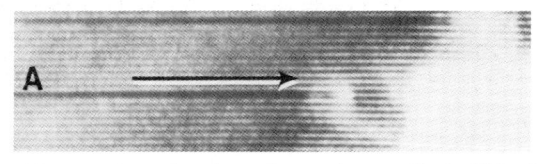

FIG. 8. The collapse of conduction planes (arrowed) in β''-alumina.

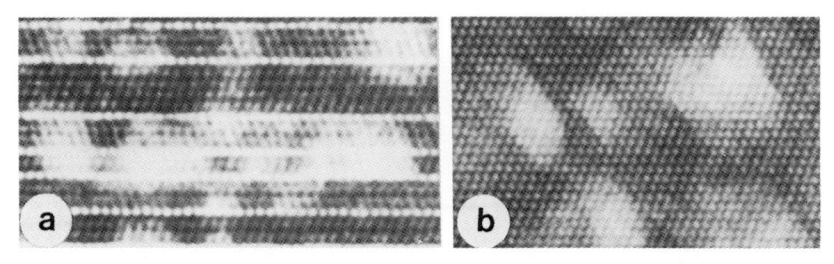

FIG. 9. a) Damage done to the spinel blocks in a second phase particle, b) a spinel sample by the electron beam.

SUMMARY

The combination of HRTEM, EDS and computer image simulation is needed for the complete understanding of the nature of the second-phase particles in alumina. Among the difficulties encountered in studying β-aluminas in second-phase particles in α-Al_2O_3 is the fact that the chemistry of the material is actually changed by ion thinning. Furthermore during examination by TEM any Na present tends to rapidly migrate away from the electron beam. Simulated images predict that the experimental image will be rapidly insensitive to such changes in composition until the structure itself begins to collapse.

ACKNOWLEDGMENTS

The authors would like to thank Mr. Ray Coles and Mr. John Hunt for maintaining the electron microscopes and analytical equipment. The electron microscope facility is supported by the NSF through the Materials Science Center at Cornell University. The multislice image simulation program was adapted from that developed at Arizona State University. This research is supported by the NSF under grant no. DMR81-02294.

REFERENCES

1. K. J. Morrissey and C. B. Carter, "Analysis of Second-Phase Particles in Al_2O_3," Adv. in Materials Characterization, 297-307 (1983).
2. K. J. Morrissey and C. B. Carter, "Second-Phase Particles in Al_2O_3," in preparation.
3. M. Blanc, A. Mocellun and J. L. Strudel, "Observations of Potassium β'''-Alumina in Sintered Alumina," J. Am. Ceram. Soc. 60 (9-10), 403-409 (1977).
4. L. C. De Jonghe, "Fast Ion Conductors," J. Am. Ceram. Soc. 62 (5-6), 289-293 (1979).
5. J. Bovin, "High Resolution Electron Microscopy Images of Defects in Mg- and Li-stabilized β''-aluminas," Acta Cryst. A35, 572-580 (1979).
6. Y. Matsui and S. Horiuchi, "Irradiation-Induced Defects in β''-Alumina Examined by IMV High-Resolution Electron Microscopy," Acta Cryst., A37, 51-61 (1981).
7. M. Bettman and L. L. Turner, On the Structure of Na_2O·4Mgo·15Al_2O_3, a Variant of β-Alumina, Inorg. Chem. 10:(7), 1442-1446 (1971).
8. R. Hull, D. Smith and C. J. Humphreys, "A high-resolution electron microscopic study of defects in sodium β'''-alumina," J. of Microscopy, 130, Pt. 2, 203-214 (1983).

THE $\gamma \to \alpha$ PHASE TRANSFORMATION IN Al_2O_3

D. S. TUCKER* AND J. J. HREN**
*Atlantic Richfield, Materials Development Lab,
27017 Prairie Street, Chatsworth, CA 91311
**Department of Materials Science and Engineering,
University of Florida, Gainesville, FL 32611

INTRODUCTION

   The primary objective of this research was to deduce the mechanisms of the γ to α transformation of Al_2O_3. The specimens studied were from spherical powder precipitated from sulfate solutions and thin films prepared by low temperature oxidation of Al. The primary experimental technique employed was the transmission electron microscope. Since alumina is so important commercially, there is a body of literature on both the mechanism and kinetics of the transformation. A brief review of the most relevant publications is thus essential.

QUALITATIVE STUDIES

   The first reported investigation of the $\gamma \to \alpha$ transformation was by Biltz and Lemke [1] who observed that cubic alumina converted into the stable α-modification at about 1000°C. Later, Stumpf et al. [2] found that corundum appeared between 1150°C and 1250°C, depending on the starting material and the experimental conditions. Saalfield [3] showed that a crystallographic relationship persisted from hydrargillite up to δ-Al_2O_3 during hydrothermal decomposition, indicating a succession of displacive changes. However, when α-Al_2O_3 formed, this relationship was destroyed, implying that the final transformation was reconstructive. Ervin [4] and Kalinina [5] suggested that the intermediate phases of alumina were not discrete but merely transitions from a disordered to a more ordered state. However, neither postulated a mechanism for this conversion. Iler [6] observed the transformation sequence as follows:

   boehmite :   500-700°C γ 1000°C θ 1100°C α

He shared Kalinina's view [5] that θ was merely a better ordered form of γ-Al_2O_3 and claimed that θ-Al_2O_3 consisted of finely divided particles whereas α-Al_2O_3 was much coarser. After observing what appeared to be an amorphous phase surrounding the alpha particle, he hypothesized that something similar to a eutectic liquid existed between the two crystalline forms, providing the means by which Al^{3+} and O^{2-} were transported from the initial to the final phase. However, Badkar and Bailey [7], studying the transformation of boehmite alumina gels, did not observe any amorphous regions and suggested that the transformation from θ to α-Al_2O_3 occurred by nucleation and growth. They also suggested that α nucleates uniformly in the θ matrix on a scale of one α grain per 10^5 θ grains. After partial transformation of θ, they observed that the α grains showed fingerlike projections into the θ matrix. This they attributed to pinning of the advancing α interphase by fine pores.

   A synchroshear mechanism in the formation of corundum was first suggested by Kronberg [8]. His mechanism involved the simultaneous shear of oxygen and alumina ions with these shear directions differing by 60°. Bye and Simpkin [9] found the amorphous alumina hydroxide doped with 2 to 5 wt.% Cr and Fe transformed by a combination of sintering and synchroshear.

Synchroshear was also suggested by Kachi et al. [10] for the transformation of γ-Fe$_2$O$_3$ to α-Fe$_2$O, which is analogous to the γ-Al$_2$O$_3$ to α-Al$_2$O$_3$ transformation. On the other hand, Ridge [11] concluded that the α forms of Fe$_2$O$_3$ and Al$_2$O$_3$ are produced from the γ forms by volume diffusion rather than synchroshear.

KINETIC STUDIES

The first thorough kinetic study of the $\gamma \to \alpha$ Al$_2$O$_3$ transformation was concuded by Clark and White [12]. They prepared γ-Al$_2$O$_3$ by calcining hydrated alumina at 840°C and followed the transformation isothermally by comparing the density measurements of green compacts with that separate from compacts of pure γ and pure α-Al$_2$O$_3$, as a function of time and temperature between 1195°C and 1500°C. They found the reaction to obey first order kinetics and obtained an activation energy of 79 Kcal/mole. Ching et al. [13] prepared γ-Al$_2$O$_3$ by melting α-Al$_2$O$_3$ followed by rapid cooling. They obtained an activation energy of 153 Kcal/mole and proposed a mechanism also involving first order kinetics. The discrepancy between their values of activation energy and those of Clark and White were attributed to the difference in particle size.

Yanagida et al. [14] investigated the transformation of both θ and γ-Al$_2$O$_3$ into corundum by x-ray diffraction. They found activation energies, respectively, of 101.6 and 144 Kcal/mole. They also noted that a high atmospheric moisture content decreased the activation energy.

Using high purity γ-Al$_2$O$_3$ powder as a starting material, Steiner [15] obtained an activation energy of 116,000 Kcal/mole for the transformation to α. He reported that the transformation was thermally activated, followed zero order kinetics, and appeared to be reconstructive. Steiner proposed that the transformation could be considered a special case of an interface controlled massive transformation beginning at existing nuclei. Kato et. al. [16] studied the $\gamma \to \alpha$ Al$_2$O$_3$ transformation and confirmed that an activation energy of 116 Kcal/mole and zero order kinetics was correct. They used alumina powder prepared from anhydrous aluminum sulfate whereas Steiner used a commercial grade high purity γ alumina powder.

Yoldas' [17] study of the transformation of an alumina solution prepared from alumina alkoxide fould that γ-Al$_2$O$_3$ transformed to α-Al$_2$O$_3$ at 1200°C. Using differential thermal analysis he found an activation energy of 133 Kcal/mole, but did not postulate a mechanism. Dynys and Halloran [18] found that γ-Al$_2$O$_3$ prepared by calcining NH$_4$Al(SO$_4$)$_2$ · 12 H$_2$O did not follow zero or first order reaction kinetics, but transformed by nucleation and growth. After observing isolated colonies of single crystal α-Al$_2$O$_3$ in the γ-Al$_2$O$_3$ matrix, they proposed that each colony grew by surface diffusion from a single nucleus.

IMPURITY EFFECTS

The diversity of data on the transformation kinetics suggests that impurities (or dopants) may be of critical importance. There is good evidence to support this view. Iler [19] found that addition of about 8 wt.% silica to fibrillar boehmite retarded the conversion of γ to θ to α-Al$_2$O$_3$ and created an intermediate kappa phase. Bye and Simpkin [9] found that additions of Fe to amorphous Al hydroxide increased the rate of formation of α-Al$_2$O$_3$ while additions of Cr decreased the transformation rate. The decreased rate was attributed to the retarding effect of Cr^{+6} on synchroshear mechanism because of the resistance to change from tetrahedral bonding. The increased transformation rate to α in the presence of Fe was attributed to the formation and segregation of α-Fe_2O_3 nuclei which did not hinder synchroshear. Wakao and Hibino [20] surveyed the effects of additions of 1 to 10% MgO, CuO, NiO, MnO, Fe_2O_3 and TiO_2 on the temperature dependence of the γ to α transformation. They found that as the cationic radius increased, the transformation temperature decreased. Fink [21] discovered that above 700°, 1% V_2O could greatly accelerate the kinetics of the $\gamma \rightarrow \alpha$ Al_2O_3 transformation, bypassing the θ phase. He proposed that the impurities acted as mineralizers, destroying the metastable Al-O-Al bond in γ or θ Al_2O_3 and reforming these metastable phases into the thermodynamically stable γ-Al_2O_3 form. Even water vapor was found to lower the activation energy for the Al_2O_3 transformation by Yanagida et al. [14] who suggested that the water vapor also acted as a mineralizer by breaking Al-O-Al bonds on the surface to form an Al-O(H)-Al bond, in turn lowering the activation energy for the formation of α-Al_2O_3.

PREPARATION AND CHARACTERIZATION

It is clear that any study of the $\gamma \rightarrow \alpha$ phase transformation must begin by carefully characterizing the starting conditions. In the present work, both powders and thin films were used as specimens. A number of analytical methods were employed to characterize the powders. These included spectrochemical analysis, scanning electron microscopy (SEM); analytical electron microscopy (TEM, SEM, STEM); high voltage electron microscopy (HVEM); x-ray diffraction (XRD); infrared reflection spectroscopy (IRRS); and electron spectroscopy for chemical analysis (ESCA). In addition, data concerning porosity, particle size and specific surface area of these powders was available [22]. The thin films were characterized only by analytical electron microscopy.

Powders

The powders used were prepared as follows: aluminum sulfate hydrate, $Al_2(SO_3)_4 \cdot 18\ H_2O$, was dissolved in deionized water to form a 0.4 M solution. The solution was then passed through a 0.22 micrometer filter. Urea, $Co(NH_2)_2$, was separately dissolved in deionized/distilled water to form a 6.0 M solution, and filtered in the same manner as the sulfate solution. These two solutions were mixed in a 1:1 ratio with an equivalent amount of deionized/distilled water. This total solution was loaded into a thermostatted water bath at 94 ± 1°C and aged for three hours during which time precipitation of spherical hydrated basic aluminum sulfate occurred. After aging, the liquid above the precipitated product was decanted to prevent dissolution during cooling. The precipitates were then repeatedly washed in distilled water using a pressure filtration unit. Finally, the spherical powders were calcined for two hours at temperatures of 450, 600, 700, 800, 1175 and 1300°C in static air using a heating rate of 2°C/min.

Chemical analysis of the uncalcined powders showed the following impurities in weight percent (w/o): Fe 0.004, Si 0.002, Na 0.001, Ti < 0.005, Pb < 0.005, NH_4 < 0.001, and CO_3 < 0.01. In addition, the powders contained 18 w/o SO_3.

Fourier transform infrared reflection spectroscopy (FTIR) was used to follow the decomposition of the sulfate phases and the dehydration of the powders as a function of calcining temperatures. ESCA (electron spectroscopy for chemical analysis) was used to detect the presence of sulfur in selected samples. A JEOL 35CX scanning electron microscope was used to study the morphology of individual particles and to observe porosity. Phase transitions as a result of calcination were monitored by x-ray diffraction.

Samples for electron microscopy were prepared as follows. Powders were allowed to settle in deionized water. The topmost solution was decanted and the fines allowed to settle onto carbon coated grids. The solution was then drained and the grids allowed to dry. Average particles sizes of less than ~ 1 µm were obtained in this way. TEM and SEM observation of these specimens were made in a JEOL 200CX STEM. For larger particles (~ 1 µm) the 200CX was used in the SEM mode and transmission images were obtained with a 1 meV transmission microscope.

In order to obtain information about the internal structure of the particles a microtoming technique was developed [23]. Powders which were calcined at 1050°C and 1175°C were selected for these studies because they were mixed γ and α as deduced from x-ray diffraction. Some serial sections were obtained with the samples lined up on a single grid.

Thin Films

Because the γ-Al$_2$O$_3$ formed during calcination had such a small grain size, the grains overlapped in the TEM specimens. Specimens were thus obtained by an alternate method, oxidation of a thin Al film. Pure aluminum foils were oxidized in static oxygen for five days at 600°C. Any remaining metallic aluminum was dissolved using 1 molar NaOH. The remaining γ-Al$_2$O$_3$ films were collected on copper grids, washed in deionized water, and allowed to dry. A few other Al foils were heated to 1000°C for 1 h to bring about a partial transformation of the γ-Al$_2$O$_3$ to α-Al$_2$O$_3$.

A direct study of the $\gamma \rightarrow \alpha$ transformation in the precipitated powders was made with a 1 meV transmission electron microscope [24]. Tungsten grids were chosen since they are refractory, have a relatively high thermal conductivity, and are relatively inert in contact with the platinum heater supports. Support films of "holey" platinum were prepared by a method described elsewhere [25]. The grids were dipped into precalcined powder and the excess powder blown off with a stream of air. The temperature was varied by pulse heating and then reduced approximately 150°C, to freeze in the microstructure.

MICROSTRUCTURE OF THE POWDERS

Powders which were calcined up to 800°C (for 2 hrs.) showed no obvious crystallinity by TEM. This seemed to contradict the x-ray diffraction data which show some γ-Al$_2$O$_3$ at 600°C, 700°C and 800°C. However, the volume sampled by the TEM is very small and the degree of crystallinity low. Further, the mean powder size was 5.5 µm, whereas the largest particle observed in the TEM was approximately 1 µm. At 900°C powders gave a distinct ring which indexed as γ-Al$_2$O$_3$. Sectioning of these particles by microtoming confirmed that they were fine polycrystals with a mean diameter of 10-20 µm. Powders calcined at 1050°C appeared qualitatively similar to those calcined at 900°C. The larger particles, which were sectioned with a microtome, revealed a duplex structure, as shown in Figure 1a. Figure 1b is a SADP from one such microtomed particle. It reveals a single crystal α-Al$_2$O$_3$ diffraction pattern superimposed on the γ-Al$_2$O$_3$ ring pattern.

After calcining at 1175°C for two hours, the SADP shows the powder particles to be nearly perfect single crystals of α-Al$_2$O$_3$. Dark field imaging demonstrates that individual particles actually have slight internal misorientations (Figure 1c).

Combined TEM data from serial sections and hot stage studies show that the α-Al$_2$O$_3$ grows from the surface into the particle when the particles are not sintered (Figure 2a). When particles are sintered, serial

(a) (b)

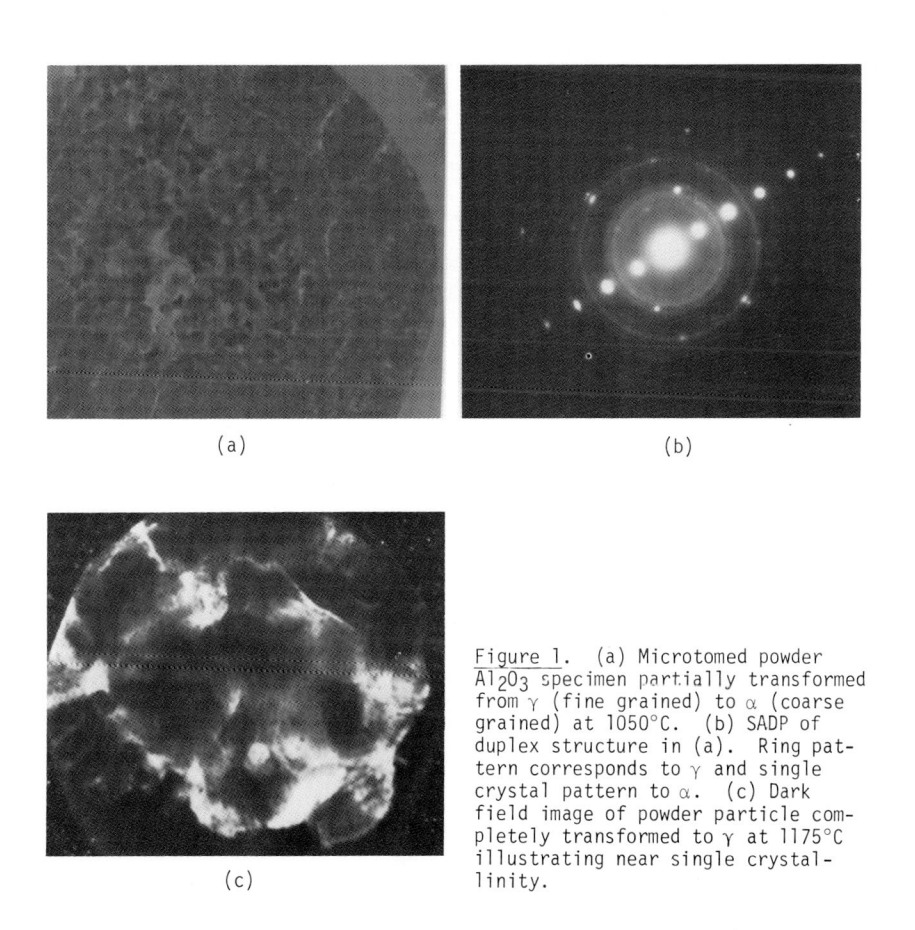

(c)

Figure 1. (a) Microtomed powder Al_2O_3 specimen partially transformed from γ (fine grained) to α (coarse grained) at 1050°C. (b) SADP of duplex structure in (a). Ring pattern corresponds to γ and single crystal pattern to α. (c) Dark field image of powder particle completely transformed to γ at 1175°C illustrating near single crystallinity.

sectioning shows that the large grained α phase spreads into both particles from the neck (Figure 2b). Typically, there is very little misorientation between the α-Al_2O_3 in each particle and the neck. Growth apparently begins in the necked region and spreads into both particles.

Thin Films

Thin films of γ-Al_2O_3 were produced by oxidation of Al metal and had a typical grain size of 10-50 μm after subsequent heat treatment. The measured lattice parameter of 7.90 Å was the same as the calcined powder. Some streaking in the diffraction pattern (Figure 3a) is apparent. This result and double diffraction around (220) are symptomatic of twinning. When this film is imaged in the dark field the twins are obvious (Figure 3b). If a trace analysis is performed, one concludes that the twins are of the {111} <11$\bar{2}$> type. Films that were partially transformed to α at 1000°C for 1 hour were also studied and showed a duplex structure consisting of a twinned region and large γ-Al_2O_3 grains. Some films were then heated to ~ 1100°C in situ. Small spherical regions of α-Al_2O_3 appeared (Figure 3c) which became distinctly faceted as the transformation proceeded. After

(a)   (b)

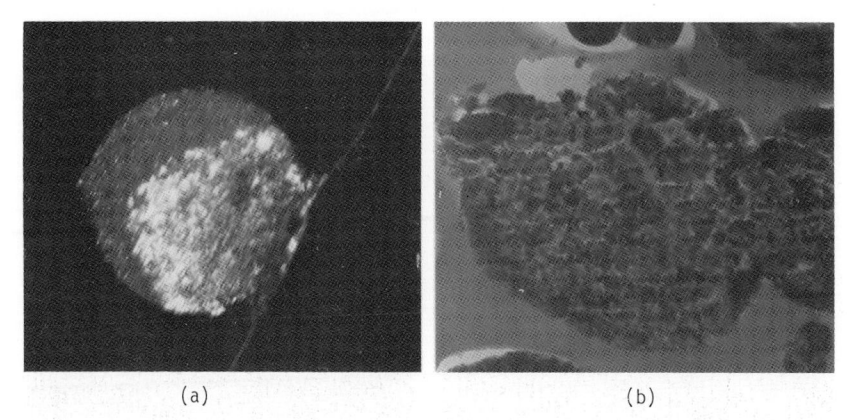

Figure 2. (a) HVEM dark field image of powder particles of Al_2O_3 partially transformed from $\gamma \rightarrow \alpha$ showing large crystal α advancing into γ. In situ heating at approximately 1100°C. (b) Microtomed section of partially transformed $\gamma \rightarrow \alpha$ powders sintered together illustrating growth of α across both particles.

further in situ growth, the twins were less numerous. Dark field imaging from a portion of the (220) ring of a calcined and microtomed section confirmed that twins were present in the fine grained γ-Al_2O_3 powders as well (Figure 3d).

TRANSFORMATION MECHANISM

The mechanism for the transformation of $\gamma \rightarrow \alpha$ Al_2O_3 apparently involves the growth of stacking faults and twins. It is well known that stacking faults transform the FCC structure into HCP locally; however, in Al_2O_3 this simple mechanism accounts for the (111) oxygen layers only. The stacking sequence of the Al cations must be accounted for as well. Following the proposal by Verwey, γ-Al_2O_3 can be said to have two types of cation layers. In the O layers, all of the aluminum atoms are in octahedral sites. In the T layers, two-thirds of the aluminum atoms are in tetrahedral sites and one-third are in octahedral sites. The O and T layers alternate between the oxygen layers, repeating every sixth layer. The aluminum atoms are already in octahedral sites so that if a stacking fault occurs about the O layer they will already be correct; however, only two-thirds of the aluminum sites are normally occupied in γ-Al_2O_3. The extra aluminum atoms must be replaced with vacancies to maintain stoichiometry. If the fault is centered about the T layer, tetrahedral sites are occupied as well as octahedral sites. To preserve stoichiometry in this case, one-third of the aluminum atoms on tetrahedral sites must be replaced with vacancies while the other tetrahedral aluminum atoms diffuse to octahedral sites.

The transformation of the thin film of γ-Al_2O_3 to α-Al_2O_3 probably occurs by a stacking fault growth mechanism. Numerous microtwins were observed, all bounded by stacking faults. Each would have an equal probability of nucleating α-Al_2O_3 and this would explain why the thin film was ultimately transformed to polycrystalline α-Al_2O_3. On the other hand, in the spherical γ-Al_2O_3 particles, there are other nucleation sites and a unique description of the transformation remains elusive. The fact that nucleation and growth begins in the necked regions is consistent with the

(a) (b)

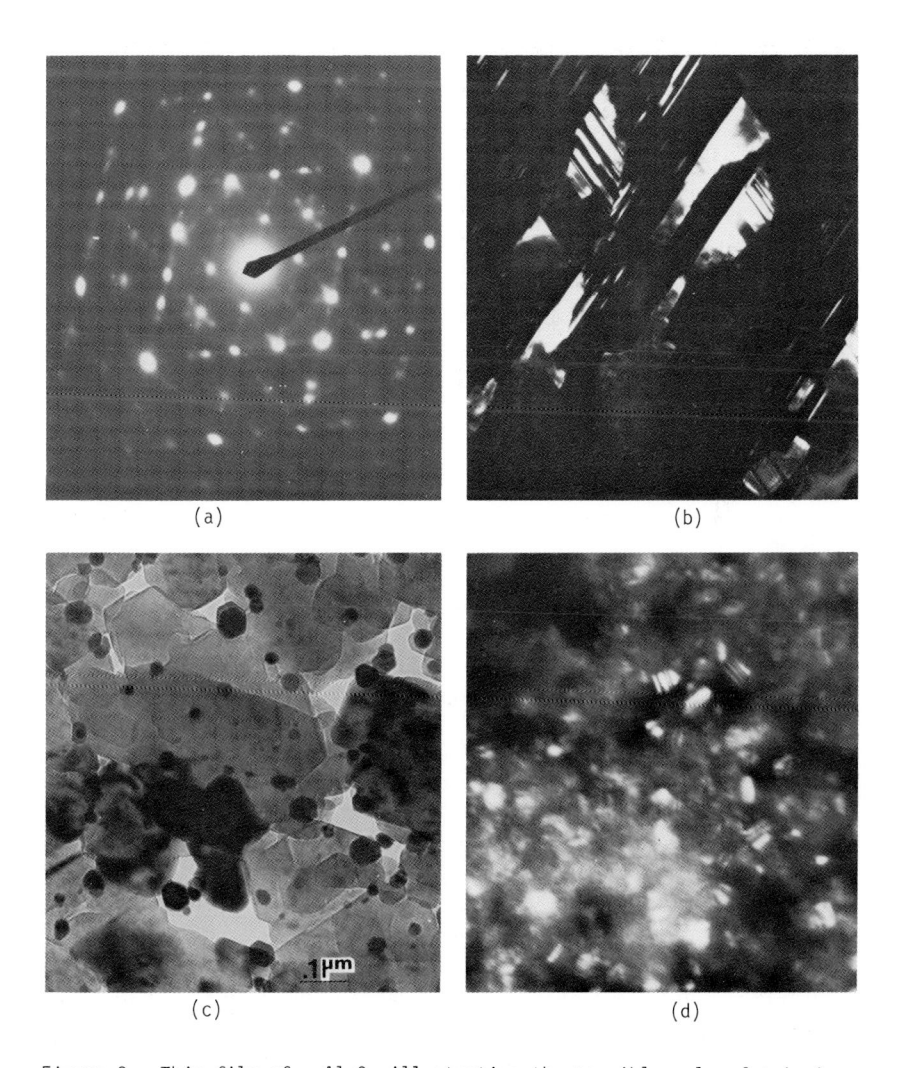

(c) (d)

Figure 3. Thin film of γ-Al_2O_3 illustrating the possible role of twinning in the $\gamma \rightarrow \alpha$ transformation. (a) SADP in (110) orientation showing streaking and double diffraction typical of twinning. (b) Dark field image using twin spots. (c) Nucleation of α at 1100°C as small nearly spherical grains. (d) Dark field image from twin spots in γ-Al_2O_3 powder.

lower activation expected there. A reaction involving short-range diffusion and stacking fault growth at the interface is a possible model for the movement of the α interface. This mechanism would require only short-range diffusion of cations and vacancies and some realignment of the twins at the $\gamma \rightarrow \alpha$ interface to achieve the nearly perfect single crystal transformation product of α-Al_2O_3.

ACKNOWLEDGEMENTS

Partial support of this work was provided by the Major Analytical Instrumentation Center of the University of Florida. The cooperation and assistance of Professor M. Sacks and Mssrs. E. J. Jenkins and E. Clausen are gratefully acknowledged.

REFERENCES

1. H. C. Stumpf, A. S. Russell, J. W. Newsome and C. M. Tucker, Ind. Eng. Chem., 42, 1398-1403 (1950).
2. H. Saalfield, N. Jaheb. F. Min. Ab., 95, 1-88 (1960).
3. G. Ervin, Acta. Cryst., 5, 103 (1952).
4. A. M. Kalinina, Russ. Jnl. of Inorg. Chem., 4, 568-573 (1959).
5. R. K. Iler, J. Am. Cer. Soc., 44, 618-624 (1961).
6. M. J. Buerger, Pergammon Press, Oxford, 54 (1965).
7. P. A. Badkar, J. E. Bailey, J. Mat. Sci., 11, 1794-1806 (1976).
8. M. L. Kronberg, Acta. Met., 5, 507 (1957).
9. G. C. Bye, G. T. Simpkin, J. Am. Cer. Soc., 57(8) (1974).
10. S. Kachi, K. Momiyama, S. Shimizu, J. Phys. Soc., Jap., 18(1), 106-116 (1963).
11. M. J. Ridge, B. Molony and G. R. Boell, J. Chem. Soc. Cal., 4, 594-597 (1967).
12. P. W. Clark and J. White, Trans. Brit. Ceram. Soc., 49, 305-333 (1950).
13. Yangjuo-Ching and Yentungsheng, Kwei Suan. Hsueh. Pao, 5, 1-11 (1966).
14. H. Yangida, G. Yamaguchi and J. Kubota, J. Cer. Soc. Jap., 74, 371-377 (1966).
15. C. Steiner, Ph.D. Thesis, Lehigh University (1972).
16. E. Kato and K. Daimon, J. Am. Cer. Soc., 62, 313 (1979).
17. B. E. Yoldas, Am. Cer. Soc. Bull., 54(3), 289-290 (1975).
18. F. W. Dynys, J. W. Halloran, J. Am. Cer. Soc., 65(9), 442-448 (1982).
19. R. K. Iler, J. Am. Cer. Soc., 47, 339-341 (1964).
20. Y. Wakao, J. Hibino, Nagoya. Kogyo. Gijutsu. Shikenko, Hahoku, 11, 588-595 (1962).
21. G. Fink, J. Inorg. Nucl. Chem., 30, 59-61 (1968).
22. M. Sacks, University of Florida, private communication.
23. D. Tucker, E. A. Kenik and J. J. Hren, Proc. 41st Ann. EMSA Meeting, 76 (1983).
24. Oak Ridge National Laboratories, Oak Ridge, TN 37830.
25. E. J. Jenkins, D. S. Tucker and J. J. Hren, Proc. 41st Ann. EMSA Meeting, 74 (1983).

CALCIUM SITE OCCUPANCY IN BaTiO$_3$

H. M. Chan, M. P. Harmer, M. Lal, and D. M. Smyth
Materials Research Center, Lehigh University, Bethlehem, PA 18015 U.S.A.

ABSTRACT

Measurements of the equilibrium electrical conductivity of Ca-doped BaTiO$_3$ indicate a highly acceptor-doped behavior when y > 1 in the general formula (Ba$_{1-x}$Ca$_x$O)$_y$TiO$_2$. It has been suggested that when there is an excess of alkaline earth oxides, some Ca^{++} occupies Ti-sites, where it acts as a doubly charged acceptor, Ca$_{Ti}^{''}$. This possibility has been checked with a recently developed technique for determining the crystallographic site occupancy of impurity atoms. Known as ALCHEMI (Atom Location Using Channeling-Enhanced Microanalysis), the technique relies on the dependence of the characteristic X-ray emission on the incident electron beam direction. A significant channeling effect was observed for slightly positive and negative excitations of the [100] reflection. The results indicate that for the composition (Ba$_{0.85}$Ca$_{0.15}$O)$_{1.02}$TiO$_{2.00}$, a substantial fraction of the total calcium ($\geq$ 20%) occupies Ti sites.

INTRODUCTION

A technique capable of determining the crystallographic site occupancy of impurity atoms has been recently developed [1]. The technique, known as ALCHEMI (Atom Location Using Channeling-Enhanced Microanalysis), relies on the dependence of the characteristic X-ray emission on the incident electron beam direction. ALCHEMI is well-suited to investigating the site occupancy of impurity atoms in BaTiO$_3$, because in this material, the barium and titanium atoms lie on alternate, parallel planes (<h00>). Thus depending on the incident beam orientation, the electron intensity will be concentrated at either the A or B planes, hence X-ray emission from one element will be enhanced relative to the other. By comparing the ratio of the characteristic X-ray intensity of an impurity element to that of either Ba or Ti for the two different orientations, it is possible to determine the fraction of impurity atoms on each type of site.

The interest in calcium-doped BaTiO$_3$ stems from its widespread use as a ceramic capacitor. Many commercial BaTiO$_3$ based multilayer capacitors contain significant amounts of calcium added by way of curie point shifters such as CaZrO$_3$. It has generally been assumed that when calcium is added to BaTiO$_3$, it substitutes for barium. However, during the course of a recent study of the high temperature (1000°C) equilibrium electrical conductivity of calcium-doped BaTiO$_3$, it was found that for A site excess compositions ((Ca+Ba)/Ti >1), calcium doping promoted acceptor type behavior [2]. Acceptor doping in BaTiO$_3$ leads to the formation of oxygen vacancies (V$_O^{°°}$) and a characteristic shift in the position of the minima on a plot of log conductivity vs. log P$_{O_2}$(g) (see Fig. 1). The implication, therefore, is that for calcium-doped BaTiO$_3$, acceptor type behavior is caused by the presence of calcium on titanium sites (Ca$_{Ti}^{''}$) and the resulting oxygen vacancy formation. Further support for the proposed mechanism comes from electrical degradation measurements [3]. It has been shown that calcium excess samples degrade more rapidly than their stoichiometric counterparts under conditions where degradation is believed to be controlled by V$_O^{°°}$ migration.

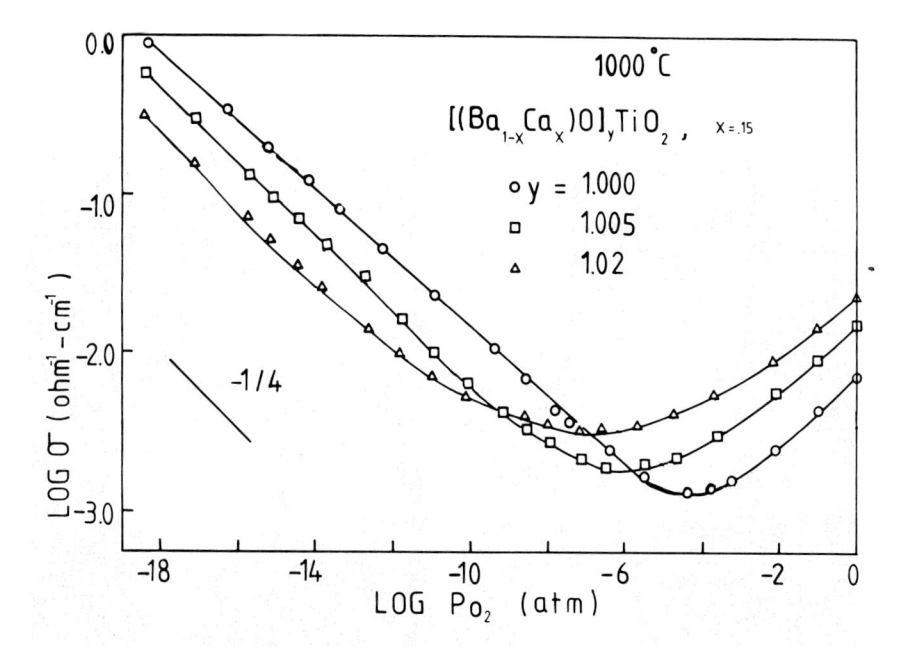

FIG. 1. Equilibrium electrical conductivity of Ca-doped BaTiO3 at 1000°C.

All of the above measurements provide indirect evidence for the occupation of Ti atom sites by Ca atoms in BaTiO3. Accordingly, the purpose of the present work has been to exploit the technique of ALCHEMI to obtain direct evidence and further quantitative information concerning the site occupancy for calcium ions in BaTiO3.

EXPERIMENTAL PROCEDURE

Powders of precisely determined chemical composition were made by the liquid mix process which is described elsewhere [4]. Samples for electrical conductivity measurements were prepared by die pressing the powder into rectangular bars at 350 MPa followed by sintering in air at 1420°C for 2½ hours. All samples sintered well into the range of impermeability ($\gtrsim$ 95% theoretical density). The equilibrium electrical conductivity was measured using a four point d.c. technique which is described in detail elsewhere [4].

Thin foil specimens were prepared by ion beam milling, and examined in a Philips 400T electron microscope equipped with an EDAX energy dispersive X-ray analyzer. The specimen was tilted so that a systematic row of <h00> reflections was visible in the diffraction pattern. It was found that the most pronounced channeling effect occurred for slightly positive and slightly negative excitations of the [100] reflection (see Figs. 2a and b, respectively). These two orientations were used for collecting the X-ray spectra. It is worth noting, however, that in the ALCHEMI technique, the precise orientations within the systematic row are unimportant, providing sufficient channeling is obtained. The following two compositions were studied $(Ba_{1-x}Ca_x O)_y TiO_2$: y = 1.02 and 0.98; x = 0.15.

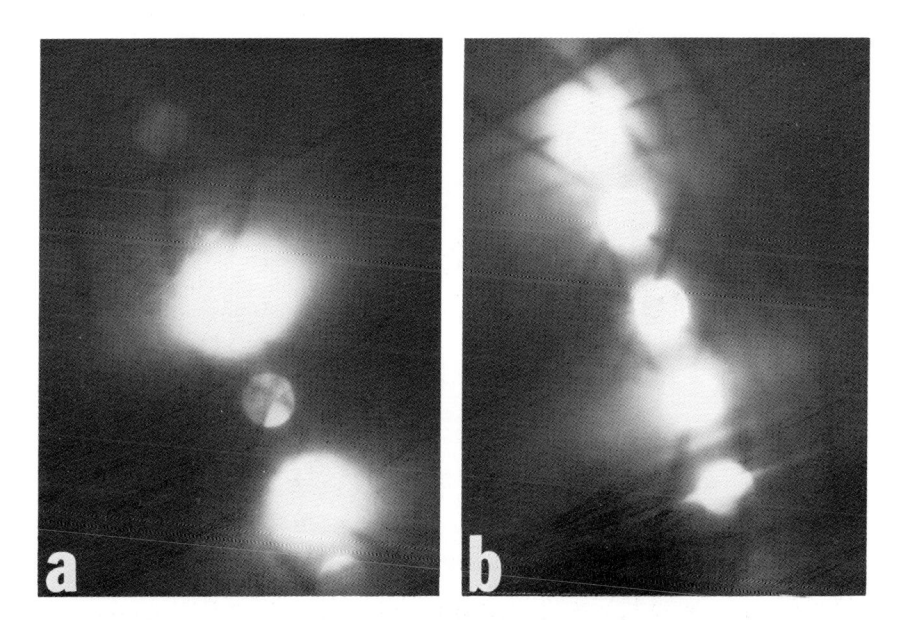

FIG. 2. Orientations used for ALCHEMI determination. a) S+, b) S-.

RESULTS

The microstructure of both specimens was similar, and consisted of fine grains ∼5-10 μm in diameter. A large number of ferroelectric domains could also be observed (see Fig. 3).

The results of the ALCHEMI determination are shown in Table 1. C_x represents the fraction of Ca atoms occupying Ti sites, and was calculated from the following expression [1]:

$$C_x = (R-1)/(R-1 + \gamma - \beta R)$$

where
$$R = (N_{Ba}^{(1)}/N_{Ca}^{(1)})/(N_{Ba}^{(2)}/N_{Ca}^{(2)}) \qquad (1)$$
$$\beta = N_{Ti}^{(1)}/kN_{Ba}^{(1)}$$
$$\gamma = N_{Ti}^{(2)}/kN_{Ba}^{(2)}$$
$$k = N_{Ti}/N_{Ba} \quad \text{For an orientation in which there is no channeling.}$$

The superscripts (1) and (2) represent the two orientations, and N is the number of counts in the X-ray peak.

It can be seen that the spread in C_x values was quite high, the reasons for this are discussed later. By applying a Student's t-test to the two sets

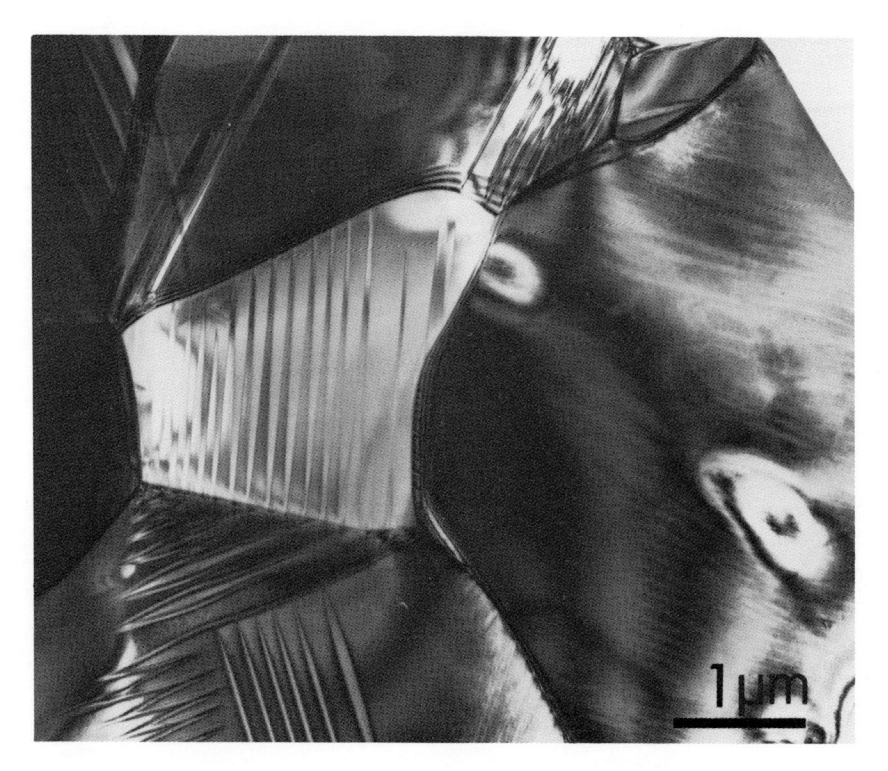

FIG. 3. Bright field TEM image of $(Ba_{0.85}Ca_{0.15}O)_{1.02}TiO_2$.

TABLE 1. Results of ALCHEMI Determination

Specimen	C_x (%)	n
$(Ba_{0.85}Ca_{0.15}O)_{0.98}TiO_2$	9.0 ± 9.7	10
$(Ba_{0.85}Ca_{0.15}O)_{1.02}TiO_2$	26.3 ± 13.2	18

n is the number of determinations. The uncertainty in the C_x value was calculated as described in the following section.

of data, however, it can be shown that the difference in C_x values is indeed "significant" (for a 95% confidence limit), and cannot be accounted for by the spread in the data alone. Thus, it can be concluded that for the A site excess specimen, a definite fraction of calcium atoms is occupying Ti sites.

DISCUSSION

The ALCHEMI results show good qualitative agreement with the predictions of the equilibrium conductivity data, namely that a larger fraction of Ca atoms occupy Ti sites in the A site excess sample. Quantitatively, however, the ALCHEMI data appear to be less satisfactory. Assuming that for the A site excess sample, whose composition can be expressed as $(Ba_{0.867}Ca_{0.133})TiO_3 + 0.02\ CaO$, the excess CaO is distributed evenly among the Ba and Ti sites, thus:

$$0.02\ CaO \rightarrow 0.01\ Ca_{Ba}^{x} + 0.01\ Ca_{Ti}'' + 0.02\ O_o^{x} + 0.01\ V_o^{\circ\circ} \qquad (2)$$

the fraction of Ca atoms on Ti atoms sites (C_x) can be calculated to be 6.54%. This value may be compared with the value of $26.3 \pm 13.2\%$ obtained using the ALCHEMI technique. We now follow with a discussion of the errors involved in the ALCHEMI procedure.

X-ray production is a statistical process, so that even under ideal conditions, the number of counts in a given peak for a large number of trials will lie on a Gaussian curve [5]. Let $\bar{N}$ be the most probable value of the number of counts, then the standard deviation (σ) of the curve is given by $\bar{N}^{\frac{1}{2}}$. 95% of the readings will lie within 2σ of the mean, so that for a given determination of the number of counts N, the uncertainty is given by

$$\pm 2\sigma = \pm 2N^{\frac{1}{2}} \qquad (3)$$

Expressed as a relative uncertainty (%), this gives

$$\frac{2N^{\frac{1}{2}}}{N} \times 100\% \qquad (4)$$

Thus the lower the value of N, the higher is the relative error, so that >10000 counts are required to keep the error below ±2%. To see how this affects the ALCHEMI determination, it is useful to consider a particular example. In a given spectra, the number of counts in the calcium peak (N_{Ca}) will be significantly lower than in the barium and titanium peaks, thus for simplicity, we can assume that the error in N_{Ca} will be the largest contribution to the error in C_x.

The following table shows the actual number of X-ray counts for each element for an individual determination. The error in N_{Ca} is calculated from eqn. (3).

TABLE II.

Orientation	N_{Ba}	N_{Ti}	N_{Ca}
S−	53066	58426	7256 ± 170
S+	56734	73305	8068 ± 180

Substituting these values of N_{Ca} into eqn. (1) gives a maximum possible value of C_x of 45.8% and a minimum value of −3.9%. Fortunately, the uncertainty for a large number of readings is considerably less, and is given by

$$\pm \frac{t_{95}^{n-1}}{\sqrt{n}} \sigma \qquad (5)$$

where σ is the standard deviation for n determinations, and t_{95}^{n-1} is the student t value at the 95% confidence level.

Clearly the accuracy in C_x can be improved by increasing N_{Ca}, i.e., counting for longer times. However, errors due to specimen drift and contamination become more significant for long acquire times, thus a compromise must be made. Because the calculation of C_x is highly sensitive to the number of X-ray counts from the dopant element, the technique is limited to compositions in which sufficient counts can be obtained in reasonable times. Otherwise it becomes impossible to distinguish whether a change in the number of counts is due to channeling or statistical fluctuations.

The strength of the channeling effect which can be obtained is also an important factor, and this is dependent on the crystal structure and chemical composition [6]. The channeling which can be obtained in $BaTiO_3$ is relatively weak (~30% increase in the N_{Ti}/N_{Ba} ratio for the two orientations), hence the sensitivity of the C_x value to the count rate. Clearly for materials where the channeling is more pronounced, the accuracy will be improved.

CONCLUSIONS

The technique of ALCHEMI was used to determine the fraction of Ca atoms on Ti atom sites in Ca-doped $BaTiO_3$. The fractions were determined to be $9.0 \pm 9.7\%$ and $26.3 \pm 13.2\%$ for A site deficient ((Ba + Ca)/Ti = 0.98)) and A site excess ((Ba + Ca)/Ti = 1.02)) samples, respectively. The findings were consistent with the results of recent equilibrium conductivity studies which suggest that Ca can occupy Ti sites in $BaTiO_3$ and act as a doubly charged acceptor (Ca_{Ti}'').

ACKNOWLEDGEMENTS

We are grateful to Dr. J. Spence and Dr. J. Taftø for advice and helpful discussions. Financial support from the Office of Naval Research under contract number N00014-83-K-0190 is also gratefully acknowledged.

REFERENCES

1a. J.C.H. Spence and J. Taftø, Scanning Electron Microscopy, Vol. II, SEM Inc., 523-531 (1982).
1b. J.C.H. Spence and J. Taftø, J. Micros. 130, 147-154 (1983).
2. J. Appleby, Y.H. Han and D.M. Smyth, 85th Ann. Meeting of the American Ceramic Society, Chicago, April 25, Abstract 13-E-83 (1983).
3. M. Lal, M.P. Harmer and D.M. Smyth, to be presented at the 86th Ann. Meeting of the American Ceramic Society, Pittsburgh, May 1984.
4. N.H. Chan, R.K. Sharma and D.M. Smyth, J. Electrochem. Soc. 128, 1762-1769 (1981).
5. A.D. Romig, Jr. and J.I. Goldstein, Met. Trans A. 11A, 1151 (1980).
6. J. Taftø, Z. Naturforsch 34a, 452 (1979).

CHARACTERISATION OF PORTLAND CEMENT HYDRATION BY ELECTRON OPTICAL TECHNIQUES

KAREN L. SCRIVENER AND P. L. PRATT
Department of Metallurgy & Materials Science, Imperial College,
Prince Consort Road, London SW7 2BP, UK.

ABSTRACT

In the past the application of electron microscopy to cement hydration has been limited largely to the study of fracture surfaces in SEM, of ground cement in TEM and of polished sections in EMPA. Consequently the microstructure of bulk cement pastes is, as yet, ambiguous.

The present work shows how techniques not previously used in the study of cement, an environmental cell on HVEM, preparation of ion beam thinned foils for STEM and BEI of polished surfaces in SEM, can give a more detailed view of the development of morphology in hydrating cement. Using a combination of electron optical techniques together with other methods a detailed characterisation of cement hydration can be made.

INTRODUCTION

The study of the hydration reaction of cement poses many problems for the electron microscopist. Cement is non-conducting, and so must be coated with a conducting material, i.e. carbon or gold; it is very brittle, making the preparation of thin sections difficult; and it contains large amounts of water in its normal state, both as pore fluid and combined in hydration products.

Despite these problems electron microscopy is an invaluable technique for determining the structure and morphology of the hydration products and their spatial distribution, which are essential to an understanding of the mechanisms of hydration and strength development.

Portland cement consists of a mixture of calcium silicates, calcium aluminate and aluminoferrite. The hydration products of the calcium silicates are calcium hydroxide, and an amorphous calcium silicate hydrate of indeterminate composition generally referred to as C-S-H. The calcium aluminate and aluminoferrite react with water and calcium sulphate, added when the cement is ground, to give hexagonal rods, similar in composition to the mineral ettringite (also known as AFt phase); and later hexagonal plates are formed with a lower sulphate content generally called "monosulphate" or AFm phase.

EARLY TRANSMISSION ELECTRON MICROSCOPE STUDIES

Early electron optical observations (such as reported by Grudemo [1]) were made by grinding hydrated cement pastes. The resulting powder was then dispersed on carbon films and examined in the transmission electron microscope. These studies enabled the crystal structure of those hydration products which were crystalline to be determined by electron diffraction. Many of the hydration products, however, were amorphous or cryptocrystalline and were very susceptible to dehydration under the electron beam; even the crystalline products such as ettringite and calcium hydroxide tended to dehydrate, whilst retaining their morphology. Despite the considerable knowledge gained from this technique, little or nothing could be deduced regarding the mechanism by which strength development takes place; as spatial relationships existing in the original paste were destroyed during the specimen preparation.

MICROSCOPY OF DRIED FRACTURE SURFACES

This problem was avoided to some extent by observation of fracture surfaces, originally by preparing carbon replicas for TEM and later by observing gold coated fracture surfaces directly in the SEM. Secondary electrons are generated by the incident beam fairly near to the surface, their intensity increasing with the inclination of the surface from the normal. Thus a shadowless image of the surface is produced with fairly good resolution (Fig.1).

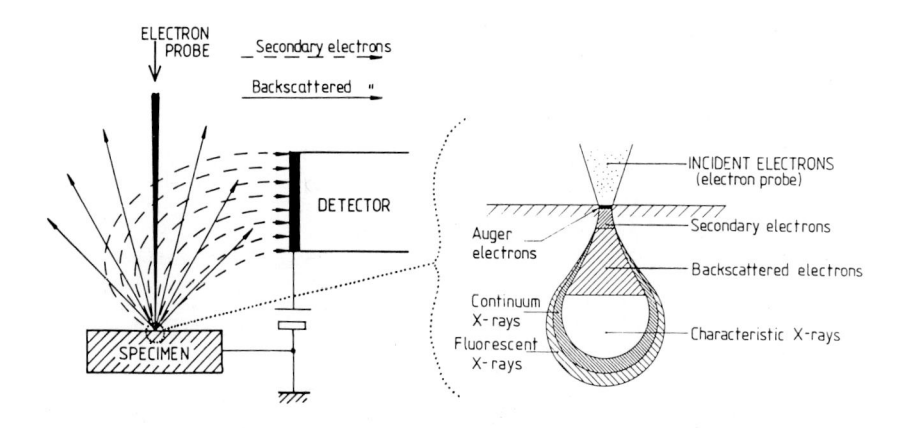

FIG.1. Secondary electron imaging in the SEM to give a shadowless image and (inset) depth of quantum generation.

Using this technique the development of microstructure during hydration can be studied, by freeze drying samples of pastes of various ages [2], [3].

The microstructure observed, however, is that of cement paste which has been previously dried. Given the amorphous nature of the hydration products, it is likely that the microstructure will be altered during the drying process. Also, only fracture paths are revealed; in young pastes this is interparticular, only showing the outer surface of the hydrated layer; in older pastes the fracture surface is dominated by areas of cleaved calcium hydroxide, which has grown to engulf the outer hydration products.

MICROSCOPY OF "WET" CEMENT PASTE

The use of an environmental cell in a high voltage electron microscope (first used to study cement by Double et al. [4] and by Jennings [5]) allows the study of hydrated cement without pre-drying. In the 1MEV AEI EM7 there is a comparatively large space between the objective pole pieces into which an environmental cell can be incorporated (Fig.2). Thus a pressure of some 200-300 torr of nitrogen, saturated with water vapour, can be maintained around the specimen. The cement is mixed in small plastic bags at a realistic water to solids ratio (∿0.5) into which copper films supported on carbon grids are then introduced. After the desired hydration time the grids are extracted from the paste and placed directly into the microscope.

Using this technique the formation of C-S-H is observed as early as 30 minutes and after 1½ hours (Fig.3) can be seen to have a foil-like structure in the "wet state". These foils can be seen to crumple and roll up on drying. In very young (10 min) tricalciumaluminate pastes, ettringite rods are observed to form away from the particle surface, the surface itself being surrounded by a filmy amorphous product (Fig.4).

This technique, however, is still selective as the paste adhering to the carbon film is only electron transparent in a few areas where the cement grains are small and well dispersed. In addition it is still only the outer hydration surface which can be seen.

FIG.2. Differentially pumped aperture environmental cell in the HVEM.

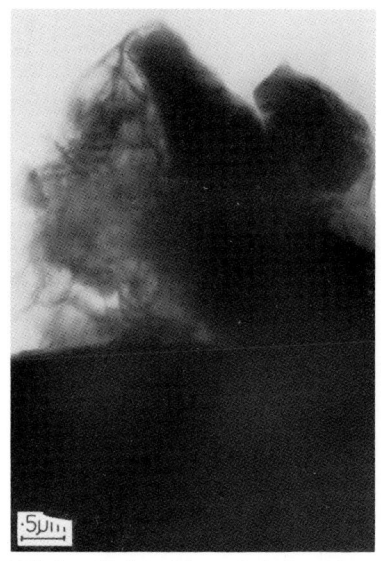

FIG.3. 1½ hr old, undried, alite paste in the HVEM.

FIG.4. 10 min old, undried, tricalciumaluminate paste in the HVEM.

STEM OF ION BEAM THINNED FOILS

The preparation of electron transparent foils by ion-beam thinning [6], [7], [8] enables the observation of the hydrated structure in cross-section. As cement is very brittle the preparation of thin foils is very difficult and young pastes must first be impregnated with epoxy resin. To minimise damage by the electron beam the foils are observed in scanning transmission electron microscopy (STEM).

Even at very early ages the formation of hydration products slightly away from the surface of the cement particle can be seen (Fig.5). The short ettringite rods noted on fracture surfaces (Fig.6) are observed together with a more gelatinous product not distinguishable on the fracture surface.

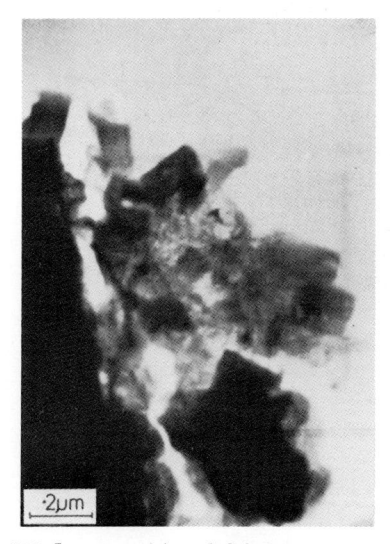

FIG.5. Ion thinned 2 hr cement paste in STEM.

FIG.6. Fracture surface of 1 hr cement paste (SEI in SEM).

Grains where the hydrated shell is separate from the anhydrous grain can be seen on fracture surfaces of pastes after 12-24 hrs (Fig.7). In thin foils, however, a definite "shell" structure can be seen around some grains at 5 hrs (Fig.8) and this becomes a common feature at 12 hrs (Fig.9).

FIG.7. Fracture surface of cement paste hydrated for 1 day (SEI in SEM).

FIG.8. Ion thinned 5 hr cement paste in STEM.

FIG.9. Ion thinned 12 hr cement paste in STEM.

Chemical analysis of the hydration products is also possible in thin foils by EDXA, with better spatial resolution and accuracy (due to simplification of the correction factors) than in EMPA.

With careful thinning at low angles fairly extensive electron transparent areas can be obtained, nevertheless a certain amount of selectivity is inevitable.

BACK SCATTERED ELECTRON IMAGES OF POLISHED CEMENT SURFACES

With polished cement sections large cross-sectional areas can be observed in the SEM. Using a pair of back scattered electron detectors images can be produced in which the intensity is dependent solely on the average atomic number of the scanned area, and independent of small fluctuations in topography (Fig.10). Thus in polished unetched cement sections, not only the anhydrous material but also areas of massive calcium hydroxide and other hydration products can be clearly distinguished.

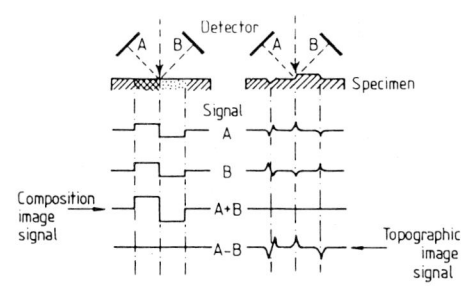

FIG.10. Signal processing for BEI in JEOL JSM 35CF SEM.

Regions of high atomic number back scatter electrons more strongly than those of low atomic number. In cement the anhydrous grains appear brighter than the hydration products, which have a lower average atomic number, due to the presence of water. This is the inverse of the case in STEM images of thin foils where the denser anhydrous areas absorb more electrons, and so appear darker. Thus a negative back scattered electron image (Fig.11) shows comparable features, though at lower resolution, to those seen in STEM (Fig.9).

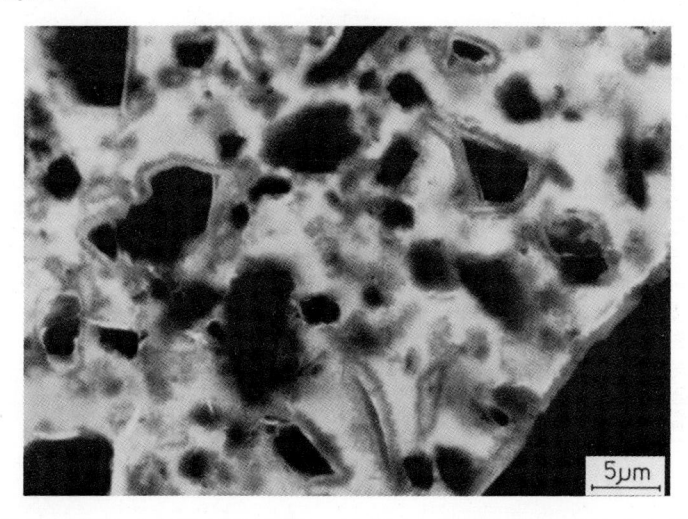

FIG.11. BEI of polished surface of 12 hr cement (negative image).

The contrast between anhydrous and hydrated material also allows the proportion of unreacted material to be quantified by image analysis, so giving a direct measure of the degree of hydration.

The examples presented here illustrate how a single observation method may be misleading. The use of a combination of electron optical techniques with other methods leads to more complete picture of the development of microstructure during cement hydration.

REFERENCES

1. A. Grudemo in: The Chemistry of Cements, H.F.W. Taylor (Academic Press 1964) pp 371-389.
2. S. Diamond in: Hydraulic Cement Pastes: their structure and properties (Cement & Concrete Association 1976) pp 2-30.
3. B.J. Dalgleish, P.L. Pratt, and E. Toulson, J. Mat. Sci. 17, 2199-2207 (1982).
4. D.D. Double, A. Hellawell, and S.J. Perry, Proc. Roy. Soc. Lond. A359, 435-451 (1975).
5. H.M. Jennings and P.L. Pratt, J. Mat. Sci. 15, 250-253 (1980).
6. B.J. Dalgleish and K. Ibe, Cement & Concrete Res. 11, 729-739 (1981).
7. R.L. Berger, unpublished.
8. J.P. Ollivier, R. Javelas, J.C. Masco, and B. Thenoz, Cement & Concrete Res. 5, 285-294 (1975).

ORDERED DEFECT-FLUORITE COMPOUNDS IN ZrO_2 ALLOYS

S. Farmer, J. Hangas, V. Lanteri, T.E. Mitchell and A.H. Heuer, Case Western Reserve University, Department of Metallurgy and Materials Science, Cleveland Ohio 44106

ABSTRACT

Intermediate ordered-defect fluorite compounds are an important component of ZrO_2 alloys containing the aliovalent solutes MgO, CaO, or Y_2O_3. These compounds also form in oxygen-deficient rare earth oxides and in other ternary oxides based on the fluorite structure, and can best be studied by TEM. Examples are given of such phases in the $MgO-ZrO_2$, $Y_2O_3-ZrO_2$, and $CaO-ZrO_2$ systems.

I. INTRODUCTION AND LITERATURE REVIEW

The high temperature cubic (c) polymorph of ZrO_2 with the fluorite structure (space group Fm3m) transforms on cooling below 2370°C to a tetragonally distorted fluorite-type structure (space group $P4_2/nmc$) (1). The subsequent tetragonal (t) → monoclinic (m) transformation (below 1000°C) can have untoward consequences in ZrO_2-containing ceramics, and so considerable effort, extending over half a century (2), has been expended in "stabilizing" the cubic fluorite phase. For this purpose, MgO, CaO, Y_2O_3, Sc_2O_3, or any of a host of rare earth oxides are suitable; all show extensive solid solubility in ZrO_2 and act as "fluorite-stabilizers", in that they lower the temperature of the c → t transformation (3).

One consequence of the dissolution of considerable quantities of such aliovalent cation solutes is the generation of massive quantities of charge-compensating oxygen vacancies, one for each divalent cation or two trivalent cations. Dissolution of ZrN in ZrO_2 also stabilizes the c polymorph (4) and again produces one oxygen vacancy for every two nitrogen anions.

It has been recognized for some time (5) that the massive concentration of point defects thus engendered is not randomly distributed ans non-interacting; rather, such systems exhibit significant short-range and long-range ordering (6), resulting in many cases in the formation of ordered defect-fluorite compounds.

The fluorite structure is also notable for its ability to accommodate non-stoichiometry, even in the absence of aliovalent solutes. For example, UO_{2-x} exists at high temperatures from x=0.1 on the hypostoichiometric side to x=0.25 on the hyperstoichiometric side of UO_2 (7). In rare earth oxides such as CeO_2, PrO_2, and TbO_2, oxygen-deficiency is prevalent and results in a series of ordered fluorite superstructures (8), which are members of a homologous series M_nO_{2n-2}, with n=7, 9, 10, 11, and 12. While all the superstructures are structurally related, only the n=7 example is isostructural in the Ce, Pr, and Tb systems. This M_7O_{12} structure is very common, and is the structural type for $Mg_2Zr_5O_{12}$, $Zr_3Sc_4O_{12}$, $Zr_4Y_4O_{12}$, and $Zr_7O_8N_4$ (and some of their HfO_2 analogues). (The crystal structure of this group of M_7O_{12} compounds was first determined for UY_6O_{12} (9), the isomorphous phase in which the cations are ordered.)

These M_7O_{12} structures are rhombohedral, all cations having either six or seven oxygen near neighbors. All the 6-coordinated cations lie along the 3-fold axis, forming so-called "Bevan clusters" (one 6-coordinated cation and six 7-coordinated cation (Fig. 1)) which constitute infinite strings in the structure (10). However, not all members of the rare earth homologous series

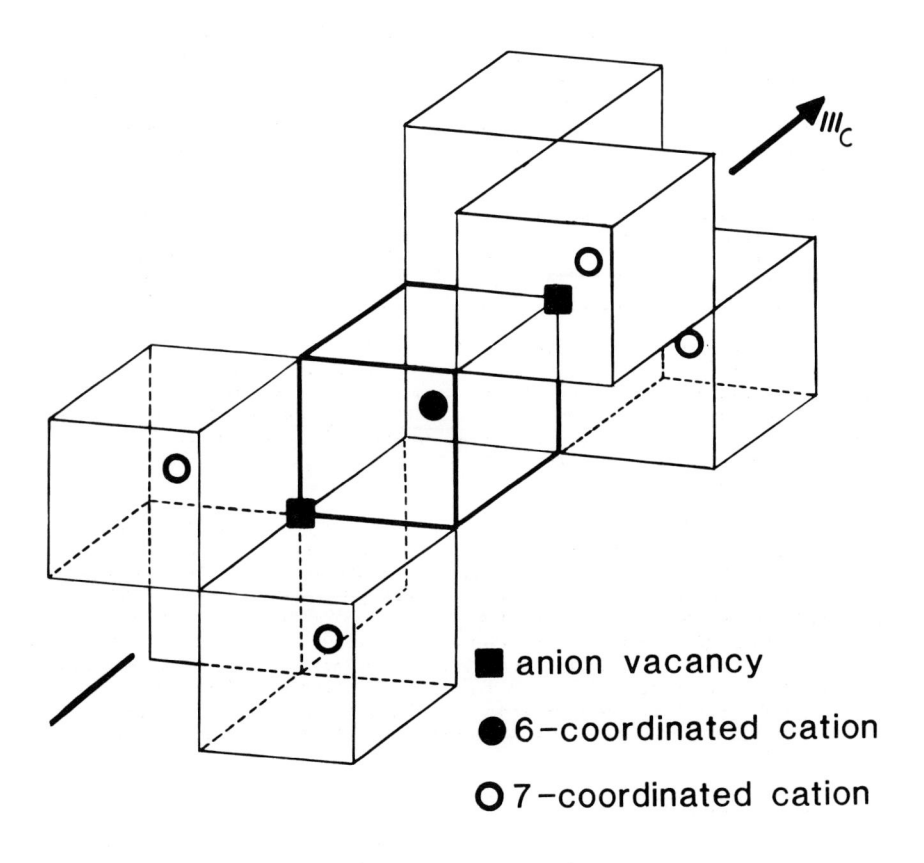

FIG. 1: Bevan cluster of one 6-coordinated cation, two oxygen vacancies, and six 7-coordinated cations. (Only four of the six 7-coordinated cations are indicated).

contain these strings of "Bevan clusters" along $<111>_F$ directions; rather the oxygen deficiency is accommodated by the clusters aligning on $\{135\}$ planes along $<211>_F$ directions (11); the $\{135\}$ planar spacing determines the stoichiometry, being every nth plane for $M_N O_{2n-n}$. It is of interest that infinite strings of 6-coordinated cations and two adjacent oxygen vacancies along all 3-fold axes are also found in the C-type rare earth sesquioxides (M_2O_3).

Although numerous intermediate phases have been identified in the binary rare earth oxides, isostructural phases are absent in most mixed rare earth oxide systems. This has been attributed to the sluggish diffusion in the cation sublattice, and the relatively low temperature of the order-disorder transformation so that equilibrium is rarely attained.

In the $Sc_2O_3-ZrO_2$ and $Sc_2O_3-HfO_2$ systems, three rhombohedral intermediate phases with disordered cations sublattices form (10). The ordered intermediate phases in the ZrN-, Y_2O_3-, and $MgO-ZrO_2$ binaries are similar to those observed in the $Sc_2O_3-ZrO_2$ system. In the $CaO-ZrO_2$ and $CaO-HfO_2$ binaries, the larger Ca ion causes a different packing of the 6-fold cations and the associated oxygen vacancies; three ordered phases form and all show some degree of cation order.

TEM has been an important, and in some cases an essential, tool in the study of these defect-fluorite oxides and oxynitrides, in part because the x-ray diffraction patterns are so similar, in part because of the incomplete "ordering" which is so common in these systems, and in part because the occurrence of these ordered compounds as fine precipitates requires a technique with good spatial resolution. It is this last advantage of TEM that has motivated our studies, in which the formation of these intermediate compounds as precipitate phases in $c-ZrO_2$ solid solutions alloyed with MgO, CaO, or Y_2O_3, has been investigated. Our studies also have relevance to the understanding of the good mechanical properties of partially stabilized ZrO_2's (PSZ's) (1).

II. The MgO ZrO$_2$ System

There are no stable intermediate compounds in the currently accepted phase diagram for the $MgO-ZrO_2$ binary system (12). However, the existence of the compound $Mg_2Zr_5O_{12}$ (and its analogue in the $MgO-HfO_2$ system) was reported some years ago (13); as noted already, it is isostructural with $Zr_7O_8N_4$ and $Zr_3Sc_4O_{12}$.

The early literature (13, 14) suggested that $Mg_2Zr_5O_{12}$ could form from a $c-ZrO_2$ solid solution of the appropriate composition (28.57 m/o MgO) but decomposed to $c-ZrO_2$ and MgO below 1850°C. More recently, this compound has been formed as a precipitate phase in Mg-PSZ's containing 8-14 m/o MgO (15-17).

Our studies have focussed on two compositions, 8.1 and 11.3 m/o MgO, initially formed as a $c-ZrO_2$ solid solution and progressively decomposed (diffusionally) by annealing between 800 and 1100°C. Decomposition was found to be a sensitive function of previous thermal history.

Heat treatment of either composition (as a single phase solid solution) at 800°C for one hour, followed by 1100°C for 5 hours, induced growth of coherent $t-ZrO_2$ precipitates and precipitates of the ordered phase $Mg_2Zr_5O_{12}$ within the $c-ZrO_2$ matrix (Figs. 2a and b). The $t-ZrO_2$ phase actually nucleated and grew to ~5 nm on cooling from a previous heat treatment at 1800°C, during which the solid solution was homogenized; it further coarsened to 50-80 nm during the 800/1100°C "aging". The equilibrium phases for these temperatures, $m-ZrO_2$ and MgO, are not much in evidence after such short

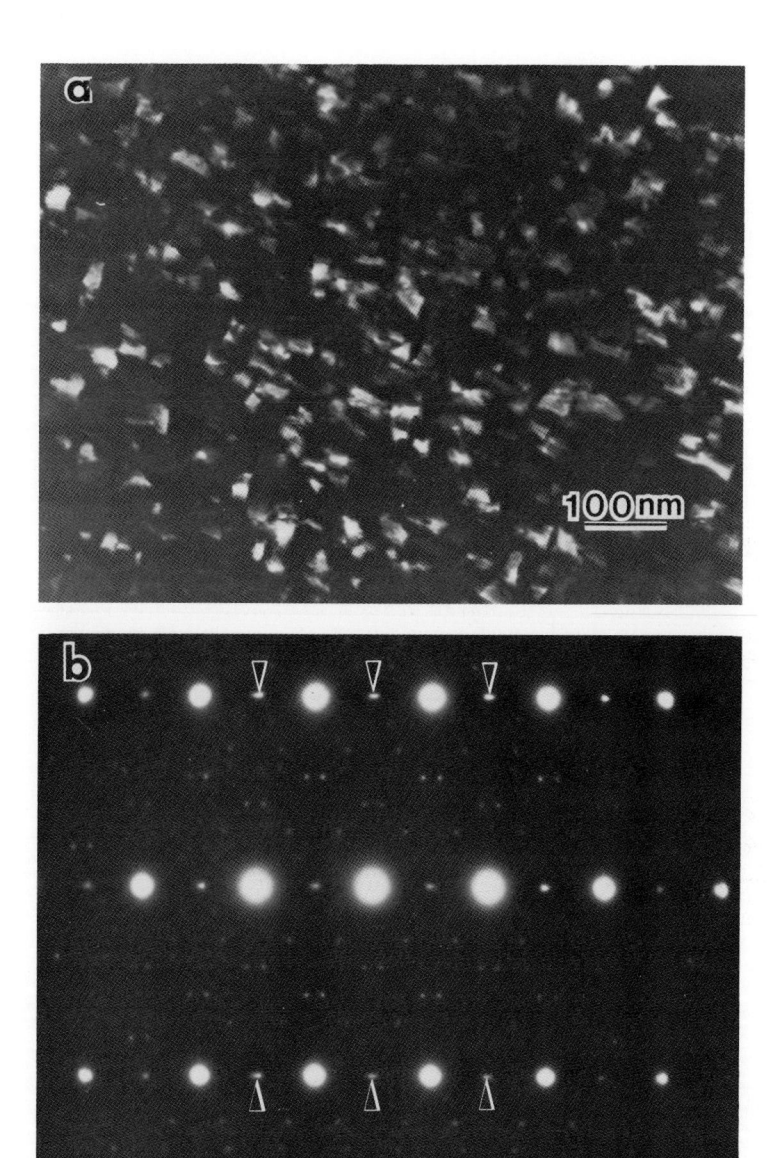

FIG. 2: $Mg_2Zr_5O_{12}$ precipitate in Mg-PSZ. (a) is a dark field image of a single precipitate variant in a c-ZrO_2 matrix and (b) is a <310>$_F$ zone axis diffraction pattern. The most intense ("fundamental") reflections are from the c-ZrO_2 matrix, while the weakest reflections are from two variants of $Mg_2Zr_5O_{12}$. The arrowed reflections are due to t-ZrO_2 precipitates, which are out of contrast in (a).

aging times.

If the aging time at 1100°C is increased to 10 hours, the $Mg_2Zr_5O_{12}$ does not coarsen appreciably, but a good deal of m-ZrO_2 is found at grain boundaries of the polycrystalline Mg-PSZ. This grain boundary "phase" is actually a eutectoid decomposition product consisting of m-ZrO_2 plus MgO, which advances into the surrounding grains as a coupled growth product (Fig. 3), and consumes the c-ZrO_2 and its t-ZrO_2 and $Mg_2Zr_5O_{12}$ precipitates. The ordered compound $Mg_2Zr_5O_{12}$ is therefore clearly a metastable decomposition product, although it may have a region of stability at some other temperature on the equilibrium phase diagram.

The conditions under which the $Mg_2Zr_5O_{12}$ formed were quite specific-- solution annealing at T>1800°C, followed by the 800/1100°C heat treatment. The lower temperature heat treatment is necessary to nucleate this phase; 800°C was usually used although 900°C was also found to be suitable. It was of interest that nucleation and growth of $Mg_2Zr_5O_{12}$ could be suppressed by an intermediate heat treatment. For example, prior heating at 1600°C for 1 hr, followed by furnace cooling, prevented the formation of $Mg_2Zr_5O_{12}$ during the 800/1100°C aging. If such material is subsequently given a second solution annealing of 2 hrs at 2100°C in an oxy-acetylene furnace, followed by rapid cooling (∼ 5 min) to room temperature, aging at 1100°C leads to the ready formation of $Mg_2Zr_5O_{12}$ without the need for a lower temperature nucleation heat treatment. Similar behavior has previously been noted concerning the formation of ordered defect-fluorite phases in some rare earth oxide systems (18), in which a suitable quenched-in high temperature cation arrangement is necessary to permit nucleation of particular ordered phases.

III. The ZrO_2-Y_2O_3 System

Two intermediate ordered compounds are known in the ZrO_2-Y_2O_3 system, an M_7O_{12} compound at 40 m/o Y_2O_3 and an M_7O_{11} phase at 75 m/o Y_2O_3. The most recent phase diagrams for this system (19-21) suggest that the former phase should be ubiquitous in high ZrO_2 compositions. However, a number of TEM studies have been reported of technologically interesting compositions containing between 2 and 8 m/o Y_2O_3 (22, 23); only m, t, and c-ZrO_2 polymorphs have been reported. This is probably due to the low temperature of the eutectoid in this system.

Recently, however, we have found evidence of an unusual ordering reaction in a 6.3 m/o Y-PSZ (the composition was determined by calibrated EDS analysis) (24). This material is a commercial furnace tube; we presume it was fired at a temperature above 1500°C and cooled slowly to room temperature.

Most grains have t-symmetry but contain anti-phase domain boundaries (Fig. 4), a result of a diffusionless c → t transformation (1). Such microstructures are common in other Y_2O_3-ZrO_2 ceramics (23). Some t-ZrO_2 "colonies" are also present, each colony consisting of stacked plates sharing a {101} habit plane, alternate plates being twin-related (their c-axes are rotated by 90°). Such colonies are also common in Y_2O_3-ZrO_2 samples heat treated in the two-phase t plus c field (22, 25).

A few grains in this polycrystalline ceramic possessed a different structure, one not observed previously in any prior study of lower Y_2O_3 material. The fundamental reflections in SAD patterns (Fig. 5) show tetragonal symmetry, and satellite reflections are seen about each of the strong fundamental reflections. Very weak superlattice reflections are also visible in $<111>_F$ orientations.

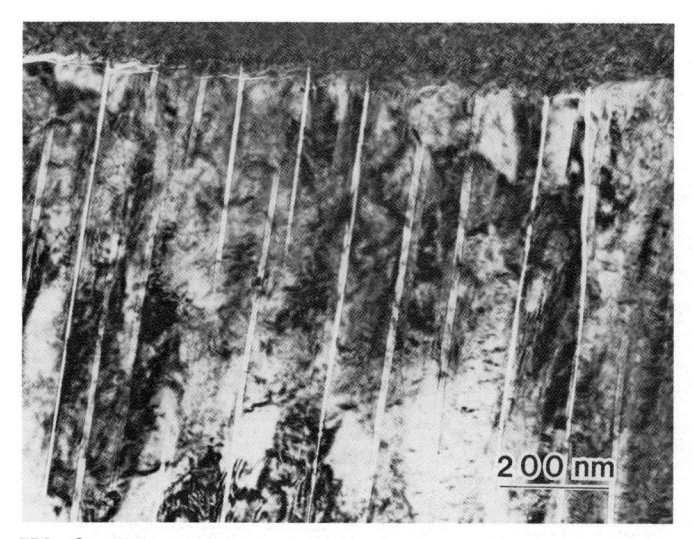

FIG. 3: Eutectoid decomposition product of $\underline{m}$-ZrO_2 and MgO in Mg-PSZ. The grain being consumed is at the top of the micrograph.

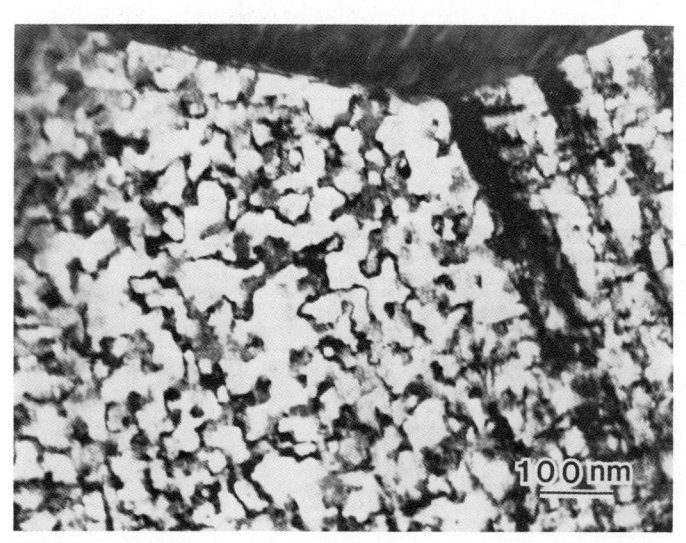

FIG. 4: Antiphase domain boundaries arising from a diffusionless but non-martensitic $\underline{c} \rightarrow \underline{t}$ transformation in 6.3 m/0 Y-PSZ.

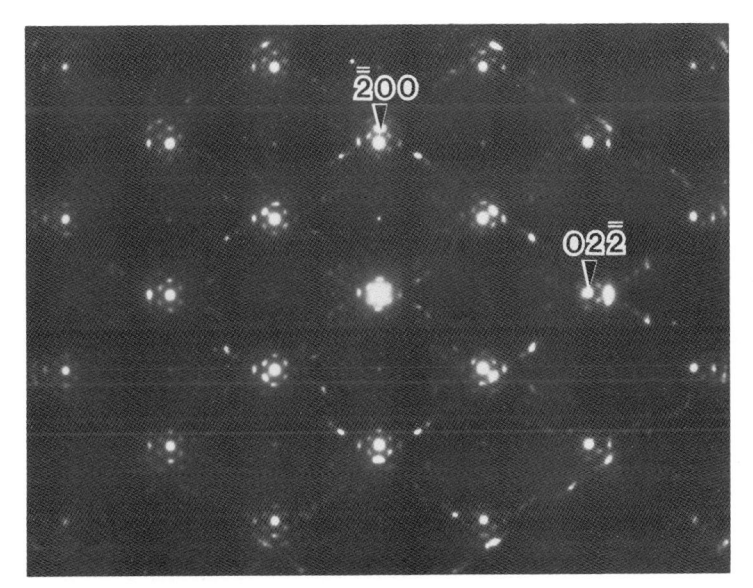

FIG. 5: <011>$_F$ zone axis diffraction pattern of possible ordered phase in 6.3 m/o Y-PSZ.

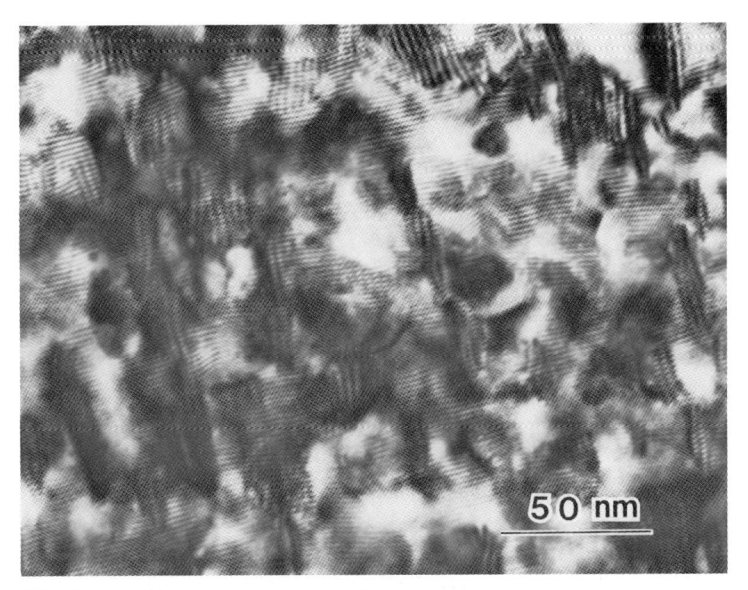

FIG. 6: Bright field image of 6.3 m/o Y-PSZ oriented to a <111>$_F$ zone axis orientation.

A bright field image of the region giving rise to the diffraction pattern in Figure 5 appears in Figure 6. The fringes correspond in spacing and direction to the satellite spacing in the diffraction pattern. EDS analysis confirmed that the grains showing this superstructure were richer in Y_2O_3 than the 6.3 m/o matrix. Further work is continuing on the identification of this phase, and particularly on the thermal history required for its formation.

IV. The CaO-ZrO$_2$ System

Three ordered compounds are known in the CaO-HfO$_2$ system, two of which have analogues in the CaO-ZrO$_2$ system, namely ϕ_1 (CaHO$_4$), ϕ(Ca$_2$Hf$_7$O$_{16}$) (the ZrO$_2$ analogue apparently does not form), and ϕ_2(Ca$_6$M$_{19}$O$_{44}$). All involve different arrangements of 6-coordinated cations. ϕ_1 contains linked clusters of three such cations (26), ϕ_2 has the cations linked in helical chains (26), and ϕ has isolated cations (27). In some senses, the degree of cation order increases from ϕ_1 to ϕ_2 to ϕ. When present as precipitate phases in CaO-ZrO$_2$ samples, they can readily be distinguished by electron diffraction (Fig. 7).

The recently determined phase diagram for this system (28) is shown in Figure 8. The data in this figure were obtained from 18-22 m/o CaO skull-melted single crystals heat treated at 1100-1300°C for up to 336 hours (29, 30); the compositions shown were determined by EDS, using a CaZrO$_3$/c-ZrO$_2$ eutectic as a standard for chemical calibration standards; the compositions are thought to be accurate to better than $\pm$ 0.5 m/o. The absence of appreciable quantities of ϕ_1 in compositions between 20 and 24 m/o CaO aged for long times suggests that ϕ_1 may be a metastable phase, in spite of the fact that it coarsened from elongated striated particles into the blocky

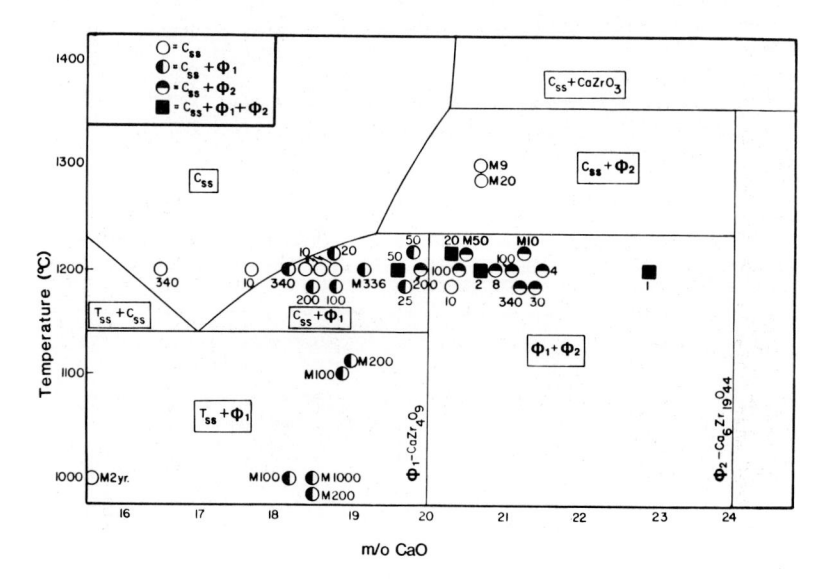

FIG. 8: Phase diagram for CaO-ZrO$_2$ system (28-30) showing compositions that precipitated either ϕ_1 or ϕ_2 on aging at elevated temperatures.

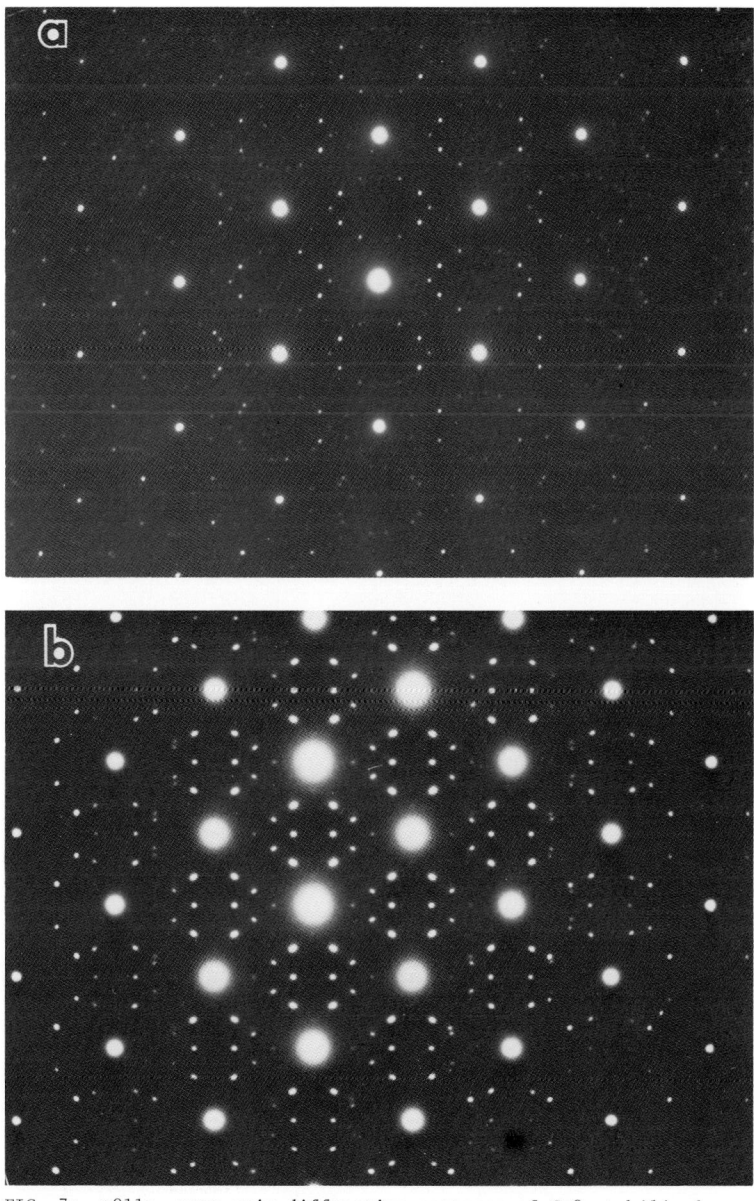

FIG. 7: $<011>_F$ zone axis diffraction patterns of CaO-stabilized ZrO_2 containing (a) $\emptyset_1$ precipitate (19 m/o CaO) and (b) $\emptyset_2$ precipitates (21 m/o CaO).

precipitates in Figure 9 in compositions leaner in CaO than 20 m/o aged for long times (30).

The morphology of $\emptyset_2$ precipitates in the CaO-rich cyrstals studied is shown in Figure 10. The $\emptyset_2$ particles are equiaxed and exhibit higher coherency strains in bright field images (not shown here) than do $\emptyset_1$, in keeping with the notion that $\emptyset_1$ is metastable (31) but nucleates readily. If this is correct, the assemblage c-ZrO$_2$ plus $\emptyset_2$ must be only modestly more stable than the assemblage c-ZrO$_2$ plus $\emptyset_1$, which combined with the high coherency strains, renders $\emptyset_2$ more difficult to nucleate than $\emptyset_1$.

ACKNOWLEDGMENT

Our research has been supported by the National Science Foundation (the MgO-ZrO$_2$ and CaO-ZrO$_2$ systems) and the AFOSR (the Y$_2$O$_3$-ZrO$_2$ system).

1. A.H. Heuer and M. Rühle, Adv. in Ceramics, Science and Technology of Zirconia, eds. N. Claussen M. Rühle and A.H. Heuer, The Amer. Cer. Soc., (1984). To be published.
2. O. Ruff and F. Ebert; Z. anorg. allg. Chem., 180, 19 (1929).
3. A.H. Heuer, "Advances in Ceramics 3, Science and Technology of Zirconia", ed. A.H. Heuer and L.W. Hobbs, Amer. Cer. Soc., 98-115 (1981).
4. N. Claussen, R. Wagner, L.J. Gouckler and G. Petzow, J. Amer. Cer. Soc., 61, 369-370 (1968).
5. R.E. Carter and W.L. Roth; in "Electromotive Force Measurements in High Temperature Systems. Institution of Mining and Metallurgy, 125-144 (1968).
6. M. Morinaga, J.B. Cohen and J. Faber, Jr., Acta Cryst., A-36, 520-530 (1980).
7. K. Hagamark and M. Broli, J. Inorg. Nucl. Chem., 28, 3632 (1966).
8. L. Eyring, in "Nonstoichiometric Oxides", ed. O.T. Sorensen, Acad. Press Inc., 337-398 (1981).
9. S.F. Bartram, Inorg. Chem., 5, 749-754 (1966).
10. M.R. Thornber, D.J.M. Bevan and J. Graham, Acta. Cryst., B24, 118-1191 (1968).
11. R.T. Tvenge and L. Eyring, J. of Solid State Chem., 29, 165-179 (1979).
12. C.F. Grain, J. Amer. Cer. Soc., 50 (6), 288-290 (1967).
13. C. Delamarre, "Hafnium Dioxide - MO Systems. Comparison with Corresponding Systems Based on Zirconia", Rev. Int. Hautes Temper. et. Refract. 9 (2), 209-224 (1972).
14. O. Yoranovitch and C. Delamarre, Mat. Res. Bull., 11 (8), 1005-1010 (1976).
15. R.H.J. Hannink and R.C. Garvie, J. Mat. Sci. 17, 2637-2643 (1982).
16. R.H.J. Hannink, J. Mat. Sci. 18, 457-470 (1983).
17. S.C. Farmer, L.H. Schoenlein and A.H. Heuer, J. Amer. Ceram. Soc. 66, 107-109 (1983).
18. D.J.M. Bevan and E. Summerville, "Handbook of the Physics and Chemistry of Rare Earths", Vol. 3, ed. C.A. Gschneider and L. Eyring, North-Holland Publishing Co, 401-523 (1979).
19. H.G. Scott, J. Mat. Sci. 10 (1975) 1527.
20. V.S. Stubican and S.P. Ray, J. Amer. Ceram. Soc. 60 (1977) 534.
21. C. Pascual and P. Duran, J. Amer. Ceram. Soc. 66 (1983) 23.
22. V. Lanteri, T.E. Mitchell and A.H. Heuer, in Ref. 1.
23. M. Rühle, N. Claussen and A.H. Heuer, in Ref. 1.
24. V. Lanteri, T.E. Mitchell, A.H. Heuer and M. Rühle, to be published.
25. R. Chaim, M. Rühle and A.H. Heuer, to be published.
26. J.G. Allpress, H.J. Rossell and H.G. Scott, J. Solid State Chem., 14, 264-273 (1975).
27. H.J. Rossell and H.G. Scott, J. Solid State Chem., 13, 345-350 (1975).
28. J.R. Hellmann and V.S. Stubucan, J. Amer. Cer. Soc., 66, 260-264 (1983).
29. J.M. Marder, T.E. Mitchell and A.H. Heuer, Acta Metall. 31, 387 (1983).
30. J. Hangas, T.E. Mitchell and A.H. Heuer, in Ref. 1.
31. V.S. Stubican, G.S. Corman, J.R. Hellmann and G. Senft, in Ref. 1.

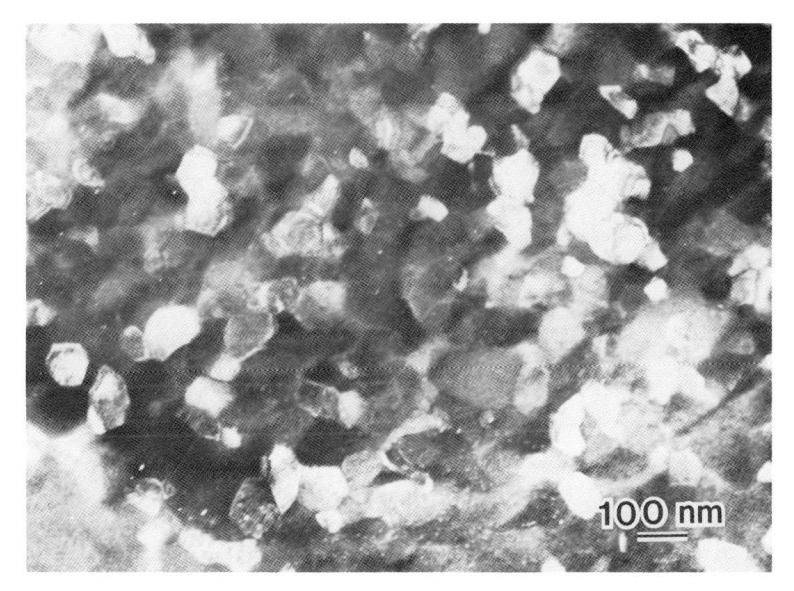

FIG. 9: $\emptyset_1$ precipitates in 19 m/o CaO-stabilized ZrO_2 aged 336 hours at 1200°C.

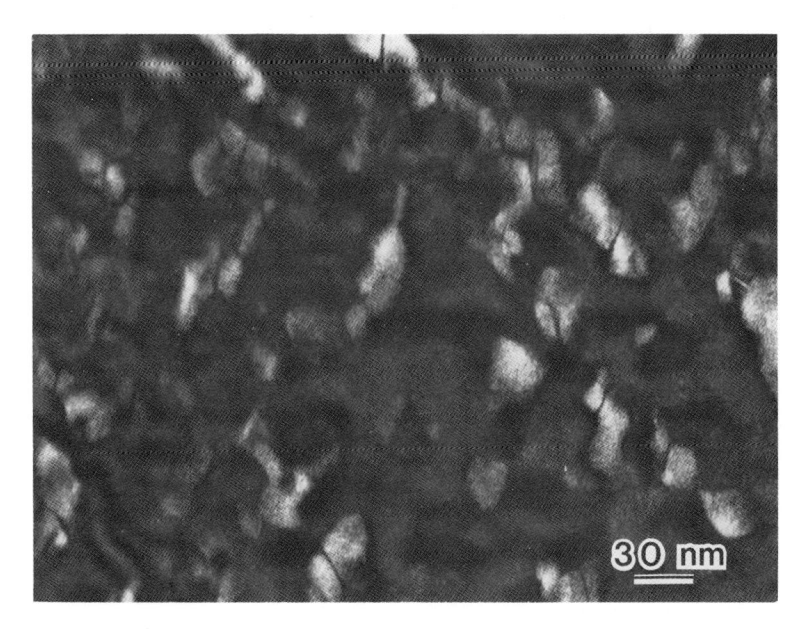

FIG. 10: $\emptyset_2$ precipitates in 21 m/o CaO-stabilized ZrO_2 aged 100 hours at 1200°C.

Author Index

Ast, D.G., 85
Batson, P.E., 33
Carpenter, G.J.C., 255
Carter, C.B., 267, 331
Chan, H.M., 345
Chen, L.J., 165
Clark, W.A.T., 211
Clarke, D.R., 325
Cowley, J.M., 177
Cunningham, B., 85
Elgat, Z., 331
Fahy, J.S., 71
Farmer, S., 357
Gleichmann, R., 85
Gomez, A., 233
Gronsky, R., 1, 241
Grovenor, C.R.M., 33, 247
Hagemann, P.N., 71
Hahn, S., 153
Hangas, J., 357
Harada, Y., 51
Harmer, M.P., 345
Heuer, A.H., 303, 357
Herd, S.R., 201
Hobbs, L.W., 291
Holland, O.W., 97
Honda, T., 51
Hosoi, J., 23
Houghton, D.C., 255
Hren, J.J., 337
Hsieh, Y.F., 165

Isakozawa, S., 63
Isozumi, S., 171
Kamimura, S., 63
Kokubo, Y., 23, 51
Komiya, S., 171
Kouh, Y., 331
Krakow, W., 39, 189
Krishnan, K.M., 79
Kuan, T.S., 143
Kusunoki, T., 171
Lagerlof, K.P.D., 303
Lal, M., 345
Lanteri, V., 357
Matsui, I., 63
Mazur, J.H., 105
Mishra, R., 79
Mitchell, T.E., 303, 357
Moriguchi, S., 23
Morrissey, K.J., 331
Narayan, J., 97
Nash, J., 23
Nguyen, T.A., 291
Otsuka, N., 261
Pennycook, S.J. 97
Petroff, P.M., 117
Ponce, F.A., 153
Pratt, P.L., 351
Rajeswaran, G., 159
Rez, P., 79
Ruhle, M., 317
Sabatini, R.L., 159

Sands, T., 241
Sato, H., 261
Scrivener, K.L., 351
Shaw, T.M., 325
Smith, D.A., 33, 223, 247
Smyth, D.M., 345
Steeds, J.W., 325
Sullivan, T.D., 111
Suzuki, S., 51
Tafto, J., 159
Tan, T.Y., 127
Thomas, G., 79
Tien, T., 201
Tonomura, A., 63

Truszkowska, K., 233
Tu, K.N., 201
Tucker, D.S., 337
Ueda, O., 171
Umebu, I., 171
Vanier, P.E., 159
Van Tendeloo, G., 279
Washburn, J., 105, 241
Watanabe, E., 23
Williams, D.B., 11
Wong, C., 33
Yacaman, M.J., 233

Subject Index

Alcemi, 345
Alumina, 267, 303, 317, 331, 337
Amorphous materials, 39
Amorphous silicide, 201
Analytical electron microscopy, 11, 317, 331
Antiphase domains, 291
Arsenic segregation, 33
Automatic diffraction analysis, 23
$BaTiO_3$, 345
Boolean processor, 39
Boron doping, 159
Cement, 351
Ceramics, 279
Channeling, 345
Chemical Analysis, 11
Chemical composition, 111
Computers, 23, 39
Contrast enhancement, 23
Convergent beam diffraction, 51
Copper sulfied, 241
Cross sections, 143
Crystalline clusters, 159
Current collection, 111
CVD, 325
Defects, 85
Differentiation, 23
Diffusion in silicon, 127
Digital diffraction, 23
Digital reconstruction, 23
Digital image processing, 39
Digitization, 39
Dislocation, 211
Dislocation loops, 171
Displacement faults, 325
Dopant profiles, 97
EBIC, 85, 111
Electron beam holography, 63
Electron channeling, 97
Electron energy loss, 33
Elemental analysis, 33
Energy dispersive x-ray analysis, 171
Environmental cell, 351
Epitaxial relationships, 247
Feature analysis, 39
Fe_9S_{10}, 291
Field ion microscopy, 39
Fourier analysis, 39
Gallium arsenide, 117, 143, 171
Gold diffusion
Gold films, 189
Gold particles, 233
Goniometer, 71
Grain boundary structure, 211, 267 317
Grain boundary width, 317
Grey scale expansion, 23
Hard copy from images, 39
Height measurement, 23
Heterojunctions, 143
High resolution microscopy, 51, 63, 153, 241, 331

High resolution objective lens, 71
Hollow cone illumination, 23
HVEM, 351
Image motion, 39
Image processing, 23
Image simulation, 189, 331
Incoherent defects, 153
Indium, 171
Inorganic crystals, 1
Interfaces, 105, 143
Interfaces, 211
   Alumina/spinel, 267
   Wustite/spinel, 267
   Spinel/alumina, 267
Interfacial defects, 223
Interface migration, 223
Interfacial reactions, 165
Interphase boundaries, 247
Interstitial oxygen, 153
Ion implanted silicon, 97
Lanthanum Hexaboride, 71
Letetium, 79
Lattice images, 291
Line width measurement, 23
Liquid phase epitaxy, 171
Low pass filtering, 23
LPCVD, 255
Magnetic materials, 63
Metal spheres, 33
Mg-Si-O, 279
Microanalysis 71, 177
Microcrystalline silicon, 159
Microdiffraction, 11, 177, 233
Microprocessor control, 1
Misfit, 241
Misfit dislocations, 261
Molecular beam epitaxy, 117, 143
Nickel films, 165
NiO, 317
Objective lens, 63
Octahedral occupation, 79
Optical diffractogram, 63, 291

Ordered defect compounds, 357
Oxidation, 105
Particle size analysis, 39
Phase boundary structure, 211
Plasmons, 33
Point defects, 127
Polycrystalline silicon, 85
Polymorphs, 325
Polysilicon, 255
Precipitates, 97
Probe specimen interaction, 33
Process control, 39
Rare earths, 79
Reconstructed surfaces, 117
Reflection electron microscopy, 117
Resolution limit, 63
RHEED, 177
Ribbon material, 85
Rotation faults, 325
Samarium, 79
Sapphire, 303
Si-Al-O-N, 279
Silica, 153
Silicide, amorphous, 201
Silicon, 85, 105, 117, 159
Silicon-antimony alloys, 97
Silicon-boron doped, 105
Silicon-Czochralski grown, 153
Silicon nitride, 325
Silicon polycrystalline, 33
Silicon/silicon dioxide, 39
Silicon-silicon dioxide interface, 105
Silicon (7x7), 117
Silver, oxidized, 189
Small dimension probes, 71
Software for image processing, 39
Solar cells, 85
Solute distribution, 11
Spatial resolution, 11
Specimen drift, 33

Specimen height control, 51
Specimen tilt, 1
Spinel, 267
STEM, 177, 351
Stacking faults, 171, 291, 303
STEM, 143, 171
Stigmator control, 23
Structure analysis, 23
Superstructures, 291
Surface dislocations, 117
Surface effects, 223
Surface steps, 177
Surface structure, 39, 177
Surface topography, 117, 189
Twin boundaries, 85, 267, 303
Twin lens, 71
Ultra high vacuum, 71
Vacancy and interstitial diffusion, 1
V_2O_3, 261
Weak beam technique, 261
Wustite, 267
X-rays, 79, 255
Yttrium, 79
Zoom and pan, 39
ZrO_2, 317
ZrO_2 Alloys, 357
ZrO_2-ZrN, 279